The Minerals, Metals & Materials Series

Ramana G. Reddy · Alexandra Anderson ·
Corby G. Anderson · Camille Fleuriault ·
Erik D. Spiller · Mark Strauss · Edgar E. Vidal ·
Mingming Zhang
Editors

New Directions in Mineral Processing, Extractive Metallurgy, Recycling and Waste Minimization

An EPD Symposium in Honor of Patrick R. Taylor

TMS

Springer

Editors
Ramana G. Reddy
The University of Alabama
Tuscaloosa, AL, USA

Alexandra Anderson
Gopher Resource
Tampa, FL, USA

Corby G. Anderson
Colorado School of Mines
Golden, CO, USA

Camille Fleuriault
Eramet Norway
Sauda, Norway

Erik D. Spiller
Colorado School of Mines
Golden, CO, USA

Mark Strauss
Aqua Metals
Reno, NV, USA

Edgar E. Vidal
NobelClad
Broomfield, CO, USA

Mingming Zhang
Baowu Ouyeel
Shanghai, China

ISSN 2367-1181 ISSN 2367-1696 (electronic)
The Minerals, Metals & Materials Series
ISBN 978-3-031-22764-6 ISBN 978-3-031-22765-3 (eBook)
https://doi.org/10.1007/978-3-031-22765-3

This Springer imprint is published by the registered company Springer Nature Switzerland AG
The registered company address is: Gewerbestrasse 11, 6330 Cham, Switzerland

Preface

Professor Patrick R. Taylor is the George S. Ansell Endowed Chair Professor of Chemical Metallurgy and Director of the Kroll Institute for Extractive Metallurgy at the Colorado School of Mines. He is an active member of TMS and has over 45 years of experience in mineral processing, extractive metallurgy, recycling and waste minimization, engineering, research, teaching, and consulting. He has published numerous papers and given invited presentations in over 13 nations and holds 9 patents. He has advised and mentored over 100 graduate students and postdoctoral research associates and served as a consultant for more than 20 companies. Professor Taylor is a Fellow of ASM International and a Distinguished Member (Fellow) of SME. He has received several outstanding awards from several societies and organizations including the Milton E. Wadsworth Award from SME, the EPD Distinguished Lecturer Award, the TMS EPD Distinguished Service Award, and the AIME James Douglas Gold Medal.

To honor Professor Taylor and his vast contributions to extractive metallurgy, The Professor Patrick R. Taylor International symposium titled "New Directions in Mineral Processing, Extractive Metallurgy, Recycling, and Waste Minimization" was held during the TMS 2023 Annual Meeting in San Diego, California, USA, from March 19-23, 2023.

The new concepts and fundamentals, update on reactor design and processes, industrial practices and developments, and environmental issues which influence the selection of metallurgical processes are discussed in the scientific papers published in this proceeding. These issues are discussed relative to new directions in pyrometallurgy, hydrometallurgy, electrometallurgy, mineral processing, metals and e-waste recycling, waste minimization and innovations in metallurgical engineering education and curriculum development.

The editors thank the symposium sponsors Society for Mining, Metallurgy & Exploration (SME), TMS Extraction and Processing Division, and its committees of Pyrometallurgy, Hydrometallurgy and Electrometallurgy, Materials Characterization, Energy, and Recycling and Environmental Technologies. We also thank the authors for providing the invaluable scientific and technical information contained in this book, and session chairs for their keen interest and support in making the symposium a success.

The lead organizer and editor Professor Ramana Reddy would like to acknowledge the assistance and encouragement of Corby G. Anderson, Colorado School of Mines, Erik D. Spiller, Colorado School of Mines, Edgar E. Vidal, NobelClad, Camille Fleuriault, Eramet Norway, Alexandra E. Anderson, Gopher Resource, Mingming Zhang, Baowu Ouyeel, Christina Meskers, SINTEF, and Mark Strauss, Aqua Metals, Inc., who were very helpful in numerous areas, particularly in soliciting invited papers, organizing sessions, and reviewing and editing the manuscripts.

Finally, we thank TMS Headquarters, particularly Patricia Warren, Kelly Markel, Trudi Dunlap, and James Robinson who made it possible to organize the symposium and publish the papers in the book *New Directions in Mineral Processing, Extractive Metallurgy, Recycling and Waste Minimization: An EPD Symposium in Honor of Patrick R. Taylor*.

Dr. Ramana G. Reddy
Lead Organizer

Contents

About the Honoree

Patrick R. Taylor is a registered professional engineer with over 45 years of experience in mineral processing, extractive metallurgy, recycling, and waste minimization; engineering, research, teaching, and consulting. He grew up in West Denver and served in the US Army in Vietnam. He then attended the Colorado School of Mines (B.S. Mathematics 1974, B.S. Metallurgical Engineering 1974, and Ph.D. Metallurgical Engineering 1978). He began his professional career at the University of Idaho in 1977. He progressed through the various positions and was the Department Head his last five years (1995–2000). He then joined the University of Tennessee where he was head of the Material Science and Engineering Department (2000–2002). In 2002, he accepted the offer to be the George S. Ansell Endowed Chair at the Colorado School of Mines (2002–2022). He directed the Kroll Institute for Extractive Metallurgy. He was part of the team that helped develop the Center for Resource Recovery and Recycling as well as the CSM part of the Critical Materials Institute.

Dr. Taylor has worked or given invited presentations in Canada, Mexico, Peru, Venezuela, Argentina, Chile, Bolivia, Colombia, Brazil, India, England, Turkey, and Egypt. He is experienced and trained in pyrometallurgy, hydrometallurgy, and mineral processing. He has been responsible for lab work, pilot plant work, research, and process development for mineral processing, and extractive metallurgy processes related to a wide variety of metals. He has authored or co-authored numerous papers and presentations and holds nine patents. He has served as a consultant for more than 20 companies and has been an expert witness. He has directed research for more than 100 graduate students and postdocs. He has taught extractive metallurgy and mineral processing university courses for the past 45 years. He has developed and taught ten short courses to industry. He wrote professional engineering exam questions for 25 years. He is active in many professional organizations including participation in SME (fellow), TMS, ASM (fellow), and MMSA.

In 1982, he was named Outstanding Faculty Member in the College of Mines at the University of Idaho. In 1985, he was the organizing chairman, TMS Fall Meeting for Process and Extractive Metallurgy: Intl. Symposium on the Recycle and Secondary Recovery of Metals. In 1990, he was named Distinguished Faculty Member at the, University of Idaho. In 1994 he received the Research Excellence Award from the University of Idaho. In 1996, he was named Fellow of ASM International. In 2003, he received the Milton E. Wadsworth Award from SME. In 2004, he was the TMS Extraction and Processing Division Luncheon Speaker. In May 2006, he was the CSM Alumni Association MME Graduating Senior Outstanding Faculty Member. In December 2006, he was the CSM Alumni Association MME Graduating Graduate Student Outstanding Faculty Member. In 2006, he was the TMS Extraction and Processing Division Distinguished Lecturer. In 2008, he was a co-organizer and co-editor of the SME 5th International Hydrometallurgy Symposium. In 2008, he was named a Distinguished Member (Fellow) of SME. In 2009, he was a co-organizer and co-editor of the SME Fall meeting, Mineral Processing Plant Design 2009. In 2010, he received the TMS EPD Distinguished Service Award. He received the AIME James Douglas Gold Medal in 2013. In 2017, he was a co-organizer of the symposium "Applications of Process Engineering Principles in Materials Processing, Energy, and Environmental Technologies" TMS. He was elected to the University of Idaho's Academy of Engineers in 2017. He was selected to be interviewed for the AIME Oral Histories program in 2018. Along with his graduate students (Tom Boundy), he received the TMS Light Metals/Extraction and Processing Subject Award—Recycling, 2019. Along with his graduate student (Vivek

Kashyap), he received the Taggart Award for the best paper in mineral processing from SME/MPD in 2021. He received the Industry Involvement award from CSM in 2021. He was named Emeritus Ansell Distinguished Professor of Chemical Metallurgy at CSM in 2022.

Dr. Taylor has been married to his high school sweetheart (Peggy) since 1969. He has two sons (Dylan and Travis) and two granddaughters (Brooke and Morgan).

About the Editors

Ramana G. Reddy is an ACIPCO Endowed Professor of Metallurgical and Materials Engineering. He served as the Head of the Department of Metallurgical and Materials Engineering and Associate Director of Center for Green Manufacturing at The University of Alabama, Tuscaloosa, Alabama, USA. He was a visiting professor at the Lawrence Berkeley Laboratory, University of California, Berkeley; Indian Institute of Technology, Bombay; and Argonne National Laboratory, Chicago and Renewable National Energy Laboratory, Golden USA. He was appointed as the University of Alabama Coordinator for the National Space Science and Technology Center (NSSTC) and NASA and served as Council Member for the Alabama State Committee for Department of Defense-EPSCoR programs.

Dr. Reddy's teaching and research experience are in the field of chemical and materials engineering, particular in the areas of thermodynamics, materials synthesis, energy materials, fuel cells, ionic liquids, and renewable energy. He has published over 433 research papers and 26 books, including an undergraduate textbook on thermodynamics. He has also delivered more than 305 invited lectures and research presentations in 26 nations. Dr. Reddy advised and mentored over 115 graduate students and worked with postdoctoral and research associates. He is the recipient of 5 USA patents.

Dr. Reddy is an Associate Editor of several national and international journals. He has received several awards such as the Service Award and the Best Research Paper Awards from several Materials Societies. He is the recipient of the TMS Extraction & Processing Distinguished Lecture Award, the Milton E. Wordsworth Award for Extractive Metallurgy of SME, ATA Award for Excellence in Engineering, and Henry Krumb Lecturer of SME. He is the recipient of the EPD-TMS Science Award, LMD-TMS Energy Committee Energy Paper Award, TMS Distinguished Service Award, Distinguished Alumni Professional Achievement Award (NITW), TMS Educator Award, SEC Faculty Achievement Award, AIME Mineral Industry Education Award, and Burnum Distinguished Faculty Award (University of Alabama, USA). Dr. Reddy was honored with the Professor Ramana Reddy Honorary symposium on Applications of Process Engineering Principles in Materials Processing, Energy, and Environmental Technologies, TMS.

Dr. Reddy is a Distinguished Member (Fellow) of SME and a Fellow of ASM International.

Alexandra Anderson Ph.D., PMP, is an R&D manager at Gopher Resource, LLC, an environmental solutions company specializing in lead battery recycling. Her work focuses on driving furnace productivity and efficiency initiatives through computational fluid dynamic (CFD) modeling and implementing novel equipment designs. Currently, she is also the principal investigator for a DOE HPC4Manufacturing partnership between Gopher Resource and Oak Ridge National Lab investigating high-fidelity multiphase furnace modeling. Alexandra obtained her B.S. in Mechanical Engineering from Gonzaga University and her M.S. and Ph.D. in Metallurgical and Materials Engineering from the Colorado School of Mines. Her dissertation investigated fluid flow and thermal profiles within secondary lead reverberatory furnaces using CFD techniques. Alexandra is active in The Minerals, Metals & Materials Society (TMS), where she serves as the vice-chair of the Process Technology and Modeling Committee; she was also the recipient of the 2021 TMS Extraction and Processing Division (EPD) Young Leader Award. Her scholarly activities include nine peer-reviewed publications, co-editorships of seven special topics for *JOM*, as well as several podium presentations at national conferences.

Corby G. Anderson is a Licensed Professional Chemical Engineer with over 40 years of global experience in industrial operations, corporate level management, engineering, design, consulting, teaching, research, and professional service. He is a native of Butte, Montana, USA. His career includes positions with Thiokol Chemical Corporation, Key Tronic Corporation, Sunshine Mining and Refining Company, H. A. Simons Ltd., and at CAMP-Montana Tech. He holds a B.Sc. in Chemical Engineering from Montana State University, an M.Sc. from Montana Tech in Metallurgical Engineering, and a Ph.D. from the University of Idaho in Mining

Engineering-Metallurgy. He is a Fellow of both the Institution of Chemical Engineers and of the Institute of Materials, Minerals, and Mining. He has directed or co-directed over 40 graduate students. He shares 16 international patents and 3 new patent applications covering several innovative technologies, 6 of which were successfully reduced to industrial practice. He currently directs the Kroll Institute for Extractive Metallurgy as the Harrison Western Professor as part of both the Mining Engineering Department and the George S. Ansell Department of Metallurgical and Materials Engineering at the Colorado School of Mines. He is also the CSM Director for the Center for Resource Recovery and Recycling. In 2009 he was honored by the Society for Mining, Metallurgy & Exploration with the Milton E. Wadsworth Extractive Metallurgy Award for his contributions in hydrometallurgical research. In 2015 he was awarded the International Precious Metals Institute's Tanaka Distinguished Achievement Award. In 2016 he received the Distinguished Member Award from the Society for Mining, Metallurgy & Exploration, and became a Distinguished Member of the U of Idaho Academy of Engineering. In 2017 he received the EPD Distinguished Lecturer Award from The Minerals, Metals & Materials Society. In 2019 he was named as a Henry Krumb Distinguished SME Lecturer. In 2019 he was also appointed and serves now as a Visiting Faculty within the Minerals Engineering Department of Central South University in China, the largest program of Mineral Processing and Extractive Metallurgy in the world. In both 2016 and in 2021 he received an Outstanding Faculty Award from the Colorado School of Mines. He was also elected in 2021 to the Sigma Xi Scientific Research Honor Society. In 2022 he received the TMS EPD Distinguished Service Award for his career contributions. Finally, in 2022 he and his coauthors received the SME Taggart Award for a notable contribution to the science of mineral processing.

Camille Fleuriault is senior project manager at Eramet Norway AS in Sauda, Norway. Her work focuses on identifying and enabling zero carbon strategies for the production of manganese alloys. She previously worked on developing innovative and environmentally friendly recycling processes for the secondary metals industry. She holds a B.S. in Geological Engineering and an M.Eng. in Mineral Engineering from the National School of Geological Engineering in Nancy, France, and a M.Sc. in Metallurgical Engineering from Colorado School of Mines, USA. She is chair of the TMS Pyrometallurgy Committee and a former *JOM* advisor for the same committee.

Erik D. Spiller has more than 45 years of experience managing research, process development, technology, engineering, sales, and operations in the minerals industries. For the past 30 years he combined that successful career with a second career in academia lecturing and conducting research at the Colorado School of Mines. He contributes technical expertise in processing from mineral liberation (comminution) through separations using various physico-chemical techniques based on particle attributes such as size, magnetic susceptibility, specific gravity, surface chemistry (flotation), conductivity, leaching, and liquid/solid separation. He received the AIME-SME prestigious Richards Award in 2009 and was a SME Fellow Member in 2018.

Mark Strauss is passionate about creating and implementing extractive metallurgical innovations to advance decarbonized and electrified clean energy technology, waste minimization, and critical materials independence. As a post-doctoral fellow at Idaho National Laboratory, he developed innovative recycling strategies to separate and purify lithium batteries into saleable lithium, copper, manganese cobalt, and nickel products. His research experience involves the primary and secondary production of critical materials using pyro-electro-hydrometallurgical techniques. His undergraduate research involved by-product recovery of rare earths from copper leach solutions at the University of Arizona. His master's and Ph.D. research investigated hydrometallurgical strategies for recycling fluorescent lights at the Colorado School of Mines and Worcester Polytechnic

Institute. He is currently developing and implementing electro-hydrometallurgical and pyrometallurgical innovations to optimize lithium-ion battery recycling at Aqua Metals.

Edgar E. Vidal holds a doctorate in Metallurgy-Mining Engineering and Bachelor's and Master's degrees in Materials Engineering. For the last four years he has been responsible for NobelClad's strategy in developing new composite metal products, applications, and markets. Prior to NobelClad, for ten years Dr. Vidal developed beryllium-related materials and processes at Materion (former Brush Wellman) and became Global Director of Sales, Marketing and Business Development. He holds four patents with an additional two pending. Dr. Vidal has over 50 publications in diverse areas of materials, from mineral processing to nuclear applications. He continues to be involved in academia as an Affiliate Faculty at Colorado School of Mines.

Mingming Zhang is currently a principal consultant at Wood Mackenzie. He has more than 20 years of experience in the field of mining, mineral processing, smelting and refining, and materials engineering. Before joining Wood Mackenzie, Dr. Zhang held lead engineer position at ArcelorMittal Global R&D in East Chicago, Indiana. He obtained his Ph.D. in Metallurgical Engineering from The University of Alabama and his master's degree in Mineral Processing from the General Research Institute for Non-ferrous Metals in China. Prior to joining ArcelorMittal, he worked with Nucor Steel in Tuscaloosa, Alabama where he was a metallurgical engineer leading the development of models for simulating slab solidification and secondary cooling process. Dr. Zhang has conducted a number of research projects involving mineral beneficiation, thermodynamics and kinetics of metallurgical reactions, electrochemical processing of light metals, metal recycling, and energy-efficient and environmentally cleaner technologies. He has published more than 50 peer-reviewed research papers and he is the recipient of several U.S. patents. Dr. Zhang also serves as editor and reviewer for a number of prestigious journals including *Metallurgical and Materials Transactions A and B*, *JOM*, *Journal of Phase Equilibria and Diffusion*, and

Mineral Processing and Extractive Metallurgy Review. Dr. Zhang has made more than 30 research presentations at national and international conferences including more than 10 keynote presentations. He was the recipient of the 2015 TMS Young Leaders Professional Development Award. He has served as conference/symposium organizer and technical committee chair in several international professional organizations including The Minerals, Metals &Materials Society (TMS), the Association for Iron & Steel Technology (AIST), and the Society for Mining Metallurgy & Exploration (SME).

Development of Ironmaking Technology by the Direct Gaseous Reduction of Iron Concentrate

Hong Yong Sohn

Abstract Considering the two most important issues the ironmaking industry faces today, i.e. energy consumption and greenhouse gas emissions, it would be advantageous to utilize the concentrate-size raw materials directly without pelletizing or sintering, especially without the use of coke. This plenary lecture describes two such processes developed at the University of Utah. One is the Flash Ironmaking Technology (FIT), and the other is a moving-bed process for continuous ironmaking with gaseous reduction of iron ore concentrate (MBIT). These technologies are designed to produce iron directly from iron concentrate without requiring pelletization/sintering and cokemaking. They take advantage of the high reactivity of the concentrate particles and will significantly reduce energy consumption and carbon dioxide emissions compared with the current processes. The process of the development from the conception of the idea, to kinetic feasibility establishment, and to the operation of a prototype facility will be discussed.

Keywords Flash ironmaking technology (FIT) · Moving-bed ironmaking · Carbon dioxide · Concentrate · Hydrogen · Magnetite · Natural gas · Pilot plant · Reformerless · Energy consumption · Sohn's law

Introduction

Primary steel is produced from iron oxide minerals through the integrated production using blast furnace (BF), combined with subsequent refining steps such as basic oxygen furnace (BOF), or by direct reduction (DR) process, combined with subsequent steelmaking steps such as electric arc furnace (EAF) treatment. Steel is also made from scraps through the EAF steelmaking process.

The integrated steelmaking process involves many steps, emits a great deal of carbon dioxide, and requires a large amount of energy compared with steelmaking

H. Y. Sohn (✉)
Department of Materials Science and Engineering, University of Utah, Salt Lake City, UT 84112, USA
e-mail: h.y.sohn@utah.edu

R. G. Reddy et al. (eds.), *New Directions in Mineral Processing, Extractive Metallurgy, Recycling and Waste Minimization*, The Minerals, Metals & Materials Series,
https://doi.org/10.1007/978-3-031-22765-3_1

from scraps. However, integrated steelmaking will remain the major steel producing route for a long time. This is because most steel products last for decades, before they become scrap and there are insufficient amounts of scrap to meet the rapidly increasing demand for steel.

Raw materials preparation consumes much energy and emits a large amount of CO_2 gas. The U.S. emitted 55 MMT CO_2 from iron and steel production in 2014 [1]. Pelletizing, sintering, and cokemaking processes emit ~20% of the total CO_2 in the BF-BOF route and the BF contributes ~70%.

According to the above discussion, the two most critical issues that the steel industry currently faces are energy consumption and greenhouse gas emissions. Large amounts of energy are consumed in the sintering/pelletizing steps and cokemaking, which are also responsible for much carbon dioxide emissions. Up to ~70% of iron production in the U.S. currently depends on magnetite concentrate produced from the low-grade taconite ore. Further, there is a trend to upgrade even reasonably high-grade iron ores up front to remove bulk of impurities before reduction. This required grinding the ores to concentrate-size particles.

It would then be advantageous to utilize the concentrate-size raw materials directly without pelletizing or sintering, especially without the use of coke. This plenary lecture describes two such processes developed at the University of Utah. One is the Flash Ironmaking Technology (FIT) [2–8], and the other is a moving-bed process for continuous ironmaking with gaseous reduction of iron ore concentrate (MBIT) [9].

Flash Ironmaking Technology (FIT)

The Flash Ironmaking Technology (FIT) is based on the reduction of iron oxide concentrates by gas in a flash reactor. This technology utilizes hydrogen or natural gas as a fuel as well as reducing agent. The development of FIT started from the kinetic feasibility determination to the laboratory flash furnace work and finally the tests in a pilot plant. Process simulation and economic analysis on the new process were also performed.

A sketch of the Flash Ironmaking process is shown in Fig. 1. A gaseous fuel is partially oxidized with industrial oxygen to generate a reducing gas at 1600–1900 K. Iron ore concentrate is fed from the top, and the reduced iron product can be collected as a solid powder or as a molten bath for direct steelmaking.

Natural gas is plentiful in the U.S. and could easily supply potential ironmaking based on the FIT. Hydrogen would be cleaner once the hydrogen economy is developed.

The development of the FIT started with the establishment of sufficient kinetic feasibility, considering the fact that there are only a few seconds of residence time available in a typical flash reactor. Upon the establishment of the kinetic feasibility, a laboratory flash reactor was tested, which was followed by a pilot-scale reactor test program.

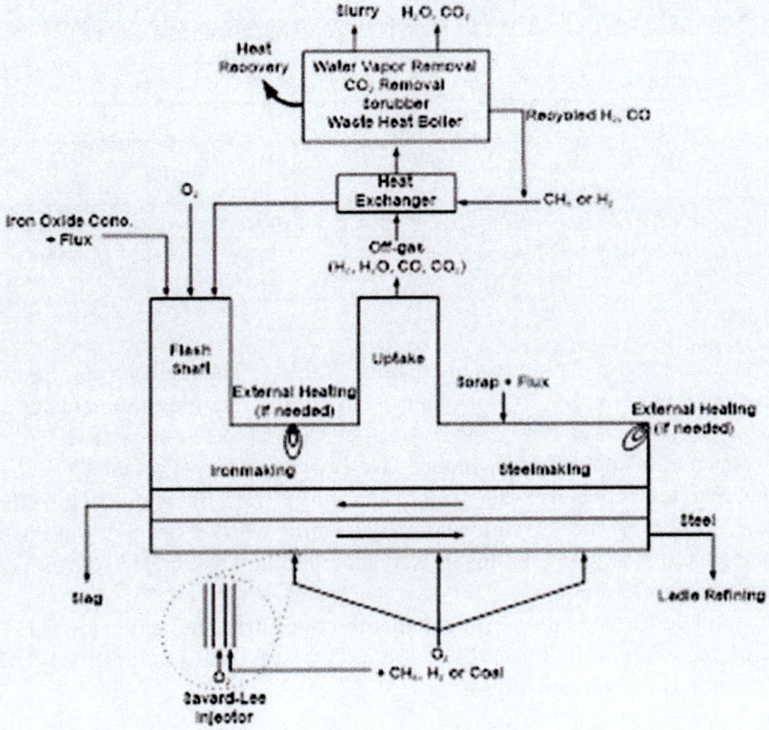

Fig. 1 A sketch of a direct steelmaking process based on the Flash Ironmaking Technology (FIT) (Adapted from Sohn and Choi [10])

Reduction Kinetics of Magnetite Concentrate Particles

Magnetite concentrate from a taconite ore of the Mesabi Range was used in this study. Sohn and coworkers [11–16] investigated the reduction rates of magnetite concentrate under the conditions of the FIT. The results were expressed by the following equation for component gases H_2 or CO:

$$\left.\frac{dX}{dt}\right|_j = k_j \cdot \left[p_j^{m_j} - \left(\frac{p_{jO}}{K_j}\right)^{m_j}\right] \cdot n_j(1 - X)[-Ln(1 - X)]^{1-\frac{1}{n_j}} \cdot d_p^{-s_j}; \quad j = H_2 \ or \ CO \tag{1}$$

where k_j is the reaction rate constant for gas j, $k_j = k_{o,j}\exp\left(-\frac{E_j}{RT}\right)$; p_j is the partial pressure of gas j; K_j is the equilibrium constant for the reduction of FeO by gas j; m_j is the reaction order with respect to gas j; n_j is the Avrami parameter; $d_p^{-s_j}$ is the particle size dependence; and X is the fraction of the total removable oxygen in the concentrate particles removed by the reaction.

Table 1 Kinetic parameters for reduction of magnetite concentrate by each component gas

Reducing gas, j	Temperature range (K)	$k_{o,j}$	E_j (kJ/mol)	m_j	n_j	s_j
H_2	1423–1623	1.23×10^7 atm^{-1} s^{-1}	196	1	1	0
	1623–1873	6.07×10^7 atm^{-1}·s^{-1}. µm	180	1	1	1
CO	1423–1623	1.07×10^{14} atm^{-1} s^{-1}	451	1	0.5	0
	1623–1873	6.45×10^3 atm^{-1}·s^{-1}. µm	88	1	0.5	1

The relevant kinetic parameters are given in Table 1. The reader is referred to the original papers [11, 13, 14, 16] for other details of the rate measurements and data analyses.

When magnetite concentrate is reduced by a mixture of H_2+CO, the CO enhances the rate of reaction between H_2 and iron oxide. This is most likely due to the effect of CO on the morphology of the reduced iron by forming whiskers, which was observed in a separate study [17]. Taking this into consideration, Fan et al. [13] developed the following rate expression.

The complete rate equations for magnetite concentrate reduction by a H_2+CO mixture at 1423 K (1150 °C)–1623 K (1350 °C) and 1623 K (1350 °C)–1873 K (1600 °C) are given, respectively, as:

$$\frac{dX}{dt} = \left(1 + 1.3 \cdot \frac{p_{co}}{p_{co} + p_{H_2}}\right) \cdot \left.\frac{dX}{dt}\right|_{H_2} + \left.\frac{dX}{dt}\right|_{CO} \quad 1423\text{K} < T < 1623\text{K} \tag{2}$$

$$\frac{dX}{dt} = \left[1 + (-0.01T + 19.65) \cdot \frac{p_{co}}{p_{co} + p_{H_2}}\right] \cdot \left.\frac{dX}{dt}\right|_{H_2} + \left.\frac{dX}{dt}\right|_{CO} \quad 1623\text{K} < T < 1873\text{K} \tag{3}$$

where $\left.\frac{dX}{dt}\right|_{H_2}$ and $\left.\frac{dX}{dt}\right|_{CO}$ represent the rates of reduction individually by H_2 and CO, respectively, obtained from Eq. (1) with the parameters listed in Table 1.

These kinetics measurements confirmed the fact that a few seconds of residence time in a flash reactor at temperatures of 1473 K or higher are sufficient to reduce iron ore concentrate.

Tests in a Laboratory Flash Reactor

Tests were then performed in a laboratory flash reactor shown in Fig. 2 [18, 19], based on the partial oxidation of methane, generating H_2+CO mixtures, and/or hydrogen with oxygen. The heating was supplemented by electrical heating.

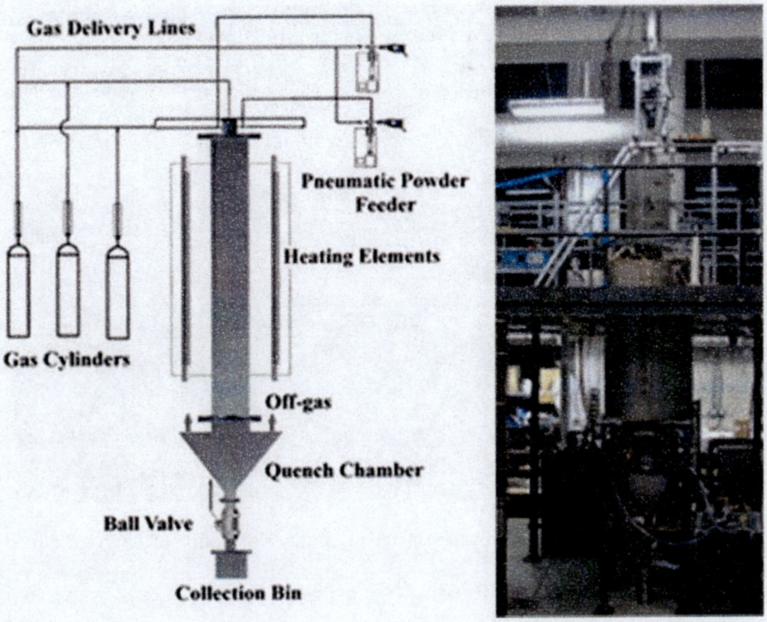

Fig. 2 The Utah laboratory flash ironmaking reactor (I.D. 0.19 m and height 2.13 m) (Adapted from Sohn et al. [3])

In these tests, a number of aspects that are relevant to the operation of an industrial flash ironmaking reactor were investigated. In addition to the effects of the main operating conditions of temperature and gas composition, other important features such as the position of concentrate feeding relative to the flame and the configuration of the flame in terms of the injection of fuel *vis-à-vis* that of oxygen were evaluated. For details of the significant test results, the reader is referred to the published articles by the author and coworkers [3, 18, 19].

In addition to the experimental test work, a CFD simulation model of the laboratory flash reactor operation was also developed to analyze and interpret the experimental results [18, 19].

The test work in the laboratory flash reactor confirmed the technical feasibility of ironmaking by the *in-situ* partial oxidation of natural gas or hydrogen in a flash reactor and established a number of useful operating conditions.

Design, Installation, and Operation of a Pilot-Plant Scale Flash Reactor

The laboratory flash reactor work was followed by the installation and operation of a Pilot Flash Reactor run at 1200–1550 °C with a solid feed rate of 1–7 kg/h, shown

Fig. 3 The pilot plant with a flash reactor installed at the University of Utah (Adapted from Elzohiery [20])

in Fig. 3. In this reactor, natural gas was used as the fuel/reducing gas instead of methane, as would be in an industrial flash ironmaking process.

This Pilot Flash Reactor system was designed to be fully computer-controlled with an emphasis on numerous safety aspects as well as high levels of operational control features. The facility consisted of a main reactor vessel, various burners, a quench tank for product collection, a stack, a concentrate feeding system, an off-gas analyzer, and human-machine interface for computer control and inputs from various sensors.

The reactor roof housed an opening for the main burner and a separate preheat burner with a pilot flame and provided a number of ports for gas and solid feeding as well as an emergency off-gas conduit equipped with a rupture disc. The configuration of the main burner in terms of injecting natural gas and oxygen was designed based on what was learned from the laboratory flash reactor.

The Human Machine Interface (HMI) consisted of the main PLC and a PC. The operator monitored all parts of the facility and ran the reactor through the computer to which the main PLC continuously fed information. All the safety and emergency procedures relied on the PLC. Details of the facilities and operation of the Pilot Reactor are described elsewhere [3].

Results from Pilot Flash Reactor Runs

The experiments in the pilot plant were performed at different temperatures and reducing powers of the gas with the aim of obtaining enough data for designing the industrial flash reactor. The results from the Pilot Flash Reactor runs were reported in Elzohiery [20]. The results showed good reproducibility within ±5% of the average reduction degree by repeating the same experiment at least 3 times. This represents a very high degree of reproducibility, considering the complexity of the operation and design of this large unit.

These results were used to develop a CFD model that would be used to optimize the operating conditions and reactor sizes to be used in an industrial reactor [5]. Different experimental runs were designed and made in this reactor to yield a range of reduction degrees, deliberately at less than complete reduction, to better examine the effects of the operating conditions and validate the CFD model.

CFD Simulation of the Pilot Reactor Operation

The same CFD model for the laboratory flash reactor discussed above was used for the pilot reactor runs [5]. The run conditions used for CFD simulation and the results are compared with the experimental results in Table 2.

The CFD model predicted the H_2 and CO concentrations within 93% accuracy for most runs. The experimental values of % reduction are compared with the CFD results in Table 2. The reduction degrees agree well for the first three runs. The agreement is not as good for the last three runs. The reason for this is likely to be because of the neglect of particle interactions for Runs 4–6. The temperature of the particle-gas stream in the main reaction zone was largely uniform, and this value is used to represent the reactor temperature. These runs had higher ratios of oxygen to natural gas and thus higher temperatures than the other runs, above 1577 °C, which is higher than the melting point of iron at 1538 °C. Particle agglomerates together more readily at these high temperatures, as shown previously during flash smelting of copper [21]. This might have caused lower reduction rates in the actual cases than in the simulation. This points to the need for improving the CFD model to account for particle coalescence at high temperatures.

The temperature of the particle-gas stream in Run 1 was also above 1577 °C, but the solid feed rate in this run was only about one-half of the values in Runs 4–6. The lower solid feed rate in Run 1 together with the fact that portions of the particles usually get stuck on the wall significantly lowered the possibility of particles in the gas stream to agglomerate in Run 1.

Process and Economic Analysis of Flash Ironmaking Technology

Based on the potential advantages of the new technology and the results of the process feasibility studies, process and economic analyses have been performed [7, 8, 22,

Table 2 Run conditions for the pilot reactor and CFD simulation results

Run #	Inner wall temperature[b] (°C)		Magnetite concentrate feeding rate (kg/h)	Main burner gas flow rate (SLPM)[a]		Particle size (μm) (Mass average used for simulation)	O_2 to natural gas molar ratio	Experimental RD (%) (± 5%)	CFD simulation RD (%)
	Average used for simulation	Range during run		NG	O_2				
1	1253	1483–1563	2.5	419	331	45	0.79	94	99.8
2	1275	1483–1563	4.3	509	328	32	0.64	80	84.5
3	1353	1483–1563	5	339	238	32	0.7	94.5	99.6
4	1167	1483–1563	5	413	359	45	0.87	74	99.8
5	1321	1483–1563	4.6	289	273	32	0.94	72.5	99.5
6	1326	1483–1563	4	264	247	32	0.93	50	85

Oxygen and natural gas input temperature was 25 °C

[a]Flow rates are calculated at 25°C and 0.85 atm, the barometric pressure at Salt Lake City

[b]Temperatures of inner wall were recorded during the main experiment

23]. The results of these analyses indicated that the new ironmaking technology will consume up to 44% less energy than the blast furnace when the former is run in the reformerless mode, i.e. direct partial combustion in the reactor, and it will emit up to 51% less carbon dioxide. When hydrogen is used, the proposed process would consume up to 60% less energy with little CO_2 emissions. However, it is noted that a more accurate comparison must include the energy consumption and CO_2 emissions for the production of natural gas, hydrogen, or coal.

An economic feasibility analysis [23] indicated that the new technology using natural gas with in-situ partial combustion would be economical at this time, owing to the small capital and operating costs as well as the low price of natural gas. The sensitivity analysis performed on the estimated NPV indicated that the price of natural gas affects the NPV most strongly. These economic analyses point to the fact that the proposed ironmaking technology would be economically feasible at this time if it is operated using natural gas.

Summary on Flash Ironmaking Technology

The overall process of developing a novel Flash Ironmaking Technology (FIT) has been described in this article.

Rate equations for the reduction of iron ore concentrate by hydrogen, carbon monoxide, and a mixture of the two formulated in this work established the fundamental feasibility of the concept of the flash ironmaking and form the basis of the design of a reactor to realize the process. Scale-up experiments were performed in a laboratory flash reactor that operated at conditions similar to those of the industrial flash ironmaking reactor, and >90% reduction degree was obtained at temperature as low as 1175 °C.

A pilot reactor that operated in the temperature range 1200–1550 °C was installed and operated to collect the data necessary for scaling up the process to an industrial scale. This reactor was used to validate the design concept of the Flash Ironmaking in terms of heat supply, residence time, reduction degree, and the determination of optimum operating conditions. These tests also identified a number of technical hurdles. This investigation proved the technical feasibility of the flash ironmaking technology for large-scale iron production. The results of this work will facilitate the complete design for the industrial flash ironmaking reactor.

The new technology does not require pellets, sinters, and coke. Instead, it would produce iron directly from concentrates using natural gas or hydrogen. As a result, the energy consumption is expected to be up to 44% less than that for the average blast furnace process when the Flash Ironmaking Technology (FIT) is operated with natural gas in the reformerless mode, i.e. in-situ partial combustion, and it will emit up to 51% less carbon dioxide. When hydrogen is used, the proposed process would consume up to 60% less energy with little carbon dioxide emissions.

Moving-Bed Ironmaking Technology (MBIT)

A novel horizontal moving-bed ironmaking process has been developed at the University of Utah. This process uses iron concentrate directly without sintering or pelletization and reduces it with hydrogen in the temperature range of 500–1000 °C. The work started with the determination of the particle kinetics, which was then combined with diffusional effect to analyze the reduction rate of a particle bed. Based on the kinetics formulation, a mathematical model of the furnace for the proposed technology was developed for a modest industrial ironmaking operation designed to produce iron at a rate of 0.1 million tonnes per year (Mtpy).

Process Concept

The Flash Ironmaking Technology (FIT) described above requires an operating temperature of 1300–1600 °C and could produce either solid-phase iron or molten hot metal. Further, it is thought to be more suitable for a large-scale ironmaking plant. For medium-level steelmaking operations, sponge iron produced at a lower temperature would make a suitable feed. Even in the latter case, direct use of iron concentrate with its high reactivity would be advantageous. Based on this reasoning, a process using a moving-bed reactor has been developed [9].

Configuration of a Horizontal Moving-Bed Furnace

The furnace to be used for the proposed process would continuously carry iron concentrate placed as a layer on a moving grate in a counter-current flow with the reducing gas. A sketch of such a counter-current moving-bed reactor is shown in Fig. 4.

The evaluation of the feasibility of such a moving bed for direct reduction of iron concentrate requires a quantitative information on the kinetics of reduction of concentrate particles in the anticipated temperature range, that of a particle bed including diffusional effects, and finally a mathematical model of the moving-bed reactor.

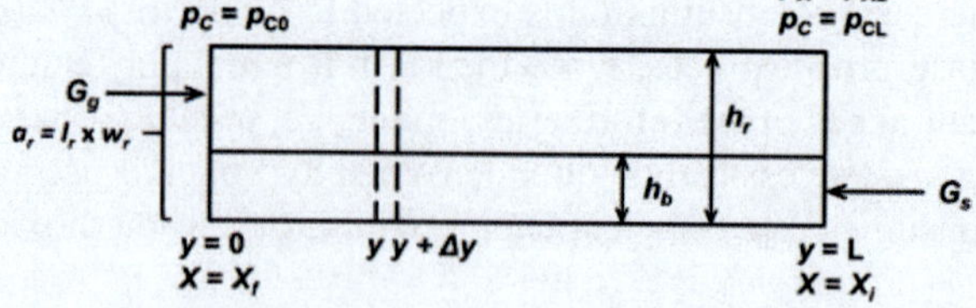

Fig. 4 Sketch of a counter-current horizontal moving-bed reactor (Adapted from Roy [24])

Hydrogen Reduction Kinetics of Concentrate Particles

The reduction kinetics of iron ore concentrate particles were measured in the temperature range expected for the proposed process using a thermogravimetric analysis (TGA) unit [24].

The rate equations for the hydrogen reduction of iron concentrate, formulated from the experimental data, are given for different temperature ranges, as follows:

In the range 800–1000 °C,

$$[-\mathrm{Ln}(1-X)]^{1/1.5} = 9.99 \times 10^3 \times \exp\left(-\frac{125,100}{\mathrm{RT}}\right) \times \left(p_{H_2} - p_{H_2O}/K\right) \times t \tag{4}$$

In the range 650–800 °C,

$$[-\mathrm{Ln}(1-X)]^{1/1.5} = 2.25 \times 10^{-5} \times \exp\left(\frac{6430}{T}\right) \times \left(p_{H_2} - p_{H_2O}/K\right) \times t \tag{5}$$

In the range 500–650 °C,

$$[-\mathrm{Ln}(1-X)]^{1/1.5} = 1.79 \times \exp\left(-\frac{33,500}{RT}\right) \times \left(p_{H_2} - p_{H_2O}/K\right) \times t \tag{6}$$

where X is fractional conversion of iron oxide, R is 8.314 J mol^{-1} K^{-1}, T is in K, p is in atm, and t is in seconds. It is noted that the rate decreased with temperature in the range 650–800 °C. This phenomenon has been observed previously [25] for iron oxide reduction.

The iron is refined in the steelmaking step, and thus, the iron just needs to be reduced to a degree of 90% or higher.

Incorporation of Interparticle Diffusion in the Rate Analysis

To determine and incorporate such diffusional effects, experiments were performed with deeper beds of concentrate of predetermined depths [24]. The experimental data were converted to a closed-form *X*-*t* relation using Sohn's law of additive reaction times [26–28] that is applicable under the general conditions of mixed control.

A good agreement was found [24] between the experimental data and the results from the application of Sohn's Law, which renders unnecessary numerical solution of the differential equations.

Design of a Horizontal Moving-Bed Furnace

Model Formulation

The model is formulated based on the configuration of the counter-current moving-bed reactor presented in Fig. 4. The details of the model formulation can be found elsewhere [24]. Here, the final result of the formulation that relates the reactor length to all other reactor parameters and operating conditions is presented, as follows:

$$L = C_1 C_4 \int_0^{X_0} \frac{C_2(1-X)^{-1}(-\mathrm{Ln}(1-X))^{1/n-1} + C_3 X}{C_1 + X - X_0} dX \tag{7}$$

where $C_1 = \frac{G_g}{4G_s}\frac{(x_{A0}-x_{C0}/K)}{1+1/K}$; $C_2 = \frac{1}{nk_0 P(x_{A0}-x_{C0}/K)}$; $C_3 = \frac{4(1-\varepsilon_P)\rho_B}{D_e}\left(\frac{V_p}{A_p}\right)^2 \frac{RT(1+1/K)}{P(x_{A0}-x_{C0}/K)}$; and $C_4 = \frac{G_s}{\rho_B(1-\varepsilon_P)(1-\varepsilon_v)}$.

In the above equations, X_o is the final value of X at the gas entrance; G_g and G_s are, respectively, the total input rate per unit-cross-sectional area of the reactor of gas (including hydrogen, water vapor, and inert gas, if any) and the input molar rate per unit-cross-sectional area of the reactor of solid B; K is the equilibrium constant for wüstite reduction; x is the mole fraction of gaseous species; $n = 1.5$ is the Avrami parameter in the nucleation-and-growth kinetics equation; k_o is the preexponential factor in the rate equation; P is the total pressure of the gas phase; ε_p is the porosity of the product iron layer; and ρ_B is the true molar density of B.

Design of Industrial Reactors

An industrial ironmaking reactor, producing 0.1 million metric tons of iron per year (Mtpy), equivalent to 12.66 metric ton/h assuming operation of 24 h/day and 330 days/year, was designed. The reactor will produce iron of $X_0 = 0.95$. The normalized driving force of reducing gas, θ, is set at 0.3 at the gas outlet. The reactor was assumed to have a width of 5 m and a height of 3 m.

From the kinetics of magnetite reduction, it was found that between 650 and 1000 °C, a bed with a thickness of 1 cm would largely be controlled by diffusion. Thus, this reactor design selected a bed thickness equal to or greater than 1 cm. For the purpose of this work, the reactor will have at most 10 layers. For the dimensions of a horizontal moving-bed reactor to be comparable with other industrial reactors, the length of the reactor should be under 50 m.

For 0.1 Mtpy production rate, the effects of temperature and bed thickness on residence time, reactor length, bed speed, and gas velocity are shown in Table 3. For a given bed thickness, the residence time and the reactor length decreased on increasing temperature. The bed speed was unaffected by temperature as expected from the diffusion-controlled reaction conditions. For a given temperature, increasing

Table 3 Effect of bed thickness and temperature on residence time, reactor length, speed of bed, and gas velocity for a moving-bed reactor producing 0.1 million tonnes of iron per year

Temperature (°C)	Bed thickness (cm)	Residence time (min)	Reactor length (m)	Bed speed (cm/min)	Gas velocity (cm/s)
1000	1	23.5	5.76	24.5	182
	2	92	11.3	12.3	188
	5	571	28.1	4.91	211
900	1	28.7	7.03	24.5	182
	2	108	13.3	12.3	188
	5	661	32.4	4.91	211
650	1	95	23.3	24.5	196
	2	371	45.6	12.3	203
	5	2303	113.1	4.91	227

the bed thickness resulted in longer residence time, longer reactor length, and slower speed of moving bed. The demand for gas increased with decreasing temperature as the composition of hydrogen-water vapor mixture in equilibrium with a wüstite-iron system became more hydrogen rich. It was found that a reactor with a bed thickness of 2 cm and 45.6 m long can operate at a temperature as low as 650 °C.

Concluding Remarks on Moving-Bed Ironmaking Technology (MBIT)

Discussion on the horizontal moving-bed reactor for commercializing the proposed technology can be summarized as follows:

1. The proposed technology for a modest-scale ironmaking operation producing 0.1 million tonnes per year can be operated at temperatures between 650 and 1000 °C.
2. The design parameters and the operating conditions for the horizontal moving-bed reactor were established. Due to low operating temperature and elimination of the pelletization step, lower energy consumption, CO_2 emissions, and maintenance costs are expected.
3. A simple model for a moving-bed reactor was formulated.

Acknowledgements Financial support from the U.S. Department of Energy under Award Number DE-EE0005751, with cost share by the American Iron and Steel Institute (AISI) and the University of Utah, is gratefully acknowledged.

References

1. Bains P, Psarras P, Wilcox J (2017) Prog Energy Combust Sci 63:146–172
2. Sohn HY (2007) Steel Times Int 31(3):68–72
3. Sohn HY, Elzohiery M, Fan D-Q (2021) J Energy Power Technol 3(3):25
4. Sohn HY, Fan D-Q, Abdelghany A (2021) Metals 11:332
5. Abdelghany A, Fan D-Q, Elzohiery M, Sohn HY (2019) Steel Res Int 90(9):1900126(10)
6. Abdelghany A, Fan D-Q, Sohn HY (2020) Metall Mater Trans B 51:2046–2056
7. Pinegar HK, Moats MS, Sohn HY (2011) Steel Res Int 82:951–963
8. Pinegar HK, Moats MS, Sohn HY (2012) Ironmak Steelmak 39:398–408
9. Sohn HY, Roy S (2022) Moving-bed reactor for continuous ironmaking with gaseous reduction of iron ore concentrate, U-6961, May 19, 2020; Provisional U.S. Patent application, Serial no 63/389,795, July 15, 2022
10. Sohn HY, Choi ME (2012). In: Carpenter M, Shelton EJ (eds) Carbon dioxide emissions: new research. Nova Science Publishers Inc., Hauppauge, NY, p 31
11. Fan DQ, Mohassab Y, Elzohiery M, Sohn HY (2016) Metall Mater Trans B 47:1669–1680
12. Fan DQ, Sohn HY, Elzohiery M (2017) Metall Mater Trans B 48:2677–2684
13. Fan DQ, Elzohiery M, Mohassab Y, Sohn HY (2021) Ironmak Steelmak 48(9):1064–1075
14. Fan DQ, Elzohiery M, Mohassab Y, Sohn HY (2021) Ironmak Steelmak 48:769–778
15. Elzohiery M, Sohn HY, Mohassab Y (2016) Steel Res Int 88:1600133
16. Elzohiery M, Fan DQ, Mohassab Y, Sohn HY (2021) Ironmak Steelmak 48:485–492
17. Wang H, Sohn HY (2012) Steel Res Int 83:903–909
18. Fan DQ, Sohn HY, Mohassab Y, Elzohiery M (2016) Metall Mater Trans B 47:3489–3500
19. Elzohiery M, Fan DQ, Mohassab Y, Sohn HY (2020) Metall Mater Trans B 51:1003–1015
20. Elzohiery M (2018) PhD dissertation, University of Utah, Salt Lake City, Utah
21. Kimura T, Ojima Y, Mori Y, Ishii Y (1986) In: Gaskell DR et al (eds) Reinhardt Schuhmann International symposium on innovative technology and reactor design in extraction metallurgy, TMS–AIME, Warrendale, Pennsylvania, p 403
22. Pinegar HK, Moats MS, Sohn HY (2013) Ironmak Steelmak 40:32–43
23. Pinegar HK, Moats MS, Sohn HY (2013) Ironmak Steelmak 40:44–49
24. Roy S (2022) PhD dissertation, University of Utah, Salt Lake City, Utah
25. Lien HO, El-Mehairy AE, Ross HU (1971) J Iron Steel Inst 209:541–545
26. Sohn HY (2020) Fluid-solid reactions. Elsevier, Cambridge, MA, pp 241–233
27. Sohn HY, Roy S (2020) Metall Mater Trans B 51(2):601–610
28. Sohn HY (1978) Metall Mater Trans B 9(1):89–96. https://doi.org/10.1007/BF02822675

Evaporation of Phosphorus from $FeO\text{-}CaO\text{-}SiO_2\text{-}P_2O_5$ Synthesized Steelmaking Slag by Carbothermic Reduction

Huafang Yu, Miki Takahiro, Yasushi Sasaki, and Tetsuya Nagasaka

Abstract Phosphorus is an important element for both the agricultural and industrial sectors, especially within semiconductor and pharmaceutical industries. However, the resource of phosphorus, and high grade phosphate rock deposits are rapidly being depleted and finding phosphorus secondary resources turns to be more and more crucial. Steelmaking slag is considered an attractive phosphorus secondary resource as it contains relatively high phosphorus content and has a constant supply, relatively fixed composition with minor contamination. The present study used graphite powder as a reducing agent to reduce $FeO\text{-}CaO\text{-}SiO_2\text{-}P_2O_5$ synthesized slag into phosphorus. And in this study, it was clarified that phosphorus could be evaporated from steelmaking slag by carbothermic reduction at 1573 K. During the carbothermic reduction process of synthesized slag, phosphorus was detected to be distributed into gas phase, metal phase, and part of it kept in residual slag phase. And partial phosphorus transferred into the gas phase could be recovered.

Keywords Carbothermic reduction · Steelmaking slag · Phosphorus

---

H. Yu (✉) · M. Takahiro · Y. Sasaki · T. Nagasaka
Graduate School of Engineering, Tohoku University, Miyagi 980-8579, Japan
e-mail: huafang.yu.e8@tohoku.ac.jp

Y. Sasaki
Institute of Multidisciplinary Research for Advanced Materials, Tohoku University, Miyagi 980-8577, Japan

T. Nagasaka
New Industry Creation Hatchery Center (NICHe), Tohoku University, Sendai 980-8579, Japan

R. G. Reddy et al. (eds.), *New Directions in Mineral Processing, Extractive Metallurgy, Recycling and Waste Minimization*, The Minerals, Metals & Materials Series,
https://doi.org/10.1007/978-3-031-22765-3_2

Introduction

Industrial steelmaking slag generally contains CaO, SiO_2, Fe_tO, P_2O_5, and some other minor compounds. These compounds distribute into two phases in steelmaking slag. One phase is a matrix phase containing Fe_xO with a small amount of other slag components, and the other phase is solid solution phase consisting of Ca_2SiO_4-$Ca_3P_2O_8$, referred to in this study as $(C_2S\text{-}C_3P)_{ss}$. Phosphorus always exhibits remarkable segregation in the form of $(C_2S\text{-}C_3P)_{ss}$; at the same time, the iron is found to be enriched in the Fe_tO-based matrix phase (Fe_tO-CaO-SiO_2 matrix phase) [1–6].

Many efforts have been made to recover phosphorus from steelmaking slag by the carbothermic reduction method, such as the work described by Shiomi et al. [7]. It is found that most of the phosphorus is recovered in the form of Fe-P alloy instead of P_2 vapor. Meanwhile, they noticed that a large amount of iron oxide is reduced until the reduction of P_2O_5 starts and indicated that the recovery of pure iron instead of Fe-P alloy might be possible. Later, Takeuchi et al. [8] attempted to recover iron and phosphorus separately from BOF slags through the carbothermic reduction method by the addition of Fe-Si alloy. The results showed that more than 95% of the iron and phosphorus were removed from the slags, and about 60% of the reduced phosphorus was removed to the gas phase as P_2, while the iron was recovered by Fe-Si alloy addition. Morita et al. [9, 10] investigated the carbothermal reduction behavior of both synthesized and factory steelmaking slag mixed with graphite powder at a temperature higher than 1873 K using microwave irradiation. After the microwave treatment, most phosphorus was recovered as Fe-P-Csatd (Carbon saturated) alloy. And they found that a higher fractional reduction of iron and phosphorus was achieved with a larger amount of carbon used, and an increase of SiO_2 content could lead to a considerable improvement in the reduction of both iron and phosphorus. Nakase et al. [11] further systemically investigated the effect of the slag composition, including basicity and FetO content in slag, on the phosphorus separation using a similar carbothermal reduction method in an induction furnace.

Currently, most of the steelmaking slag is used for the construction of roads, bridges, ports, and other civil engineering applications [12]. The carbothermic reduction of solid solution and synthesized slag has been researched; however, the phosphorus in the steelmaking slag has not been studied systematically yet [13, 14]. Understanding the effect of Fe_tO on the phosphorus evaporation behavior using the carbothermic reduction method is of great importance in promoting the recovery and reuse of phosphorus contained in steelmaking slag.

Therefore, in this research, the recovery of pure phosphorus from pretreatment slag was verified, and the distribution of phosphorus in the gas phase, metal phase, and residual slag after carbothermic reduction of steelmaking slag was clarified.

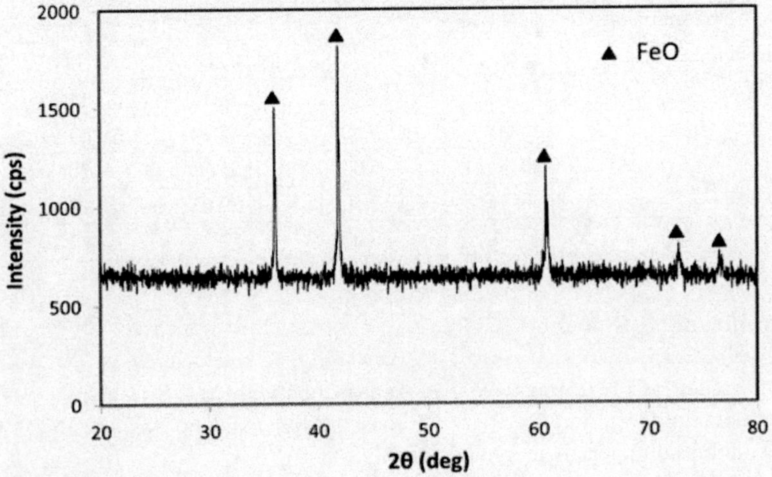

Fig. 1 XRD peaks of the synthesized FeO

Experimental

Preparation of Synthesized Slag

Preparation of FeO

Reagent Fe_2O_3 (Wako 1st Grade) was put in an alumina boat and heated in a horizontal resistance furnace. A pure iron sheet was placed in an alumina boat crucible to prevent Al_2O_3 from dissolving into raw material. Reagent Fe_2O_3 was heated up to 1200 K in an argon atmosphere. Then, a mixture of CO and CO_2 gas with a flow rate of 100ccm for each was used as a reducing agent at 1200 K for 2 h. The generated material was quenched, and then, XRD analysis was conducted to confirm the generation of FeO, as shown in Fig. 1. This procedure was repeated several times to satisfy the amount of synthesized slag.

Preparation of CaO

Reagent $CaCO_3$ (Wako 1st Grade) was put in the Platinum crucible and heated in the lift furnace. $CaCO_3$ was calcined at 1373 K for 4 h in air. Generated CaO was kept in a desiccator.

Table 1 Composition of synthesized slag

Composition/mass%	FeO	P_2O_5	CaO	SiO_2
Slag A	33.33	16.66	38.12	11.89
Slag B	20.00	5.00	57.14	17.86
Slag C	20.00	5.00	55.00	20.00
Slag D	20.00	2.50	59.05	18.45
Slag E	20.00	5.00	52.27	22.73

Preparation of (C_2S-40 mol% C_3P)$_{ss}$

Reagent-grade $CaCO_3$ (>mass 99.99%, Kanto Chemical Co.), SiO_2 (>mass 99.9%, Kanto Chemical Co.), and $Ca_3P_2O_8$ (≥ mass 98.0%, Nacalai Tesque, INC.) were used as the initial materials to synthesize (C_2S-C_3P)$_{ss}$.

Ca_2SiO_4 powder was prepared by mixing $CaCO_3$ and SiO_2 powder at a molar ratio of 2 to 1 and pressed into a pellet. The pellet of the mixture was put in a platinum crucible and heated at 1223 K for 4 h in the air using a $MoSi_2$ resistance furnace. Then, the generated material was confirmed to be Ca_2SiO_4 by XRD.

To synthesize (C_2S-40 mol%C_3P)$_{ss}$, generated Ca_2SiO_4 was mixed thoroughly with $Ca_3P_2O_8$ at a molar ratio of 6:4 [15]. The mixture was pressed into a disk pellet and put in a platinum crucible, and heated at 1773 K in air for 48 h with quenching in helium gas. The procedure, including crushing, mixing, pressing, heating, and quenching, was repeated twice and conducted to XRD to confirm the formation and homogeneity of solid solutions.

Preparation of Synthesized FeO-CaO-P_2O_5-SiO_2 Slag

5 kinds of FeO-CaO-SiO_2-P_2O_5 slags were synthesized, as shown in Table 1. Slag A was prepared based on the previous study [14]. Slag A, B, and D were synthesized with the same CaO/SiO_2, while slag B, C, and E were synthesized with different Fe/P ratio. The synthesized slags were prepared from a mixture of reagent $Ca_3P_2O_8$, reagent SiO_2, synthesized FeO, CaO, and a solid solution of C_2S-40 mol%C_3P. The mixtures were put in a pure iron crucible inside of a MgO crucible and then heated at 1673 K for 20 h in a vertical resistance furnace under an argon gas atmosphere. Synthesized slags were quenched in Helium gas flow.

Carbothermic Reduction Experiment of Synthesized Slag

A 0.5 g synthesized slag was mixed with graphite powder of 2 carbon equivalence (required carbon amount to reduce the total amount of P_2O_5 and FeO) and then pressed into a pellet (φ: 10 mm, h: 4 mm). Then, the sample was put into Al_2O_3

crucible and heated at 1573 K under 200 ccm of Ar gas flow in a horizontal resistance furnace. Samples were placed in the center of the heating zone of the horizontal resistance furnace, and after heating, the samples were pulled away from the center zone for quenching. Each sample was quenched after being heated for 1, 3, 5, and 7 h, respectively. The analysis methods were as same as the previous research [14]. Based on that, the distribution of P in the gas phase, metal phase, and slag phase of 5 kinds of synthesized slags could be calculated.

Results and Discussion

Distribution of P

Based on the results of weight change, ICP-AES, and SEM–EDS analyses, and material balance calculation of phosphorus, the phosphorus distribution in the gas phase, metal phase, slag phase, and the P% in the metal phase of 5 kinds of synthesized pretreatment slags are shown in Fig. 2.

After the carbothermic reduction experiment, phosphorus gas evaporated from all 5 kinds of synthesized slags, and the phosphorus distribution in the gas phase ranged from 6.83% to 55.26%.

Among these 5 kinds of synthesized slags, slag A, which has the highest basicity and lowest molar ratio of [Fe]/[P], has the highest amount of P distribution in the gas phase. This means about 55.26% phosphorus in the slag was transformed into phosphorus gas by carbothermic reduction at 1573 K.

Influence of P_{P2} of Initial Slag

In the previous study [14], it was found that Equilibrium partial pressure of P_2 gas in CaO-SiO_2-FeO-P_2O_5 slag in the reduction with CO-CO_2 gas mixture is larger than other kinds of gases such as Pgas, P_4 gas, and PO gas. Thus, most of the produced phosphorus gas is also assumed to be P_2 as a first approximation in this study.

The P_{P2} of slag with different Fe/P could be achieved. It was assumed that all of the Fe and P were reduced from the initial slag and dissolved into the metal phase. In order to clarify the relationship between the evaporation of P and the property of the initial slag, the P distribution in the gas phase based on P_{P2} of the initial slag was made (Fig. 3). According to the results, a higher P_{P2} of initial slag is beneficial for phoshorus recovery.

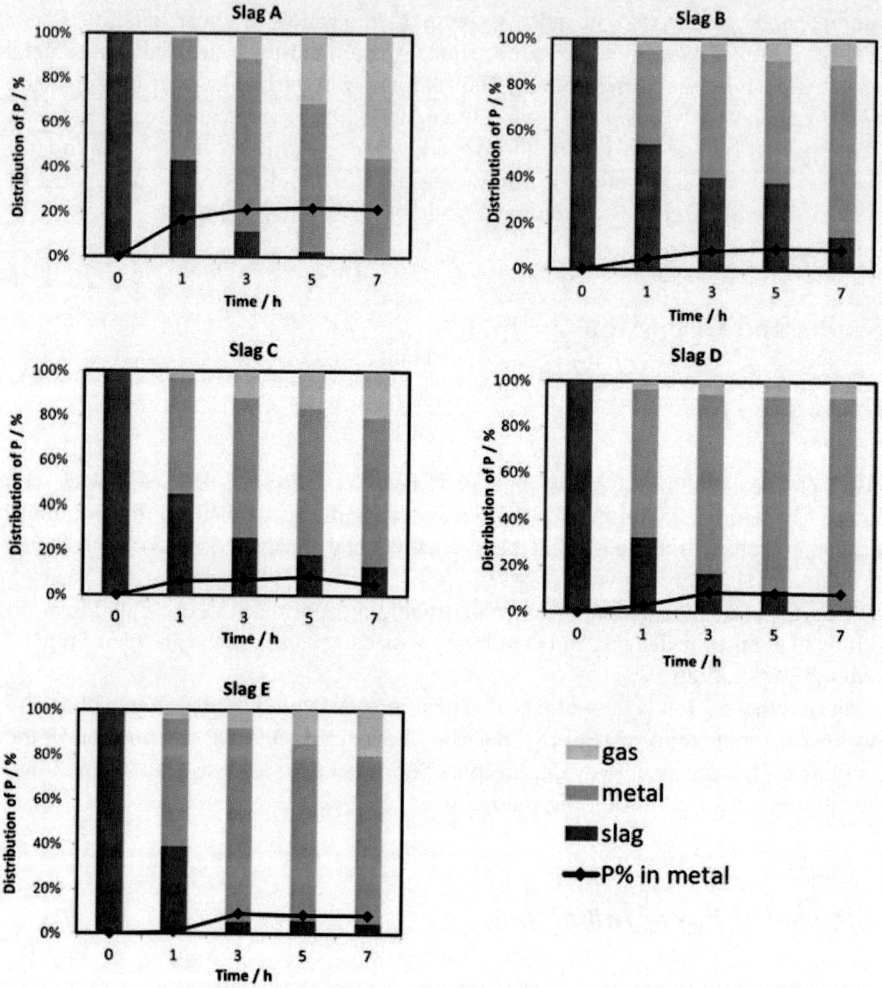

Fig. 2 Phosphorus distribution in the gas, metal, and slag phases and P/Fe in the metal phase

Conclusions

In the present work, the mechanism of carbothermic reduction experiement of synthesized slag was clarified. It was found that phosphorus could be evaporated from steelmaking slag by carbothermic reduction at 1573K, and during the carbothermic reduction process of synthesized slag, generated P was found to be dissolved into the metal phase, and then, part of it converted into gas phase. However, further investigations, especially the quantitative kinetics approach, are certainly required to implement this process.

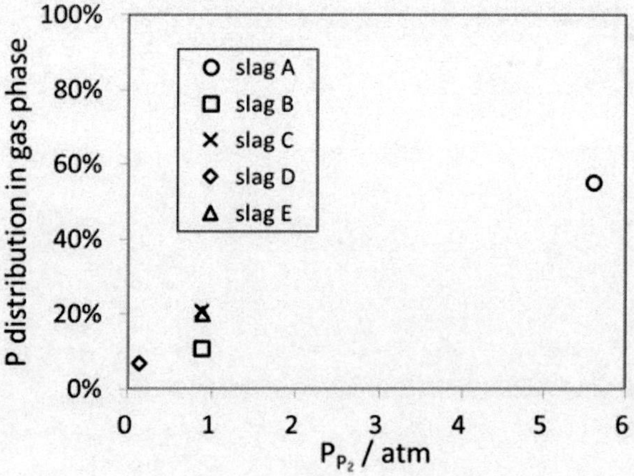

Fig. 3 P distribution in the gas phase based on P_{P2} of the initial slag

References

1. Matsubae-Yokoyama K, Kubo H, Nagasaka T (2010) ISIJ Int 50:65–70
2. Kubo H, Matsubae-Yokoyama K, Nagasaka T (2010) ISIJ Int 50:6
3. Yu H, Lu X, Miki T, Matsubae K, Sasaki Y, Nagasaka T (2022) Resour Conserv Recycl 180:106203
4. Uchida Y, Sasaki N, Miki Y (2018) ISIJ Int 58:869–875
5. Fushi-tani K, Matsu-suye M, Hasegawa M, Naito M, Iwase M (2007) Steel Res Int 78:96–101
6. Futatsuka T, Shitogiden K, Miki T, Nagasaka T, Hino M (2004) ISIJ Int 44:753–761
7. Shiomi S, Sano N, Matsushita Y (1977) Tetsu-to-Hagane 63:1520–1528
8. Takeuchi S, Sano N, Matsushita Y (1980) Tetsu-to-Hagane 66:2050–2057
9. Morita K, Guo M, Miyazaki Y, Sano N (2001) ISIJ Int 41:716–721
10. Morita K, Guo M, Oka N, Sano N (2002) J Mater Cycles Waste Manage 4:93–101
11. Nakase K, Matsui A, Kikuchi N, Miki Y (2016) Tetsu-to-Hagane 102:485–491
12. Yamasue E, Matsubae K, Nakajima K, Hashimoto S, Nagasaka T (2013) J Ind Ecol
13. Yu H, Lu X, Miki T, Sasaki Y, Nagasaka T (2021) J Sustain Metall 7:459–469
14. Yu H, Miki T, Sasaki Y, Nagasaka T (2022) Metall Mater Trans B 53:1806–1815
15. Yu H, Miki T, Sasaki Y, Nagasaka T (2020) Metall Mater Trans B 51:3007–3015

Lead Bleed-Off from Dust of Copper Smelter

Kifu Matsuura, Satoshi Shibata, and Fumito Tanaka

Abstract There has been a decrease in the grade of copper concentrate, leading to an increase in impurities load of copper smelter globally. Among such impurities, lead has a detrimental impact on electrorefining operation causing the passivation of the anode. Although reductive smelting of dust to produce lead bullion has been known to be an option of lead bleeding off the copper production circuit, impurities other than lead included in "dirty concentrate" in recent years may suffer the operations of lead smelting and successive refining of the bullion, as well. The present paper will suggest a process from pretreatment of feed to the lead smelter to pyro-refining of bullion to ensure satisfactory quality of bullion while managing impurities other than lead and discuss process factors that are crucial for the process design of the present process.

Keywords Copper smelter · Dust treatment · Impurities management · Lead

Introduction

The grade of copper concentrates has been decreasing while the content of impurities has been increasing in recent years. Among such impurities, lead has adverse effects on the tankhouse operation causing the passivation of anode [1]. In copper smelters, a substantial part of lead evaporates to dust, which is recycled to a smelter to recover copper. The dust is bled off occasionally from the copper smelting circuit to reduce the lead content of the feed [2, 3]. In the conventional process, the dust bled off is leached with sulfuric acid to recover copper and sent to lead smelters to produce

K. Matsuura (✉) · S. Shibata · F. Tanaka
Mitsubishi Materials Corporation, Tokyo, Japan
e-mail: mkifu@mmc.co.jp

S. Shibata
e-mail: ssatosi@mmc.co.jp

F. Tanaka
e-mail: f-tanaka@mmc.co.jp

R. G. Reddy et al. (eds.), *New Directions in Mineral Processing, Extractive Metallurgy, Recycling and Waste Minimization*, The Minerals, Metals & Materials Series,
https://doi.org/10.1007/978-3-031-22765-3_3

bullion. "Dirty concentrate" may, however, suffer the operation of lead smelting due to impurities other than lead. The present paper aims to suggest ferric leaching for the treatment of the dust to enhance the dissolution of the dust components and discuss a pyro-refining of bullion to ensure satisfactory quality of bullion while managing impurities other than lead.

Experimental

Leaching

Ferric leaching was applied on dust sampled from an electric precipitator of a copper smelter to remove copper to a satisfactorily low level. Figure 1 shows a schematic diagram of the experimental apparatus used for the leaching. The dust and a ferrous sulfate solution were mixed to a pulp of 300 g/L. The pulp was stirred magnetically, heated up to 90 °C, and aerated with air bubbling to accelerate oxidation of the dust for 24 h. The residue was separated from the leachate by filtration and dried subsequently in an oven. The residue dried was assayed by X-ray fluorescence analysis (XRF).

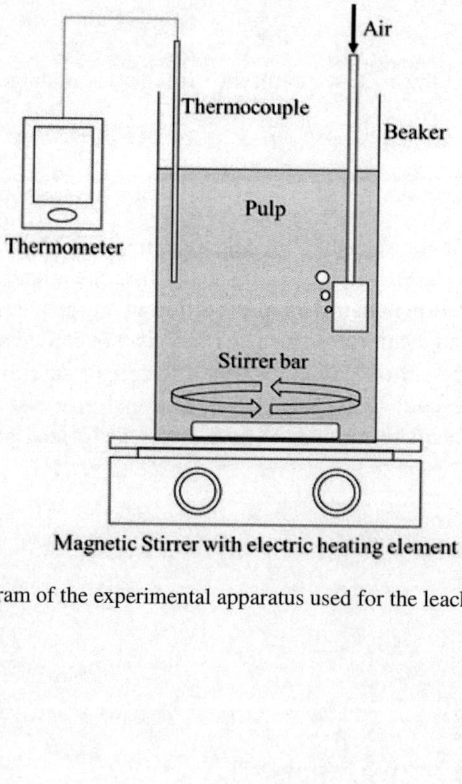

Fig. 1 Schematic diagram of the experimental apparatus used for the leaching

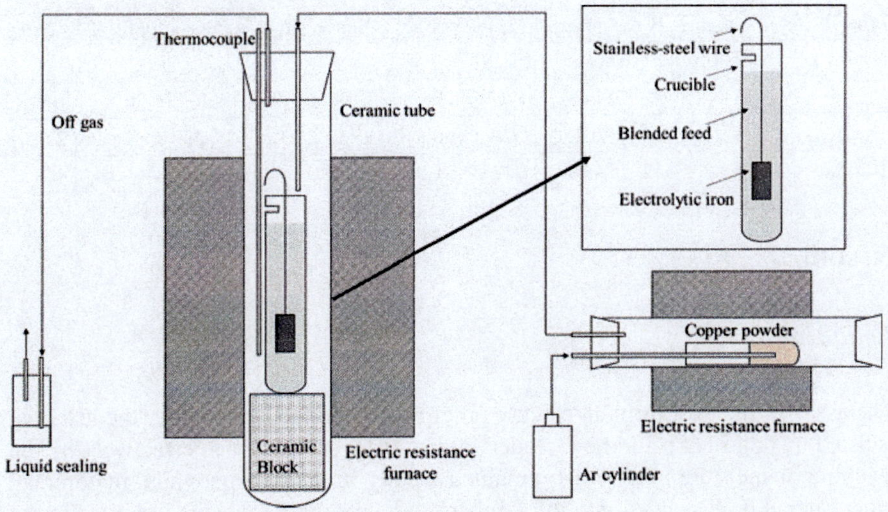

Fig. 2 Schematic view of the experimental apparatus used for the smelting experiment

Reductive Smelting

Figure 2 shows a schematic diagram of the experimental apparatus used for the smelting experiment. The dried residue of 15.0 g, granulated copper smelting slag of 1.5 g, limestone of 2.8 g, and activated carbon of 0.8 g were contained in a MgO crucible. A lump of electrolytic iron was hung in the crucible to react with PbS while maintaining the experimental system coexisting with iron. The crucible was placed in an electric-resistance furnace and then heated to 1200 °C. Argon deoxidized with preheated copper powder was used for the purging of the residual air in the furnace at the flow rate of 200 mL/min. After holding for 4 h, the lump of iron was removed from the crucible, and then, the crucible was taken out of the furnace. The crucible was quenched by continuous water spray. The slag was pulverized and assayed by XRF. The matte and bullion were dissolved into acid solution and assayed by inductively coupled plasma atomic emission spectroscopy (ICP-AES).

Liquation

Pb-based alloy was synthesized in an induction furnace to the same composition as the bullion obtained in the reductive smelting experiment. The Pb-based alloy was remelted in an electrical resistance furnace, then cooled down to 350 °C. The dross floating on the melt surface was skimmed off, and the refined bullion was cooled to room temperature. The dross and refined bullion were dissolved into acid solution and assayed by ICP-AES.

Table 1 Assay and mass of the dust from an electric precipitator and its residue after ferric leaching

	Mass (dry g)	Assay (mass%)							
		SiO_2	Ca	Fe	Cu	Zn	Sn	Pb	Bi
Dust	150	6.8	N/A	8.3	17.9	3.5	0.5	8.8	2.0
Residue	34.8	24.0	0.4	1.9	1.6	1.1	0.2	42.3	4.7

Results

Leaching

The assay of the dust from an electric precipitator and its residue after the leaching is listed in Table 1. Pb and Fe included in the dust were separated effectively by the leaching, namely Pb in the dust remained wholly in the residue while, in contrast, most Cu and Fe dissolved into the leachate. As a result, the residue to be fed to the reductive smelting was reduced in dry mass to 25% of the dust.

Reductive Smelting

Table 2 shows the average assay and mass of the experiments conducted twice because both results were close to each other. Slag, matte, and bullion obtained in the experiments were estimated homogeneous at the smelting temperature based on the microscopic observation of frozen samples. The mass of slag phase was calculated assuming that CaO included in the residue was completely distributed to the slag. Then, the mass of the matte and bullion was calculated by simultaneous linear equations for the Pb and Bi input/output balance. Most of the impurities were removed to slag and matte, and over 95% of Pb included in residue was distributed in the bullion. Over 30% of Cu and almost all Bi was distributed in the bullion too.

Table 2 Assay and mass of slag, matte, and bullion obtained by the reductive smelting

	Mass (g)	Assay (mass%)							
		SiO_2	CaO	Fe	Cu	Zn	Sn	Pb	Bi
Slag	10.2	36.7	17.2	21.8	0.1	1.1	0.0	0.5	0.0
Matte	2.5	N/A	N/A	50.8	6.1	0.4	0.0	9.8	0.5
Bullion	6.8	N/A	N/A	0.3	1.2	0.0	0.1	89.4	9.1

Table 3 Assay and mass of the samples fed to and obtained by the liquation

	Mass (g)	Assay (mass%)				
		Pb	Cu	Sn	Fe	Bi
Bullion (feed)	1,000	89.4	1.2	0.1	0.3	9.1
Dross	173	79.2	5.2	0.6	1.7	13.9
Liquated bullion	827	91.7	0.4	0.0	0.0	8.0

Liquation

Table 3 shows the assay and mass of samples fed to and obtained by the liquation. The mass balance demonstrates that liquation is a necessary step before casting because substantial dross forms during cooling to 350 °C. All the Fe and Sn as well as 75% of Cu included in bullion was removed to dross. In contrast, over 70% of Bi remained in the liquated bullion.

Suggested Flowsheet

Figure 3 suggests the flowsheet of the present lead bleeding-off process indicating mass and elemental distribution based on the dust to feed to the leaching step. Lead recovery from the dust to the liquated bullion may be 83%, and most lead loss happens in the liquation. Since the liquated metal is mainly copper, the decopperization in the leaching is decisive in the performance of the process. Although most impurities are eliminated from the feed, 40% of the bismuth in the feed is distributed to the liquated bullion.

Table 3 suggests that residual impurities of the liquated bullion are copper and bismuth. Those impurities can be removed in several lead smelters effectively rather than in copper smelters. The liquated bullion can be processed by the Hulst process to remove copper, subsequently electrorefined by the Betts process in which bismuth is distributed to the anode slime. Bismuth included in the slime can be separated by partial oxidation in a smelting furnace into bismuth litharge, which is further smelted and electrorefined to obtain a bismuth cathode [4]. Pyrorefining by Kroll-Betterton process, which has been adopted more commonly than partial oxidation, is not applicable to remove bismuth from the liquated bullion due to its excessive level of bismuth content.

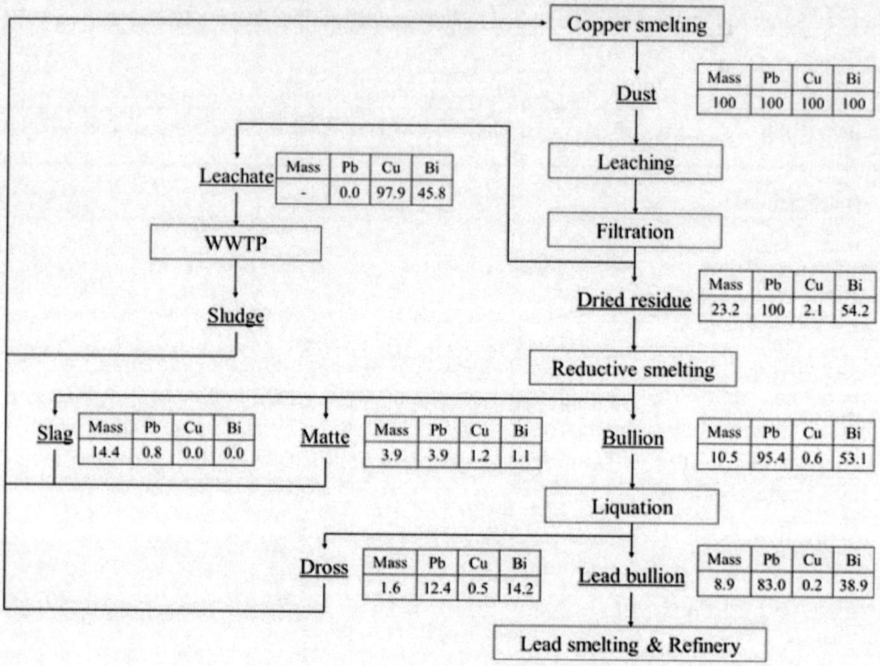

Fig. 3 The flowsheet of lead bleed-off and material balance of major elements (Tables in the figure indicate the distribution ratio in percentage from the dust to the designated product)

Mass and Energy Balances of the Reductive Smelting

Since reductive smelting is a high-temperature process, the energy balance of the reductive smelting was evaluated in addition to the mass balance in the case that lead of 2,000 tons/year is removed from a copper smelting circuit to a lead smelting circuit by the use of a Héroult furnace. In this case, the dried residue fed to the Héroult furnace was set to 5,703 tons/year under the condition which the Pb content of the residue is the same value as listed in Table 1. The energy balance of such smelting operation during the period from one tap to the next of the bullion other than the heat loss from the furnace was calculated referring to thermodynamic data from HSC Chemistry ver 9.6.1 and assuming that:

- The tap-to-tap period of the bullion is 1 day and the availability of the furnace is 90%, (based on the assumption the cumulative amount of the residue to be fed becomes 17.4 tons/day);
- Every reactant charged to the furnace has the temperature of 25 °C and every product discharged from the furnace has the temperature of 1200 °C;
- The residue comprises $PbSO_4$, Cu_2S, SnO_2, $CaSO_4$, SiO_2, Fe_2O_3, ZnO, Bi_2O_3, and other gangue oxides;

- Copper smelting slag comprises FeO, Fe3O4, CaO, SiO_2, ZnO, and other gangue oxides;
- Coke contains solely fixed carbon, and limestone and iron scrap contain negligible impurities;
- Reductive smelting slag comprises PbO, FeO, CaO, SiO_2, ZnO, and other gangue oxides;
- Reductive smelting matte comprises PbS, FeS, Cu_2S, ZnS, and Bi_2S_3;
- Reductive smelting bullion comprises Pb, Fe, Sn, Cu, and Bi; and
- Furnace off-gas comprises CO, CO_2, SO_2, Pb(g), and Zn(g).

Based on a rule of thumb for designing metallurgical furnaces, the throughput per hearth area was set to the same value as an existing furnace used for a similar operation to the present process. Referring to the Héroult furnace at Kosaka smelter for smelting 1.79 tons/(m^2·day) of Pb-bearing residue [5], the inner diameter of the present furnace was set to 3.51 m. Equations (1)–(3) were applied on the calculation of the heat loss through the hearth [6] so as to meet both Q in Eqs. (1) and (2), neglecting the temperature gradient between bulk bullion and brick surface and assuming the ambient temperature to be 30 °C. The overall heat loss from the furnace was presumed twice the heat loss from the furnace in the present calculation.

$$Q = A \cdot \frac{(1473 - T)}{\frac{l_{brick}}{\lambda_{brick}} + \frac{l_{shell}}{\lambda_{shell}}} \quad (1)$$

$$Q = h \cdot A \cdot [T - 303] + \varepsilon \cdot \sigma \cdot A \cdot \left[T^4 - 303^4\right] \quad (2)$$

$$\frac{h \cdot D}{\lambda_{air}} = 0.27 \cdot (Gr \cdot Pr)^{0.25} \quad (3)$$

where

Q Heat flux, (W/m^2)
T Surface temperature of shell, (K)
A Hearth area, (m^2)
l Thickness of layer, (Brick = 0.4 m, Shell = 0.045 m)
λ Thermal conductivity, (Brick = 3 W/(m · K), Shell = 43 W/(m · K))
h Heat transfer coefficient, (W/(m · K))
ε Emissivity of shell, (=0.79)
σ Stefan–Boltzmann constant, $\left(= 5.67 \times 10^8 \mathrm{W}/\left(\mathrm{m}^2 \cdot \mathrm{K}^4\right)\right)$
D Inner diameter of furnace, (m)
Gr Grashof Number
Pr Prandtl number

Table 4 shows the mass and energy balances per a cycle from a tap to the next tap of the bullion. The difference in energy between input and output was balanced by the Joule's heat originating from electricity. Considerable endothermic heat of the reaction heat stems from thermal decomposition of lead sulfate and calcium

Table 4 Mass and energy balances of the reductive smelting

Input			Output		
	Mass (kg/cycle)	Heat (MJ/cycle)		Mass (kg/cycle)	Heat (MJ/cycle)
Dried residue	17,360	–	Off-gas (Nm^3/cycle)	4,884 (2,317)	6,464
Limestone	3,258	–	Slag	10,769	15,502
Copper smelting slag	1,741	–	Matte	2,900	3,111
Coke	938	–	Bullion	7,843	1,557
Iron scrap	3,098	–	Heat of reactions	–	10,782
Joule's heat	–	48,228	Heat loss	–	10,813
Total	26,396	48,228	Total	26,396	48,228

sulfate. The off-gas composition calculated was $CO_2/CO = 1.52$ which corresponds to $log p_{O_2} = -10.6$ at 1200 °C. Therefore, the present experiment was conducted presumably in close conditions to lead smelting operations [7].

Conclusions

The present paper has suggested an alternative lead bleeding-off process from copper smelting circuits in which lead-bearing dust collected from the electrostatic precipitator was leached in an acid ferric solution followed by the reductive smelting and the liquation to obtain bullion containing over 90%Pb. The liquated bullion requires the separation and recovery of residual bismuth which can be achieved by the litharge process having been operated economically and solely in several lead smelter and refinery. Therefore, bleeding-off impurities from "dirty concentrate" requires metallurgical network of copper production circuit and lead production circuit.

References

1. Schlesinger ME, Sole KC, Davenport WG, Flores GRFA (2022) Extractive metallurgy of copper, 6th edn. Elsevier, pp 331–359
2. Tomita M, Higashi M and Okamoto H (1992) Hydrometallurgical process for copper converter dust at the Saganoseki smelter & refinery. In: Proceedings of the international symposium on residues and effluent—processing and environmental considerations, pp 283–294
3. Moldabayava GZ, Akilbekova SK, Mamyrbayeva KK, Mishra B (2015) Electrosmelting of lead-containing dusts from copper smelters. J Sustain Metall 1:286–296

4. Kuwabara M, Oba H (1993) Lead smelting and refining at Hosokura. J Min Metall Inst Japan 109:999–1002
5. Wakamatsu N, Maeda Y, Suzuki S (1987) Behavior of impurities in the Kosaka lead smelting works. In: Proceedings of the annual meeting of the Mining and Metallurgical Institute of Japan, M-3
6. Themelis NJ (1995) Transport and chemical rate phenomena. Gordon and Breach Publishing, pp 145–175
7. Moon N, Hino M, Lee Y, Itagaki K (1999) Composition region of homogeneous liquid phase in lead-containing slags related to QSL process. J MMIJ 115:263–268

Simulation and Post-mortem Studies: The Holistic Approach for Optimized and Engineered Lining Concepts

D. Gregurek, G. Unterreiter, C. Lind, and A. Spanring

Abstract Copper smelting furnaces are typically lined with magnesia-chromite refractories. This paper evaluates the common refractory wear mechanisms of infiltration, spalling, and chemical attack as observed in the copper anode furnace. All these wear parameters lead to severe degeneration of the brick microstructure and to a decreased lining life. Additionally, the influence of infiltration on the thermal conductivity and temperature field on a simplified furnace lining was investigated by means of numerical simulation. A steady state heat transfer finite element analysis was conducted using a theoretical model of a cross-section of the furnace. Two cases with different material properties were considered. The first case considered virgin lining and the second a lining with deep slag infiltrations. Accordingly, a detailed investigation and understanding of the wear mechanisms through "post-mortem studies" in conjunction with numerical simulations is an important prerequisite for the refractory producer.

Keywords Copper anode furnace · Numerical simulation · Finite element analysis · Wear phenomena

Introduction

During the copper production process, so-called white metal or matte, with a copper content of approximately 75%, can be upgraded in a Peirce-Smith converter to blister copper which has a copper content of 98%. This process consists mainly of blowing air through tuyeres for oxidizing and slagging of the iron in the matte [1]. In the next step, blister copper is refined in anode furnaces before it is cast in anodes for further refining by electrolysis.

D. Gregurek · G. Unterreiter (✉)
RHI Magnesita, TC Leoben, Magnesitstrasse 2, A-8700 Leoben, Austria
e-mail: Guenter.unterreiter@rhimagnesita.com

C. Lind · A. Spanring
RHI Magnesita, Kranichberggasse 6, A-1120 Vienna, Austria

R. G. Reddy et al. (eds.), *New Directions in Mineral Processing, Extractive Metallurgy, Recycling and Waste Minimization*, The Minerals, Metals & Materials Series,
https://doi.org/10.1007/978-3-031-22765-3_4

Fig. 1 Anode furnace for the copper industry

The anode furnace shown in Fig. 1 is a horizontal cylindrical rotary furnace which operates in batches. In the first step, sulphur is removed by oxidation followed by a reduction step to remove the oxygen. Impurities such as arsenic lead and antimony can also be removed from the blister copper. In the field of copper production as in an anode furnace, magnesia-chromite bricks are mainly used for the linings [2]. The so-called mag-chrome bricks contain a variety of refractory materials which differ both in cost and in effectiveness in copper making. These materials can be distinguished by (a) the type of bond, (b) composition, and (c) purity level [3]. Thermal conductivity of a ceramic or a refractory is strongly determined by its microstructure. The presence of pores, grain boundaries, impurities, etc., inhibits heat transfer through the material and consequently reduces the thermal conductivity [4]. The heat conductivity of pure copper is about 400 $W{\cdot}m^{-1}\ K^{-1}$ of copper oxide 80 $W{\cdot}m^{-1}\ K^{-1}$ [5] and 4 $W\ m^{-1}{\cdot}K^{-1}$ for the un-infiltrated (virgin) magnesia-chromite brick. Within the reaction zone, the microstructure was described as completely dissolved by numerous Cu-Fe-Cr oxides. Hence, this will lead to a significant change of the material properties such as thermal conductivity and thermal expansion. Based on material measurements and finite element analysis, the aim of this study is a qualitative comparison of the temperature field in virgin and infiltrated conditions.

The paper at hand describes the main wear phenomena of spent magnesia-chromite bricks from the copper anode furnace based on material measurements and finite element analysis. The wear phenomena discussed include corrosion caused by slag and by copper infiltration. The refractory sample was collected as shown in the schematic sketch of the anode furnace (see Figs. 2 and 3).

The investigations were done by chemical analysis (X-ray fluorescence), mineralogical investigation (reflected light microscopy and scanning electron microscope), and laser flash method for thermal conductivity.

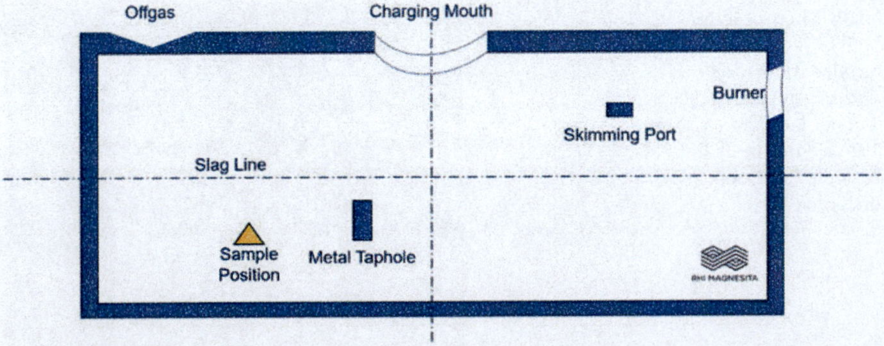

Fig. 2 Schematic sketch of the cylindrical copper anode furnace showing the location of sampling

Fig. 3 Cut section of used magnesia-chromite brick from the copper anode furnace (sample position from Fig. 2)

Wear Phenomena

Slag Corrosion

The main driving force behind corrosion is dissolution reaction between refractory components (mainly oxide phases) and slag. The dissolution is always diffusion controlled. It is governed by the difference in concentration gradients, i.e., in concentration directly at the interface slag/refractory versus bulk concentration farther from the interface (Fick's First Law of Diffusion) [6]. In addition to dissolution reaction at the hot face, the slag also infiltrates the refractory pore space. The chemistry of the slag is an additional crucial parameter to be considered. In the infiltrated refractory microstructure, there is a chemical reaction within the inherent components (e.g.,

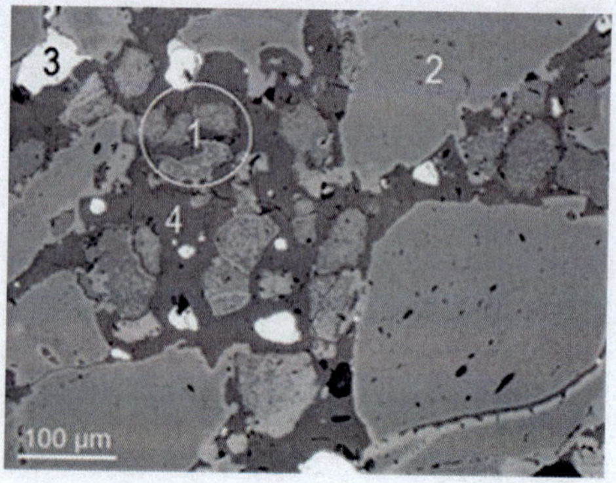

Fig. 4 Microstructural detail of infiltrated and corroded refractory microstructure. Distance from hot face approx. 2 mm. Corroded magnesia fines (1). Chromite (2). Copper (3). Forsterite (4)

dissolution of magnesia), according to the respective phase equilibrium. In particular, the magnesia fines responsible for the brick bonding are the most vulnerable fraction in the refractory. The infiltrating slag will decrease the critical share and prepare hot erosion due to loss of brick bonding. An example of the infiltrated and corroded refractory microstructure is shown in Fig. 4.

Within the infiltrated brick microstructure, the main reaction products include Mg–silicate (forsterite, Mg_2SiO_4) due to corrosion of sintered magnesia and the interstitial phase. In general, the second main brick component chromite is not corroded compared to the magnesia.

Copper Infiltration

The slag infiltration of the brick microstructure is frequently followed by copper/copper oxide infiltration (Fig. 5). Hereby, the kinetic aspects, such as viscosity, pore size and pore distribution in the refractory, temperature during the furnace operation, and wetting angle/surface tension between the refractory and the metal, play a dominant role in the infiltration behaviour [7]. Generally, non-oxide infiltration only densifies the bricks' microstructure without any corrosion of the brick inherent phases.

Numerical Simulation and Finite Element Analysis (FEA)

Simulation Model

Two different thermal steady state scenarios (infiltrated and virgin) were modelled by using a FEA based software package [8]. A representative three-dimensional section of the geometry (see Fig. 6) was used for the geometrical model. The sample which

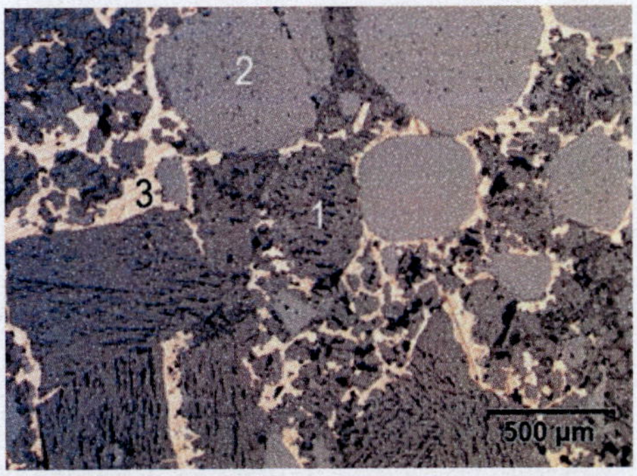

Fig. 5 Microstructural detail of infiltrated microstructure. Distance from hot face approx. 20 mm. Fused MgCr material (1). Chromite (2). Copper (3)

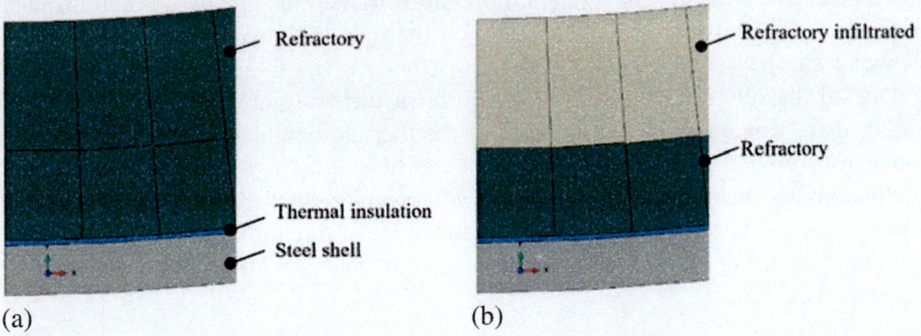

Fig. 6 Geometrical domain **a** without infiltration and **b** with infiltration representing a part of the anode furnace

was used for measuring material data used in the finite element analysis was taken from the sample position shown in Fig. 2.

A theoretical model for an infiltrated and a non-infiltrated scenario was set up in order to investigate the influence of infiltration. To highlight the differences caused by the modified thermal conductivity, the same wall thickness was used for both the infiltrated and the non-infiltrated scenario. Firstly, the non-infiltrated refractory was taken under consideration. Therefore, material data for bulk density and thermal conductivity of the virgin magnesia bricks were used. Then, an infiltrated refractory sample was analysed to measure density and thermal conductivity. Due to wear phenomena, the length of the brick under research was reduced from 450 to 220 mm with an infiltration depth of 130 mm at the sample position. In our model, we were assuming a theoretical lining where the lining thickness was constant; hence, no wear profile has been developed.

The model represents a steady state condition for the hot faces above and below the slag line at a constant temperature of 1280 °C. The thermal boundary condition for the outer surface of the steel shell was considered with a constant heat transfer

Table 1 Temperature dependent thermal conductivity of materials in W·m^{-1}·K^{-1}

Material	25 °C	100 °C	200 °C	400 °C	500 °C	600 °C	750 °C	800 °C	1000 °C	1200 °C
Refractory	4.95		4.73	4.22		3.93		3.8	3.67	3.15
Refractory infiltrated	14.9		14.6	13.6		13.4		13.4	12.5	
Insulation			0.02	0.03	0.03	0.04		0.05		
Steel	55	55	51	44	41					

coefficient of 7.5 W·m^{-2}·K^{-1}, a temperature of 25 °C, and an emissivity of 0.9 for surface radiation. These boundary conditions resulted in temperature of about 200 °C which is in agreement with temperature measurements of the steel shell temperature of the furnace in operation. In Table 1, the temperature dependent physical properties of the considered materials are provided.

The change of the thermal conductivity in the transition zone between infiltrated and virgin material was modelled based on the mineralogical observations which showed a sharp change.

Due to the observed infiltration of the refractories by copper oxide shown in Fig. 7, there was a significant increase of the thermal conductivity. The refractory under research had an apparent porosity of 15 vol%.

Phenomena such as chemical reaction which would alter the material properties were neglected. The goal was to show the influence of the infiltration on the temperature field resulting in a comparison of the infiltrated versus virgin lining based on a theoretical design.

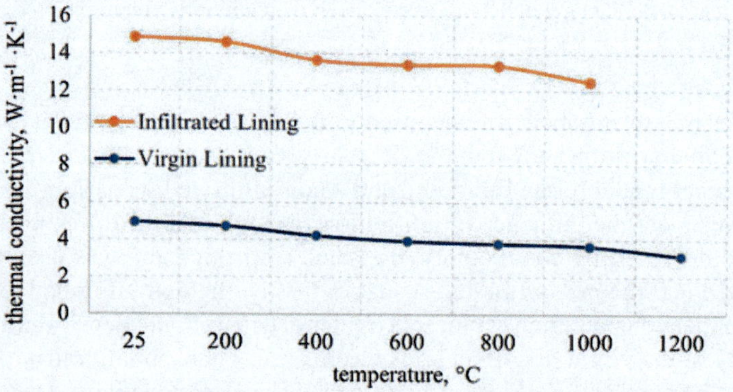

Fig. 7 Measured thermal conductivity of the mag-chrome refractories

Results

The thermal conduction of the virgin lining and of the infiltrated lining shows a decline between the temperature range of 25 °C and 1200 °C.

The steady state temperature distribution after heat-up is shown in Fig. 8. The temperature profile of the infiltrated sample (plotted along path 1, see Figs. 8 and 9) shows a lower gradient at the beginning and higher temperatures in general, which cause higher heat losses, and consequently, higher energy consumption. An individual solution based on calculations and numerical simulations in order to minimise the heat losses can be found.

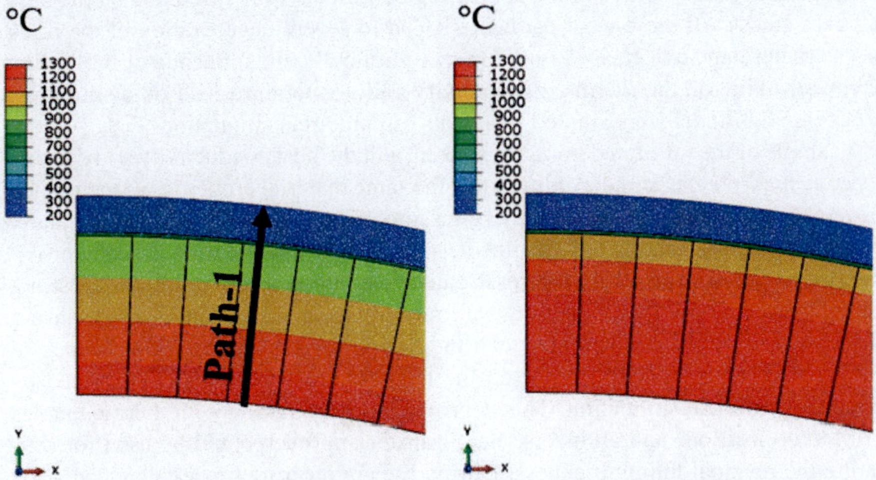

Fig. 8 Steady state temperature distribution of the **a** virgin and **b** infiltrated lining

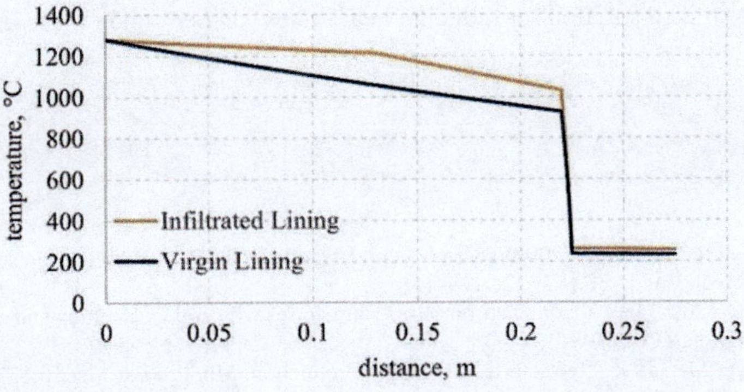

Fig. 9 Temperature along path 1

In view of future work, the stresses and strains for both scenarios will be simulated and compared based on a linear-elastic approach. It is furthermore of interest to know the magnitude of the Mises stresses in the steel shell. In the next step, creep effects will also be under consideration, what will be of interest here is the extent to which creep phenomena in the lining reduce stresses.

Conclusions

In this research, common refractory wear mechanisms of infiltration, spalling, and chemical attack as observed in the copper anode furnace were discussed on the base of a case study. All these wear parameters lead to severe degeneration of the brick microstructure and a decreased lining life. Additionally, the influence of infiltration by copper oxide on the thermal conductivity and temperature field on a simplified furnace lining model were studied by means of numerical simulation.

A sample of the infiltrated lining was taken, and the heat conductivity and density, respectively, were measured. Additionally, the same material properties as mentioned above were analyzed on a so-called virgin sample. Due to the infiltration by mainly copper oxides and metallic copper, the thermal conductivity was increased approximately by a factor of four. This increase can be explained by the porous microstructure of the refractory and the strong infiltration with copper oxides which have a much higher thermal conductivity than the non-infiltrated brick. The influence of the increased thermal conductivity on the temperature field was demonstrated by means of numerical simulation. This approach in combination with thermocouples at different positions and an inverse heat transfer approach could be used for estimating the residual lining thickness. During the engineering process, a calculation with the material data of the virgin lining followed by measured material parameters of the infiltrated lining can be made. Hence, the in-service performance of the lining can be described more realistically by means of a numerical simulation such as FEA. Thus, an optimized solution for reducing heat losses can be found for the individual process situation.

References

1. Aminizadeh N, Mansouri S (2007) Thermochemical model of the Peirce-Smith copper converter. Proc Inst Mech Eng E-J P M E 221:129–138
2. Crites MEA (2000) Interaction of chrome-free refractories with copper smelting and converting slags. Can Metall Q 39:129–134
3. Schlessinger ME (1996) Refractories for copper production. Min Pro Ext Met 16:125–146
4. Vitiello D (2021) Thermo-physical properties of insulating refractory materials. Universite' de Limoges, vol NNT: 2021LIMO0034, p 23
5. Dhivyaa Anandan KR (2012) Synthesis and stability of cupric oxide-based nanofluid: a novel coolant for efficient cooling. Asian J Sci Res 1

6. Harmuth H, Vollmann S (2014) Refractory corrosion by dissolution in slags - challenges and trends of present fundamental research. Iron Steel Rev 58(4):157–170
7. Barthel H (1981) Wear of chrome magnesite bricks in copper smelting furnaces. Interceram 30:250–255
8. Dassault Systems. https://www.3ds.com/products-services/simulia/products/abaqus/. Accessed 16 Sept 2021

A Kinetic Description of Physico-Chemical Processes Taking Place in the Burden of HCFeMn Submerged Arc Furnaces

Ainur Nigmetova, Bertil Farjaudon, Haoxue Han, Astrid Hecquet, and Gilles Nussbaum

Abstract The pre-reduction reactions of high Mn-oxides in Comilog and Nchwaning manganese ores have long been investigated because of their potential to reduce both the specific energy and coke consumptions of FeMn alloys. However, the performances of most industrial furnaces do not beneficiate from fully completed pre-reduction reactions. In this context, a new incremental approach, using an experimental and numerical tool, is developed to investigate the mechanisms involved from the microscale to the industrial scale using experimental tools from the laboratory set-ups to pilot equipment. The pre-reduction optimization study was started at the microscale. Results of the lab-scale trials on both ores at different temperatures, CO/CO_2 ratios, and the numerical model simulating the kinetics and heat transfer of the pre-reduction reactions will be presented. The main kinetic parameters have been identified from the lab-scale trials using the numerical model, compared to the literature, and interpreted to explain the physico-chemical behavior of pre-reduction.

Keywords Multi-scale approach · Mn ore pre-reduction · Experimental tool · Kinetic model · Physico-chemical parameters

Introduction

As a significant Mn alloys producer, Eramet has strong ambitions regarding climate change and is committed to reduce its climate footprint by reducing CO_2 emissions. The production of Mn alloys is carried out by smelting Mn ore in an electrical furnace using carbonaceous material as a reducing agent. This process ultimately creates CO_2 gas emissions [1]. The most promising way to reduce GHG emissions is to improve the efficiency of the ore pre-reduction by CO gas generated in the HCFeMn furnace.

Mn ores are progressively reduced to metal: (1) reduction to lower Mn-oxide by carbon monoxide in the pre-reduction zone (Fig. 1) leading to exothermic reactions:

A. Nigmetova (✉) · B. Farjaudon · H. Han · A. Hecquet · G. Nussbaum
Eramet Ideas, Trappes, France
e-mail: ainur.nigmetova@eramet.com

R. G. Reddy et al. (eds.), *New Directions in Mineral Processing, Extractive Metallurgy, Recycling and Waste Minimization*, The Minerals, Metals & Materials Series,
https://doi.org/10.1007/978-3-031-22765-3_5

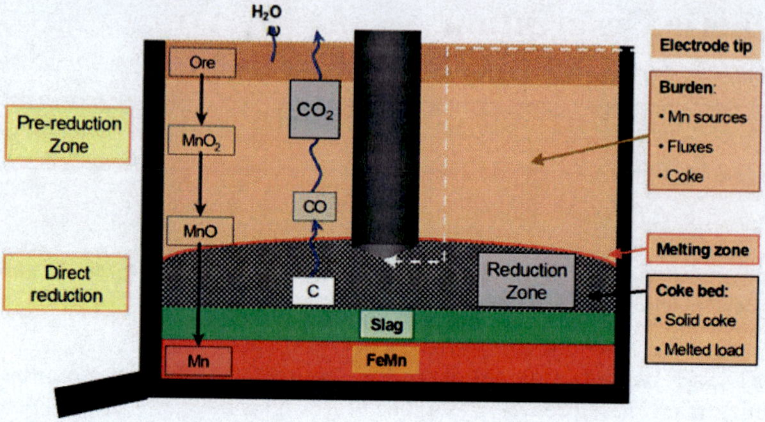

Fig. 1 FeMn alloys production operation

(a) $2MnO_{2(s)} + CO_{(g)} \rightarrow Mn_2O_{3(s)} + CO_{2(g)}, \Delta H = -199.9kJ.mol^{-1}Mn$

(b) $3Mn_2O_{3(s)} + CO_{(g)} \rightarrow 2Mn_3O_{4(s)} + CO_{2(g)}, \Delta H = -187.8kJ.mol^{-1}Mn$

(c) $Mn_3O_{4(s)} + CO_{(g)} \rightarrow 3MnO_{(s)} + CO_{2(g)}, \Delta H = -50.7kJ.mol^{-1}Mn$ (1)

(2) MnO is eventually reduced to metal by C at the bottom of the furnace, also called direct reduction zone (Fig. 1) leading to endothermic reaction:

$$MnO_{(l)} + C_{(s)} = Mn + CO, \Delta H = 275kJ.mol^{-1}Mn \quad (2)$$

CO_2 is produced by the Mn-oxides pre-reduction reactions (Eq. 1). If the materials are not entirely pre-reduced (MnO_x, x > 1), when reaching 800 °C, a significant amount of CO_2 may react with the coke particles to produce CO according to Boudouard reaction: $CO_2 + C \rightarrow 2CO$ [2]. Besides consuming additional carbon, it is highly endothermic, thus increasing electric energy consumption. Today, the pre-reduction yield in industrial furnaces is estimated to be low (about 30%). Increasing the pre-reduction reactions kinetics will therefore result in large gains in specific electricity and coke consumptions. The impact of the physical and chemical properties of the pre-reduction zone (burden) and its components on the main parameters that control kinetics of pre-reduction reactions are studied by many authors [2–4]. The Mn ore pre-reduction represents a system involving the different phenomena in competition and depending on the different conditions at different scales.

At the Ore Scale

At the scale of one ore particle, gas-solid reactions can be controlled by gaseous diffusion, chemical kinetics, or both [5]. The diffusion of gas within the solid depends on the reaction temperature, and the ore particle microstructure changes all along the pre-reduction reaction. The chemical reaction takes place on a boundary layer between an unreacted core and a solid product shell and depends on the gas composition. In the case of pre-reduction, the gas diffusion and chemical reaction are mostly controlled by the following parameters:

- The specific surface area of the particle accessible to the gas, which in turn depends on:
  - Particle size: the finer the particle, the faster the reaction. The particle size may change as the reactions proceed and the temperature increases.
  - Particle microporosity: more microporosity will facilitate the access of CO gas to Mn-oxides via gas diffusion. The microporosity may change as the reactions proceed and the temperature increases.
- The reactivity of Mn-oxides in the minerals, which acts as the driving force for the chemical reactions between the minerals and CO gas.
- The partial pressure of CO: the higher this pressure, the faster the reaction.
- Gas temperatures, which essentially locally depend on the endothermic and exothermic reactions that take place in the burden and on the intensity of the pre-reduction reactions.
- Particle temperature, which depends on their nature and physical properties.

At the Burden Scale

The burden can be regarded as a multi-particle system composed of single particles. Mentioned above parameters are valid to each particle of the multi-particle system, but in addition, at the burden scale the following factors are considered to control the burden pre-reduction behavior:

- Macroporosity between ore particles: higher macroporosity will facilitate the gas flow in the burden.
- Permeability: a high permeability of the burden is related to the open connection between macropores, which in turn provides a good gas flow distribution.
- The moisture: as the moisture content increases, the width of the low temperature zone ($T < 200$ °C) increases. At such low temperatures, the kinetics of the pre-reduction reactions are very low.
- Position of the particle in the burden: temperature difference on the upper or lower part of the burden affects the pre-reduction behavior by causing endothermic reactions.

In this article, we present our incremental approach to study the kinetics of prereduction of the Mn ores.

Incremental Approach

This multi-scale approach is presented in Fig. 2. It follows an incremental approach aiming at increasing gradually the scope of the studies and the complexity of the investigated mechanisms, from the lab-scale to the industrial scale. Both the experimental work and the numerical modeling work reflect this approach.

As one sees, pre-reduction involves the coupling of numerous chemical reactions and phase changes, multi-phase material flow, and heat transfer. The optimization of pre-reduction should hence be built upon a clear understanding of the above-mentioned phenomena. This is the reason why we adopted a multi-scale and incremental approach which consists of decomposing pre-reduction into phenomena studied at different scales and combining experiment and modeling work.

Alongside these studies, we are building a numerical model of the burden, the "Burden Model". The model simulates the simultaneous heat and mass transfers between a gas circulating in a porous bed coupled with the kinetics of chemical reactions. It shall be a numerical twin of the burden found in an industrial furnace, as it is not only following a rigorous and coherent mathematical description of the physico-chemical phenomena but is also calibrated with solid experimental results from laboratory and pilot trials. The "Burden Model", once accomplished, will be a helpful tool: to capitalize on the knowledge and in-depth understanding of the burden operation, to identify key levers and evaluate their impact on the pre-reduction yield, and to guide experimental and industrial trials for pre-reduction optimization.

At the lab-scale, the focus is the kinetics of different reactions in the ore particles; at pilot scale, the pre-reduction efficiency is largely determined by the reductant gas

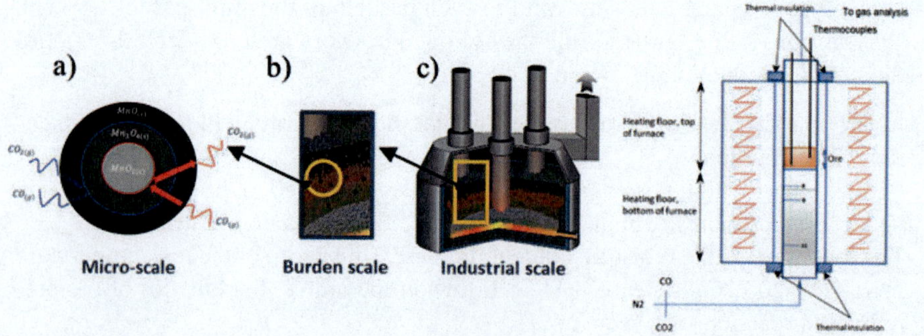

Fig. 2 Left: Multi-scale problem applied to the FeMn furnace: from micro to industrial scale. Right: Schematic representation of the lab-scale furnace

flow in the burden porous network, whereas in an industrial furnace, the spatial distributions of temperature and flow resistance are the important influencing parameters, and meanwhile a result of the chemical reactions in the burden. In the present work, the lab-scale results are presented. The pilot equipment is currently being built, and a short description of the equipment is given below.

Lab-Scale

The aim of the lab-scale trials is twofold. The first one is to deepen our understanding of the Mn-oxides reduction and to reveal important points before starting pilot tests. The second is to identify kinetic parameters, specific to the ores and their properties, to provide a predictive model at the microscale (Fig. 2 left a). To achieve the latter, the strategy based on the work of [6] is applied: cover the whole temperature range for reductions of Mn-oxides; vary the conditions: CO concentration (%CO), temperature; decouple the Mn-oxide reduction reactions (Eq. 1). To this end, the ore is reduced at successive isothermal plateaus, to obtain the kinetics in well-defined conditions.

At the lab-scale, a small bed of Mn ore particles is introduced in a resistance furnace, above a bed of alumina beads. The materials are heated to a pre-defined temperature under an inert gas flow (N_2). Then a mix of CO/CO_2 gas is fed at the bottom of the furnace. Thermocouples in the Mn ore beds record the evolution of the temperature (see Fig. 2 right). The evolution of the chemical analysis of the outlet gas is an image of the evolution of the conversion rate of the Mn-oxides inside the furnace.

Pilot Shaft Furnace

An intermediary scale between the laboratory microscale and the industrial scale is the burden scale (Fig. 2 left b). A new pilot tool is currently being designed to represent the burden operation only and will allow to investigate in detail the pre-reduction zone, without any interaction with the fusion zone and the liquid baths. To do so, the principle of a Shaft Furnace has been used. Eramet IDEAS new pilot tool will consist of a vertical tubular furnace fed with CO gas to simulate the gas generated by the coke bed in a HCFeMn furnace; it will be operated continuously thanks to a device discharging the pre-reduced ore. The furnace will have a 4 m height reactive zone and an internal diameter of 500 mm. The feed rate will be ranging from 100 to 400 kg/h with a gas flow up to 150 Nm^3/h of CO gas to meet the industrial conditions. The furnace will be able to treat ore grain sizes up to 30 mm. It will be fully instrumented with thermocouples to characterize the vertical temperature gradient, but also with gas analysis and pressure at different heights to characterize the vertical chemical gradient and the evolution of the charge permeability.

This pilot will be particularly helpful to study the interactions between the various reactions that take place in a HCFeMn: pre-reduction reactions in a mix of different ores, vaporization of water from the ore, and decarburization of some minerals found in some ores or furnace feed. It will also simulate the conditions prevailing in the various zones of an industrial furnace burden: along the furnace shell as opposed to the zone inside the electrodes.

Numerical Approach

Numerically, pre-reduction kinetic model based on the physico-chemical and structural properties of the ores is fitted with the lab-scale test results. Computational Fluid Dynamics (CFD) is used to compute mass and heat transfers in a porous bed and coupled with the kinetic model. Here, the gas flow and burden characteristics can be controlled through a set of input parameters such as chemical composition, temperature, downward velocity, particle size distribution, porosity, and permeability for the burden.

The pre-reduction kinetics study was started at the microscale. The lab-scale trials and kinetic model are developed and applied for Comilog and Nchwaning ores. In the following paragraph, the numerical model simulating the kinetics and thermics of the pre-reduction reactions will be presented.

Kinetic Model

The first step of the incremental approach is to correctly model the kinetics of the reaction between CO and Mn-oxides. To provide better predictions, a kinetic model based on the grain model [5] is developed. For studied Mn ores, such as Comilog which is highly porous, an advantage of this model compared with the shrinking-core model [5] is that particle microporosity is considered. The current model describes the kinetics of the chemical reactions between CO gas and Mn-oxides (Eq. 1) and provides a rate of reaction progress X_{MnO_x} (Eq. 3). Secondly, thermal effects due to the reaction exothermicity and gas-ore heat exchange are considered by coupling grain model with heat balance equation (Eq. 4) providing the ore particle temperature evolution during reaction.

$$\frac{dX_{MnO_x}}{dt} = \frac{3bC_{co}}{R_p \rho_{0,MnO_x}} \frac{1}{\frac{R_p^2(1-\varepsilon)}{R_g D_e}\left(\left(1 - X_{MnO_x}\right)^{-\frac{1}{3}} - 1\right) + \frac{1}{k}\left(1 - X_{MnO_x}\right)^{-2/3}}, \quad (3)$$

$$\rho_s C_{ps} \frac{dT}{dt} = -\lambda \nabla T - \rho_{MnO_x} \Delta H \frac{dX_{MnO_x}}{dt} - \frac{4R_p^2 h_t}{V_p(1-\varepsilon)}\left(T - T_{gas}\right) \quad (4)$$

Table 1 List of physico-chemical, structural, and thermophysical parameters

Physico-chemical parameters	Structural parameters	Thermophysical
D_e—gas diffusion coefficient in the solid k—rate constant of the reaction ρ_{MnO_x}—molar density of the solid reactant C_{CO}—concentration of the CO gas b—stoichiometric number	R_p, R_g—radius of the individual particle and of the grains ε—particle microporosity ρ_s—particle density	ΔH—molar heat of reaction C_{ps}—solid particle heat capacity h_t—heat transfer coefficient

This system of equations involves three sets of parameters presented in Table 1. The rate constant of the pre-reduction reaction is based on the Arrhenius law with the pre-exponential factor k_o and the activation energy E_a. Particle and grain shapes are assumed spherical.

The evaluation of the structural parameters of the particles of each ore is out of the scope of this paper. The thermophysical parameters can be found in databases. In this work, we therefore focus on the identification of D_e, k_0, E_a, which are intrinsic to the studied Mn ores. The chemical parameters k_0, E_a were identified by many authors, and they were reviewed by [7, 8]. However, there have been very few studies on the gaseous diffusion coefficient identification, and the existing study involves the diffusion coefficient identification on the reduction of the crushed Mn ores in which the diffusion mechanism is extremely weak due to the removal of the microporosity.

The identification algorithm is based on an optimization algorithm [9] by minimization of a continuous function in a multi-dimensional space. We propose to identify the kinetic parameters D_e, k_0, E_a by using the time-dependent conversion curve $X_{MnO_x}^{\exp}(t)$ and ore temperature evolution $T^{exp}(t)$ obtained from the laboratory experiments. In this approach, the kinetic model is used in an inverse manner to deduce the input parameters from the output.

Results of the Lab-Scale Tests

Figure 3 represents a set of laboratory tests carried out under varying conditions to ensure their reproducibility at the same conditions and to identify the effect of particle sizes, temperature, and %CO. This sensitivity analysis is advantageous to provide confident parameter identification. Two ore particle diameters are investigated: d_p = 10–12.5 mm, d_p = 1–3.15 mm.

Figure 4 (left) and (right) represents the oxidation degree evolution during reduction under different conditions for Nchwaning and Comilog ores, respectively. It can be noticed that Comilog is found to reduce at lower temperature than Nchwaning, which can be explained by higher oxidation level of Comilog (MnO_2) compared to Nchwaning (Mn_2O_3). Tangstad et al. [10] presented a review of the CO reactivity of

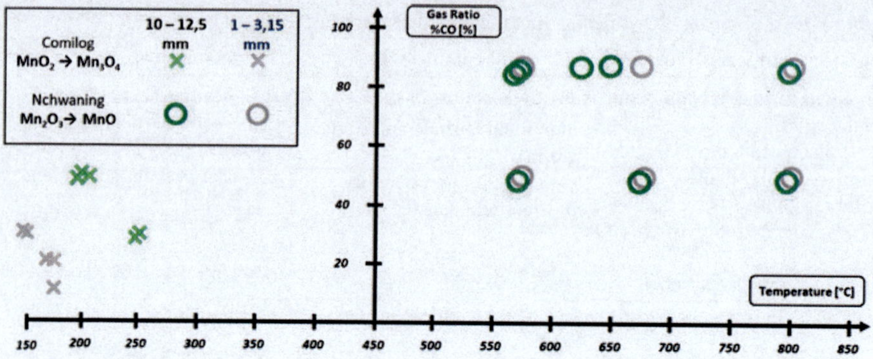

Fig. 3 All realized lab-scale tests for both Comilog and Nchwaning ores

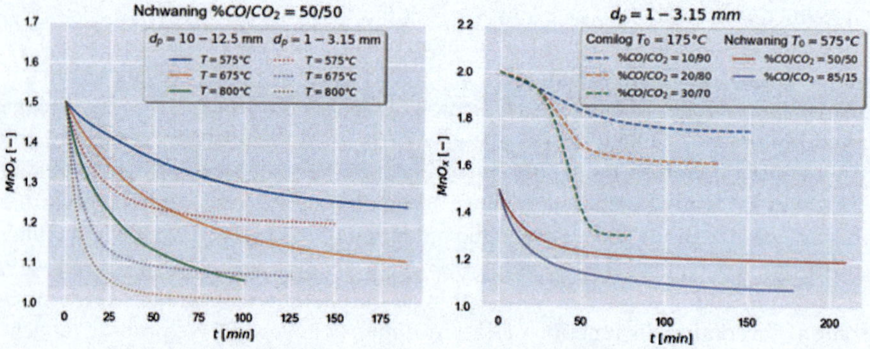

Fig. 4 Left: Particle size and temperature effect on the oxidation level of Nchwaning under CO/CO_2 = 50/50. Right: CO/CO_2 ratio effect on the oxidation level of Comilog and Nchwaning

different Mn ores including two ores of interest. It was found that CO reactivity of higher oxides is stronger like for Comilog, which leads to faster reaction.

Larssen et al. [8] and Schanche et al. [11] studied the effect of the particle size and gas composition on the reduction behavior in CO/CO_2 atmosphere of same ores. They concluded that both ores reduce more quickly with decreasing particle size and increased %CO. Figure 4 (left) shows experimental results showing that four times decrease in the particle size increases more than twice the reduction rate of Nchwaning ore. The reaction rate is faster with rising temperature (Fig. 4 (left)). Furthermore, for both ores the reaction rate can be improved with increasing %CO as shown in Fig. 4 (right) as confirmed by mentioned above authors. Larssen [12] explains that the increased reaction rate observed for increasing %CO is due to an increased partial pressure of CO (decreased oxygen pressure) providing larger driving force for the reduction.

Evaluation of the Model Parameters

The main kinetic parameters used in the model were calculated from the lab-scale trials. Figure 5 (left) and (right) presents the evolution of the reaction rate constants for Comilog and Nchwaning ores and of the diffusion coefficients with temperature respectively. Dashed lines correspond to the Arrhenius law evolution over temperature based on mean value of identified E_a and k_0 for each ore. The confidence interval is also estimated. It is found to be narrower for Comilog which is reduced at lower temperature and larger for Nchwaning, particularly at higher temperature (>700 °C). For now, it is supposed to be the effect of the carbonate's decomposition which is present in Mn ores in small quantities. For Comilog, identified E_a, k_0 parameters are in good agreement with available literature results for the Brazilian ore [6], which is supposed to have a quite close reduction behavior. For Nchwaning, lower E_a values were found in [13], but identified E_a values in present work cannot be compared to these values, since they correspond to the reduction of calcinated Nchwaning.

Figure 5 (right) shows the identified gas diffusion coefficients D_e which confirm the difference in microstructure of two ores. Less porous Nchwaning reduction provides a lower D_e at temperature lower than 800 °C compared to Comilog reducing at low temperature. The low D_e values confirm the slow reaction rate of Nchwaning pre-reduction (diffusion control) and the higher D_e values for Comilog illustrate its faster pre-reduction kinetics (chemical control). From 800 °C, D_e values for Nchwaning are close to that of Comilog. It can be explained by structural evolution which may occur at high temperature. It is known from [14] that microporosity increases with temperature during the reduction for both ores. The same authors found that CO reactivity is correlated to the high initial microporosity of the Mn ore. Indeed, highly porous Comilog has higher CO reactivity compared to Nchwaning.

The particle size effect on the diffusion can be also observed in Fig. 5 (right). Identified D_e for smaller particles is lower than that identified for larger particles. It is known that within small particles, the chemical process is very quick compared

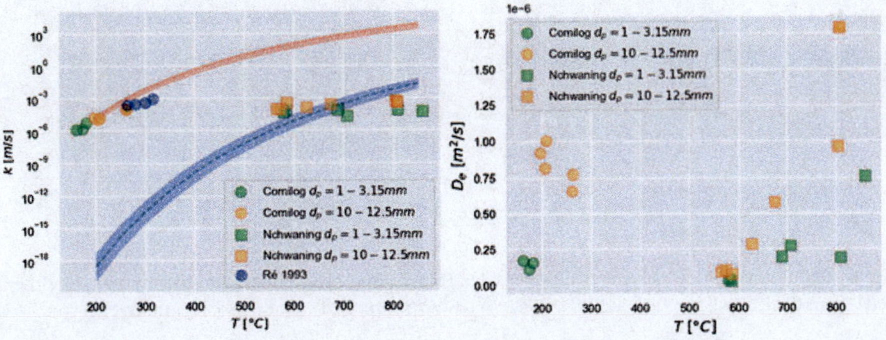

Fig. 5 Left: Identified reaction rate constant over temperature for Comilog and Nchwaning. Right: identified gas diffusion coefficient over temperature for Comilog and Nchwaning

to the diffusion processes which provides a low value for D_e. A comparison for D_e with literature is also performed. Ré et al. [6] provided very low values for D_e around 10^{-10} m^2/s that is several orders of magnitude lower than in present work. The reason is that they studied crushed mineral powders which have a characteristic size of several hundreds of microns, while we work with the larger particle size (d_p = 1–12.5 mm), in which the reduction reaction can be controlled by both chemical and diffusional processes simultaneously.

Figures 6 and 7 present the comparison of two experimental and numerical approaches for the evolution of oxidation degree of Mn-oxides and ore temperature over time. Numerical results are obtained by solving the system of Eqs. (3) and (4) with identified parameters from Fig. 5. The improved model by microporosity and thermal effect has led to very satisfactory results as the model is in a good agreement with the laboratory tests. Figure 6 (left) provides Comilog reduction from MnO_2 to Mn_3O_4 for d_p = 10–12.5 mm under $\%CO/CO_2 = 50/50$. For smaller particles d_p = 1–3.15 mm at the same temperature the direct (MnO_2 to Mn_3O_4) and fast reduction is obtained even with decreasing %CO to 30/70 as can be seen in Fig. 6 (right). This fact may confirm the particle size effect mentioned before and its much higher effect on the reduction rate comparing to %CO.

Figure 7 (left) and (right) provide Nchwaning reduction ($Mn_2O_3 \rightarrow MnO$) under $\%CO/CO_2 = 50/50$ for d_p = 10–12.5 mm and d_p = 1–3.15 mm, respectively. A decreasing of particle size accelerates the reduction. In fact, after 100 min the reduction continues for larger particles while for smaller particles MnO is reached within 50 min.

Regarding temperature profiles, first from Figs. 6 and 7 higher exothermicity of Comilog is confirmed comparing to Nchwaning. With decreasing particle size, for Comilog, the temperature peak increases from about 100–300 °C, while for Nchwaning it increases from about 20–100 °C.

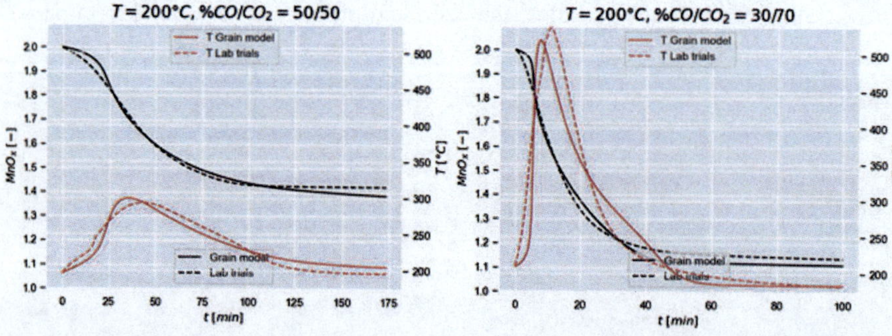

Fig. 6 Oxidation degree of Mn-oxide over time (black) and ore temperature evolution over time (red) from Comilog reduction at 200 °C: d_p = 10–12.5 mm (left), d_p = 1–3.15 mm (right)

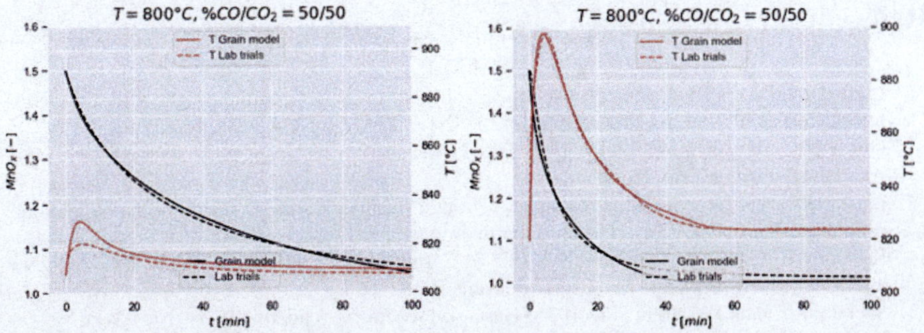

Fig. 7 Oxidation degree of Mn-oxide over time (black) and ore temperature evolution over time (red) from Nchwaning at 800 °C reduction: $d_p = 10–12.5$ mm (left), $d_p = 1–3.15$ mm (right)

Conclusions and Perspectives

An incremental strategy based on both experimental and numerical approaches was proposed. The study conducted at the microscale was presented. At this scale, the pre-reduction behavior of Comilog and Nchwaning ores was investigated in isothermal conditions in CO/CO_2 atmosphere. The effect of particle size, CO composition, temperature was evaluated and confirmed by different literature sources.

A kinetic model well-adapted for the porous Mn ores was presented. The model is coupled to the thermal model considering the particle heating due to the exothermicity and the convection. This improvement provided very satisfactory results as the model reproduces well the experimental data. This has allowed a better understanding of the physico-chemical parameters on the overall kinetics for each ore. Identified diffusion coefficients confirm the different porous microstructures of both ores. A higher activation energy was identified for Nchwaning. It means that a high chemical energy, i.e., high temperature, is needed to cause the pre-reduction of Nchwaning, as compared to Comilog for which the activation energy is almost two times lower.

In this perspective, identification of kinetic parameters will also be done on more conditions to complete and solidify the sensitivity analysis. Furthermore, the dependence of the diffusion coefficient on temperature and structural changes during reductions will be investigated. The reduction of Nchwaning ore at high temperature will be studied to decouple potentiel endothermic reactions and pre-reduction reactions. The last step consists of comparing the results of the model with industrial data, making it possible to assess the impact of operational levers on pre-reduction in industrial furnace.

References

1. Lindstad T, Olsen SE, Tranell G, Færden T, Lubetsky J (2007) Greenhouse gas emissions from ferroalloy production. In: Proceedings of the INFACON XI. New Delhi, India
2. Olsen SE, Tangstad M, Lindstad T (2007) Production of manganese ferroalloys. Tapir Akademisk Forlag, Trondheim
3. Ishak RJ (2002) Reaction kinetics for reduction of manganese ore with caron monoxide in the presence of carbon. Dr. Ing. Thesis, Trondheim
4. Kononov R, Ostrovski O, Ganguly S (2008) Carbothermal reduction of manganese oxide in different gas atmospheres. Metall Mater Trans B 662–668
5. Szekely J, Evans JW, Sohn HY (1976) Gas-solid reaction. Academic Press, New York
6. Ré M (1993) Décomposition thermique et réduction des oxydes de Managanèse par le monoxyde de carbone: étude cinétique en réacteur différentiel et mise en oeuvre de lit fluidisé
7. Cheraghi A, Yoozbashizadeh H, Safarian J (2020) Gaseous reduction of manganese ores: a review and theoretical insight. Mineral Process Extract Metall Rev 41
8. Larssen TA (2020) Production of Comilog- and N'chwaning ore. PhD thesis, Trondheim
9. Saša S, Nelder J (2009) Nelder-mead algorithm. Scholarpedia 4(7):2009
10. Tangstad M, Sibony M, Wasbø S, Tronstad R (2001) Kinetics of the prereduction of manganese ores. In: Conference proceedings, Infacon IX (Quebec)
11. Schanche Trygve L, Tangstad M (2021) Isothermal reduction of Nchwaning Manganese ore in CO/CO2/H2 atmospheres. In: Infacon XVI: International ferro-alloys congress, Trondheim
12. Larssen TA, Senk D, Tangstad M (2020) Reduction of manganese ores in CO-CO_2 atmospheres. Metall Mater Trans B 1–19
13. Berg KL, Olsen SE (2000) Kinetics of manganese ore reduction by carbon monoxide. Metall Mater Trans B 477
14. Turkova K (2014) CO reactivity and porosity of manganese materials. ISIJ Int 54 :1204–1208

Lanthanum-Light Metal Alloys Production Using Secondary Resources—Thermodynamic Analysis

Ahmad Rizky Rhamdani, M. Akbar Rhamdhani, Geoffrey Brooks, Mark I. Pownceby, Yudi Nugraha Thaha, Trevor B. Abbott, John Grandfield, and Chris Hartley

Abstract Rare-earth-elements (REE) can be used as alloying agents to improve the properties of aluminum alloys. REE are often added in the metallic form to produce the alloys, but the high cost of REE metals makes the alloys expensive thus limiting their application. An alternative way to make REE-Al alloys is by reacting REE oxides with light metals. For this route, secondary light-metal resources, such as aluminum dross, have the potential to be used to reduce the cost of alloy production and at the same time limit waste generation. In this study, a systematic thermodynamic evaluation of aluminum-lanthanum (Al-La) alloy production using lanthanum oxide (La_2O_3) and different light-metal secondary resources was carried out. Three different Al dross compositions were evaluated, each with different metal/metal oxide ratios. Based on the analysis, Al dross can successfully be used as the starting material for this process. The La_2O_3 (10 wt% of total charge) was completely reduced in all simulated compositions. Additional calculations were carried out to simulate a process in a Rotary Salt Furnace (RSF). From this analysis, the use of chloride salts

A. R. Rhamdani (✉) · Y. N. Thaha
Research Center for Metallurgy, National Innovation and Research Agency, Tangerang Selatan, Banten, Indonesia
e-mail: drhamdani@swin.edu.au

A. R. Rhamdani · M. A. Rhamdhani · G. Brooks · T. B. Abbott · J. Grandfield
Fluid and Process Dynamics (FPD) Group, Department of Mechanical and Product Design Engineering, Swinburne University of Technology, Hawthorn, VIC 3122, Australia
e-mail: arhamdhani@swin.edu.au

M. I. Pownceby
CSIRO Mineral Resources, Private Bag 10, Clayton South, Victoria 3169, Australia

T. B. Abbott
Magontec. Ltd., Sydney, NSW 2000, Australia

J. Grandfield
Grandfield Technology Pty. Ltd., Moreland West, Melbourne, VIC 3005, Australia

C. Hartley
Platina Resources Ltd., Mt Hawthorn, WA 6915, Australia

R. G. Reddy et al. (eds.), *New Directions in Mineral Processing, Extractive Metallurgy, Recycling and Waste Minimization*, The Minerals, Metals & Materials Series,
https://doi.org/10.1007/978-3-031-22765-3_6

is predicted to only slightly reduce the yield, but the use of fluoride salts should be avoided.

Keywords Aluminum · Dross · Recycling · Secondary recovery · REE · Lanthanum

Introduction

Aluminum is a light metal which has a high strength to weight ratio. These properties make aluminum and its alloys a leading choice as a structural material especially when the weight of the equipment/component must be minimized, such as in automotive and aerospace applications. Aluminum can be alloyed with Rare Earth (RE) metals to improve its strength, corrosion, and other properties. For example, the addition of lanthanum (La) to generate Al-La alloy has been shown to have an effect on grain refinement as well as improvements in its creep resistance, tensile properties, and electrical conductivity [1, 2]. The addition of RE could also give Al alloy good castability compared Al-Si alloys, as shown in cerium alloyed with aluminum [3]. This could expand the applications of Al-RE alloys when some elements are not compatible with silicon, such as scandium.

Al-RE alloys are usually produced by mixing a molten base alloy with a RE metal. This process can be quite inefficient because of various phenomena such as La losses, complex melting-casting process, and high differences between the melting temperature of the base alloy and the RE [4–6]. The high price of RE metal also makes the price of the alloy expensive and limits its commercialization [7, 8]. One alternative way of making an Al-RE alloy is by direct metallothermic reaction of the base alloy with RE oxides. In this process, Al will reduce the RE oxides to RE intermetallics and oxides of Al will form. This route of Al-RE alloy production could potentially be cheaper because the price of RE oxides can be more than half of the equivalent metal price. For this process, secondary sources of Al have the potential to be used as the reductant to make the alloy. The use of secondary sources could further reduce the price of the alloy significantly and also potentially solve existing waste generation problems. A secondary resource that could be used is a by-product from aluminum smelting process known as aluminum dross. Dross is typically divided into two types, white and black dross, with white dross having a higher Al content compared to black dross. The general composition of white dross is typically 15–80 wt.% of metallic Al, 20–85 wt.% Al_2O_3 and 0–1 wt.% salt, while black dross is typically composed of 7–50 wt.% of metallic Al, 30–50 wt.% Al_2O_3, and 30–50 salt [9]. White dross is currently processed through several different methods to recover valuable Al metal. The most common process is by melting in a Rotary Salt Furnace (RSF). In this process, the white dross is fed to a RSF and salt fluxes (mainly chlorides) are added. Salt flux addition provides a protective cover for the molten aluminum, wets the oxide film, and initiates disintegration of the film, stripping it

Table 1 Composition of aluminum dross used for simulation

Phase	Composition (wt.%)		
	WDH	WDL	WDS
Al	60	15	80
Al_2O_3	40	84	20
Flux	0	1	0

from the surface of molten aluminum droplets, and helps promote the coalescence of small aluminum droplets [9].

These secondary resources, which can still have high Al content, could potentially be used as a reductant for reducing RE oxides while also making Al-RE alloy in the process. In this study, a systematic thermodynamic evaluation of Al-RE alloy production using Al dross has been carried out. Pure Al was also investigated as a reductant for comparison. Lanthanum oxide was chosen as the RE oxide because of its thermodynamic properties of being the easiest to reduce based on standard Gibbs energy of formation. Additional calculations were also carried out with the additional of salts to simulate a possible alloy production during dross processing in a Rotary Salt Furnace (RSF).

Methodology

Three different Al dross compositions (Table 1) were selected for use in the current study. These compositions represent white dross which have high (WDH—60% Al) and low (WDL—15% Al) Al metal concentration and a white dross that has been processed through grinding and sieving and has the highest metal concentration (WDS—80% Al).

Thermodynamic modelling of the system was performed using FactSage 8.2 thermochemical software [10]. Five different databases were used, which are FactPS, FTLite, FTOxid, FTSalt, and FTHall, to represent the pure substances, alloy system, oxide system, chloride salts system, and fluoride salts system, respectively. The La_2O_3 amount was fixed at 10 wt.% of total input material. The effect of using a cover flux (56 wt.% NaCl—44 wt.% KCl) and additives (AlF_3 or Na_3AlF_6), with the ratio of 55 wt.% Al Dross—30 wt.% cover flux—10 wt.% La_2O_3—5 wt.% additives, was also studied for evaluating a possible alloy production during dross processing in an RSF.

Results and Discussion

Based on the standard Gibbs free energy of formation ($\Delta G°$), La_2O_3 is more stable compared to Al_2O_3 (Fig. 1a), which means that metallothermic reaction between Al

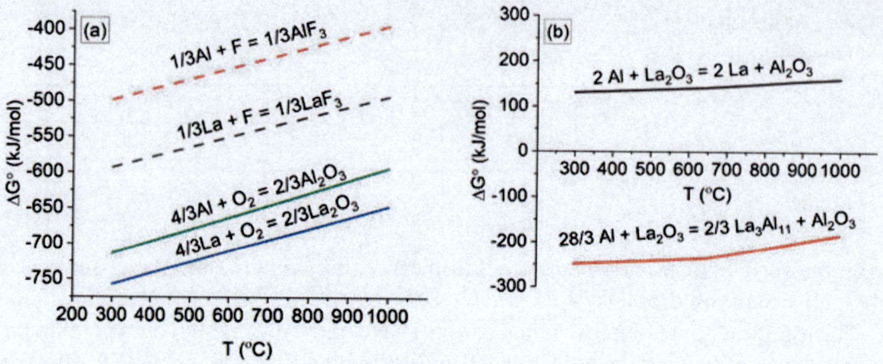

Fig. 1 **a** Comparison of standard Gibbs free energy of formation of Al_2O_3, La_2O_3, AlF_3, and LaF_3 and **b** standard Gibbs free energy of metallothermic reduction and intermetallic formation

and La_2O_3 (Eq. 1) to produce pure La is not favored thermodynamically. However, La metal could be dissolved into liquid Al resulting in lowering the activity of La compared to pure La, which would shift the reaction to the right. Reduced La would also form an intermetallic with Al which is more stable compared to its oxide (Eq. 2). Standard Gibbs free energy of formation of the intermetallic, i.e. reaction in (Eq. 2), is negative throughout the simulated temperature, which means that the reaction is favored thermodynamically compared to the metallothermic reaction (Fig. 1b).

$$2\mathrm{Al} + \mathrm{La_2O_3} = 2[\mathrm{La}]_{\mathrm{Al}} + \mathrm{Al_2O_3} \tag{1}$$

$$\frac{28}{3}Al + La_2O_3 = \frac{2}{3}La_3Al_{11} + Al_2O_3 \tag{2}$$

Effect of M/MO Ratio in Al Dross

The analysis of equilibrium states between La_2O_3 in pure Al at 850 °C can be seen in Fig. 2a. In the pure Al system, La_2O_3 was predicted to be reduced completely up to 56 wt.% La_2O_3, before $AlLaO_3$ starts to form. All reduced La will be dissolved in liquid Al metal at lower concentrations (up to ~30 wt%). Starting at 32 wt.% La_2O_3, La_3Al_{11} was predicted to form; and at much higher La_2O_3 concentrations, Al_2La and Al_3La were predicted to form. These results show that thermodynamically, an Al-La alloy could be made by reacting Al and La_2O_3.

The equivalent calculation of equilibrium of La_2O_3 mixtures with Al dross, which contains 20 wt.% Al metal and 80 wt.% Al_2O_3, instead of pure Al, was also conducted and these results are shown in Fig. 2b. At low La_2O_3 concentrations, all La_2O_3 was predicted to be reduced and dissolved in liquid metal up until 9 wt.% concentration

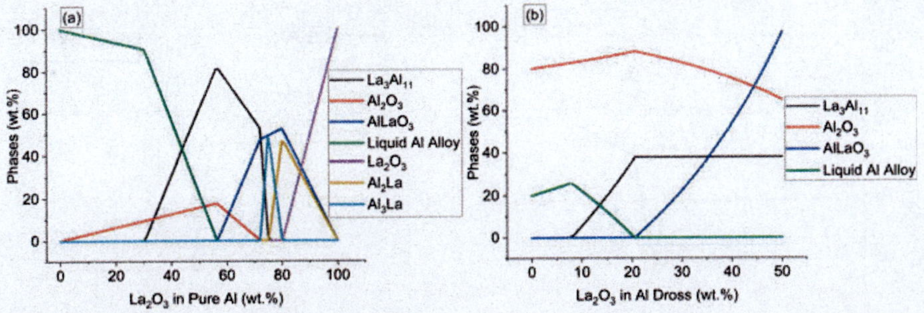

Fig. 2 Predicted phases/compounds at equilibrium of reactions **a** La_2O_3 in pure Al, and **b** La_2O_3 in Al dross containing 20 wt.% Al metal at 850 °C

before La_3Al_{11} starts to form. At 20 wt.% La_2O_3 in the dross, no liquid alloy was predicted to form, rather all La was predicted to form intermetallic La_3Al_{11}. Beyond that, La_2O_3 reacted with Al to form $AlLaO_3$.

Effect of Temperature

Figure 3 shows the results of equilibrium reactions between 10% La_2O_3 with different dross compositions (WDH, WDL, and WDS) at different temperatures. At 850 °C, when dross with high Al concentration is used (i.e. WDH and WDS), all La was predicted to dissolve in the liquid Al. As the temperature was decreased, the La_3Al_{11} intermetallic phase started to form, i.e. at 750 °C and 650 °C in the case of reactions with WDH and WDS, respectively. At about 650 °C, solid Al alloy (FCC) was predicted to form. As more liquid was solidified, more solid Al alloy and intermetallic were formed. At 600 °C, no more liquid alloy was predicted. At this point, almost all La was predicted to form La_3Al_{11}, as the solubility of La in the Al FCC matrix is only ≈0.02 wt.% La at 600 °C. When dross with low Al concentration was used (WDL), the La_3Al_{11} intermetallic phase was predicted to form even at high temperatures. This was because only a relatively small amount of molten Al was available to dissolve the reduced La, and therefore, the rest of La forms intermetallic (La_3Al_{11}) with Al.

Effect of Salts During Melting in Rotary Salt Furnace

A possibility of Al-La alloy production (by adding La_2O_3) during dross processing through Rotary Salt Furnace (RSF) processing was evaluated from a thermodynamic perspective. In a typical process for Al recovery from dross in an RSF, about 30 wt.% of a chloride salts mixture is commonly used as cover flux and as a media for oxides removal. Additives to improve coalescence of molten alloy droplets and to increase

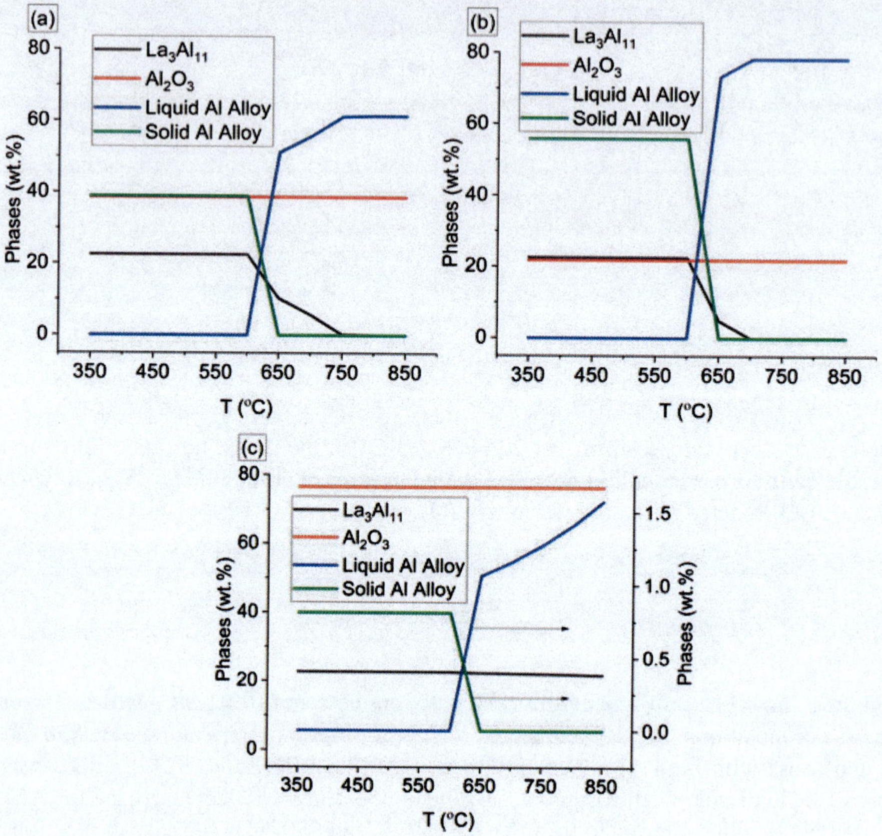

Fig. 3 Effect of temperature on the phases formation at equilibrium when 10 wt.% La_2O_3 is reacted with **a** WDH dross, **b** WDS dross, and **c** WDL dross

Al recovery in the RSF process are also commonly added to the system in the form of 5 wt.% of AlF_3 or 5 wt.% Na_3AlF_6.

The results show that the presence of fluoride salt as an additive appeared to inhibit the process of Al-La alloy formation. In the case of AlF_3 additive addition (Fig. 4), almost all La_2O_3 reacts with AlF_3 to form LaF_3 (Eq. 3), instead of reacting with Al and becoming reduced to form the alloy and the intermetallic. Unlike La_2O_3, LaF_3 could not directly react with Al to form the alloy because the difference in $\Delta G°$ of formation of AlF_3 with LaF_3 is higher compared to Al_2O_3 with La_2O_3 (Fig. 1). This model also predicts trace amount of RE halide liquid phases (<0.003 wt.%) at 850 °C (not shown in Fig. 1).

$$2AlF_3 + La_2O_3 = 2LaF_3 + Al_2O_3 \tag{3}$$

$$Na_3AlF_6 + La_2O_3 = 2NaF + NaLaF_4 + LaAlO_3 \tag{4}$$

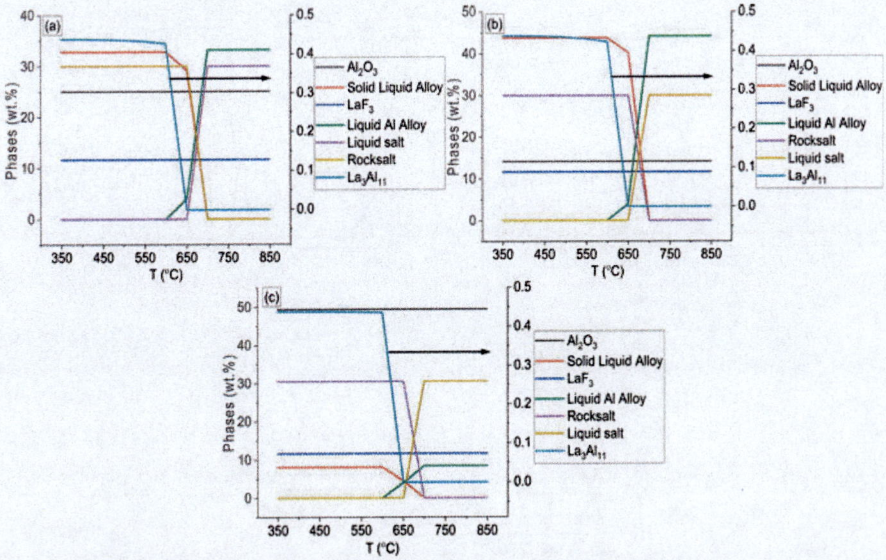

Fig. 4 Effect of temperature on phase formation at equilibrium when 10 wt.% of La_2O_3, 30 wt.% chloride salts mixture, 5 wt.% AlF_3 additives, and 55 wt.% dross are reacted. **a** With WDH dross, **b** with WDS dross, and **c** with WDL dross

$$2NaF + La_2O_3 = 2LaOF + Na_2O \quad (5)$$

$$3LaOF + 3Al = Al_2O_3 + AlF_3 + 3[La]_{Al} \quad (6)$$

On the other hand, the use of cryolite as an additive slightly improved the yield, with higher La_3Al_{11} concentration (Fig. 5). This is because cryolite will dissociate into Na^+ and Al_6^3 [11]. The reaction between La_2O_3 and Na_3AlF_6 (Eq. 4) is more complex compared to those with AlF_3 (Eq. 3) [12]. NaF formed from this reaction could further react with La_2O_3 (Eq. 5) (this reaction has been studied by Jang et al. [13]), and the formed LaOF will be reduced by Al to form the alloy (Eq. 6). The use of cryolite also resulted in the formation of $NaAl_9O_{14}$, which appeared to be formed from the reaction between Na^{2+} in the salt and Al_2O_3 to form $NaAl_9O_{14}$. This result may suggest that the use of AlF_3 as additive should be avoided and the use of cryolite as additives should only be conducted with very high cryolite ratio (>3).

Equilibrium calculations were also performed for the system without fluorides addition, i.e. 10wt.% La_2O_3—35 wt.% chlorides salts—65 wt.% Al dross, and the results are presented in Fig. 6. The results show that the chloride salts appeared to only react slightly with La_2O_3. Only a trace amount of K_2LaCl_5 ($<10^{-6}$ wt.%) was predicted from this equilibrium reaction (not shown in Fig. 6). This result may suggest that chlorides salts could be used in the process without significantly affecting the alloy formation.

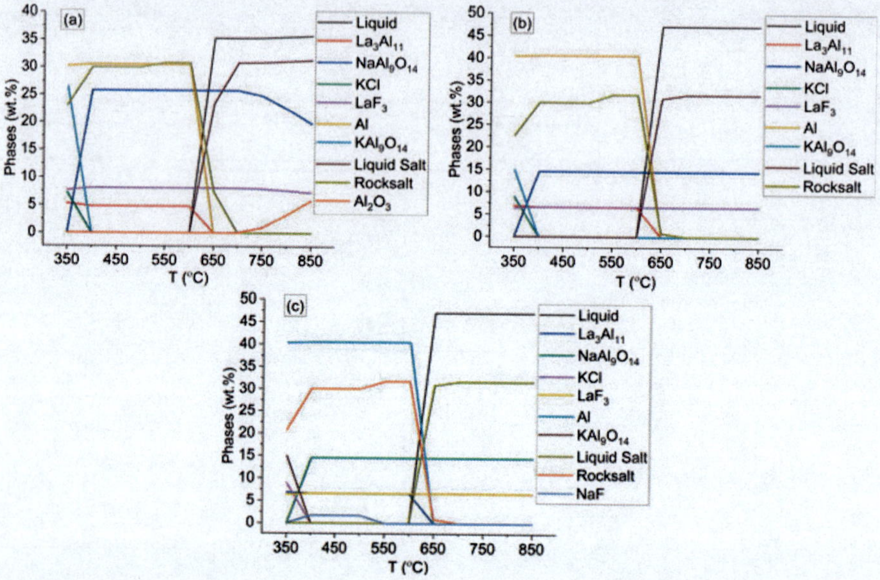

Fig. 5 Effect of temperature on phase formation at equilibrium when 10 wt.% of La_2O_3, 30 wt.% chloride salts mixture, 5 wt.% Na_3AlF_3 additives, and 55 wt.% Dross are reacted. **a** With WDH dross, **b** with WDS dross, and **c** with WDL dross

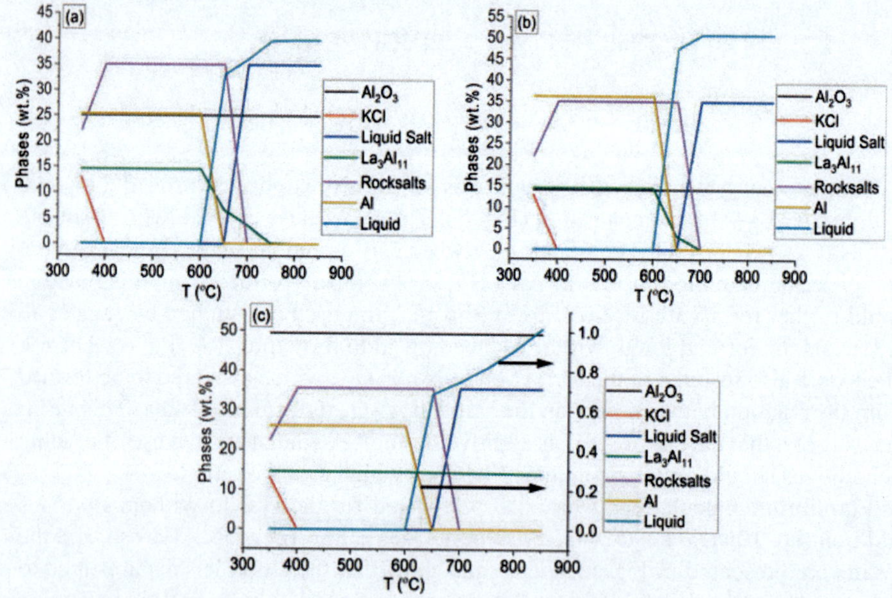

Fig. 6 Effect of temperature on phase formation at equilibrium when 10 wt.% of La_2O_3, 35 wt.% chloride salts mixture, and 55 wt.% dross are reacted. **a** With WDH dross, **b** with WDS dross, and **c** with WDL dross

Conclusions

Systematic thermodynamic calculations were carried out examining the potential use of Al dross and La_2O_3 as the starting material for making Al-La alloy. The results show that Al dross could be used to reduce La_2O_3 to produce Al-La alloy. The maximum concentration of La_2O_3 in the starting material was predicted to be 20 wt.% before the yield starts to decrease.

Addition of La_2O_3 (and other salt additives) during Al recovery from dross through melting in an RSF was also simulated to evaluate the possibility of Al-La alloy making during this process. The results predicted that the use of fluoride salts as additives decreased the yield of La_2O_3 reduction. This is because La_2O_3 is predicted to react with fluoride salt instead of forming an alloy with aluminum. A process with only chloride salts addition, appeared to provide a similar yield as that of without any salt additions.

Overall, the results from the current study show that the production of Al-La Alloy through reaction between La_2O_3 and Al dross is thermodynamically possible. This, however, does not take into account possible kinetic limitations. Hence, further study and experimental trials are needed to verify this simulation, which will be the next focus of our activities.

Acknowledgements This work is part of a PhD project of Mr Ahmad Rizky Rhamdani under Joint Swinburne-BRIN (National Research and Innovation Agency, Indonesia) Research Degree Program with additional funding support from Industry Consortium (Magontec, Grandfield Technology, and Platina Resources), and the Commonwealth Scientific and Industrial Research Organisation, Australia (CSIRO).

Conflict of Interest The authors declare that they have no conflict of interest.

References

1. Jiang H, Li S, Zheng Q, Zhang L, He J, Song Y, Deng C, Zhao J (2020) Effect of minor lanthanum on the microstructures, tensile and electrical properties of Al-Fe alloys. Mater Des 195. https://doi.org/10.1016/j.matdes.2020.108991
2. Yao D, Qiu F, Jiang Q, Li Y, Arnberg L (2013) Effect of lanthanum on grain refinement of casting aluminum-copper alloy. Int J Met 7:49–54. https://doi.org/10.1007/BF03355544
3. Plotkowski A, Rios O, Sridharan N, Sims Z, Unocic K, Ott RT, Dehoff RR, Babu SS (2017) Evaluation of an Al-Ce alloy for laser additive manufacturing. Acta Mater 126:507–519. https://doi.org/10.1016/j.actamat.2016.12.065
4. Shtefanyuk Y, Mann V, Pingin V, Vinogradov D, Zaikov Y, Tkacheva O, Nikolaev A, Suzdaltsev A (2015) Production of Al-Sc alloy by electrolysis of cryolite-scandium oxide melts. Light metals 2015. Wiley, Hoboken, pp 589–593
5. Polat Ç, Erdoğan M, İplikçioğlu AS, Karakaya İ (2018) Electrochemical formation of alloys of scandium in molten salts. In: Rare metals, pp 1555–1558
6. Liu X, Guo Z, Xue J, Zhang C (2019) Effects of synergetic ultrasound on the Sc yield and primary Al3Sc in the Al-Sc alloy prepared by the molten salts electrolysis. Ultrason Sonochem 52:33–40. https://doi.org/10.1016/j.ultsonch.2018.09.009

7. Savchenkov S, Kosov Y, Bazhin V, Krylov K, Kawalla R (2021) Microstructural master alloys features of aluminum–erbium system. Curr Comput-Aided Drug Des 11:1–13. https://doi.org/10.3390/cryst11111353
8. Xiao J, Ding W, Peng Y, Chen T, Zou K (2020) Preparing Sc-bearing master alloy using aluminum–magnesium thermoreduction method. Metals (Basel) 10:1–14. https://doi.org/10.3390/met10070960
9. Peterson RD, Newton L (2002) Review of aluminum dross processing. In: Light metals 2002. TMS, Warrendale, pp 1029–1037
10. Bale CW, Bélisle E, Chartrand P, Decterov SA, Eriksson G, Gheribi AE, Hack K, Jung IH, Kang YB, Melançon J, Pelton AD, Petersen S, Robelin C, Sangster J, Spencer P, Van Ende MA (2016) FactSage thermochemical software and databases, 2010–2016. Calphad Comput Coupling Phase Diagrams Thermochem 54:35–53. https://doi.org/10.1016/j.calphad.2016.05.002
11. Kvande H (2011) Production of primary aluminium. In: Fundamentals of aluminium metallurgy. Elsevier, pp 49–69
12. Ambrová M, Jurisová J (2006) Solubilities of lanthanum oxide in fluoride melts: Part I. Solubility in M3AlF6 (M = Li, Na, K). Thermochim Acta 443:105–108. https://doi.org/10.1016/j.tca.2006.01.007
13. Jang P, Li H, Kim W, Wang Z, Liu F (2015) Preparation of Al-La master alloy by thermite reaction in NaF-NaCl-KCl molten salt. JOM 67:1130–1136. https://doi.org/10.1007/s11837-015-1392-x

Selective Chlorination as an Innovative Extraction Method for Valuable Metals from Iron Containing Matrix

S. Steinlechner, L. Höber, and K. Witt

Abstract In different nonferrous metal producing industry sectors, the impurity element iron has to be removed from the process solution. Examples are the jarosite or goethite from nickel or zinc production but also the red mud from aluminum production. Regardless of environmental concerns, the material is land filled in almost any case, although valuables such as indium, silver, gold, nickel, or zinc are present in considerable amounts. Within the presented research, a low carbon dioxide emitting multi-metal recovery from such iron containing residues by means of a selective chlorination extraction has been fundamentally evaluated by experiments but also by comprehensive thermodynamic calculations. The paper summarizes the thermodynamic fundamental concept exploited to separate the dominating iron matrix from the valuable elements and shows verification experiments in a lab size of several grams.

Keywords Selective chlorination · Jarosite · Goethite · Iron precipitate · Nickel industry · Zinc industry · Silver · Zinc · Lead · Indium · Nickel

Introduction

Iron, being the fourth most abundant element in the earth's crust, is contained to notable extents in common minerals of base metals like zinc and nickel and must be separated during their extraction [1–3]. In hydrometallurgy, iron removal is commonly realized by precipitation methods, generating a purified solution for further processing and a precipitate as residue, which is usually named after the formed iron compound. In zinc production, mainly jarosite and goethite residues occur, while in nickel production, iron is also removed in the form of hematite or ferric hydroxide [1, 4].

S. Steinlechner (✉) · L. Höber · K. Witt
Christian Doppler Laboratory for Selective Recovery of Minor Metals Using Innovative Process Concepts, Montanuniversität Leoben, Franz Josef-St. 18, 8700 Leoben, Austria
e-mail: stefan.steinlechner@unileoben.ac.at

R. G. Reddy et al. (eds.), *New Directions in Mineral Processing, Extractive Metallurgy, Recycling and Waste Minimization*, The Minerals, Metals & Materials Series,
https://doi.org/10.1007/978-3-031-22765-3_7

With the climate crisis as one of the most pressing issues of our time, the need for cleaner technologies is on the rise and so is the demand for raw materials required for the transition. Zinc, being mainly applied in the galvanization of steel products, plays an important role in the corrosion protection of wind turbines while nickel demand increases due to its use in electric vehicle batteries, geothermal power plants, and hydrogen electrolysis [5, 6]. With zinc and nickel production growing, so do the amounts of generated residual materials, like iron precipitates. Nowadays, they are primarily discarded in tailing ponds, which not only poses an environmental hazard on their surroundings, but also results in an unexploited extraction potential of contained valuable metals [7–10]. Iron precipitation residues can contain considerable amounts of valuable elements, that are partially regarded critical raw materials. The base metals copper, lead, nickel, and zinc can be found in such residues, but also technology metals like gallium or indium and the precious metal silver [10]. The extraction of such elements out of the iron containing matrix could help save primary resources, reduce waste volumes, and mitigate negative effects on the environment caused by their disposal [11].

Numerous investigations on possible processing strategies for iron precipitation residues have been conducted in the past decades, targeting the treatment of residual materials for environmentally safe disposal and the recovery of economically valuable elements. Nonetheless, the only industrially applied methods are in the field of immobilization and pyrometallurgy. Immobilization still leads to vast amounts of residues being landfilled, while contained metals are not recycled and therefore lost. Pyrometallurgical processes, on the other hand, enable the recovery of various elements, but are usually associated with the emission of greenhouse gases [12].

A novel approach consists of the chlorination of valuable elements and the subsequent volatilization of the formed chlorides. Chlorination is an increasingly important way of extracting nonferrous metals not only from ores or concentrates but also from industrial residues [13]. Chlorinating agents like gaseous chlorine, hydrogen chloride as well as chlorides of alkali and alkaline earth metals possess a high reactivity at moderate temperatures, selectivity in chlorination of the desired metal values, and an easy availability with low cost [14]. These properties have led to numerous applications in metallurgy on an industrial scale, like the extraction of titanium, raw earth elements, tantalum, niobium, or tin [15]. Additionally, chlorination—volatilization processes have been examined as a means of recycling metals from a wide variety of wastes and residues, like, e.g., LCD powder, PVC wires, or pyrite cinder [16–18]. In chlorination—volatilization approaches, a chlorinating agent is added to the metal-containing material, enabling the separation of volatile metal chlorides from the non-volatile matter [19]. To volatilize metals, the formation of the respective metal chloride must be thermodynamically favorable, and its vapor pressure must be high enough at the given temperature for volatilization. These two requirements are at the same time the basis for the selectivity of such processes: Various materials like Al_2O_3, SiO_2, or Fe_2O_3 fail to fulfill the first one [17]. In the case of Fe_2O_3, for example, it would be necessary to provide a reducing agent like carbon in order to lower the oxygen potential sufficiently for a successful chlorination [14]. The second one, on the other hand, is often not met by chlorides like NaCl, KCl, or $CaCl_2$, which makes the separation of valuable metals, whose chlorides do show the necessary vapor pressure, from the matrix possible [17].

Materials and Methods

The approach described in this paper aims at processing different iron precipitation residues with chloridic compounds in a thermal treatment process at temperatures in the range of up to 1100 °C, leading to a selective removal from valuable metals as chlorine compounds. The fact that iron chlorinates poorly while other volatile metal chlorides are formed can be exploited for this innovative separation of valuable metals from jarosite or goethite residues. The thermodynamic principles and the behaviour of investigated residues under thermal treatment are discussed in more detail below.

The basic idea is based on the one hand on the low process temperature compared to other metallurgical processes required for chlorination and on the other hand on the carbon addition and thus CO_2 emission that is not required due to the direct chlorination reaction. Figure 1 shows the evaporation temperatures of a variety of relevant metals in their metallic, oxidic, and as well their chloridic form. The metals are found to form chlorides in different states of valence, which are considered in the comparison. The right plot shows the average temperature for metallic and oxidic compounds as well as for all depicted chlorides. This comparison illustrates that all investigated metals are more easily vaporized as chlorides and remain non-volatile in their oxidic state. The mean temperature for the evaporation of chlorides is significantly lower than that of the metals or oxides. This circumstance together with

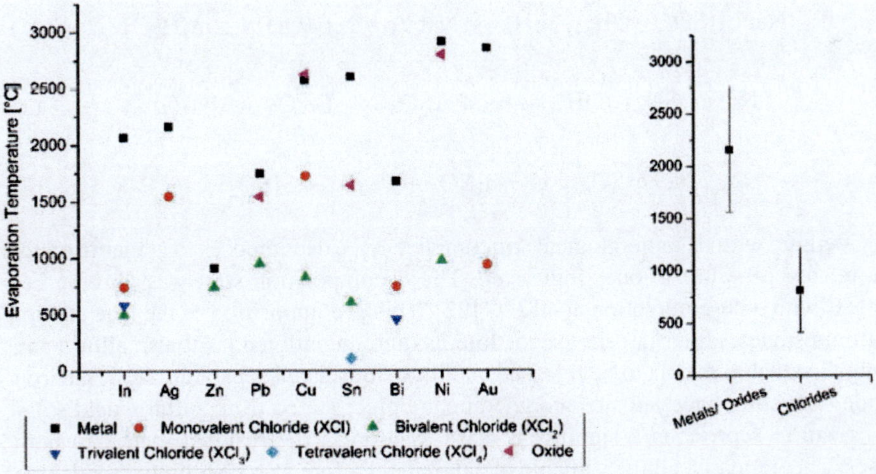

Fig. 1 Evaporation temperatures of a range of metals of interest in their metallic, oxidic, and chloridic form

the instability of iron chloride, described later, is exploited for selective separation of valuables from iron precipitation residues.

Jarosite is the mineralogic term for a certain group of basic iron sulfates with the general formula $XFe_3(SO_4)_2(OH)_6$. The placeholder X is generally represented by monovalent cations such as Na^+, K^+, $NH_4{}^+$, H_3O^+, Ag^+, or ½ Pb^+. Similar than in nature also the precipitate forms different kinds of jarosite depending on what cation is added to the process solution or what impurities or also valuable elements are present in the process solution. Alternatively, iron can be precipitated from sulfatic solutions also as goethite, with the formula FeO·OH. Both precipitation residues, either from zinc or nickel industry, can incorporate other valuable metals in their structure or co-precipitate it as sulfates, oxides, or hydroxides, due to decreasing solubility of the individual valuable element compounds during the iron precipitation process. Notables are concentrations of zinc, lead, silver, and indium, as well as nickel, the latter two being among the world's critical elements in terms of security of supply [20].

During a possible thermal treatment of these materials, a decomposition takes place before or in parallel of the actual treatment, leading to a splitting of the hydroxide and sulfate groups. This results in an easier processability of the simpler oxidic or sulfatic compounds as well as in an enrichment of the valuable metals remaining in the matrix. In jarosite with its structural formula $XFe_3(SO_4)_2(OH)_6$, water is bonded in the form of OH^- and H_3O^+ (in the case of hydronium jarosite). The evaporation of moisture can be observed below 200 °C, the following split and removal of OH-group follow up to 450 °C and between 600–850 °C the SO_3 is removed [21], illustrated by Eqs. (1)–(3) exemplarily for a sodium jarosite. This decomposition can take place either in parallel with the chlorination in a process step or upstream in order to obtain the SO_3 separately from the chlorine compounds formed.

$$NaFe_3(SO_4)_2(OH)_6 \cdot mH_2 \rightarrow NaFe_3(SO_4)_2(OH)_6 + m\{H_2O\} \tag{1}$$

$$NaFe_3(SO_4)_2(OH)_6 \rightarrow NaFe(SO_4)_2 + Fe_2O_3 + 3\{H_2O\} \tag{2}$$

$$NaFe(SO_4)_2 \rightarrow Na_2SO_4 + Fe_2O_3 + 3\{SO_3\} \tag{3}$$

Goethite, with its mineralogical structure FeO·OH, decomposes to hematite under the release of water in one single step. The decomposition start was observed at 200 °C with 96% completion at 400 °C [22]. This decomposition of the iron precipitation residues, either jarosite or goethite, is quite intentional, as it also allows any trapped valuable metals to be released and thus chlorinated separately from the iron oxide. Anyhow, since jarosite and goethite are precipitated from sulfuric acid solutions, sulfur represents a significant mass fraction in the residual matrix in both cases. Theoretically, all the valuable metals can therefore also be present as sulphate compounds. So, calculations were also carried out for the presence of valuable metal sulfates.

For the following thermodynamic calculations, mainly the software FactSage but partly also HSC Chemistry ($CuCl_2$) were used. The FactSage Reaction Module offers the possibility of automatically considering elements and compounds in their most stable form at certain temperatures. Accordingly, any changes in the slopes of the various curves are due to the melting and evaporation of the components present. As a pre-stage to temperature-induced evaporation, it is necessary to convert the metals from their present form into chlorides. In the course of this work, thermodynamic calculations were used to determine chlorination using chlorine gas, gaseous hydrochloric acid, and the three metal chlorides $MgCl_2$, $AlCl_3$, and $FeCl_3$ in their most stable form. The reason for this is the possible dissociation of the chloride compound, the previous reaction with split-off hydrate water, or the direct reaction. Sodium chloride, potassium chloride, and calcium chloride were excluded by tentative experiments but also thermodynamic calculations due to their high stability and therefore poor suitability to act as chlorination agent for others. For easier comparison, all reactions are normalized to a total amount of two atoms of chlorine participating in the reaction as given in Eqs. (4)–(6), exemplarily for oxidic compounds. The term "Sca" stands for solid chlorination agent, for example, Mg in $MgCl_2$. All calculations were performed with the FactSage Reaction Module (Database FactPS), taking into account all compounds in their most stable form at the respective temperature

$$1/y\,Me_xO_y + 2HCl \leftrightarrow \boldsymbol{x/y}\,MeCly + H_2O \tag{4}$$

$$1/\boldsymbol{y}\,Me_xO_y + Cl_2 \leftrightarrow \boldsymbol{x/y}\,MeCly + 1/2O_2 \tag{5}$$

$$1/\boldsymbol{y}\,Me_xO_y + 2/z\ ScaCl_2 \leftrightarrow \boldsymbol{x/y}\,MeCl_{2y/z} + 2/z\ MO_{z/2} \tag{6}$$

Figure 2a, b is illustrating the results of the performed calculations, respectively, the Gibbs Energy of the reactions of various metal oxides (a) and sulphates (b) of interest with hydrochloric acid in the most stable form as a function of temperature. Most metal oxides are transformed to their corresponding chlorides, while iron oxide theoretically does not react to iron chloride above ~100 °C (red dotted line). In the case of present sulphate compounds, a temperature above 800–1100 °C is required, to shift the equilibrium to the product side, respectively, chlorine compound. However, this thermodynamic consideration does not take into account any other furnace atmosphere and with that the stability of the sulphates themselves in, for example, oxidizing conditions, whereby a possible decomposition could take place and thus result in a reaction of chloride compound with the oxide again. This fact applies to all subsequent considerations.

The comparison of Fig. 2a, b with Fig. 3a, b—illustrating the reaction with chlorine gas instead - draws a somewhat similar picture. In the case of metal oxides, the decrease in driving force with increasing temperature is less pronounced than for the reaction with HCl. Generally, copper and nickel oxide show a lower tendency to react than other metal oxides. The iron oxide is again not stable below 1200 °C in

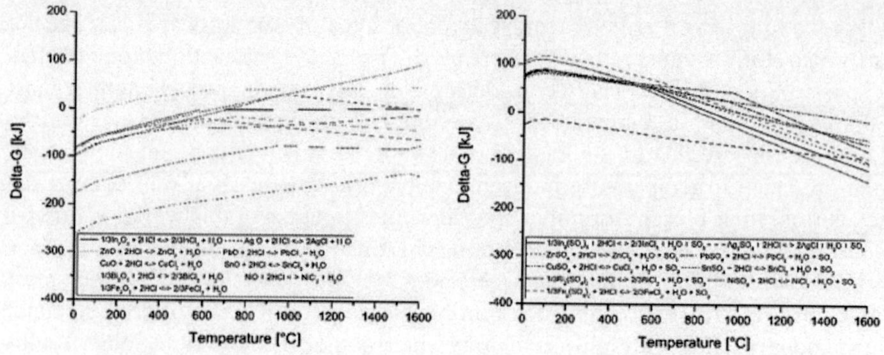

(a) (b)

Fig. 2 Gibbs energy of the reactions of valuable metal oxides (**a**) or metal sulphates (**b**) and hydrochloric acid

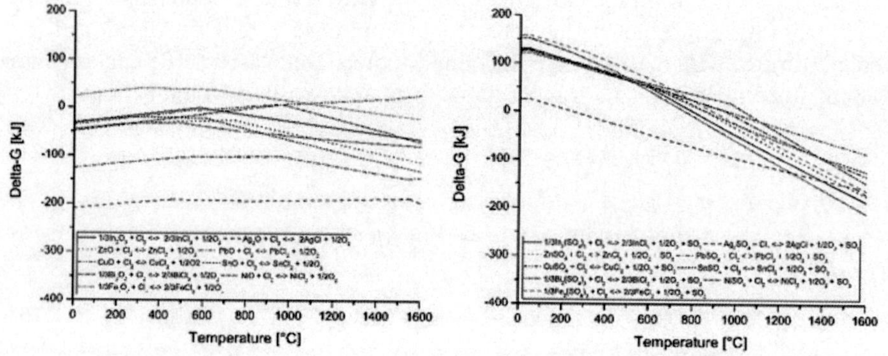

(a) (b)

Fig. 3 Gibbs energy of the reactions of valuable metal oxides (**a**) or metal sulphates (**b**) and gaseous chlorine

the case of oxides and 900 °C in the case of sulphates. All in all, the reaction either with gaseous chlorine or gaseous HCl shows very similar characteristics.

More interesting than the reaction with Cl_2 or HCl seems to be the reaction with solid chlorine carriers, as these are by far easier to handle. Nevertheless, the previously mentioned reactions are of great importance as well since they can take place through a previous reaction of the solid chlorine carriers with hydrate water or moisture of the residual materials. Based on stability calculations of various metal chlorides, the decision was taken for the three chlorides $MgCl_2$, $AlCl_3$, and $FeCl_3$. All three are typically present in their hydrated form and upon exposure to heat, react with the release of either gaseous chlorine or hydrochloric acid. This would lead to aforementioned reactions. Nevertheless, also the unhydrated forms can be present as well and therefore directly react with valuable metal compounds.

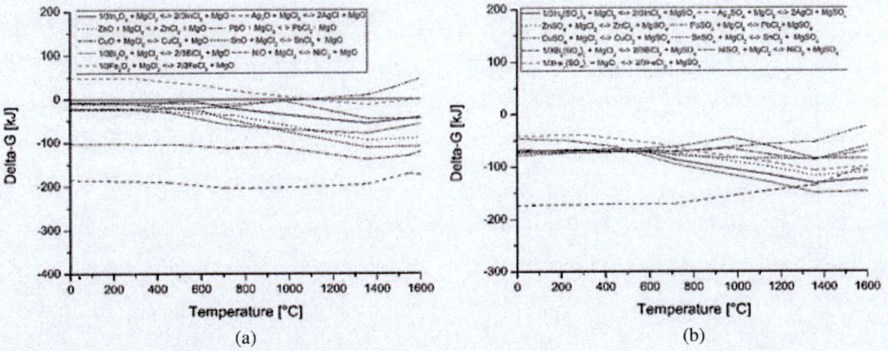

Fig. 4 Gibbs energy of the reactions of valuable metal oxides (**a**) or metal sulfates (**b**) and magnesium chloride

Figure 4a, b show the reactions with the solid chlorine carrier $MgCl_2$. All reactions take place up to a temperature of 1100 °C, except the one with iron oxide. In the case of sulphate compounds, even iron sulphate reacts to its chloride, being strongly influenced by the stability of the second reaction product, which is magnesium sulphate instead of magnesium oxide.

With $AlCl_3$, all metals, including iron, can be chlorinated in the complete investigated temperature range, in some cases with high driving forces, as shown in Fig. 5a, b. Here, the conclusion is obvious that $AlCl_3$ would not be suitable due to the possible formation of volatile ferric chloride, but formed ferric chloride can also react further as it is illustrated in the following Fig. 6.

The utilization of iron chloride as solid chlorination agent has two possible advantages. The first is that its formation when using other solid chlorination agents is unavoidable anyway due to the huge amount of iron dominating the matrix. The second reason is that the formed reaction product, iron oxide, is not forming any new

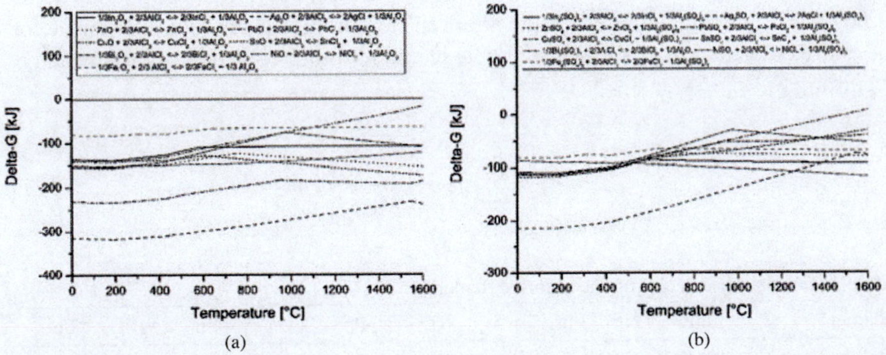

Fig. 5 Gibbs energy of the reactions of valuable metal oxides (**a**) or metal sulfates (**b**) and aluminium chloride

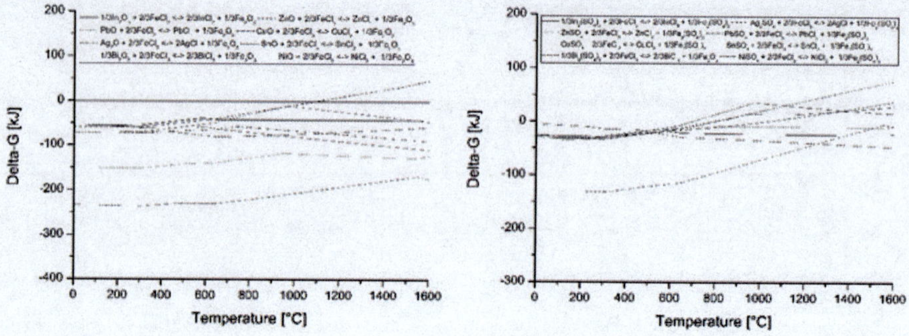

(a) (b)

Fig. 6 Gibbs energy of the reactions of valuable metal oxides (**a**) or metal sulphates (**b**) and iron chloride

impurity in the already existing iron containing matrix. The reaction of iron oxide with iron chloride is not illustrated in Fig. 6a, b as the reaction products are the same as the reactants but it can be seen that in case of metal oxides up to 1100° and in the case of sulphates up to a temperature of around 700 °C all equilibria are on the product sides.

Summarizing it can be stated that the chlorination of various valuable metals from an iron containing matrix is viable based on the carried out thermodynamic calculations. Due to the low to completely non-existent stability of ferric chloride compared to the other metal chlorides considered, a satisfactory separation effect should thus be achievable.

The thermodynamic method development for selective chlorination was verified by practical experiments utilizing industrial jarosite and goethite from the zinc industry. The required amount of chlorine was calculated based on the elements and their chlorine consumption, summarized in Table 1. By far the biggest consumer of chlorine is for both, jarosite and goethite, the metal zinc with 33% and 70%, respectively, for goethite. The amount of the residues and chlorine compounds in the mixtures was calculated based on the required stoichiometric amount of chlorine for the entire chlorination of specific valuable elements and 5% of the iron in the matrix according to Eq. (7).

$$n_{Cl} = m_r * \sum f_{Cl}(i)\frac{x_i}{M_i} * f_{El}(i) \quad (7)$$

Table 1 Factors for calculating the chlorine demand

Element	K	Na	In	Ag	Zn	Pb	Cu	Sn	Bi	Fe
Chlorine compound	KCl	NaCl	$InCl_3$	AgCl	$ZnCl_2$	$PbCl_2$	$CuCl_2$	$SnCl_2$	$BiCl_3$	$FeCl_3$
Chlorine factor f_{Cl}	1	1	3	1	2	2	2	2	3	3
Extraction factor f_{El}	1	1	1	1	1	1	1	1	1	0.05

The chlorine factor f_{Cl} is the ratio of chlorine and the valuable element in the compound ($InCl_3$: $f_{Cl} = 3$). The extraction factor $f_{El}(i)$ is the assumed extent of chlorination of the respective metal, being 1 (100%) for all elements except for iron 0.05 was assumed.

The base value of the respective addition of $MgCl_2{\cdot}6H_2O$, $AlCl_3{\cdot}6H_2O$, or $FeCl_3{\cdot}6H_2O$, being two times the stoichiometric amount for chlorination, was calculated according to Eq. (8). Furthermore, the base addition was doubled and tripled, representing four times and six times the stoichiometric calculated amount.

$$m_{Chloride(base)} = \frac{2 * n_{Cl}}{f_{Cl/Chloride}} * M_{Chloride} \tag{8}$$

The experimental evaluations consisted of four campaigns with a total of 36 trials. This included 2 materials, a dried jarosite and a dried goethite, each treated at 2 temperatures (900 and 1100 °C). The mass of the residue and chlorine carrier mixture was 80 g, and the treatment time was for all experiments 30 min in a high-temperature muffle furnace of Nabertherm. The mixtures were charged into the hot furnace at trial temperature using a silica crucible and placed on a weighing pan which is connected to the external scale. During the trial, the mass loss was tracked over the entire experimental time of 30 min. Subsequently, the remaining materials were milled, packaged, and sent for analysis. The experimental results to interpret the effectivity of the chlorination procedures are the calculated extraction rates of specific elements.

Results and Conclusion

The following Fig. 7 summarizes the realized individual extraction rates of the elements In, Ag, Zn, Pb, Cu, Sn, Au, and Bi of the experiments at 1100 °C. Due to the significantly lower extraction rates at 900 °C, these values are not presented.

It can generally be stated that the chlorination of iron, shown in Fig. 8, is significantly lower compared to all targeted valuable elements in the majority of experiments. This confirms the underlying principle of lower affinity to chlorine than the valuable elements in focus, enabling the separation and extraction from the remaining iron containing residue.

Summarizing, a novel method for the simultaneous recovery of various valuable metals via selective chlorination reactions was investigated and tested on an extended laboratory scale. A main advantage is that chlorination reactions proceed without the presence of carbon as a reducing agent and lead to effective evaporation of the targeted elements. As a result, no greenhouse gas emissions are generated in the process itself, excluding the necessary energy input for temperature control.

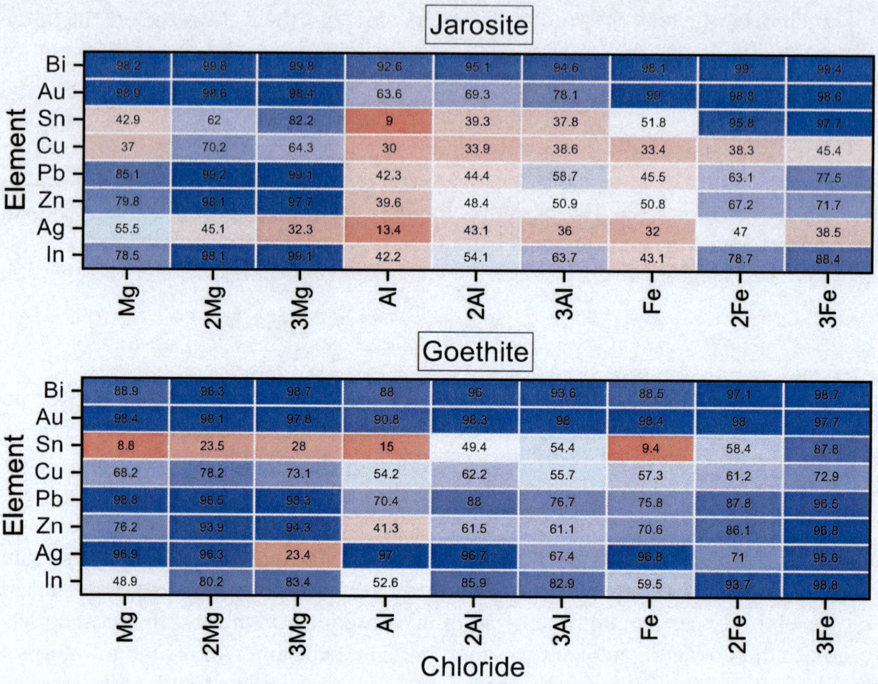

Fig. 7 Individual extraction rates of the experiments at 1100 °C with varying addition of chlorine carrier (for example: Mg … base amount $MgCl_2 \cdot 6H_2O$; 2 Mg … two times the base amount and four times the calculated stoichiometric amount, respectively; Al = $AlCl_3 \cdot 6H_2O$, Fe = $FeCl_3 \cdot 6H_2O$)

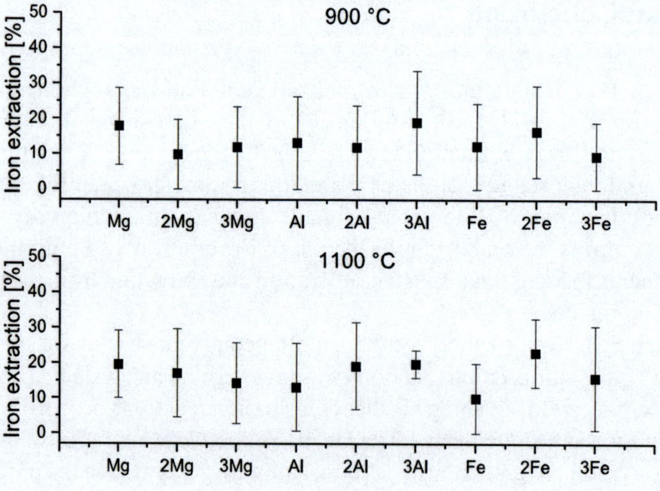

Fig. 8 Iron extraction rates averages over all assessed residues

Funding This work was funded by the Christian Doppler Research Association, the Austrian Federal Ministry for Digital and Economic Affairs, and the National Foundation for Research, Technology and Development.

References

1. Crundwell FK, Moats MS, Ramachandran V, Robinson TG, Davenport WG (2011) Extractive metallurgy of nickel, cobalt and platinum-group materials. Elsevier, Amsterdam. https://search.ebscohost.com/login.aspx?direct=true&scope=site&db=nlebk&db=nlabk&AN=383424
2. Frey PA, Reed GH (2012) The ubiquity of iron. ACS Chem Biol 7(9):1477–1481. https://doi.org/10.1021/cb300323q
3. Sinclair RJ (2005) The extractive metallurgy of zinc, 1st edn. AusIMM (Spectrum series/Australasian Institute of Mining and Metallurgy, 13), Carlton Vic
4. European Commission, Joint Research Centre, Delgado Sancho L, Roudier S, Farrell F, Cusano G, Rodrigo Gonzalo M (2017) Best available techniques (BAT) reference document for the non-ferrous metals industries: industrial Emissions Directive 2010/75/EU (integrated pollution prevention and control): Publications Office
5. Bobba S, Carrara S, Huisman J, Mathieux F, Pavel C (2020) Critical raw materials for strategic technologies and sectors in the EU. A foresight study. Luxemburg: Publications Office of the European Union. https://op.europa.eu/en/publication-detail/-/publication/8e167f11-077c-11eb-a511-01aa75ed71a1
6. International Energy Agency (2021) World energy outlook special report. The role of critical minerals in clean energy transitions. Edited by International Energy Agency. https://iea.blob.core.windows.net/assets/ffd2a83b-8c30-4e9d-980a-52b6d9a86fdc/TheRoleofCriticalMineralsinCleanEnergyTransitions.pdf. Accessed 28 Mar 2022
7. Di Maria A, van Acker K (2018) Turning industrial residues into resources: an environmental impact assessment of goethite valorization. Engineering 4(3):421–429. https://doi.org/10.1016/j.eng.2018.05.008
8. Linsong W, Peng Z, Yu F, Lu S, Yue Y, Li W, Wei S (2020) Recovery of metals from jarosite of hydrometallurgical nickel production by thermal treatment and leaching. In Hydrometallurgy 198:105493. https://doi.org/10.1016/j.hydromet.2020.105493
9. Pappu A, Saxena M, Asolekar SR (2006) Jarosite characteristics and its utilisation potentials. Sci Total Environ 359(1–3):232–243. https://doi.org/10.1016/j.scitotenv.2005.04.024
10. Reuter M (2013) Metal recycling. Opportunities, limits, infrastructure. United Nations Environment Programme (International resource panel), Nairobi. http://hdl.handle.net/20.500.11822/8423
11. Ndlovu S (2017) Waste production and utilization in the metal extraction industry. With assistance of Geoffrey S Simate, Elias Matinde, 1st edn. Taylor & Francis Group, London. https://ebookcentral.proquest.com/lib/kxp/detail.action?docID=4890676
12. Hoeber L, Steinlechner S (2021) A comprehensive review of processing strategies for iron precipitation residues from zinc hydrometallurgy. Clean Eng Technol 4(1):100214. https://doi.org/10.1016/j.clet.2021.100214
13. Kanari N, Allain E, Joussemet R, Mochón J, Ruiz-Bustinza I, Gaballah I (2009) An overview study of chlorination reactions applied to the primary extraction and recycling of metals and to the synthesis of new reagents. Thermochim Acta 495(1–2):42–50. https://doi.org/10.1016/j.tca.2009.05.013
14. Jena PK, Brocchi EA (1997) Metal extraction through chlorine metallurgy. Mineral Process Extract Metall Rev 16(4):211–237. https://doi.org/10.1080/08827509708914136

15. Gaballah I, Bonazebi A, Gaballah L, Kanari N, Menad N, Mugica JC, Coelho MC (1996) New approach for valuable elements recovery from concentrates & wastes by selective chlorination through binary chlorides. Edited by European Commission. European Commission. https://cordis.europa.eu/docs/projects/files/BRE/BRE20173/26839761-6_en.pdf. Accessed 21 May 2022
16. Ding J, Han P-W, Lü C-C, Qian P, Ye S-F, Chen Y-f (2017) Utilization of gold-bearing and iron-rich pyrite cinder via a chlorination–volatilization process. Int J Miner Metall Mater 24(11):1241–1250. https://doi.org/10.1007/s12613-017-1516-0
17. Kameda T, Fukushima S, Grause G, Yoshioka T (2013) Metal recovery from wire scrap via a chloride volatilization process: Poly(vinyl chloride) derived chlorine as volatilization agent. Thermochim Acta 562:65–69. https://doi.org/10.1016/j.tca.2013.03.012
18. Park K-S, Sato W, Grause G, Kameda T, Yoshioka T (2009) Recovery of indium from In_2O_3 and liquid crystal display powder via a chloride volatilization process using polyvinyl chloride. Thermochim Acta 493(1–2):105–108. https://doi.org/10.1016/j.tca.2009.03.003
19. Grause G, Yamamoto N, Kameda T, Yoshioka T (2014) Removal of lead from cathode ray tube funnel glass by chloride volatilization. Int J Environ Sci Technol 11(4):959–966. https://doi.org/10.1007/s13762-013-0286-0
20. Hayes SM, McCullough EA (2018) Critical minerals: a review of elemental trends in comprehensive criticality studies. Resour Policy 59:192–199
21. Steinlechner S, Antrekowitsch J (2018) Thermodynamic considerations for a pyrometallurgical extraction of indium and silver from a jarosite residue. Metals 8. https://doi.org/10.3390/met8030xxx
22. Naono H et al (1987) Porous texture in hematite derived from goethite: mechanism of thermal decomposition of geothite. J Colloid Interface Sci 120:439–450

High Vacuum Solar Thermal Dissociation for Metal and Oxide Extraction

M. G. Shaw, G. A. Brooks, M. A. Rhamdhani, A. R. Duffy, and M. I. Pownceby

Abstract The current interest in space-based mineral and metal extraction technologies, and the increased likelihood of establishing research facilities on the lunar surface, provides a strong impetus for high vacuum metallurgical research. The current work examines the viability of a thermal dissociation process for metal and metal oxide extraction from beneficiated and un-beneficiated lunar feedstocks. Thermal dissociation experiments using lunar regolith simulants and pure oxide samples were performed using a bespoke apparatus involving a vacuum reactor coupled with a solar simulator heat source. Specific focus was given to sub-liquidus operation and the sublimation of metal oxides under low temperature and low vacuum conditions. The thermodynamic and kinetic considerations, as well as the practical demonstration of such a process, are also discussed. This work demonstrates the potential of utilising the natural high vacuum conditions on the Moon for developing novel high vacuum extraction processes.

Keywords Solar thermal · Vacuum metallurgy · Astrometallurgy · Thermal dissociation

Introduction

Vacuum metallurgy as an area of study has been around for many decades [1, 2]; however, industrial implementation has been limited due to the high cost and maintenance requirements of pumping equipment rendering most potential vacuum

M. G. Shaw (✉) · G. A. Brooks · M. A. Rhamdhani
Fluid and Process Dynamics Group, Swinburne University of Technology, Hawthorn, Melbourne, VIC, Australia
e-mail: mgshaw@swin.edu.au

A. R. Duffy
Centre for Astrophysics and Supercomputing, Swinburne University of Technology, Hawthorn, Melbourne, VIC, Australia

M. G. Shaw · M. I. Pownceby
CSIRO Mineral Resources, Clayton, VIC, Australia

R. G. Reddy et al. (eds.), *New Directions in Mineral Processing, Extractive Metallurgy, Recycling and Waste Minimization*, The Minerals, Metals & Materials Series,
https://doi.org/10.1007/978-3-031-22765-3_8

metallurgical processes un-economic [3]. As a result, very few vacuum metallurgy processes have been proposed, tested, and implemented industrially. It is not unreasonable to assume that this is in part due to the difficulty in designing and operating apparatus that has the capacity to operate under high vacuum conditions. We note that the definition of 'levels' of vacuum is somewhat subjective—in the case of the current work when talking about 'high vacuum' we refer to those conditions $<10^{-10}$ atm and extending down into the ultra-high vacuum range of $<10^{-14}$ atm and below.

With multiple international consortiums aiming to send missions to the Moon in the near term [4], most notably NASA's Artemis program [5], there will soon be considerably more access to the easily maintainable high vacuum environment of space for research purposes. By operating in space or on the lunar surface, the complexities of designing high vacuum capable, high temperature reactors, and pumping systems capable of pumping rates sufficient to maintain such conditions are largely removed from the initial testing processes. While such systems would still need to be engineered for implementation of extractive processes on Earth, such advantages could significantly increase the viability of conducting larger scale testwork in the field of vacuum metallurgy.

Ellingham diagrams displaying the specific effect on metal and metal oxide system stability of varying vacuum conditions can be found elsewhere [6]. These diagrams demonstrate that, theoretically, many metal extraction processes could benefit from operation under vacuum conditions and some of these have already been partially investigated in the pursuit of oxygen production on the Moon [7, 8]. For example, thermal decomposition of metal oxides for oxygen production has been investigated previously [9–13]. Some significant benefits of these processes are the lack of reagents required for reduction as well as the possibility of using concentrated solar thermal (CST) energy as a heat source. The current work investigates the nature of this process in terms of its applicability to metal and metal oxide extraction from complex silicate ores and its thermodynamic and kinetic viability, and concludes with a discussion of results from a preliminary laboratory demonstration study.

Evaporation Kinetics

The partial pressure of a phase dictates the pressure of that phase that will be reached and maintained by evaporation in a system. In order for an oxide to continually evaporate, the saturated vapor pressure of the oxide, or indeed a sub-oxide of the material, must exceed that of the ambient or total pressure [14]. At atmospheric pressures, the evaporation and often accompanying thermal dissociation of oxide materials require extremely high temperatures, well above the liquidus [14]. However, as ambient pressure decreases, the required saturated vapor pressure for evaporation to become thermodynamically favourable decreases. With ambient pressures in space, and specifically on the Moon, having been measured at below 10^{-14} atm [15], it is predicted that the sublimation or continuous sub-liquidus evaporation of oxides is thermodynamically viable under these conditions [16].

As is well documented, thermodynamic viability cannot be interpreted as kinetic viability. For example, consider the Hertz-Knudsen-Langmuir (HKL) equation shown in Eq. 1.

$$\frac{dn_i}{dt} = -A\frac{a_e p_{i,sat} - a_c p_i}{\sqrt{2\pi RM_i T}} \tag{1}$$

In which $\frac{dn_i}{dt}$ is the evaporation rate in mol/s of substance i, A is the surface area, $p_{i,sat}$ is the equilibrium partial pressure of the gas species of element i, p_i is the elements' partial pressure at the surface in question, M_i is its molar mass, a_e the Langmuir coefficient for evaporation and a_c the same for condensation, R is the universal gas constant, and T the temperature [17–20]. It can be seen that p_i acts as a limiting factor and will only directly influence the evaporation rate when relatively close in value to $p_{i,sat}$. As p_i approaches 0 (i.e. in a vacuum), a change in this value will not result in a significant alteration to the evaporation rate. Instead, a change in rate necessitates a change in the $p_{i,sat}$ term.

Since evaporation kinetics correlate to this saturated vapor pressure and therefore temperature, the ambient pressure in space, while increasing thermodynamic viability, does not directly affect the kinetics of such a reaction to any measurable extent. In fact in practice, under high vacuum conditions the p_i in Eq. 1 can be considered 0, in this case, the equation evaluates what is known as Langmuir evaporation [17, 20]. As such, while establishing the thermodynamic viability is important, the industrial viability of sublimation as a mechanism of metal and oxide extraction will be more limited to the kinetics of sub-liquidus operation.

The thermodynamic basis and kinetic evaluation of the current work have been discussed in detail elsewhere [16, 21], and this paper will focus on the presentation of some practical demonstrations of the thermal dissociation of oxides under vacuum conditions heated using simulated concentrated solar energy.

Methods

Tests in the current work were completed using a 42 kW solar simulator developed by Ekman et al. [22]. Testing was carried out using a custom-made vacuum reactor connected to an Ilmvac™ STP-1 turbomolecular pumping station that, in a clean environment, is capable of achieving a vacuum of below 10^{-10} atm.

The custom-made reactor tube used for all tests was a silica glass closed end tube with a length of 400 mm, an outer diameter of 34 mm, a thickness 2.5 mm, and a 5 mm flange at the open end. A 304 stainless steel cooling collar was machined in-house and consisted of two parts, a water cooled collar that fitted over the silica reactor itself and a base plate that continued onto a standard KF40 fitting for connection to the vacuum pump. The reactor tube mounted into the cooling collar can be seen in

Fig. 1 Vacuum reactor tube connected to a steel cooling collar

Fig. 1. The reactor tube flange was sealed between the base plate and the collar using Viton™ o-rings.

The full experimental setup can be seen in Fig. 2. The procedure used for dissociation tests was as follows. A ~2.6 g sample (oxides, monomineralic concentrates, or complex silicate mixtures) was placed in a steel die assembly with a 13 mm diameter. The die was compressed using a manual hydraulic press for 5 min at 9 tons of pressure. The samples were placed in an alumina boat and dried at 150 °C for at least 2 h. The mass of the samples and the boat were recorded accurately. The silica glass reactor tube was inserted into the cooling collar (d) and sealed and the samples were then inserted. The collar was attached to the vacuum pump (c) and pumped down for 1 to 2 h sufficient to achieve a vacuum of below 10^{-7} atm. The electrically heated furnace (b) was then turned on and allowed to ramp up to 900 °C. The apparatus was left over night (~18 h) to de-gas and to pump down to a reaction pressure of ~10^{-8} atm. The simulator (a) reflectors were cleaned with ethanol and a non-linting wipe, and the ballast systems (e) were turned on. The simulator was then initiated. For most experiments, the sample was heated to between ~1000 and ~1150 °C for 2–6 h. It was not possible to run the solar simulator for long times (i.e. days) so for tests requiring longer run times, the simulator was 'pulsed' for four to six hours a day for two to three days—this was completed depending on the expected volatility of the sample. The reactor was designed to allow metals and oxides vaporised at high temperatures to be transported down the tube before depositing on the insides of the reactor as it exited the furnace body. Once the experiments were completed and the sample was carefully removed and weighed, the reactor tube with its deposition bands was preserved for further analysis.

Temperature and pressure readings during the tests were collected using a DT85 series 3 DataTaker data logger. The DataTaker was controlled using the dEX-2 control software communicating with the device via a USB connection.

Sample analysis was completed on the deposits coating the inside of the reactor tubes and in some cases on the reacted pellets. Deposits in the reactor tubes formed band like structures of different colours and degree of transparency. The deposits were extremely thin in thickness (estimated at up to a few 10's of nanometers) and at different positions along the inside of the tube. For analysis, the reactor tubes were placed into a plastic bag and carefully broken. Shards that contained specific

Fig. 2 Swinburne Solar Simulator connected to a turbomolecular vacuum pump. **a** simulator lamps, **b** hybrid resistance furnace with open cavity (facing away), **c** turbomolecular vacuum pump, **d** reactor cooling collar, **e** simulator ballast systems, **f** variable height apparatus stand, **g** vacuum extension pipe

deposition bands were labelled and mounted onto aluminium sample holders using carbon dag before being carbon coated for analysis (to prevent the surface of the glass, an insulator, from charging during analysis). The surface coatings were analysed using Scanning Electron Microscopy (SEM) and Energy Dispersive Spectroscopy (EDX). Reacted samples were analysed using X-ray Fluorescence Spectroscopy (XRF) and X-ray Diffraction (XRD).

Results and Discussion

The conditions and reaction details of the tests completed are listed in Table 1. Feed samples tested using the above methods included nanophase FeO/Fe_2O_3 mixture, ilmenite mineral concentrate, and LMS-1 lunar regolith simulant (supplied by the Exolith Lab). A blank test under vacuum conditions was run (#1), as was a test using LMS-1 under atmospheric conditions (#2). Under atmospheric conditions, no deposit was formed at the flange end of the reactor tube. Under vacuum conditions,

Table 1 Test conditions for vacuum thermal dissociation tests

Sample #	Feed material	Average pressure (atm)	Max temperature (°C)	High temp duration (hours)
1	None	1.29E-08	1143	3.5
2	LMS-1	1	1153	2.1
3	FeO/Fe_2O_3	2.73E-07	1127	1.8
4	FeO/Fe_2O_3	3.42E-07	1025	4.3
5	Ilmenite	2.74E-08	1116	7.1
6	LMS-1	5.01E-08	1121	4.1
7	LMS-1	8.84E-08	1147	3.5
8	Reacted sample 7	3.83E-08	1151	8.8

but with no sample, a light white deposit consisting of Si and O was formed - similar deposit bands are not mentioned in the rest of the analysis.

The composition of surficial deposit bands referred to in the current work is from EDS analysis as discussed above and omits the interference from the strong Si and associated O signals. The thin nature of the deposited samples resulted in background Si and O signals in every analysis from the silica reactor tube. The elements identified as present in the deposition bands formed in each test are listed in Table 2. It is to be noted that while accurate mass loss measurements were calculated for each sample, the estimated errors exceeded the calculated mass losses and thus the results were deemed insignificant.

Table 2 Elements present in the deposit bands formed after thermal dissociation of samples. Compositions determined by EDX analysis

Sample #	Feed material	Deposits found	Notes
1	None	None (Fractured Si and O)	General frosting on glass
2	LMS-1	None	No deposit
3	FeO/Fe_2O_3	(Fe, O)*	Distinct brown/deep red band
4	FeO/Fe_2O_3	None (Fractured Si and O)	General frosting on glass
5	Ilmenite	(Zn, O)*	Distinct brown/dark brown band
6	LMS-1	(Na, Cl, K, S, O)*, (F, P, Pb, Cu)+	Multiple faint bands
7	LMS-1	(Na, K, O) *, (P, Pb, Cu)+	Multiple faint bands
8	Reacted Sample 7	(Na, Cu)**	Faint pink band

*Distinct crystals or background layer

+Trace amounts

**Almost below detection limit of EDX scan, predominant reading was Si and O background

The temperature sensitivity of the sublimation of Fe is evident when comparing the results of samples #3 and #4, both using a FeO/Fe_2O_3 mix as a feed material. While a noticeable deposit of FeO was seen at 1127 °C after reaction for only 1.8 h, even after 4.3 h at the lower temperature of 1025 °C, no significant deposit was seen (#4).

The ilmenite sample in test #5 yielded somewhat unexpected results. The ilmenite used was a naturally-sourced mineral concentrate rather than a chemically pure sample XRF analysis on the feed material indicated a number of impurity elements (present as distinct gangue mineral phases), and it was discovered that this mineral sample contained trace amounts of zinc (0.028%). Zinc is well known for its volatility [23] and the main deposition band formed in this test consisted almost exclusively of ZnO. XRF analysis of the residue indicated that 82% of the ZnO was sublimated out of the feed sample during the test. Of even more surprise was the discovery that the ZnO deposit had in some places formed nano-sized rods or filaments. The formation of these structures is well described elsewhere [23], and further investigation of this structure formation is beyond the scope of the current work.

With the main target of the current work being the demonstration of oxide sublimation for lunar resource processing applications, the majority of tests were conducted using a feed material of lunar maria simulant 1 (LMS-1). This is a simulated lunar regolith material developed by the Exolith Lab [24]. It is of note that while some of the trace elements detected in the deposition bands (F, Pb, Cu, Cl, etc.) are not part of the bulk composition of the LMS-1 material, they are present in trace quantities [24]. As was demonstrated by the test with natural ilmenite, the concentration of these trace elements via vacuum sublimation is not surprising.

Three separate tests were run using LMS-1 material. One test was at a temperature of ~1120 °C for 4 h (#6) followed by two consecutive tests where the reacted samples from the first test (#7) were transferred to a clean reactor tube before further reaction (#8). The target temperature for both tests was 1150 °C. Test #8 was allowed to run for almost twice the duration of the previous LMS tests to allow for even slowly sublimating material to be captured. The deposition band formed in test #8 can be seen in Fig. 3.

A secondary electron (SE) image of one of the deposit bands formed during the reaction of sample # 6 is shown in Fig. 4. For accuracy, most EDX results were taken at 6000 times magnification; however, the 1500 times SE image in Fig. 4 shows well what was a common theme in the analysis, that being a general background coating (in this case predominantly Na with O and Cu), with regularly spaced crystal structures of varying types representing a concentration of different elements (Na, P, and O with traces of Cu). Since oxygen was present in the silica tube, it was often hard to quantify how much was present in the crystal/area in question. A significantly more thorough analysis would be required to characterise the nature of the individual crystals formed during the deposition process and since the target of this testwork was a demonstration of the sublimation of the feed material under vacuum conditions, it was deemed beyond the scope of the current work.

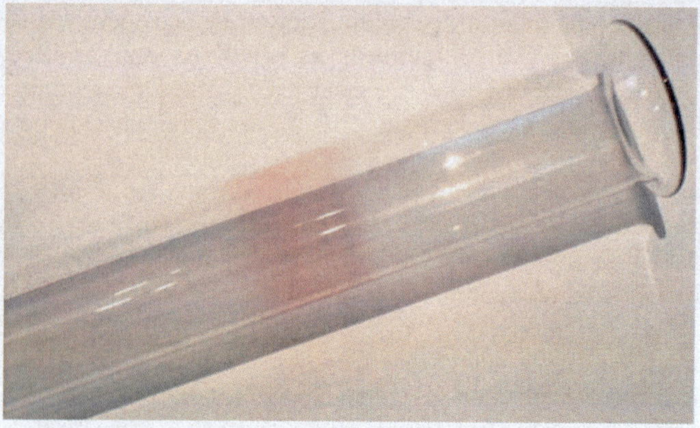

Fig. 3 Na deposition band formed from the sublimation of LMS-1 at ~1150 °C for 8.8 h at 3.83E-08 atm, after pre-processing for 3.5 h at similar conditions

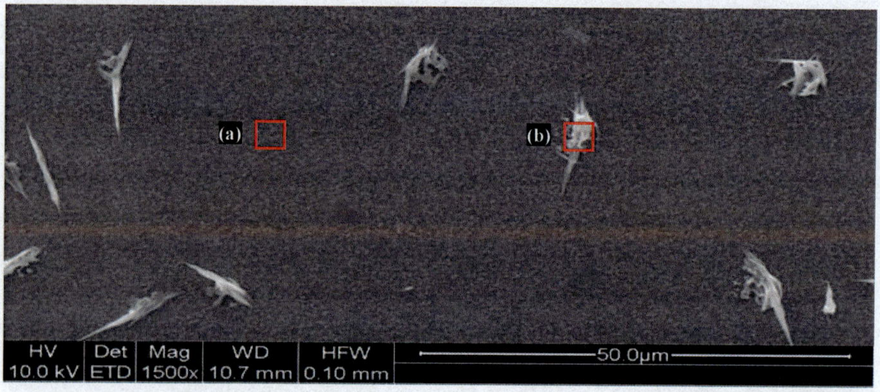

Fig. 4 Secondary electron image of deposition sample from solar reactor test # 6 (LMS-1 feed, 1121 °C, 4.1 h, ~5.01E-08 atm). EDX analysis of the red boxed regions indicated the presence of **a** Na with trace O, Cu, **b** Na, P, O, trace Cu

Conclusions

In this work, we have explored the viability of sublimation under high vacuum conditions for metal and metal oxide extraction. Thermodynamically, the dissociation of oxides is viable at sub-liquidus temperatures under high vacuum conditions such as those present in space or on the lunar surface. While thermodynamic viability is a necessary and not insignificant hurdle, of more importance for potential industrial application is the kinetics of sublimation. The laboratory tests shown in the current work suggest that the rate of sublimation of oxides is, not unexpectedly, slow at sub-liquidus temperatures. However, despite the slow reaction speeds, the sublimation and deposition of multiple oxides were demonstrated, albeit at pressures orders of magnitude higher than those expected for operation in space. Fe and O were sublimated from a FeO/Fe_2O_3 concentrate, Zn and O were sublimated from an ilmenite mineral concentrate, and Na, K, P, and S, and in some cases, trace amounts of Pb, Cl, and F were sublimated from a lunar maria regolith simulant material. It was apparent with the FeO tests that the sublimation rate and viability of sublimation at pressures of 10^{-7} to 10^{-8} atm were sensitive to temperature, with the 100 °C temperature difference between samples #3 and #4 being the difference between a significant FeO deposition band at the higher temperature (1127 °C), and no visible or measurable deposition at the lower temperature (1025 °C). In all tests involving the LMS-1 material, the deposition bands consisted of a background coating (usually Na, and O, with trace amounts of K, Cu, P, S, F, Cl, and Pb) and liberal smattering of crystal structures varying from sub-micron size up to those visible with the naked eye.

It is to be noted that in the current work, the operation of the demonstration tests under high vacuum conditions was difficult. The operation and maintenance of so-called dirty high vacuum chambers, especially when conducting high temperature testwork is a significant challenge. This highlights the potential benefits of having access to the lunar environment in terms of accessibility to testwork such as that covered here. While the predicted and measured rate of oxide sublimation under vacuum conditions is slow, there may still be ways of utilising this process in future, or even ways by which the kinetic viability can be increased. These sorts of fields of study will, in the future, be much more easily investigated when the vacuum of space as a resource can be utilised within astrometallurgical research.

References

1. Kroll W (1951) Vacuum metallurgy: its characteristics and its scope. Vacuum 1(3):163–184
2. Pidgeon L, Alexander W (1944) Thermal production of magnesium-pilot plant studies on the retort ferrosilicon process. Trans AIME 159:315–352
3. Balomenos E, Panias D, Paspaliaris İ (2012) Exergy analysis of metal oxide carbothemic reduction under vacuum–sustainability prospects. Int J Thermodyn 15(3):141–148
4. Li C et al (2019) China's present and future lunar exploration program. Science 365(6450):238–239
5. Sanders G (2019) NASA lunar ISRU strategy, NASA, Editor. Washington, DC, USA
6. Shaw MG et al (2021) High vacuum metallurgy: opportunities in lunar resource processing. In: 5th international future mining conference, AusIMM: Perth, Western Australia. p 463
7. Crawford IA (2015) Lunar resources: a review. Prog Phys Geogr 39(2):137–167
8. Shaw M et al (2021) Mineral processing and metal extraction on the lunar surface—challenges and opportunities. Mineral Processi Extract Metall Rev 1–27
9. Nakamura T, Senior CL (2005) Solar thermal power system for lunar ISRU processes. In: AIP conference proceedings 746. Albuquerque, New Mexico, USA
10. Senior C (1992) Lunar oxygen production by pyrolysis. In: Space programs and technologies conference. AIAA, Huntsville, AL, USA
11. Sauerborn M (2005) Pyrolyse von Metalloxiden und Silikaten unter Vakuum mit konzentrierter Solarstrahlung. Bonn, Germany, Universitäts-und Landesbibliothek Bonn
12. Sauerborn M et al (2004) Solar heated vacuum pyrolysis of lunar soil. In: 35th COSPAR scientific assembly. Paris, France

13. Matchett J (2006) Production of lunar oxygen through vacuum pyrolysis. In: School of Engineering and Applied Science. George Washington University Washington, DC, USA
14. Rosenqvist T (2004) Principles of extractive metallurgy. 2nd edn. Tapir Academic Press, Trondheim, Norway
15. Stern SA (1999) The lunar atmosphere: history, status, current problems, and context. Rev Geophys 37(4):453–491
16. Shaw MG et al (2021) Thermodynamic modelling of ultra-high vacuum thermal decomposition for lunar resource processing. Planet Space Sci 105272
17. Sossi PA, Fegley B Jr (2018) Thermodynamics of element volatility and its application to planetary processes. Rev Mineral Geochem 84(1):393–459
18. Hertz H (1882) Ueber die Verdunstung der Flüssigkeiten, insbesondere des Quecksilbers, im luftleeren Raume. Ann Phys 253(10):177–193
19. Knudsen M (1909) Die Gesetze der Molekularströmung und der inneren Reibungsströmung der Gase durch Röhren. Ann Phys 333(1):75–130
20. Langmuir I (1916) The evaporation, condensation and reflection of molecules and the mechanism of adsorption. Phys Rev 8(2):149
21. Shaw MG et al (2023) Unpublished thesis - Swinburne University of Technology
22. Ekman BM, Brooks G, Rhamdhani MA (2015) Development of high flux solar simulators for solar thermal research. In: Energy technology 2015. Springer, pp 149–159
23. Zhang Y et al (2012) Synthesis, characterization, and applications of ZnO nanowires. J Nanomater 2012:624520
24. Isachenkov M et al (2022) Characterization of novel lunar highland and mare simulants for ISRU research applications. Icarus 376:114873

Development of a Dynamic Model of Collision and Coalescence for Molten Matte Droplets in Copper Smelting Reaction Shaft Considering Interfacial Deformation

Yuko Goto, Shungo Natsui, and Hiroshi Nogami

Abstract In the flash furnace for copper smelting, Cu loss in the slag is an economically important topic. It is known that the size of the matte droplets, one of the significant factors of mechanically entrained Cu loss, is complicated by the local number density in the reaction shaft, initial size, and chemical composition of the concentrate. Numerical simulation based on a dynamic model can clarify the various factors' effects instead of the experimental measurements restricted in the industrial high-temperature reaction field. In this study, the collision and coalescence behavior of many droplets consisting of matte and slag with various sizes assumed in the flash smelting reaction shaft is calculated explicitly by smoothed particle hydrodynamics method considering interfacial deformation. This numerical simulation visualizes droplets' interfacial deformation, matte coalescence, and slag interference. It reveals the considerable effect of interfacial energy among matte-slag-gas on each droplet size growth.

Keywords Copper smelting · Flash furnace · Matte · SPH · Binary droplet · Collision · Coalescence

Y. Goto (✉)
Sumitomo Metal Mining Co., Ltd, 3-5, Sobiraki-Cho, Niihama, Ehime 792-0001, Japan
e-mail: yuko.goto.r4@smm-g.com

S. Natsui · H. Nogami
Institute of Multidisciplinary Research for Advanced Materials, Tohoku University, 2-1-1 Katahira, Sendai 980-8577, Japan
e-mail: natsui@tohoku.ac.jp

H. Nogami
e-mail: nogami@tohoku.ac.jp

R. G. Reddy et al. (eds.), *New Directions in Mineral Processing, Extractive Metallurgy, Recycling and Waste Minimization*, The Minerals, Metals & Materials Series,
https://doi.org/10.1007/978-3-031-22765-3_9

Introduction

A flash furnace, smelting Cu concentrates into Cu matte, increasing the Cu content from 20–30 to 50–65%, has been widely used in the pyrometallurgical production of Cu since its industrial development in the past century. The mixture of Cu concentrates and flux is continuously fed into the reaction shaft of the flash furnace with oxygen-enriched air through a concentrate burner. By the exothermic reactions of sulfide in the Cu concentrate, molten matte and slag are generated, and matte is separated from slag in the settler. It is required to reduce the Cu loss into slag and operate a flash furnace with high efficiency to realize a sustainable society.

The matte droplets obtained in the reaction shaft must be well separated from the slag in the settler. The matte droplets percolate the upper slag layer to reach the bottom matte layer; otherwise, they are stagnant on the surface of the slag. The past sampling results and numerical studies show that the separability of matte from the slag layer in the settler depends on the matte droplet diameter in the reaction shaft part [1]. Thus, many studies on particle size growth in reaction shafts have been conducted [2–7].

The reaction model, proposed by Kim and Themelis [2] and further modified by Jokilaakso et al. [3] and Yli-Penttila et al. [4], explains that the concentrate particles undergo fragmentation during the oxidization because the internal pressure raised by the sulfur gasification reaction breaks the oxide shell formed around the particles. Pérez-Tello et al. [5] confirmed that the fine concentrate particles under 45 μm behave differently from coarse ones. Namely, the tiny particles melt rapidly, increasing their mean size by collision. Although the laboratory scale studies show the effect of the initial size distribution of Cu concentrate, the particle behavior in large-scale reactors such as commercial furnaces is different and has not yet been fully clarified. Kemori et al. [7] sampled concentrate particles falling in the reaction shaft of a pilot-scale flash furnace and confirmed that the mean size and the amount of sulfur eliminated from the particles increased along the reaction shaft. In their observation, 5–800 μm droplets were eventually produced from 5–80 μm input concentrates in the reaction shaft. Thus, from observation results on a laboratory scale experiment, it is expected that a significant number of collisions and coalescence occur during the flight in the commercial furnace, as the model proposed by Kemori et al. mentioned.

An image of matte and slag droplet distribution in the reaction shaft is shown in Fig. 1. Evaluating the frequency of the collision among droplets in the reaction shaft is essential to estimate the matte droplet diameter that significantly impacts Cu loss. However, despite its importance, the behavior of the interfacial boundary between the matte and slag is not well understood.

To describe the movement of the dispersed droplets, the most instinctive approach is to use a mesh-free discrete-element type simulation. In this study, a dynamic model using the Smoothed Particle Hydrodynamics (SPH) method considering interfacial deformation [8, 9] is developed to evaluate the collision and coalescence behavior of many droplets consisting of matte and slag with various sizes assumed in the flash smelting reaction shaft.

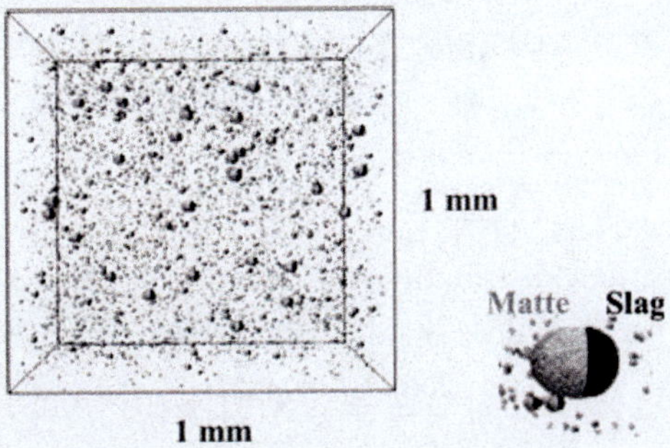

Fig. 1 Image of droplets distribution in reaction shaft in flash furnace. (Grey color shows matte, black color shows slag)

Numerical Methods

Governing Equation of Liquid Motion and Discretization

The Navier–Stokes equation under adiabatic conditions was adopted to simulate the movement of many matte-slag binary droplets in the reaction shaft.

$$\frac{\mathbf{Du}}{\mathrm{D}t} = -\frac{\nabla P}{\rho} + \nu \nabla^2 \mathbf{u} + \mathbf{g} + \frac{\mathbf{F}_\mathrm{s}}{\rho} \tag{1}$$

where, $\boldsymbol{F}_s$, $\mathbf{g}$, P, t, $\mathbf{u}$, ν, and ρ are the interfacial force applied per unit volume [kg m^{-2} s^{-2}], gravitational acceleration [m s^{-2}], pressure [kg m^{-1} s^{-2}], time [s], velocity [m s^{-1}], dynamic viscosity [m^2 s^{-1}], and density [kg m^{-3}]. On the right-hand side of Eq. (1), the first, second, third, and fourth terms denote the pressure gradient, viscous force, gravity, and interfacial force, respectively.

In the SPH method [9, 10], the governing equations are discretized and replaced by particle interaction forces. The interaction is limited to a finite range using an effective radius h, and density and the other variables in Eq. (1) are calculated using kernel function W as follows.

$$\rho_\mathrm{i} = \sum_j m_\mathrm{j} W\left(\mathbf{r}_\mathrm{ij}, h\right), \tag{2}$$

$$\phi_\mathrm{i} = \sum_j m_\mathrm{j} \frac{\phi_\mathrm{j}}{\rho_\mathrm{j}} W\left(\mathbf{r}_\mathrm{ij}, h\right), \tag{3}$$

where the subscripts i and j denote the particle indices, h is the effective radius [m], m is the mass of particle [kg], $\mathbf{r}$ is a vector obtained from the difference between particle i and j [m], W is the kernel function [m^{-3}], and ϕ is the variables in Eq. (1). In this study, Wendland's kernel [10] is adopted.

$$W\left(\mathbf{r}_{\mathrm{ij}}, h\right) = \frac{21}{16\pi h^3}\begin{cases} \left(1 - \frac{q}{2}\right)^4 (2q + 1) & q \le 2 \\ 0 & 2 < q \end{cases}, \tag{4}$$

where $q = |\mathbf{r}_{\mathrm{ij}}|/h$.

Thus the Navier–Stokes equation can be discretized as follows, and using the particle distribution, the velocity and the position of the particle are updated.

$$\begin{aligned} m_{\mathrm{i}}\frac{\mathrm{D}\mathbf{u}_{\mathrm{i}}}{\mathrm{D}t} &= -\sum_j \left(P_{\mathrm{i}} V_{\mathrm{i}}^2 + P_{\mathrm{j}} V_{\mathrm{j}}^2 + \prod_{\mathrm{ij}} \right) \nabla W_{\mathrm{ij}} \\ &+ \sum_j \frac{2\mu_{\mathrm{i}}\mu_{\mathrm{j}}}{\mu_{\mathrm{i}} + \mu_{\mathrm{j}}} \left(V_{\mathrm{i}}^2 + V_{\mathrm{j}}^2\right) \frac{\mathbf{r}_{\mathrm{ij}} \cdot \nabla W_{\mathrm{ij}}}{\left|\mathbf{r}_{\mathrm{ij}}\right| + (0.1h)^2} \mathbf{u}_{\mathrm{ij}} + m_i \mathbf{g} + \mathbf{F}_{\mathrm{s,i}}, \end{aligned} \tag{5}$$

where V is the volume [m^3], μ is the viscosity [kg m^{-1} s^{-1}] and Π is the artificial viscosity [kg m^5 s^{-2}]. Tait's equation of state is used to calculate pressure.

$$p_{\mathrm{i}} = \frac{c^2 \rho_0}{\gamma}\left\{ \left(\frac{\rho_{\mathrm{i}}}{\rho_0}\right)^{\gamma} - 1 \right\}, \tag{6}$$

where c is the sound speed [m s^{-1}], we set it to 12.24 m/s in this study [8].γ is the adiabatic exponent, and in this study, $\gamma = 7.0$. The subscript 0 denotes the initial condition.

The potential model Kondo et al. proposed [11] is adopted to calculate the interfacial tension force.

$$\mathbf{F}_{\mathrm{s,i}} = -\frac{2\sigma_{\mathrm{i}}\left|\mathbf{r}_{\mathrm{ij}}\right|_0^2}{\sum_j E\left(\left|\mathbf{r}_{\mathrm{ij}}\right|\right)} \sum_j \frac{\partial E\left(\left|\mathbf{r}_{\mathrm{ij}}\right|\right)}{\partial r} \frac{\mathbf{r}_{\mathrm{ij}}}{\left|\mathbf{r}_{\mathrm{ij}}\right|}, \tag{7}$$

$$E\left(\left|\mathbf{r}_{\mathrm{ij}}\right|\right) = \frac{1}{3}\left(\left|\mathbf{r}_{\mathrm{ij}}\right| - \frac{3}{2}\left|\mathbf{r}_{\mathrm{ij}}\right|_0 + h_{\mathrm{s}}\right)\left(\left|\mathbf{r}_{\mathrm{ij}}\right| - 2h_{\mathrm{s}}\right)^2, \tag{8}$$

where E is the potential energy [m^3], h_{s} is the effective radius [m] and σ is surface tension [kg s^{-2}], respectively.

This study introduces the multidimensional moving least squares interplant with constraint condition (CLS) method, as it allows for more accurate approximations near the boundary sampling points than the standard [12]. The CLS method improves the mass-area-density consistency and filters out small-scale pressure oscillations.

In this study, the models shown above were written in Fortran90. The source code was GPU-parallelized by OpenACC for speed, compiled with the nvfortran, the compiler for NVIDIA GPU, and executed using Nvidia A100.

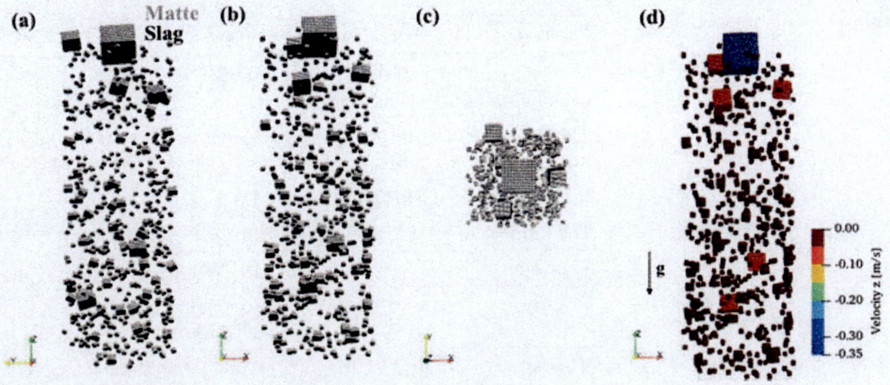

Fig. 2 **a–c** Initial placement of matte and slag from each direction (Gray color shows matte, black color shows slag). **d** Initial velocity

Calculation Conditions

This study assumes several sets of matte-slag binary droplets generated after the Cu concentrate reaction, and their collision calculations are performed. Figure 2 shows the initial placement and velocity of the calculated particles. Four droplet size groups were set. The matte and slag binary droplet sizes are determined that they are generated after the reaction of the 40, 20, 10, and 5 μm concentrate. The concentrate is assumed as $CuFeS_2$. The Cu grade of the matte generated after the reaction is set to 60%, and the Fe/SiO_2 of the slag is set to = 2.05.

$$2CuFeS_2 \rightarrow Cu_2S + 2FeS + 0.5S_2 \tag{9}$$

$$2FeS + SiO_2 + O_2 \rightarrow 2FeO \cdot SiO_2 + S_2 \tag{10}$$

Table 1 shows the parameters used in this calculation. The matte and slag were simulated with a collection of 1 μm SPH particles. The initial velocity of each droplet is set to the terminal settling velocity. Other physical properties used in the calculations are listed in Table 1.

Results

Figure 3 shows the time change of the droplet distribution. The matte-slag binary droplets are spherical due to interfacial tension, and the relative position of the droplets changes during falling due to the difference in the terminal settling velocity.

Figure 4 shows the time change of a droplet morphology. The large droplet impact to small one at $t = 20$ μs. A part of the matte in the small droplet contacts the larger

Table 1 Parameters used in the calculation [13]

Parameter	Value
Concentrate diameter used for calculation of matte and slag size	40, 20, 10, 5 μm
The initial velocity of the matte-slag binary droplet	0.343, 0.082, 0.016, 0 m/s
Initial distance of SPH particle	1 μm
The number of total SPH particles	110,848
Time step	10^{-9} s
Density	4660 kg/m^3 (Matte) 3200 kg/m^3 (Slag)
Viscosity	0.004 Pa s (Matte) 0.150 Pa s (Slag)
Surface tension	0.43 N/m (Matte) 0.40 N/m (Slag)
Interfacial tension	0.10 N/m

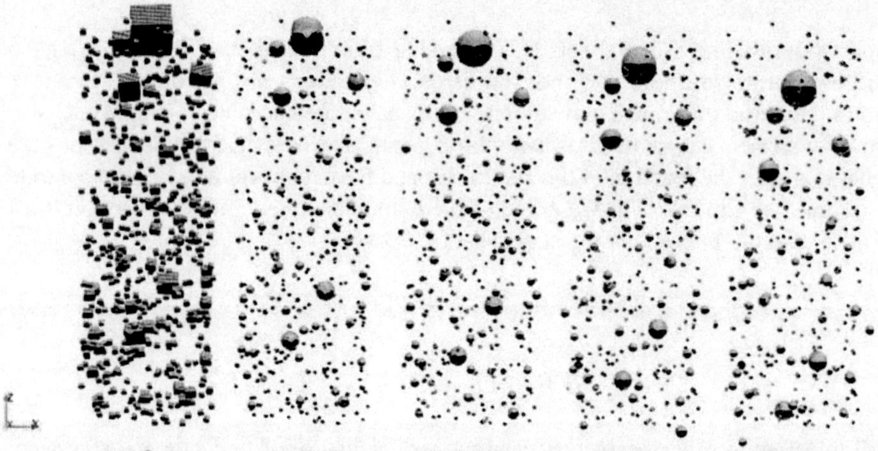

Fig. 3 Time change of droplet distribution. (Grey color shows matte, black color shows slag)

one at the effect at $t = 22$ μs. After that, most of the matte in the small droplet is captured in the large matte during falling because the force is worked, reducing the interface's size, i.e., the interfacial tension.

The dynamic model developed in this study can visualize the interfacial deformation of droplets, matte coalescence, and slag interference in the flash furnace's reaction shaft. In this study, only the difference in terminal settling velocity was considered. In the flash furnace, the collision frequency is expected to increase because of the gas-phase turbulence and the ejection of small particles generated by coarse concentrate rupture. By taking these effects into account, this analytical

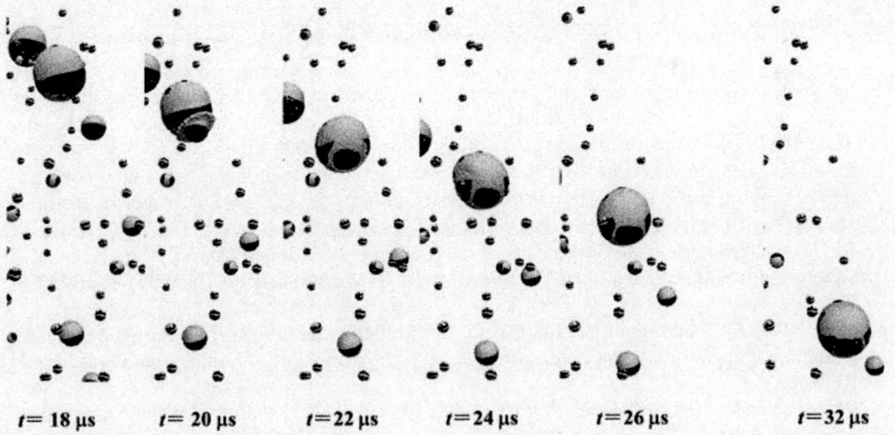

Fig. 4 Time change of droplet morphology. (Grey color shows matte, black color shows slag)

method is expected to represent in-furnace phenomena in more detail. Although the validation by comparison with phenomena in a flash furnace is a future issue, the easy-to-understand result obtained by this calculation can be expected to help consider how to operate a flash furnace.

Conclusion

In this study, the collision and coalescence behavior of many droplets consisting of matte and slag with various sizes assumed in the flash smelting reaction shaft is calculated explicitly by smoothed particle hydrodynamics method considering interfacial deformation using original code written in Fortran90. This numerical simulation visualizes interfacial deformation of a droplet, coalescence of matte, and interference of slag, and it showed that the interfacial energy significantly affects the matte's size growth. Although it is necessary to consider the effects of other forces on the droplets and compare the calculation results with actual phenomena in a flash furnace, these easy-to-understand calculation results seem to be helpful tool that provides guidelines on how to operate a flash furnace.

Acknowledgements This research was conducted using the FUJITSU Supercomputer PRIMEHPC FX1000 and FUJITSU Server PRIMERGY GX2570 (Wisteria/BDEC-01) at the Information Technology Center, The University of Tokyo.

References

1. Schmidt A, Montenegro V, Wehinger GD (2021) Transient CFD modeling of matte settling behavior and coalescence in an industrial copper flash smelting furnace settler. Metall Mater Trans B 52:405–413. https://doi.org/10.1007/s11663-020-02026-0
2. Kim YH, Themelis NJ (1986) The Reinhardt Schuhmann international symposium on innovative technology and reactor design in extractive metallurgy. In: Gaskell DR, Hager JP, Hoffman JE, Mackey PJ (eds) Warrendale, PA, TMS, pp 349–369
3. Jokilaakso A, Suominen R, Taskinen P, Lilius K (1991) Oxidation of chalcopyrite in simulated suspension smelting. Trans Instn Min Metall (Sect C: Mineral Process Extr Metall) 100:C79-C90
4. Yli-Penttilä JT, Peuraniemi EJ, Jokilaakso A, Riihilahti KM (1998) Dust formation in flash oxidation of copper matte particles. Min, Metall Explor 15(4):41–47. https://doi.org/10.1007/BF03403157
5. Pérez-Tello M, Parra-Sánchez VR, Sánchez-Corrales VM, Gómez-Álvarez A, Brown-Bojórquez F, Parra-Figueroa RA, Balladares-Varela ER, Araneda-Hernández EA (2018) Evolution of size and chemical composition of copper concentrate particles oxidized under simulated flash smelting conditions. Metall Mater Trans B 49(2):627–643. https://doi.org/10.1007/s11663-018-1183-1
6. Perez-Tello M, Sohn HY, Marie K, Jokilaakso A (2001) Experimental investigation and three-dimensional computational fluid-dynamics modeling of the flash-converting furnace shaft: Part I. Experimental observation of copper converting reactions in terms of converting rate, converting quality, changes in particle size, morphology, and mineralogy. Metall Mater Trans B 32B:847–868. https://doi.org/10.1007/s11663-001-0072-0
7. Kemori N, Ojima Y, Kondo Y (1988) Flash reaction processes. In: Proceedngs of a center for pyrometallurgy conference University of Utah, Salt Lake City, UT June 15–17
8. Natsui S, Tonya K, Nogami H, Kikuchi T, Suzuki RO (2021) SPH simulations of binary droplet deformation considering the Fowkes theory. Chem Eng Sci 229:116035. https://doi.org/10.1016/j.ces.2020.116035
9. Lucy LB (1977) Astron J 82:1013–1024. https://doi.org/10.1086/112164
10. Gingold RA, Monaghan JJ (1977) Mon Not R Astron Soc 181:375–389. https://doi.org/10.1093/mnras/181.3.375
11. Kondo M, Koshizuka S, Suzuki K, Takimoto M (2007, January) Surface tension model using inter-particle force in particle method. In: Fluids engineering division summer meeting, vol 42886, pp 93–98
12. Colagrossi A, Landrini M (2003) Numerical simulation of interfacial flows by smoothed particle hydrodynamics. J Comput Phys 191(2):448–475
13. Shibata E, Nakamura T (2013) Situation and problem of slag/matte separation in copper smelting. J MMIJ 129(5):171–176. https://doi.org/10.2473/journalofmmij.129.171

Options for Sustaining Metallurgical Engineering Education

E. Jak and P. C. Hayes

Abstract To design, develop, and efficiently operate the new technologies needed by the metallurgical industry, we need a skilled workforce with a sound understanding of the key aspects of the value chain and the metal processing cycle. That is, we need a workforce with specialist expertise in metallurgical engineering with the ability to use advanced computer-based tools to analyse, predict, and control these processes. With few exceptions, the availability of, and enrolments in, metallurgical engineering education programs around the world are low, and programs are under-resourced. There are major issues in relation to attracting students, designing and delivering programs, and ensuring financial sustainability of our university programs. In this context, we explore the approach taken to the development of the metallurgical engineering program at The University of Queensland, and the teaching of undergraduate students and engineers working in the metallurgical industry. These examples illustrate the importance of engagement with, and support from, industry in order to significantly increase the number of metallurgical engineering graduates and sustain metallurgical engineering as a discipline in our universities.

Keywords Education · Metallurgical engineering · Universities · Industry · Sustainability

Introduction

Our present societies face significant challenges such as sustainability and climate change. Major technological changes are taking place with rapid developments in renewable energy sources, energy storage, electric vehicles, and electrical and electronic devices. The growth in the production of advanced materials used in these manufactured devices has led to significantly increased demands for metals to the

E. Jak (✉) · P. C. Hayes
Pyrometallurgy Innovation Centre (Pyrosearch), School of Chemical Engineering, The University of Queensland, Brisbane, Australia
e-mail: e.jak@uq.edu.au

R. G. Reddy et al. (eds.), *New Directions in Mineral Processing, Extractive Metallurgy, Recycling and Waste Minimization*, The Minerals, Metals & Materials Series,
https://doi.org/10.1007/978-3-031-22765-3_10

extent that we are now at the start of the greatest growth in non-ferrous metal production the world has seen. This expansion can only be achieved with increased primary production from ores—from resources with decreasing ore grade. Added to this is the increase in the number of different metals and combinations of different metals used in device manufacture. Increased recycling and reuse of metals present in end of life products leads to significantly increased complexity and variability of metallurgical process feed streams.

To be able to design, develop, and efficiently operate the technologies of the future, we need a workforce skilled in metallurgical engineering with a sound understanding of the key aspects of the value chain and the metal processing cycle. That is, we need a workforce with a range of specialist expertise in mineral processing,pyrometallurgy, hydrometallurgy and electro-metallurgy with ability to use advanced computer-based tools to analyse, predict, and control these processes. Also, we need a work-force with an understanding and appreciation of the processing steps before and after metallurgical processing, and of the wider impacts of their actions on society as a whole [1–6].

The problem is, there are currently simply not enough metallurgical engineers, and the demand for the metallurgical engineers is rapidly increasing!

In Australia, enrolments in metallurgical engineering programs are low—far below what are needed to meet existing industry needs, never mind the projected significant growth. Similar situations exist around the world with few exceptions, notably in China and South America [6].

There are a number of reasons for this:

(1) Profile of metallurgy or lack of it-most high school and university students are not aware that the discipline of metallurgy even exists and what it encompasses.
(2) There is a lack of information on what metallurgical engineers do and recognition of the importance of the discipline to the functioning of our technologically based societies.
(3) The current financial structures and enrolment system of our universities.

There is an urgent need to reverse this trend of decreasing enrolments in metallurgy and to significantly increase student numbers of graduates in metallurgical engineering.

Options for Professional Education and Training

Typical academic and training programs offered by universities in Australia include:

Undergraduate Bachelors programs The standard undergraduate engineering program in Australia is a 4-year Bachelor of Engineering. The programs are accredited by Engineers Australia. The only universities currently offering Metallurgical Engineering are Curtin University in West Australia and The University of Queensland in the State of Queensland in the east of the country. Both of these universities offer these formal programs through Schools of Chemical Engineering.

The majority of BE graduates are employed by industry, only a small fraction go on to more advanced and higher degree studies.

Postgraduate Diplomas Postgraduate Diplomas are offered to graduates with Science or Engineering backgrounds who have previously not undertaken formal studies in metallurgical engineering. These can be delivered in residential or distance education modes, the latter option is intended particularly for people already in the workforce and is currently offered only by Curtin University.

Coursework Masters Coursework Masters degrees, typically 1 year full time study, provide useful pathway for broadening knowledge and providing advanced level skills. However, with the shortages in metallurgical engineering graduates and excellent job prospects, there is little incentive for domestic students to take on additional studies and again enrolment numbers are small. As is the case for all forms of formal education, Masters degrees are only financially viable for universities in the more popular disciplines where a minimum enrolment can be guaranteed.

Research Masters/PhD Research higher degrees, such as research Masters and PhDs, provide students with additional skills in investigation, analysis and innovation, and training in the use advanced experimental and/or process modelling techniques. Research higher degree students are not only the workhorses behind research undertaken at universities, enabling fundamental research to be undertaken at reduced cost, but also an important source of professionals seeking research and development roles in industry.

Continuing Professional Development (CPD)/Short courses/Micro-credentials Increasingly offered by professional societies, Continuing Professional Development and Short Courses provide support and broadening experiences to practicing engineers. These are useful for updates on latest trends but are principally aimed at information transfer and professional networking. Some universities have started to market these short courses as Micro-credentials. Whilst superficially attractive these vehicles, in general, do not lead to "deep learning" of complex, advanced skills. Deep learning comes from practice in the implementation of these new skills, and experience shows that the learning process is accelerated by ongoing support from those skilled in the art—this is the basis for teaching practice.

In cases where the need for particular skill sets has been identified by companies, special training programs have been developed and delivered by professional engineers to upskill or retrain engineers already working in the metallurgical industry. Again, to produce effective learning outcomes practice and feedback to students need to be built into these programs. Examples of these programs include MetSkill and Metcelerate [7–9] workplace-based training programs designed for professional engineers with no prior training in mineral processing operations.

Of these options, the largest demand and importance is for graduates with Bachelor of Engineering degrees. Education at the undergraduate level has a profound impact on the fundamental understanding of, and systematic approaches taken to solve, complex engineering problems. For this reason, the focus should be on the ways in which enrolments in Bachelor level metallurgical programs can be increased.

Metallurgical Engineering Undergraduate Curriculum at UQ

In Australia, whilst the principal products are mineral concentrates, there are also significant chemical or extractive metallurgical operations producing refined metals. The requirement is to produce graduates with a knowledge of the whole value chain from ore to refined metal products. In addition, the increasing use of computer-based tools to assist in design, simulation, and control of complex integrated operations has led employers to preferentially select engineering graduates with the process engineering skills associated with Chemical Engineering programs. At the University of Queensland, in 2005, following consultation with industry stakeholders, the dual major BE in Chemical and Metallurgical Engineering was introduced. The development of the program was briefly described in an earlier publication [3]. After operating successfully since its inception, to meet a university-wide restructure of undergraduate offerings, this program has, as recently as 2021, been rebadged as a Major in Metallurgical Engineering in the degree of Chemical Engineering. The courses and the current structure of the 4-year program are illustrated in Fig. 1. The program consists of a general first year of engineering mathematics, computer programming, engineering thermodynamics and chemistry with elective courses to enable students to broaden their understanding of other engineering disciplines. The second year provides a sound background in engineering sciences and process engineering fundamentals. Included here is the first course in metallurgical engineering,

UQ Metallurgical Engineering program plan				
Year 1 Sem. 1	Professional Engineering	Calculus & Linear Algebra	Thermodynamics: Energy and the Environment	General Elective
Year 1 Sem. 2	Computer Programming	Multivariate Calculus & ODEs	Chemistry	General Elective
Year 2 Sem. 3	Mass & Energy Balances	Investigation & Statistical Analysis	Fluid Dynamics	Physical Chemistry for Engineering
Year 2 Sem. 4	Process Equipment & Control Systems	Chemical Thermodynamics	Heat & Mass Transfer	Metal Production and Recycling
Year 3 Sem. 5	Unit Operations/ Process Control	Reaction Engineering	Major elective	Major elective
Year 3 Sem. 6	Process Systems Analysis	Dynamic Modelling & Control	Process Mineralogy and Comminution	Physical Separation Processes
Year 4 Sem. 7	Impact/Risk in the Process Industries	Professional Business/Practice	Pyrometallurgy	Hydrometallurgy and Electrometallurgy
Year 4 Sem. 8	Process Engineering Design Project		Major elective	Major elective

Fig. 1 The courses and the current structure of the 4-year BE program in Chemical Engineering with the Major in Metallurgical Engineering at The University of Queensland

a general introduction to the whole value chain from ore, through metal production, to metal recycling. This introductory course is an elective course for Chemical Engineers and students in other branches of engineering, e.g. mining engineering and science students, including Earth Science.

An important feature of the Metallurgical Engineering Major program is the flexible entry up until the start of year three of the program. The delay in specifying the metallurgy major has been an important factor in keeping opportunities open and allowing for articulation of students from related programs. Most students leaving high school are unaware of the metallurgical engineering discipline and the career opportunities it provides.

Years 3 and 4 of the program provide advanced courses in process engineering, and the application of physical and chemical fundamentals to metallurgical processes through specialist courses in process mineralogy and comminution, mineral processing separations, hydro- and electro-metallurgy, pyrometallurgy, and metallurgical plant design.

The major provides the core engineering science and process engineering knowledge and skills associated with conventional chemical engineering programs. The systems approach, and with it the use of process modelling tools, provide the ability to quantitatively describe and analyse material flows in complex integrated plant operations.

Currently, all students studying metallurgical engineering at Curtin and UQ are required to take a capstone plant and process design course in the 4th year of their studies. The design brief is developed in conjunction with industry, who supply real data on key characteristics of a number of different ore deposits in different locations. The students are required to use what they have learnt in their previous specialist metallurgy and process engineering courses to develop a prefeasibility report on these options, including flow sheet design and equipment selection, process costs, and societal impacts. The students work in groups during the semester and produce a joint report. At the start of the semester, a special workshop involving the students and professional engineers is held giving the students the opportunity to learn directly of issues related to industrial practice and to use this information to inform their process selection and design. The design course is financially supported by the Minerals Council of Australia (through the Minerals Tertiary Education Council, MTEC) [10].

The current metallurgy major is accredited by the Institution of Chemical Engineers (I. Chem. E., London), Engineers Australia (EA), and the Australasian Institute of Mining and Metallurgy (AusIMM).

This program structure not only allows students from other engineering programs within UQ, other tertiary institutions in Queensland or interstate to transfer without loss of credit for prior study into the program, but also provides opportunities to international students to join the program in the later years through 2 + 2 and 2 + 3 programs with other universities.

The students are attracted to the wider career options on graduation and in their future careers offered by the chemical engineering program but also view favourably the enhanced job opportunities on offer by specialisation in an identifiable industry sector.

The metallurgical engineering major provides students with the appreciation and understanding of the critical link between material microstructure and processing requirements, and the interrelationship of process steps in the metallurgical value chain, from ore through to refined metal. These concepts are equally important for the development of efficient methods of recovery of metals from end of life devices [11]. As ores and materials become more complex in physical structure and in the combination of elements present, it is increasingly important to develop engineers who can understand, integrate, and optimise combinations of physical and chemical processing operations. This combination of chemical and metallurgical engineering provides graduates with the fundamental knowledge and skills sought by the minerals industry.

Actions Needed to Increase Metallurgical Engineering Enrolments

Providing an attractive curriculum structure is on its own not enough to obtain the level of enrolments needed in these Metallurgical Engineering programs.

Profile/Social Media The explanation given most often by students as to why they did not enrol in metallurgical engineering is that they were unaware that the discipline even existed at the time they made their choices at high school or universities. They had no idea of the key role played by metallurgical engineers in delivering the metals we need to sustain our technologically based societies.

Students of today are much more aware and concerned about major issues, such as climate change, pollution, and social responsibility, and want these issues to be addressed in their learning programs [4, 5]. The image of the mining and metallurgical industry as dirty and harmful to the environment is not attractive. The widespread acceptance of the need to address the climate emergency provides the opportunity to change this narrative. The messages we post in the media to potential students should highlight the importance and need for metals to be able to manufacture the advanced materials that will be used in the emerging technologies and in supporting the transition to sustainability. The need for increased recycling and reuse of materials is recognised and accepted in the community but the significant challenges of separating and reusing metal alloys need to be further highlighted. These messages have to be conveyed to high school students as well as engineering students at the start of their university education - these are the critical times when decisions are made by young people on subject and course selection.

Flexible entry In many instances, the students enrolling university for the first time are often required to select the Department or area of specialisation at this entry point. This is a major disadvantage for metallurgical engineering since most students are unaware of the discipline and its significance at this point in their life. University administrations should be encouraged to allow flexible entry into metallurgical engineering during years 1 and 2 until the start of year 3 of a 4-year program.

This delay in selecting the area of specialisation allows the student time to come into identity with and experience what might be their future career direction.

Engagement and support At most universities, there are student activities and societies that promote the various disciplines. Academic staff and students can contribute to spreading the word about the discipline and the advantages of studying metallurgical engineering, through, for example,

- The establishment of students societies, and engagement in sponsored activities and events
- Contributing to activities organised by other university societies and departments
- Actively publicising metallurgical events and activities through Social Media networks.

Here also is an opportunity for industry to demonstrate to that there are exciting, challenging, and promising career paths in the metallurgical industry. Industry can usefully contribute to the recruitment process through,

- Presentations to students at informal social events.
- Facilitating visits by high school, and 1st and 2nd year university student cohorts, to working process plants, particularly targeting students in regional areas and those areas where there are existing or potential industrial metallurgical activities.
- Awarding fee and living scholarships to students taking up studies in the discipline.
- Providing students with paid part-time work experience/internships with professional engineers between classes during semesters.
- Providing students with paid work during the long vacations between university studies.
- Developing and encouraging participation of industry professionals in mentor schemes for undergraduate and graduate engineers.

Many professional metallurgical engineers cite their visits to industry and employment experiences as undergraduates as the key factor determining their decision on the directions of their future careers.

Resilience University enrolments in professions closely associated with the minerals closely follow the business cycle and in general reflect commodity prices and company profits. Enrolments tend to increase when industry is promoting positive messages, making new investments, and hiring workforce. Since it takes approximately 4 years for students to complete their degrees, the graduation of these students may coincide with downturn in economic activities, times when people are laid off to maintain profit bottom line. These periodic downturns do not unnoticed by Mums and Dads who have influence on the future of their sons and daughters. The minerals industry continues to have a bad reputation in this regard hiring and firing rather than strategic development and retention of key members of its workforce. Meanwhile, universities are expected to maintain the education programs through these economic cycles. To attract the best and brightest to metallurgical engineering the industry must address, this as an issue and commit to developing attractive career paths for young engineers.

Strategic development of the discipline is also needed to provide opportunities for careers within universities so that research academics can continue to create new ideas, innovate, and at the same time also support the commercial implementation of new technologies. This approach would be of benefit to universities and to industry.

This could be achieved by creating and funding a new class of research academics, having specialist metallurgy skills and expertise, who would develop and undertake teaching specifically for industry in their field of specialisation, in addition to undertaking advanced research to support industry.

Potential objectives of such a program could be to

- Provide rapid responses to industry training and educational needs in key specialist fields.
- Transfer key skills in the use and application of specialist skills and advanced predictive tools to engineers in industry.
- Integrate these educational programs into graduate training programs for industry.
- Enhance collaboration and cooperation between industry and universities on the development of research projects, and the implementation and optimisation of industrial technologies.
- Provide job certainty for metallurgy research academics and career pathways into academia.
- Strengthen the sustainability and funding of university metallurgy research.

The detailed arrangements between industry partners and the universities would likely differ with industry requirements and host universities. All parties would benefit from this type of collaboration. Industry has access to researchers with advanced skills and expertise in specialist fields, enabling them to rapidly upskill their workforce in key areas of need. Universities and research teams are able to attract and retain high quality researchers. New pathways for career development are created for postdoctoral researchers with industry focus with potential for advancement in industry or universities.

Financial sustainability In most countries, universities are under financial pressure and are increasingly forced to operate as discrete financial entities, i.e. they are run as educational business units. The educational programs offered are increasingly determined by economic considerations rather than the strategic needs of our societies. These pressures favour programs having larger student enrolments over smaller specialist programs, such as metallurgical engineering.

The predominant source of funding for most universities in Australia is through undergraduate programs – student fees and Government funding based principally on the number of students enrolled. In general, the fee income per student for the university from domestic students is lower than from international fee paying students. This led to significant proportions of enrolments of international students in the more popular disciplines. However, the recent restrictions on movement stemming from the COVID virus outbreak and geopolitical tensions have resulted in significant declines in international student enrolments. This has placed additional financial pressures on the universities that have come to rely on this source of income.

To offer accredited programs for graduates the university must provide a sound curriculum and sufficient qualified academic staff. Rationalising the teaching resources available by combining classes from related discipline areas is one way in which costs can be reduced and academic teaching loads reduced, as demonstrated in the example of the UQ major in metallurgical engineering. However, there comes a point at which further reductions leading to cancellations of key specialist courses diminish the quality of learning program. Reducing program quality leads to loss of reputation and standing in the field and further difficulties in recruiting students—this becomes a downward spiral, which cannot be reversed without a capital injection. This point is where industry needs to make a choice. It can ignore the problem and allow the progressive loss of courses, leading to further reductions in competencies of graduates and availability of academic staff in metallurgical engineering, or it can be proactive in supporting programs it sees valuable for its future workforce needs.

If there are no graduates in metallurgical engineering, then industry has some choices to make, (i) put up with decline in the competency of its workforce and the adverse impact of efficiency and productivity, (ii) spend money, time, and effort retraining its workforce internally using highly paid industry specialist engineers in the desired skills, or (iii) alternatively, invest in university education, providing financial and other contributions to ensure there are sufficient graduates in the market place.

There are examples of direct targeted industry financial support to universities in the form of

- Academic staff fixed term or endowed teaching academic positions
- Specific teaching programs/courses
- Donations of equipment in support of teaching or research.

Now is a period of growth for our industry, why not make what are modest investments in supporting metallurgical engineering education? This would be to the benefit of industry, universities, and the wider society.

Industry and Governments can also play a role by changing the language that is used to inform the general public about the key role of downstream processing. It is not all about "mining" and "critical minerals", the narrative needs to be changed to explain the whole value chain with emphasis on **critical metals, metal recovery, reuse, recycling, and sustainability**. These are the messages that the young people want to and need to hear.

The metallurgical academic staff at our universities can and should encouraged and allowed to play a key role in facilitating these actions and messages by actively working with, and coor-dinating interactions between the major parties-industry, university management, and students. Metallurgical academic staff have direct working interactions with all these key stakeholders and are ideally placed to drive these initiatives.

All of these messages should be aimed at achieving a dramatic increase in the number of metallurgical engineering graduates. This is the key to ensuring the sustainability of metallurgical engineering as a discipline in our universities and to the supply of metallurgical graduates to the society, professionals capable to tackle the current major challenges.

Summary

We live in a time when there is an increasing demand for graduates with skills in metallurgical engineering due to the transition to renewable energy technologies, the electrification of transport, and many aspects of operations in our everyday life. However, the low student numbers over recent years have left most countries with a metallurgical skills shortage. In this paper, some suggestions as to how enrolments in metallurgical engineering can be increased have been provided. Universities can assist to facilitate the increase in metallurgical engineering enrolments by creating flexible pathways and supporting programs for students; however, they need help from industry - and industry needs graduates.

A more strategic view of tertiary education is needed, the one in which there is shared responsibility between universities and the metallurgical industry for, and ownership of, the workforce development. This will enable increased awareness of opportunities in the industry, encourage increased student enrolments in metallurgical engineering, and support for the financial sustainability of key metallurgical engineering programs.

References

1. Hayes PC (2018) The changing world of metallurgy education. In: Proceedings of Extraction 2018, Ottawa, TMS-CIM, Montreal, pp 57–67
2. Hayes PC (2019) The world needs metallurgical process engineers. JOM 71(2):463–468
3. Hannah R, Hayes PC (2014) The challenges for professional metallurgical education. In: Proceedings of extraction and processing division symposium on pyrometallurgy in honor of David G.C. Robertson, Ed.: P. J. Mackey et al., TMS, Warrendale, pp 473–480
4. Marsden LD (2020) The ecosystem of research, education and community. Metall Mater Trans A 51A:4329–4340
5. Dubova IV et al (2020) Integration of sustainability in metallurgical education: analysis of educational practices CDIO. J Phys: Conf Ser 1691:012212. https://doi.org/10.1088/1742-6596/1691/1/012212
6. IMPC (2016) A roadmap for minerals processing education. In: Drinkwater D (Ed.) Special symposium on education: mineral processing for the future. XXVIII International Mineral Processing Congress (IMPC 2016), September 11–15, Quebec City, Canada
7. Drinkwater D, Napier-Munn T (2014) Why good professional development is key to profitability in the mining industry, XVII International Mineral Processing Congress (IMPC). Santiago, Chile
8. Drinkwater D, Bianco N (2014) The MetSkill program—rapidly developing effective young engineers in the workplace. In: Proceedings of extraction and processing division symposium on pyrometallurgy in honor of David G.C. Robertson, Ed.: P. J. Mackey et al. TMS, Warrendale, pp 499–506
9. www.metcelerate.com

10. Ibana D (2014) Delivering a national process design unit with industry support. In: Proceedings of extraction and processing division symposium on pyrometallurgy in honor of David G.C. Robertson, Ed.: P. J. Mackey et al., TMS, Warrendale, pp 489–498
11. Hayes PC (2021) Process principles in minerals and materials production with emphasis on metal production and recycling, 4th edn. Hayes Publishing. www.peterhayesmetallurgy.com

Separation of Co(II) Over Ni(II) from Chloride Leached Solution of Spent Li-Ion Batteries Using Cyphos IL104

Sadia Ilyas, Hyunjung Kim, and Rajiv Ranjan Srivastava

Abstract The hydrometallurgical separation of cobalt and nickel was achieved from HCl leached solution of spent Li-ion batteries, while applying Cyphos IL104 as a potential extractant. Cobalt extraction significantly increases with increasing Cyphos IL104 and Cl^- ion concentrations, and equilibrium pH (pH_{eq}). Under the optimized condition of 0.8 mol·L^{-1} Cyphos IL104, 3.0 mol·L^{-1} Cl^- ions, pH_{eq} value 5.0, and O/A ratio of 2/3 could extract >99% Co with only 3% Ni into the organic phase. The back-extraction of cobalt performed with 2.0 mol·L^{-1} H_2SO_4 solution yielded 99.95% stripping efficiency and was crystallized to obtain high-pure $CoSO_4.xH_2O$ crystals.

Keywords Hydrometallurgical recycling · Li-ion batteries · Ionic liquid

Introduction

Lithium-ion batteries (LiBs) are widespread employed in most of the new generation electronic devices for their portable use [1, 2]. At present, the electric vehicles (EVs) are the main consumer of LiBs with an estimated consumption over 700 GWh by 2030 [3–5]. It comes together with the challenges of meeting the continuous demands for strategic and rare, and the sustainable management of spent LiBs with 8–10 years of lifespan containing the metal compounds therein. Recycling of spent LiBs is therefore a necessity which can be an environmentally friendly solution to the problems. Moreover, a considerable number of metals in spent LiBs (i.e., 10–30%

S. Ilyas (✉) · H. Kim
Department of Earth Resources and Environmental Engineering, Hanyang University, Seongdong-gu, Seoul 04763, Republic of Korea
e-mail: Ilyas.sadia24@gmail.com; sadiailyas1@yahoo.com

R. R. Srivastava
Center for Advanced Chemistry, Institute of Research and Development, Duy Tan University, Da Nang 550000, Vietnam

Faculty of Natural Sciences, Duy Tan University, Da Nang 550000, Vietnam

R. G. Reddy et al. (eds.), *New Directions in Mineral Processing, Extractive Metallurgy, Recycling and Waste Minimization*, The Minerals, Metals & Materials Series,
https://doi.org/10.1007/978-3-031-22765-3_11

Cu, 5–20% Mn, 5–30% Co, 5–20% Ni, and 2–7% Li) can be regarded as a secondary source of the strategic-rare metals [6, 7].

Among the reported recycling techniques, hydrometallurgical recycling has been found to be advantageous than pyrometallurgy because of a greater yield, low energy consumption, and lower carbon footprints. It involves pre-treatment, discharging, shredding, sieving and separation, leaching, and separation and purification steps. These steps are complex and particularly to deal with a challenging issue of hydrometallurgy for the liquid–liquid separation of Co(II) and Ni(II). For isolating targeted metal ions from the leach liquor, the liquid–liquid (solvent) extraction is widely employed. Many works have been done on cobalt and nickel extraction using acidic organophosphorus extractants from the acidic solution. The separation factor between cobalt and nickel increases in the following order: phosphoric acid (D2EHPA) < phosphonic acid (PC88A) < phosphinic acid (Cyanex 272) and pH_{50} values were determined to be Co (3.68) and Ni (3.86) for D2EHPA, Co (4.2) and Ni (5.7) for PC88A, and Co (4.2) and Ni (6.2) for Cyanex 272, respectively [8, 9].

Further, the use of ionic liquids offers an environmentally friendly alternative to conventional organic solvents in several applications. The advantageous features of low vapour pressure, high chemical and thermal stability, and a greater loading capacity have gained significant attention in solvent extraction of metal ions as well [10, 11]. Flieger et al. [12] employed the aqueous biphasic system based on 1-hexyl-3-methylimidazolium tetrafluoroborate [HMIM][BF_4] and NaCl as a salting-out agent to separate cobalt over the co-existing metals including nickel. Comparative study conducted by Zhu et al. [13] observed that tri-hexyltetradecylphosphonium chloride (Cyphos IL 101) exhibited a higher cobalt extraction than that of trioctylmethylammonium chloride (Aliquat 336) and tri-octyldecyl amine (Alamine 336) from a nickel laterite leach liquor. Hoogerstraete and Binnemans [14] applied undiluted trihexyl(tetradecyl)phosphoniumnitrate to extract samarium and lanthanum over cobalt and nickel, respectively. All these ionic liquids follow the anion exchange mechanism.

To the best of our knowledge, there is no report on cobalt-nickel separation using Cyphos IL104 from the leached solution of spent LiBs. The parameters such as equilibrium pH (pH_{eq}), temperature, and organic-to-aqueous (O/A) phase ratio were varied to examine the maximum separation between Co(II) and Ni(II). The loaded organic was stripped to recover the high-pure cobalt from the spent LiBs and shows the process leading towards the circular economy of this strategic-rare metals.

Experimental

The feed solution used in this study was generated by the leaching of NMC-cathode powder (containing 20.2% Co, 0.9% Ni, 6.2% Mn, and 4.8% Li) while maintaining a pulp density of 20%. The obtained leach liquor was then undergone for Mn-removal by the oxidative precipitation using $KMnO_4$ as described by Sattar et al. [15]. The Mn-depleted solution containing 10.2 $g\cdot L^{-1}$ Co, 0.42 $g\cdot L^{-1}$ Ni, and 2.3 $g\cdot L^{-1}$ Li was further used as the stock solution to feed in the present study. The ionic liquid Cyphos IL104 (Cytec Industries Inc.) and diluent solvent C10 (SOLVENTIS) were used for the preparation of organic feed. Hydrochloric acid (Daejung) and sulfuric acid (Lab-Scan Analytical Science) were used without further purification.

All the experiments were performed at an O/A ratio of 1/1 using 30 mL volume of each phase in a 100 mL separating funnel. Until mentioned specifically, other parameters like contact time (10 min), temperature (25 ± 1 °C), and phase separation time (10 min) were maintained unchanged. After the equilibration, two phases were separately collected and the raffinate was taken for metal analysis using an inductively coupled plasma–atomic emission spectroscopy (ICP–AES, iCAP6000 series, Thermo Scientific). The metal concentration in organic phase was determined by mass balance between the aqueous feed and raffinate. Further, the distribution coefficient (D), extraction efficiency in percentage, and molar separation factor ($\beta_{(\mathrm{Co/Ni})}$) were determined as follows:

$$D = \frac{\mathrm{CE}_{org}}{\mathrm{CE}_{aq}} \tag{1}$$

where CE_{org} and CE_{aq} are post-extraction metal concentration in the organic and aqueous phases, respectively.

$$\text{Extraction, \%} = \left[\frac{D \times \left(\frac{\mathrm{V}_{org}}{\mathrm{V}_{aq}}\right)}{D \times \left(\frac{\mathrm{V}_{org}}{\mathrm{V}_{aq}}\right) + 1}\right] \times 100 \tag{2}$$

where V_{org} and V_{aq} are volumes of the organic and aqueous phases, respectively.

$$\beta_{(\mathrm{Co/Ni})} = \frac{D_{\mathrm{Co}}}{D_{\mathrm{Ni}}} \tag{3}$$

Subsequently, the stripping was carried out by contacting the loaded organic phase with H_2SO_4 solutions of different concentrations, O/A phase ratio of 1/1, and time 10 min. Both phases were separated after 10 min of settling, and then, the aqueous phase was analyzed for knowing the stripping efficiency as follows:

$$\text{Stripping, \%} = \left[\frac{\mathrm{CS}_{aq} \times \mathrm{V}_{aq}}{\mathrm{CS}_{org} \times \mathrm{V}_{org}} \times 100\right] \tag{4}$$

where CS_{org} and CS_{aq} are post-stripping metal concentration in the organic and aqueous phases, respectively.

Results and Discussion

Effect of Extractant Concentration

The extraction efficiency of Co(II) and Ni(II) was investigated at different concentrations of Cyphos IL104 in the range of 0.1–1.0 mol·L^{-1}, while keeping an O/A ratio constant at 1/1. Figure 1 shows that the extraction of Co(II) increased with increasing concentration of Cyphos IL104 molecules into the organic phase. It can be seen that Co(II) extraction improved from <10% to >50% by increasing the extractant concentration from 0.1 mol·L^{-1} to 0.5 mol·L^{-1}, respectively. The extraction reached >90% using >0.9 mol·L^{-1} Cyphos IL104, which can be ascribed to the shift in distribution curve with increasing concentration of ionic liquid in the organic phase [16, 17]. However, the co-extraction Ni(II) also increased with >0.5 mol·L^{-1} of Cyphos IL104, yielding >17% co-extraction of Ni(II). Further, the calculated molar separation factor ($\beta_{Co/Ni}$) in Fig. 1 depicts an increasing trend reaches ~90 with 0.8 mol·L^{-1} Cyphos IL104. Then after, it decreases to a value near about 50 with 1.0 mol·L^{-1} of Cyphos IL104.

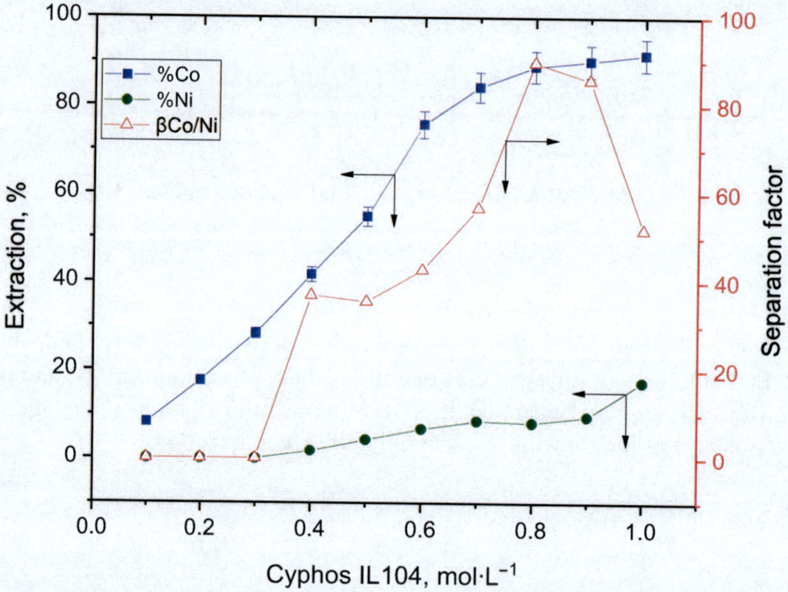

Fig. 1 Extraction behaviour of Co(II) and Ni(II) as a function of Cyphos IL104 concentration in the organic phase

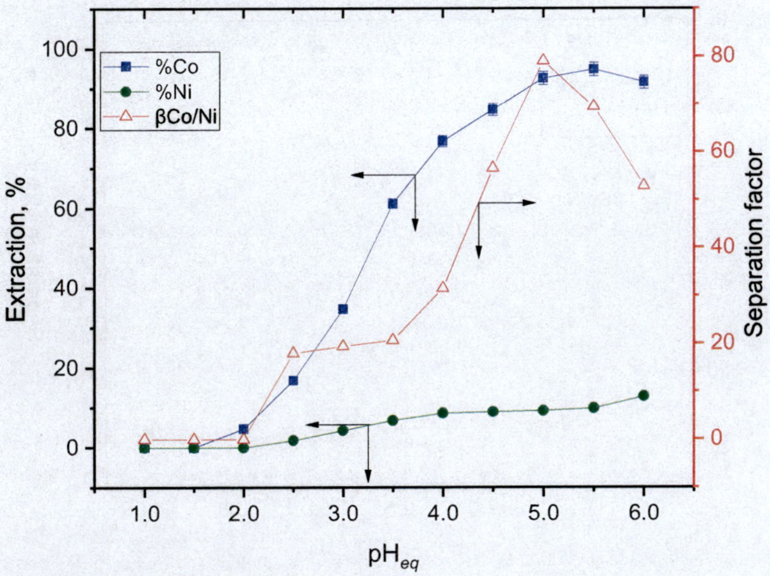

Fig. 2 Extraction behaviour of Co(II) and Ni(II) as a function of equilibrium pH

Effect of Equilibrium pH

The extraction efficiency of Co(II) and Ni(II) was investigated at different equilibrium pH (pH_{eq}) varied in the range of 2.6 to 6.0, while 0.8 mol·L^{-1} of Cyphos IL104 and O/A ratio of 1/1 were kept constant. Figure 2 shows that the extraction of Co(II) increased from 34 to 91% in between the pH_{eq} value 2.5 and 6.0. The co-extraction of Ni(II) increased from 1.8% to 12.8% within the same range of pH_{eq}. A higher selectivity over Ni(II) can be understood by a greater ratio of water exchange for Co(II) than for Ni(II) that leading to form an easy complexation between Co(II) and Cyphos IL104 [17, 18]. Further, the calculated $\beta_{Co/Ni}$ of 79.0 obtained at a pH_{eq} 5.0 was found to be optimal (Fig. 2).

Effect of Chloride Ions

The extraction efficiency of Co(II) and Ni(II) was investigated at different chloride concentration in the range of 1.0 to 4.0 mol·L^{-1} Cl^- in the aqueous phase. Figure 3 shows that the extraction of Co(II) increased from ~80% to 92% with increasing chloride ions from 1.0 to 2.5 mol·L^{-1}, respectively. After that, Co(II) extraction did not change and found about 94% at ≥3.0 mol·L^{-1} of Cl^- concentration in the aqueous phase. It can be understood by Cl^- ions driving the ion-pairing with the trihexyl(tetradecyl)phosphonium molecules of Cyphos IL104 that split at the

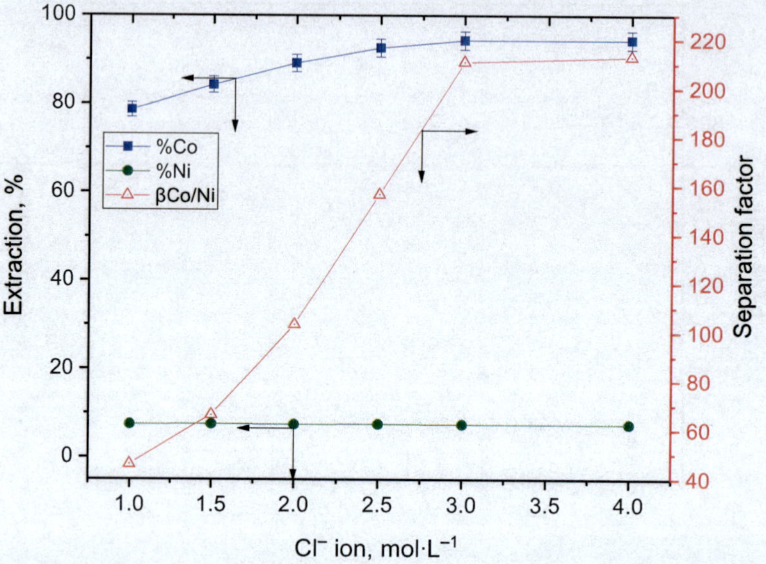

Fig. 3 Extraction behaviour of Co(II) and Ni(II) as a function of chloride ion concentration

extraction equilibrium [18, 19]. However, the co-extraction of Ni(II) did not show remarkable change with changing Cl^- ion concentration. Thus calculated $\beta_{Co/Ni}$ was found to be >210 at a Cl^- concentration of $\geq$3.0 mol·L^{-1} (Fig. 3). Based on the experimental results, the extraction reaction can be written as follows:

$$Co^{2+} + 2Cl^- + 2\overline{[R_3R'PA]} \leftrightarrow 2\overline{[R_3R'PCl]} + \overline{[CoA_2]} \quad (5)$$

Effect of Temperature

The extraction efficiency of Co(II) and Ni(II) was investigated at different temperatures ranges from 25 °C to 55 °C, and keeping 0.8 mol·L^{-1} Cyphos IL104, 3.0 mol·L^{-1} Cl^{-1}, pH$_{eq}$ 5.0, and O/A ratio of 1/1 unchanged. Figure 4 shows that raising the temperature could slightly improve Co(II) extraction from 93% to >96% (at 25 °C and 55 °C, respectively). This behaviour can be understood by the metal-binding sites of Cyphos IL104 which improves with raising the temperature [16, 17]. The co-extraction of Ni(II) was unchanged at ~7% only. Thus, the calculated $\beta_{Co/Ni}$ improved more and reached >290 at 55 °C. Further, the thermodynamic properties of Co(II) extraction were determined using the van't Hoff equation, as follows [16, 20]:

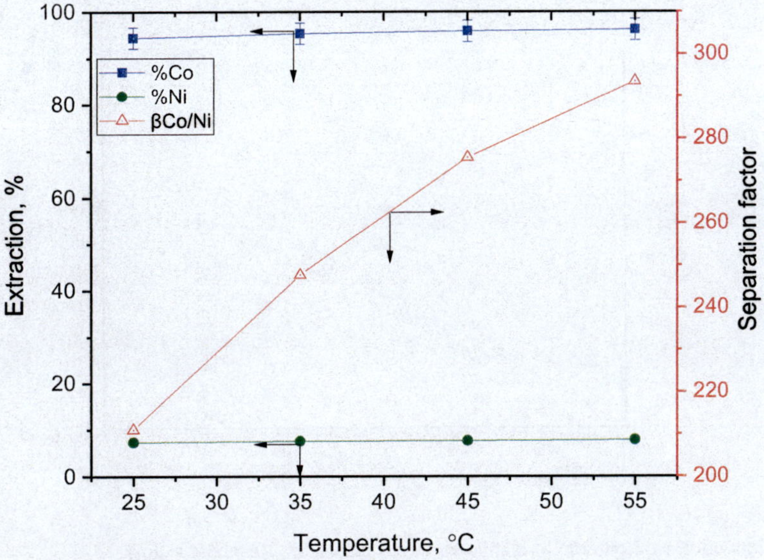

Fig. 4 Extraction behaviour of Co(II) and Ni(II) as a function of temperature

$$\ln D_{\mathrm{Co}} = -\frac{\Delta H^\circ}{\mathrm{R}T} + \frac{\Delta S^\circ}{\mathrm{R}} \tag{6}$$

The slope and intercept values of ln $[D_{\mathrm{Co}}]$ versus $1/T$ (not shown here for the sake of brevity) were used to calculate the apparent changes in both enthalpy (ΔII°) and entropy (ΔS°), which were obtained to be 10.8 $\mathrm{kJ \cdot mol^{-1}}$ and 59.9 $\mathrm{J \cdot K^{-1} \cdot mol^{-1}}$, respectively. Further, the values of ΔH° and ΔS° were used to calculate the change in standard Gibbs free energy (ΔG°) as follows:

$$\Delta G^\circ = \Delta H^\circ - T\Delta S^\circ \tag{7}$$

The obtained values of extraction thermodynamics indicated that cobalt extraction with ionic liquid is a spontaneous process of endothermic nature, which tends to disrupt the organic molecules when metal is being complexed into the immiscible phase [17].

Effect of O/A Phase Ratio

The extraction efficiency of Co(II) was investigated at different phase ratio varied from 3/1 to 1/3 by equilibrating both aqueous and organic phases at a fixed unchanged

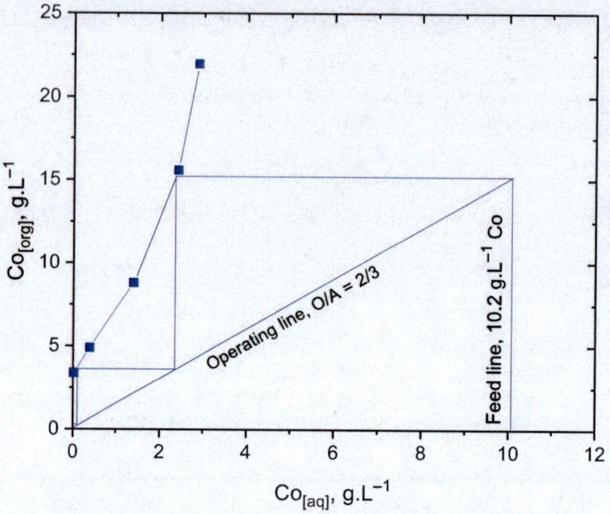

Fig. 5 Extraction isotherm of Co(II) plotted as a function of O/A phase ratio

volume. As can be seen, the increase in O/A ratio from 1/3 to 3/1 could significantly increase Co(II) extraction from 72% to about 99%, respectively. The extraction isotherm plotted in Fig. 5a shows that a 3 stage counter-current extraction at an O/A phase ratio of 2/3 can quantitatively extract Co(II) over Ni(II). Further, a batch simulation study was investigated and the steady-state data were obtained for metal distribution, which showed 20 mg·L^{-1} Co(II) corresponding >99% extraction efficiency. The extraction phase organic contains ~3% Ni(II) that yielded a separation factor ($\beta_{Co/Ni}$) of 1097.

Cobalt Recovery from the Loaded Organic Phase

Co(II) stripping from the loaded organic phase was investigated at different H_2SO_4 concentrations ranges from 1.0 to 4.0 mol·L^{-1} at an O/A ratio of 1/1. Results in Fig. 6 showed that stripping efficiency increased with increasing acid concentration in the aqueous phase. >90% Co(II) was stripped when contacted with 2.0 mol·L^{-1} H_2SO_4 solution, which was only 39% with 1.0 mol·L^{-1} H_2SO_4. Further, the stripping reached about 99% when contacting ≥3.0 mol·L^{-1} H_2SO_4 solution with the loaded organic; however, to keep a lower consumption of acid, a two-stage stripping under counter-current contact was performed using 2.0 mol·L^{-1} H_2SO_4 solution. A quantitative stripping (~99%) was achieved after a two-step contact, which was 91.6% with a single contact. The stripped solution contains 15.2 g·L^{-1} cobalt and 20 mg·L^{-1} nickel was evaporated to crystallize the sulfate salt of cobalt and analyzed to be $CoSO_4{\cdot}xH_2O$ phase as the final product.

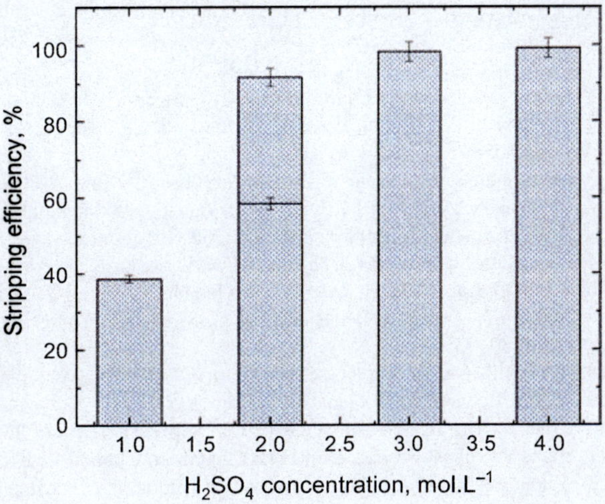

Fig. 6 Stripping behaviour of Co(II) from loaded organic phase as a function of acid concentration

Conclusions

This study demonstrated a novel liquid–liquid separation of cobalt and nickel from an HCl leached solution of spent Li-ion batteries by applying Cyphos IL104 as a potential extractant. The results showed that ~95% cobalt was extracted with 0.8 mol·L^{-1} Cyphos IL104 and 3.0 mol·L^{-1} Cl^- ions in the aqueous solution at an O/A ratio of 1/1 and a pH_{eq} value of 5.0, thereby yielding the maximum separation value of $\beta_{Co/Ni} = 254$. The extraction at a different O/A ratio of 2/3 under 3-stages of counter-current contact yielded about 99% Co(II) with only 21 mg·L^{-1} Ni(II), yielding a higher $\beta_{Co/Ni} = 1097$. The thermodynamic studies indicated that Co(II) extraction exhibited the endothermic nature with $\Delta H°$ value of 10.8 kJ·mol^{-1}. Approximately 99% cobalt was yielded by stripping using 2.0 mol·L^{-1} H_2SO_4 solution at an O/A ratio of 1/1, which was further crystalized to give high-pure $CoSO_4.xH_2O$ crystals. The study leads to present an alternative to the conventional extractants applied in Co(II)/Ni(II) separation.

Acknowledgements This work was supported by the Brain Pool Program through the National Research Foundation of Korea (NRF) funded by the Ministry of Science and ICT (Grant No. 2021H1D3A2A01100016) and Basic Science Research Program through the National Research Foundation of Korea (NRF) funded by the Ministry of Education (Project no. 2020R1I1A1A01074249), and grant funded by the Korea Government (MSIT) (No. 2022R1A5A1032539).

Conflicts of Interest There are no conflicts to declare.

References

1. Mayyas A, Steward D, Mann M (2018) Sustain Mater Technol 17:e00087
2. Takahashi VCI, Junior ABB, Espinosa DCR, Tenorio JAS (2020) J Environ Chem Eng 8(3):103801
3. Natarajan S, Aravindan V (2018) Adv Energy Mater 8(33):1802303
4. Porvali A, Aaltonen M, Ojanen S, Velazquez-Martinez O, Eronen E, Liu F, Lundström M (2019) Resour Conserv Recycl 142:257
5. The Batteries Report (2018) https://www.rechargebatteries.org. Accessed 20 July 2022
6. Chen Q, Lai X, Gu H, Tang X, Gao F, Han X, Zheng Y (2022) J Clean Prod 369:133342
7. Fahimi A, Ducoli S, Federici S, Ye G, Mousa E, Frontera P, Bontempi E (2022) J Clean Prod 338:130493
8. Devi NB, Nathsarma KC, Chakravortty V (1998) Hydrometallurgy 49:47
9. Liu Y, Lee M (2016) Geosystem Eng 19(5):214
10. Ilyas S, Kim H, Srivastava RR (2021) Sep Purif Technol 278:119577
11. Ilyas S, Srivastava RR, Kim H (2022) Green Chem 24:5204
12. Flieger J, Tatarczak-Michalewska M, Blicharska E, Madejska A, Flieger W, Adamczuk A (2019) Sep Purif Technol 209:984
13. Zhu Z, Yoko P, Cheng CY (2017) Hydrometallurgy 169:213
14. Hoogerstraete TV, Binnemans K (2014) Green Chem 16:1594
15. Sattar R, Ilyas S, Bhatti HN, Ghaffar A (2019) Sep Purif Technol 209:725
16. Muhsan MA, Ilyas S, Cheema HA, Masud S, Shabbir N (2017) Sep Purif Technol 186:90
17. Ritcey GM, Ashbrook AW (1984) Solvent extraction part I. Elsevier, Amsterdam
18. Rybka P, Regel-Rosocka M (2012) Sep Sci Technol 47(9):1296
19. Srivastava RR, Ilyas S, Kim H, Tri NLM, Hassan M, Mudassir M, Talib N (2020) JOM 72(2):839
20. Ishfaq A, Ilyas S, Yaseen A, Farhan M (2019) Sep Purif Technol 209:964

Research and Industrial Application of Selenium and Tellurium Recovery Processes

Shijie Wang

Abstract Selenium and tellurium are semiconductors. Tellurium forms inorganic and organic compounds that are superficially similar to corresponding selenium compounds yet dissimilar in terms of properties and behavior. The first part of this paper describes selenium and tellurium in terms of their chemical reactions. The second part presents applications of selenium and tellurium recovery processes that are hot topics in today's extractive metallurgy field.

Keywords Hydrometallurgy · Recovery process · Selenium · Tellurium

Introduction

Selenium (Se, atomic number 34, atomic weight 78.96) lies between arsenic and bromine and between sulfur and tellurium (Te, atomic number 52, atomic weight 127.61) in Group 6A (chalcogens) in the periodic table. It was discovered in 1817 by a Swedish chemist as a contaminant in sulfuric acid. Its name was derived from the Greek *selene*, meaning moon. Tellurium is located between selenium and polonium and between antimony and iodine in the periodic table. It was discovered in 1782 in Transylvanian gold ore. Its name was derived from the Latin *tellus*, meaning earth.

Despite the fact that selenium is a nonmetal, and tellurium is a metalloid, both elements are semiconductors. Their valence states assigned to the central atoms in their compounds are –2, 0, +2, +4, and +6. The +2 state for both selenium and tellurium is not known to occur in nature.

In terms of chemical properties, selenium reacts with active metals and gains electrons to form ionic compounds containing the selenide ion, Se^{-2}. As an essential micronutrient for mammals, it is present in biomolecules called selenoproteins. By comparison, tellurium behaves as an anion or cation, depending on the medium. Unlike selenium, it has no biological function.

S. Wang (✉)
Coeur Mining, Inc., 104 S. Michigan Ave., Chicago, IL 60603, USA
e-mail: swang@coeur.com

R. G. Reddy et al. (eds.), *New Directions in Mineral Processing, Extractive Metallurgy, Recycling and Waste Minimization*, The Minerals, Metals & Materials Series,
https://doi.org/10.1007/978-3-031-22765-3_12

Selenium is widely used in: electrical and optical applications; photoelectric and photovoltaic cells; xerography; the manufacture of metallurgical ferrous metals, copper, nickel-iron, cobalt-iron, and lead alloys, glass and ceramics, pigments, rubber, lubricants, and pharmaceuticals; chromium plating; organic chemistry; medicine; and nutrition. Tellurium has found wide application in automation, remote control, radio engineering, electronics, semiconductor manufacture, and other fields. Since cadmium-telluride photovoltaic modules have become among the lowest cost components for generating electricity from solar energy, tellurium is of keen interest in today's bull market.

Selenium and Tellurium Recovery, Control, and Removal

Selenium and tellurium are typical trace elements, also called rare metals. They are necessarily associated with many sulfide ores or heavy non-ferrous metals. During the treatment of ores and concentrates, selenium and tellurium accumulate in intermediate metallurgical products such as dusts of smelting furnaces and converters, anode slimes of copper and nickel refiners, and lead refining slag. Almost all the selenium and tellurium traded and consumed commercially are derived from the copper industry in anode slimes. Methods to recover selenium and tellurium from their bearing materials are determined by their chemical properties and compositions and by the associated metal content.

Selenium Recovery

The characteristic valences of selenium can make it difficult to remove and recover. For example, Cu_2Se is easily dissolved in a basic medium where Se^{+6} is dominant in the leachate. It is very difficult to reduce Se^{+6} to Se^{+4}, then to Se^0. Although its purity could be 4–9's Se, the coarse size of the product will carry a penalty in sales.

Regardless of the form in which selenium occurs in the raw material, it is recovered selectively by solution in water and then reduction to the element by sulfur dioxide (SO_2). The precipitate is refined by a combination of pyro- and hydrometallurgical methods to produce selenium of adequately high purity.

Typical process chemistry for selenium leaching and precipitating in acid media can be expressed as follows:

$$Se_{(s)} + Cl_{2(g)} + 3O^{=}_{(l)} \rightarrow SeO^{=}_{3(l)} + 2Cl^{-}_{(l)} \quad \text{(Leaching)}$$

$$SeO^{=}_{3(aq)} + 2SO_{2(g)} + O^{=}_{(l)} \rightarrow 2SO^{=}_{4(aq)} + Se_{(s)} \quad \text{(Precipitating)}$$

Selenium can occur in four oxidation states: selenide (−2), elemental selenium (0), selenite (+4), and selenate (+6). In general, selenate has a high solubility and is the most mobile in water, which is challenging for both selenium removal and recovery.

Selenium Control

Selenium control in solution discharge is necessary and critical for precious and platinum group metal facilities. Its presence in discharge solutions at concentrations higher than the United States Environmental Protection Agency (USEPA) requirement could result in shutdown of the plant. The USEPA has proposed daily maximum limits on the selenium concentration in some wastewater discharge streams to 76 μg/L (ppb) and monthly average limits of 31 μg/L. The Utah Pollutant Discharge Elimination System permits 54 μg/L Se in solutions discharging to Great Salt Lake.

While in charge of precious and rare metals operations, the author set up a target of 0 μg/L Se discharge from the facilities. All efforts were made to prevent selenium discharge in the final streams. First, the leach operation was optimized to maximize selenium dissolution in the circuit. Second, the different leach solutions were combined and treated in a selenium reduction circuit to maximize selenium recovery. Finally, the autoclave operation was optimized to minimize co-extracted selenium in solution and make use of the tellurium cementation circuit to co-precipitate selenium in the copper telluride byproduct. These efforts ultimately guaranteed zero discharge of selenium from the precious metals plant.

Selenium Removal

Between the smelter and concentrator tailings are the springs and wells called wetlands. Selenium removal within wetlands is challenged by the very low selenium content in a huge volume of solution. The following technologies are available and have been partially tested for wetlands.

- Ferrihydrite Adsorption removes metal ions from dilute solution by sorption onto ferrihydrite and then desorption into a more concentrated solution at lower pH, where ferrihydrite thus becomes available for reuse in subsequent sorption operations.
- Catalyzed Cementation removes heavy metals from water by cementation on the surface of iron particles. The process would have the ability to treat and remove selenium from solution regardless of its valence state (+6 or +4). Propitiate catalysts are added to increase the selenium removal efficiency.

- Biological Selenium Reduction treats selenium-containing wastewater using selenium-reducing sludge acclimatized from denitrifying sludge. Methanol was used as an electron donor in the experimental tests.
- Biological Absorption is a theoretical process whereby heavy metals are absorbed by living (e.g., the algae *Chlorella regularis*) or nonliving (e.g., Iota Chips derived from seaweed) biomass.
- Electrometallurgical Reduction involves a series of electrowinning cells with different cathodes for selenium reduction from wastewater.

Commercial application using biological absorption by Iota Chips was successfully used by the author to remove selenium from solution. The selenium discharges were effectively controlled, and the pressure of rising Se levels on the tailings pond was significantly alleviated. However, a challenge associated with biological absorption requires further study: the offensive odor of sodium sulfides from the drainage system.

Tellurium Recovery

For tellurium, hydrochemical treatment is brought about either by direct leaching of raw tellurium-bearing materials (wet method) or after a special preliminary treatment of the materials to convert the tellurium into solid, water-soluble compounds (dry method). In the dry method, tellurium-bearing materials are sintered with soda or sulfatized by heating with sulfuric acid. The resulting tellurium compounds are leached with water.

Typical process chemistry for tellurium leaching and precipitating in acid medium can be expressed by the chemical equations below:

$$\mathrm{Te} + \mathrm{Cl}_{2(g)} + 3\mathrm{O}^{=}_{(1)} \rightarrow \mathrm{TeO}_3^{=} + 2\mathrm{Cl}^- \quad \text{(Leaching)}$$

$$\mathrm{TeO}^{=}_{3(aq)} + 2\mathrm{SO}_{2(g)} + \mathrm{O}^{=}_{(1)} \rightarrow 2\mathrm{SO}^{=}_{4(aq)} + \mathrm{Te}_{(s)} \quad \text{(Precipitating)}$$

Since USEPA regulations limit the direct use of toxic gases in the process, all Cl_2 (chlorine) and SO_2 have been replaced by HCl (hydrochloric acid), H_2O_2 (hydrogen peroxide), and sodium sulfide ($NaHSO_3$) in commercial applications.

Conventional Process Flowsheet

In general, precious metals values are recovered from soda slags generated by the precious metals cupellation operation and from tellurium slags generated at the lead refinery. The production of granular selenium and tellurium is incidental to the recovery of the precious metals values. The conversion of the granulars to saleable forms is for recovery (at least partially) of the expenses of the operation and for cash flow.

In a conventional process flow diagram (Fig. 1), all selenium input is through the copper refinery circuit. The selenium-bearing filtrate containing predominantly recoverable +4 Se and some +6 Se, which is not directly recoverable, is treated by adjusting the acid content and injecting SO_2 gas. This reduces the +4 Se to the elemental form as a precipitate. This "granular" selenium is filtered out, and the filtrate is further treated by addition of a proprietary reductant and more acid plus steam heat to recover the +6 Se as an off-grade granular selenium.

In a conventional process flow diagram (Fig. 1), about 66% of the tellurium comes through the copper refinery circuit, and 34% from the lead refinery. Tellurium slag from the lead refinery is received, fused in a caustic leach, and oxidized with niter to produce soluble tellurite slag equivalent to the soda slags generated by the copper metal cupellation operation.

The fused lead refinery Te slag is treated similarly to the soda slag, except there is no significant selenium content, and the slag is given a three-stage leach with water, then acid, then caustic. The leach solutions are treated similar to soda slag to generate a tellurium dioxide (TeO_2) filter cake, which is combined with the soda slag TeO_2 in the TeO_2 circuit.

In the TeO_2 circuit, the TeO_2 is purified by repeated dissolution and reprecipitation of the TeO_2, which eliminates the majority of heavy metal impurities. The final TeO_2 precipitate, contaminated primarily by silica, is fed to the tellurium circuit.

TeO_2 precipitates from both the niter and lead refinery circuits are combined, redissolved, filtered, and reprecipitated by pH adjustment, removing the impurities and producing a purified TeO_2 feed for the tellurium circuit. The steps to produce saleable selenium and tellurium products (Fig. 2) are as follows:

1. Acid dissolving: TeO_2 filter cake from the TeO_2 circuit is treated with a mineral acid to dissolve the tellurium and leave the accompanying silica behind, generating a solution with high tellurium content.
2. Selenium removal: The above solution is treated to remove nearly all remaining tramp selenium, and then it is filtered and transferred to storage.
3. Pure tellurium solution is transferred from storage to the tellurium gas absorbers, and SO_2 gas is injected to precipitate granular elemental tellurium.
4. The granular tellurium is washed, centrifuged off, dried, sampled, and transferred to storage.

In the conventional process, precious metals in anode slimes are mainly recovered by pyrometallurgical processes. Due to required compliance with increasingly stringent environmental regulations, hydrometallurgical processing has become an attractive method to recover many metals, including relatively inert metals such as gold and platinum group metals.

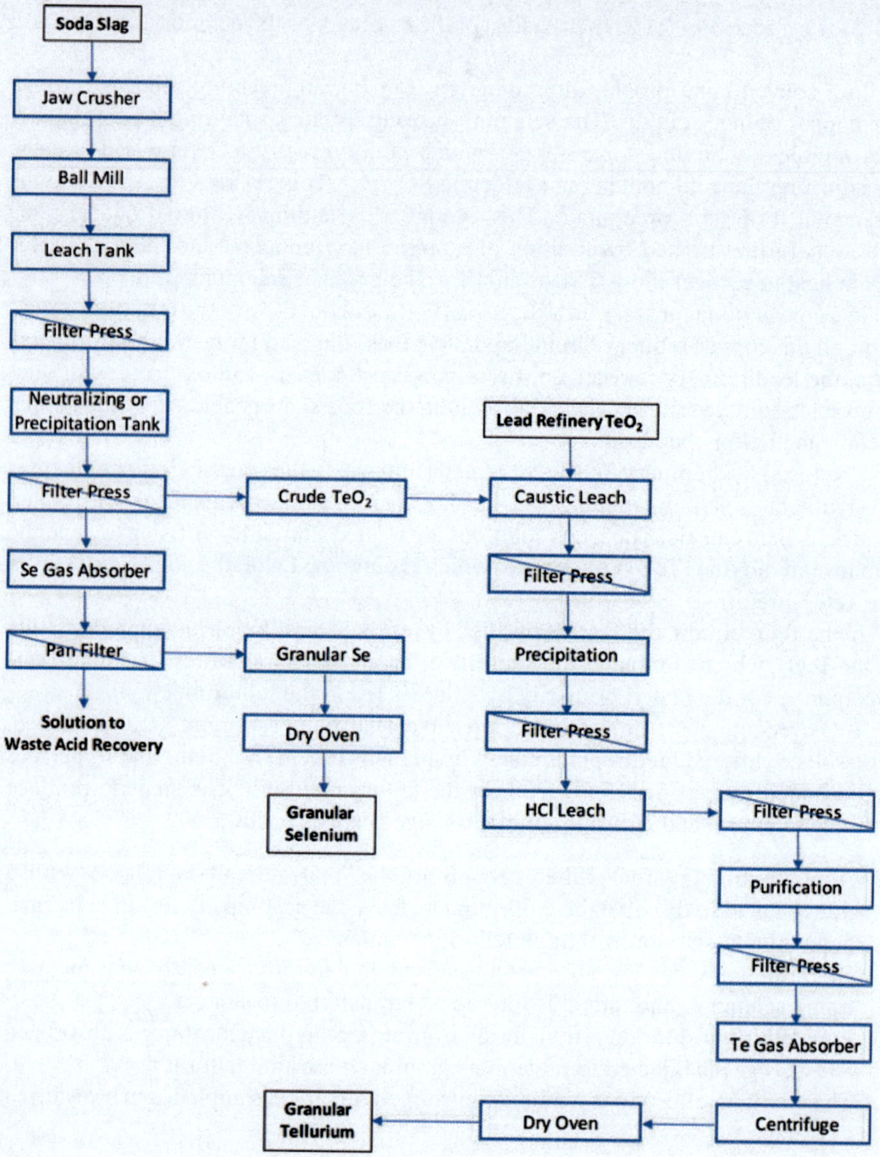

Fig. 1 Conventional process flowsheet for selenium and tellurium recovery

Newly Developed Process Flowsheet

Fully hydrometallurgical processes for extraction and recovery of gold, silver, and platinum group metals from copper anode slimes are described elsewhere. In the newly developed hydrometallurgical process for saleable crude selenium and

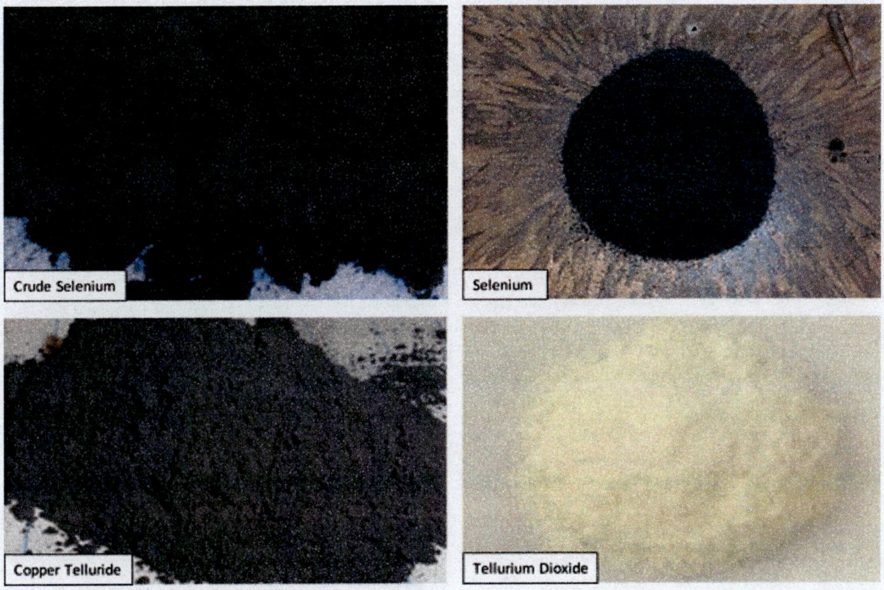

Fig. 2 Saleable selenium and tellurium products

tellurium dioxide (Fig. 3), tellurium is co-extracted with copper in the autoclave acidic pressure leach. The selenium is insoluble with gold and silver; therefore, it remains in the leach residue. The solubilized tellurium is then electrolytically cemented out as copper telluride by adding copper chips. As an intermediate product, granular copper telluride (Cu_2Te) is quickly saleable for tellurium recovery.

From the leaching gold and selenium circuit, insoluble silver is recovered from the leach residue in the silver circuit, where the solubilized gold and selenium are separated in a solvent extraction circuit. When the gold is recovered from loaded and washed organics, selenium is selectively precipitated from the raffinate as crude selenium for sale.

The newly developed hydrometallurgical process makes it possible to recover saleable tellurium and selenium along with casted gold and silver bars in only five (5) days, whereas forty-five (45) days were required with the conventional pyrometallurgical process.

Conclusions

This paper summarizes the chemical properties, applications, and challenges associated with selenium and tellurium. The newly developed hydrometallurgical process presented herein would make the recovery of these two rare elements not only desirable, but economically viable, especially for copper refineries seeking to recover selenium and tellurium from anode slimes. This process is more robust because it recovers

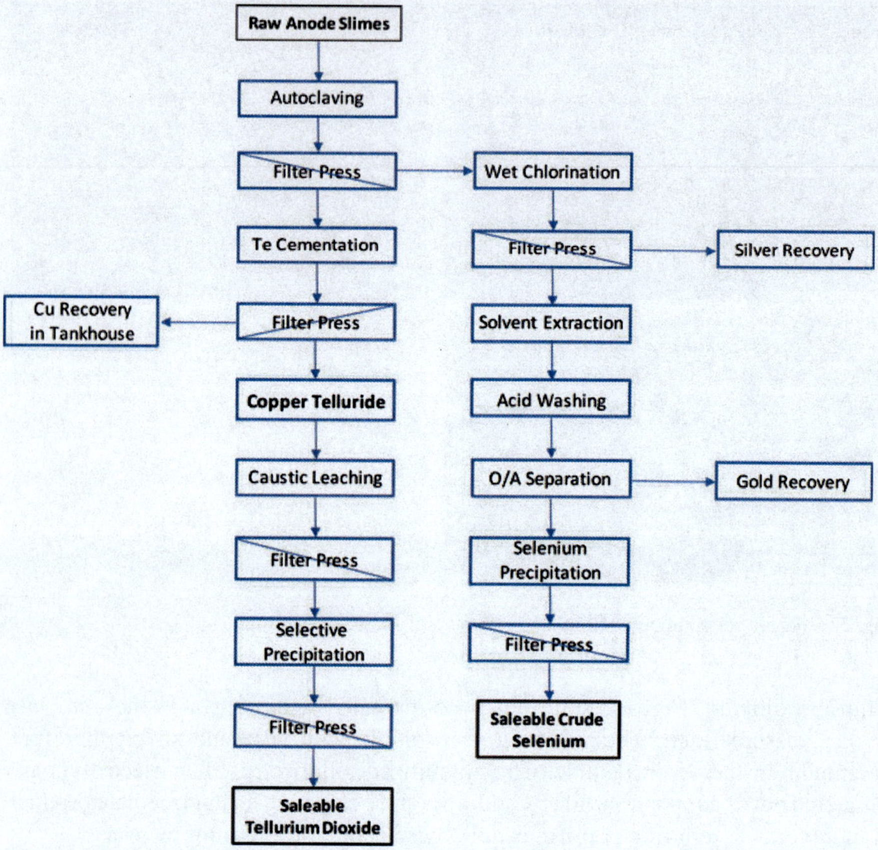

Fig. 3 Newly developed process flow diagram for selenium and tellurium recoveries

greater values and is more environmentally benign than traditional pyrometallurgical processes.

References

1. Wang S (2011) Tellurium, its resourcefulness and recovery. JOM 63(8):90–93
2. Wang S (2023) An innovative hydrometallurgical process for recovery of critical and rare metals from copper anode slimes. In: Rare metal technology 2023. Springer, Cham.

Nickel Matte as Novel Reductant in Galvanic Leaching of Spent Lithium-Ion Battery Black Mass

Erik Prasetyo and Sulalit Bandyopadhyay

Abstract Hydrometallurgy is a widely applied technology to extract valuable metals from spent lithium-ion batteries. H_2SO_4-H_2O_2 combination is the most common lixiviant used for leaching black mass. However, the use of H_2O_2 poses several problems including instability and high reactivity in high temperature leaching. In here, nickel matte (NM; Ni-Cu Sulfide) is proposed as a novel reductant in H_2SO_4 leaching. The effects of NM dosage, pulp density, H_2SO_4 concentration, temperature, and reaction time on metal recovery have been studied. The results demonstrate that NM improves metal dissolution. With leaching condition H_2SO_4 1 M, pulp density 5%, and temperature 70 °C, both Co and Mn recovery are doubled when NM is added. A significant finding was that the effect of temperature on leaching efficiency was insignificant. High leaching efficiency was attained at high pulp density (210 g/L), low leaching temperature (10 °C), and short period of time (30 min). The recovery of Li, Ni, Mn, and Co in this leaching condition was 93%, 50%, 69%, and 88%, respectively. Aside from its advantage as reductant, NM addition could increase Ni concentration in leachate, which may further facilitate direct nickel-rich cathode syntheses using precipitation and crystallization.

Keywords Lithium-ion battery · Leaching · Sulfide · Nickel matte

E. Prasetyo (✉) · S. Bandyopadhyay
Department of Chemical Engineering, Norwegian University of Science and Technology, Trondheim, Norway
e-mail: erik.prasetyo@ntnu.no

R. G. Reddy et al. (eds.), *New Directions in Mineral Processing, Extractive Metallurgy, Recycling and Waste Minimization*, The Minerals, Metals & Materials Series,
https://doi.org/10.1007/978-3-031-22765-3_13

Introduction

Lithium-ion battery (LIB) is an integral part of green technology to combat climate change and carbon emission, due to its role as energy storage device in electric vehicles (EVs). It is estimated in 2035 that more than half of the vehicles available in the market will consist of EVs [1]. This need for LIBs for EVs will boost the demand on certain elements in the form of raw material of LIB including lithium (Li), nickel (Ni), manganese (Mn), and cobalt (Co). Most of these elements are considered as critical elements since their availability in the world market is sensitively affected by geopolitical factors. To secure and sustain the supply of these critical elements, recycling process is warranted.

LIB consists of several components including steel casing, plastic separator, aluminum, and copper foils as current collectors, organic electrolytes, lithium, manganese, nickel, and cobalt metal oxides as cathode and graphite as anode. Prior to recycling, the spent battery will be discharged, dismantling, and shredded. To separate the components, physical methods such as flotation, gravity settling, and sieving are generally applied, and the cathode and anode components report together as one fraction, which is referred to as black mass (BM). BM is the main target in recycling process due to its value (>70% of total battery value) [2], and there are three basic processes to recycle BM, i.e. high temperature or smelting (pyrometallurgy), chemical dissolution (hydrometallurgy), and combination of both processes. So far, hydrometallurgical routes possess advantages over pyrometallurgical routes due to its lower energy consumption and ability to deal with low grade raw materials.

In hydrometallurgical process, the target metals in BM will be dissolved using acid-reductant combination, among which sulfuric acid-hydrogen peroxide is the most popular combination. Other types of acids such as organic acids, e.g. citric acid [3] and other polycarboxylic acids have been proposed as safer alternatives, but their use is yet to reach industrial scale. Reductants are required to enhance the metal solubility especially Co and Mn, by transforming them from insoluble species Co(III) and Mn(IV) in BM into soluble species Co(II) and Mn(II). Up until now, H_2O_2 is the most applied reductant compared to other such as ascorbic acid [4], glucose [5], ferrous iron [6], and tannic acid [7], considering the economical factor, availability and versatility (handling and application). However, H_2O_2 has limitations due to instability at high temperature leaching, which is inevitable considering the total metal recovery expected in recycling process.

Sulfide minerals, especially pyrite (FeS_2), have been reported as an effective reductant in manganese oxide reductive leaching, and its stability and efficacy in high temperature leaching have been documented [8]. Another sulfide analogue which could be proposed as reductant is nickel matte (NM), which is an intermediate product of nickel ore processing. The typical Ni content in NM may vary between 30 and 70%, while the rest is composed of Co, Cu, Fe, sulfur, and other trace elements. The addition of NM as additive hypothetically will be advantageous, aside from providing reductant in the BM dissolution, NM addition can increase Ni and Co content in pregnant leach solution. Increasing Ni and Co concentration in leachate will facilitate the direct nickel-rich cathode (NMC 811) syntheses from the leach solution using precipitation or crystallization. The objective of current study is to evaluate the efficacy of NM as reductant in BM sulfuric acid leaching including the effect of different variables, e.g. BM/NM mass ratio, pulp density, sulfuric acid concentration, leaching temperature, and leaching time.

Material and Methods

Materials

Commercial samples of LIB BM and NM powder were grinded and sieved, and the fraction with particle size less than 53 μm (270 mesh) was used further in the leaching experiments. For compositional analysis, BM and NM were acid digested using HNO_3/HCl (vol ratio 1), followed by MP-AES determination (Agilent 4210 MP-AES Optical Emission Spectrometer, Particle Engineering Core Facility, NTNU). The metal concentrations are listed in Table 1. Nitric acid, sulfuric acid, and hydrochloric acid (concentrated analytical grade) were purchased from MilliporeSigma, all used without further purification, and ultrapure water (ASTM Type 1, Sartorius Arium® Mini Ultrapure Water Systems, Particle Engineering Core Facility, NTNU) was used for solution preparation, dilution, and analysis throughout the experiments.

Leaching Experiments

The leaching was performed inside a water heated double jacketed glass reactor (250 ml volume) equipped with condenser. Julabo CORIO CD-BC4 water bath was used for circulating water in the heating jacket. Typically, 250 ml sulfuric acid solution was introduced into the reactor and heated. After the solution reached the desired temperature, BM and NM were added into reactor and the mixture was homogenized using magnetic stirring (600 rpm). The mixture was sampled, and the filtrate was separated from solid residue using a syringe filter. The filtrate was diluted using HNO_3 2% and the metal analysis was determined using MP-AES. Metal leaching efficiency (R) was calculated using equation:

$$R = \frac{C_F \times V}{(C_{BM} \times m_{BM}) + (C_{NM} \times m_{NM})}$$

Table 1 Metal concentration in black mass and nickel matte powder used in leaching studies

Sample\elements (weight %)	Li	Ni	Mn	Co	Fe	Al	Cu
Black mass	3.71 ± 0.17	9.91 ± 0.32	7.87 ± 0.25	9.56 ± 0.22	0.10 ± 0.01	0.14 ± 0.01	1.68 ± 0.09
Nickel matte		60.6 ± 1.3		1.86 ± 0.11	2.26 ± 0.05		15.12 ± 0.35

where C_F was the metal concentration in the filtrate (mg/L), C_{BM} the metal content in BM (mg/g), m_{BM} the mass of black mass (g), C_{NM} the metal content in nickel matte, m_{NM} the mass of nickel matte, and V the volume of leaching agent (L).

Results and Discussion

Effect of NM-BM Mass Ratio

The effect of NM addition dosage to the BM sulfuric acid leaching was investigated between 0 (control condition) and 50 g/L (NM/BM mass ratio 1). The constant variables include leaching temperature 70 °C, sulfuric acid 1 M, 600 rpm, leaching time 30 min, and total pulp density 100 g/L. The results in Fig. 1a show that NM addition posed positive effect on metal leaching efficiency especially Mn and Co, since both metal dissolutions depend on the reductant. Mn recovery increased from 37% at control condition to 82% at NM concentration 20 g/L (NM/BM ratio 0.4) while Co leaching efficiency almost was doubled from 49.7% at control condition to 105% at NM/BM ratio 0.4. In the case of Li, it seems that the dissolution was not dependent on the NM presence. At control condition, Li recovery was 94% which slightly increased at higher NM/BM ratio. It is hypothesized that Li dissolution from BM in acidic media was more controlled by acid attack and exchange with proton (delithiation) [9]. Ni recovery based on Fig. 1a is decreasing as the NM/BM ratio was raised. Since Ni also exists in nickel matte, increasing NM/BM ratio will increase pulp density in terms of nickel matte in leaching system. This in turn decrease nickel recovery. However, based on Fig. 1b, nickel concentration greatly increased to about 14 g/L at NM/BM mass ratio 0.8, which indicated mutual effect on nickel leaching both from BM and NM.

Effect of Sulfuric Acid Concentration

Acidic condition is required in BM leaching; hence, the effect of sulfuric acid concentration was studied in the range of 0.25–1.5 M. The constant variables include NM/BM ratio 0.4, total pulp density 70 g/L, stirring speed 600 rpm, leaching temperature 70 °C, and leaching time 30 min. The results in Fig. 2a show that increasing sulfuric acid improved the metal leaching efficiency until the sulfuric acid concentration reached 0.75 M. Higher sulfuric acid concentration did not significantly enhance both metal recovery and metal dissolution as shown in Fig. 2b.

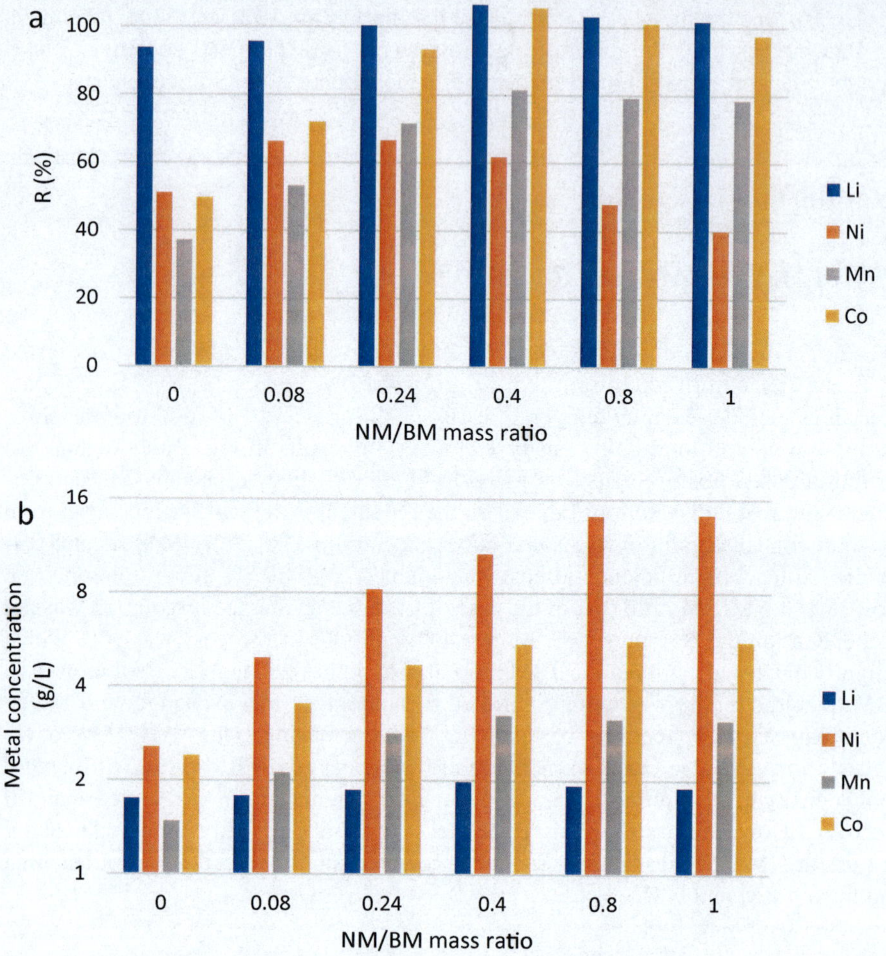

Fig. 1 The effect of NM addition as reductant in the BM sulfuric acid leaching to **a** Metal leaching efficiency, **b** Metal concentration in pregnant leach solution. Leaching condition cited in the text

Effect of Total Pulp Density

The effect of total pulp density was studied at three different solid–liquid ratios: 112, 140, and 210 g/L. The constant variable in the test includes leaching temperature 70 °C, sulfuric acid 2 M, stirring speed 600 rpm, NM/BM mass ratio 0.4, and leaching time 30 min. Increasing pulp density has negative effect on all metal recovery. Li recovery decreased from 106% (pulp density 112 g/L) to 86% (pulp density 210 g/L), and Co, Mn, and Ni show the same trends (Fig. 3a). This trend is the opposite to the metal concentration in pregnant leach solution trend, which increased as pulp density increased (Fig. 3b). Figure 3b shows the typical metal concentration in pregnant leach

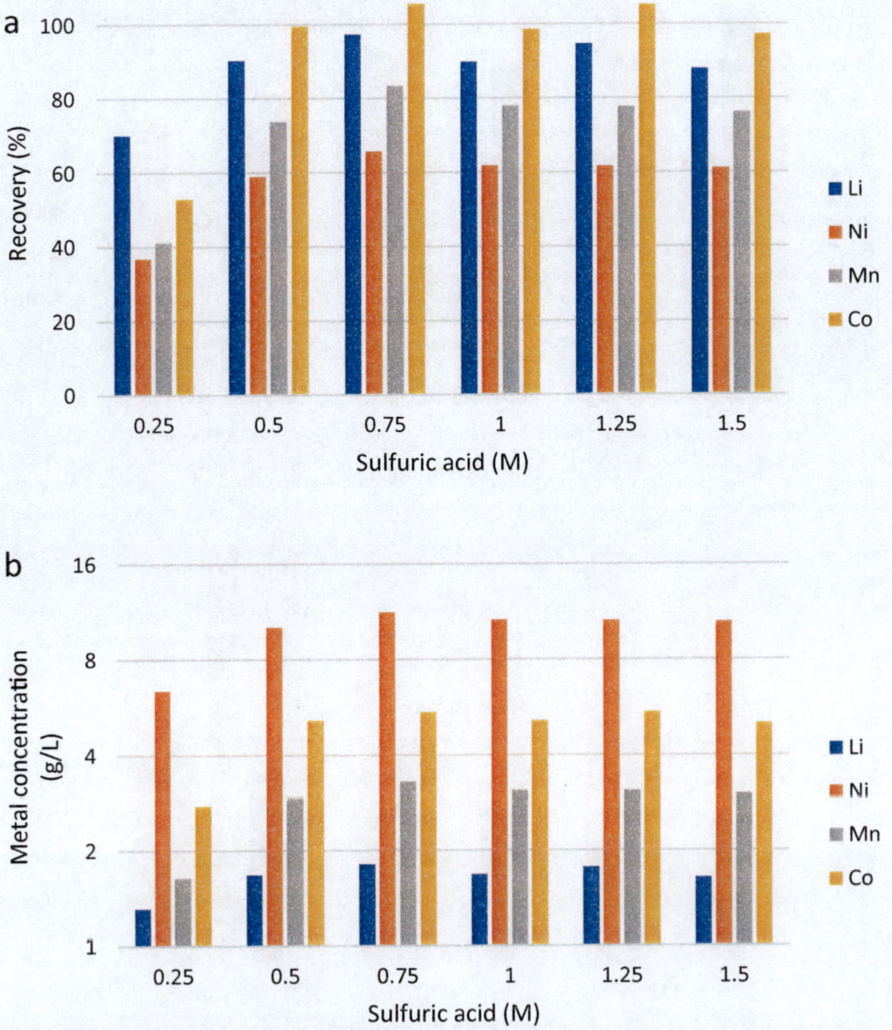

Fig. 2 The effect of sulfuric acid concentration to **a** Metal leaching efficiency, **b** Metal concentration in pregnant leach solution. Leaching condition cited in the text

solution. Ni concentration can reach about 24 g/L, which is beneficial in facilitating crystallization of nickel sulfate as battery cathode precursor. It is worth to note that in this study, typical pulp density for battery leaching was higher compared to typical pulp density used in other studies which are generally less than 100 g/L.

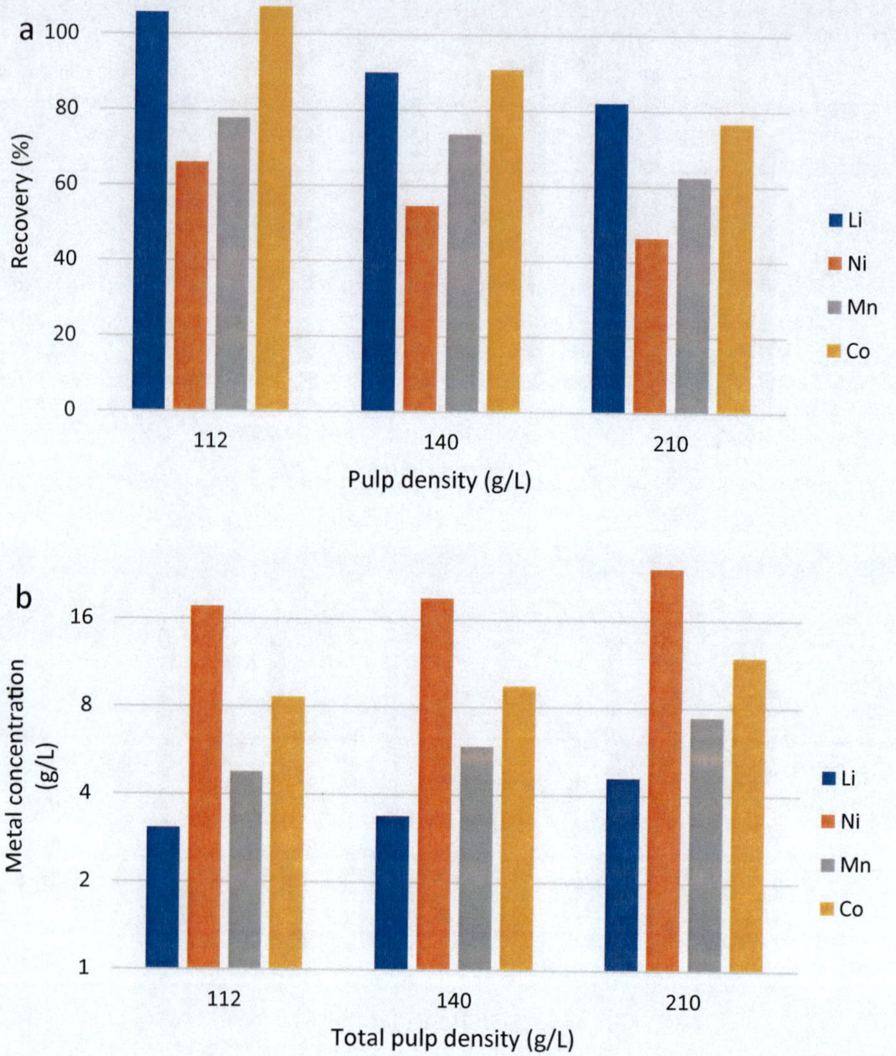

Fig. 3 Effect of pulp density on **a** Metal recovery and **b** Metal concentration in pregnant leach solution. Leaching conditions cited in the text

Effect of Leaching Temperature

The effect of leaching temperature on the metal leaching efficiency (Fig. 4a) and metal concentration in pregnant leach solution (Fig. 4b) was studied at 10, 30, 50, and 70 °C, while the constant variables included sulfuric acid 2 M, total pulp density 210 g/L, NM/BM mass ratio 0.4, stirring speed 600, and leaching time 30 min. The results in Fig. 4a, b clearly show the effect of temperature on metal leaching efficiency and metal concentration was not significant. Li recovery was within 90% at whole leaching temperature studied. Ni, Mn, and Co recovery slightly increased about 5%, when leaching temperature increased from 30 °C to 50 °C (Co from 91 to 96%, Mn from 70 to 75%, and Ni from 50 to 55%). It is still not clear how the metal leaching efficiency could reach such high value even the leaching was performed at low temperature (10 °C). There is a possibility that NM could act as catalyst in BM leaching, and this would be resolve in future investigation.

Conclusions

Nickel matte (NM) was successfully used as novel reductant in lithium-ion battery black mass leaching in sulfuric acid media. NM effectively encouraged metal dissolution, especially of Mn and Co. Li was less affected by NM since its dissolution did not require a reductant, and its dissolution occurred through mostly by acid attack. This finding was supported by studies on effect of sulfuric acid concentration, in which all metal dissolution including Li were positively correlated to acid concentration until the concentration reached 0.75 M. Current research shows that the effect of temperature on leaching efficiency was minimum under the chosen condition. Compared to most common lixiviant (sulfuric acid-hydrogen peroxide), sulfuric acid-nickel matte combination has better performance in metal leaching efficiency, in terms of higher solid–liquid ratio and shorter leaching time (Table 2). Leaching efficiency was still high even though the leaching was performed at low temperature (10 °C) and at high pulp density (210 g/L). Within 30 min, the leaching efficiency of Li, Ni, Mn, and Co reached 93%, 50%, 69%, and 88%, respectively. This leaching condition produced pregnant leach solution containing Li 5.2 g/L, Ni 26 g/L, Mn 8.2 g/L, and Co 13.7 g/L. High transition metals concentration in the pregnant leach solution is considered beneficial to facilitate crystallization process in LIB cathode resynthesis.

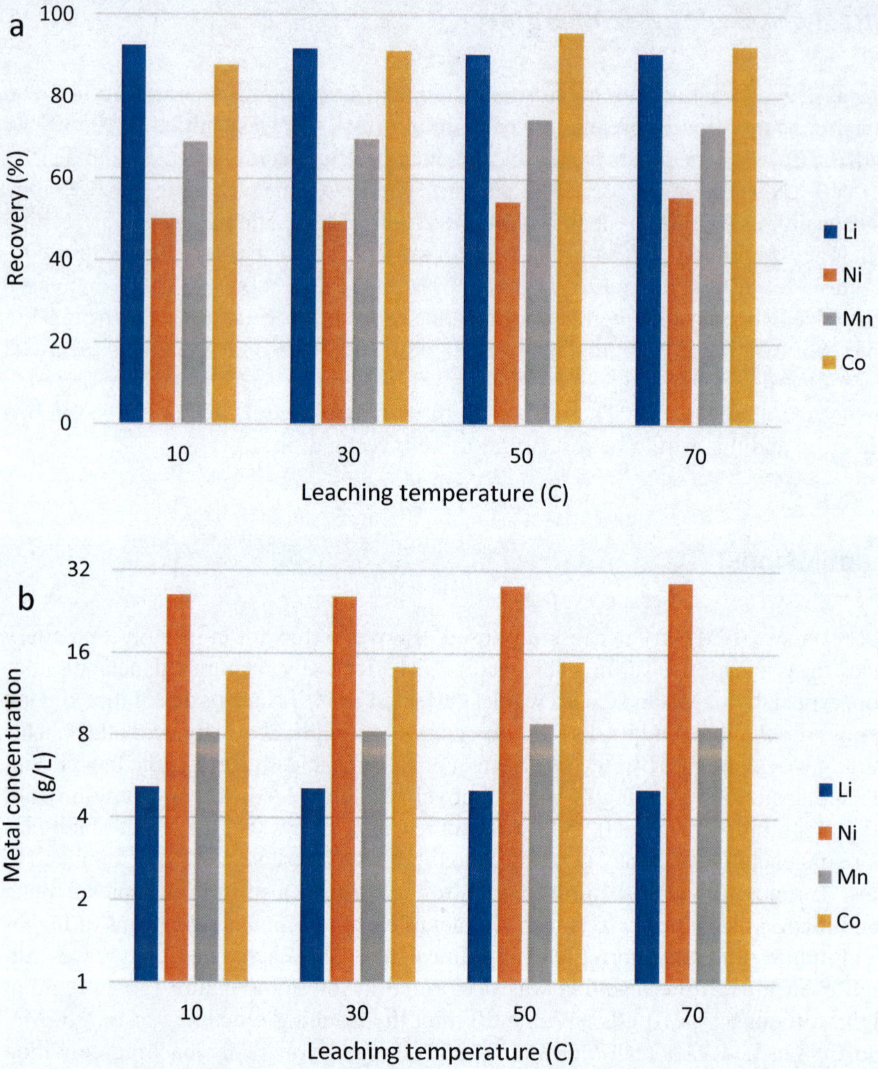

Fig. 4 Effect of leaching temperature on **a** Metal recovery and **b** Metal concentration in pregnant leach solution. Leaching conditions cited in the text

Table 2 Reports on metal recovery in previous studies using sulfuric acid-hydrogen peroxide lixiviant with current study using sulfuric acid-nickel matte

No	Lixiviants	S/L (g/L)	Temp (°C)	Leaching Time (min)	Recovery (%)				Ref
					Li	Ni	Mn	Co	
1	H_2SO_4 1 M + H_2O_2 10%	45	60	180		100	78	80	[10]
2	H_2SO_4 2 M + H_2O_2 10%	20	95	240	93	95	85	80	[11]
3	H_2SO_4 2 M + H_2O_2 10%	20	25	120	95	90	90	90	[12]
4	H_2SO_4 2 M + H_2O_2 10%	50	70	30	99			75	[13]
5	H_2SO_4 2 M + H_2O_2 2%	33	60	120	86			97	[14]
6	H_2SO_4 2 M + H_2O_2 1%	50	25	1440	100	77		87	[15]
7	H_2SO_4 2 M + Nickel Matte 60 g/L	150	70	30	91		73	93	Current study

References

1. Arambarri J, Hayden J, Elkurdy M, Meyers B, Hamatteh ZSA, Abbassi B, Omar W (2019) Lithium ion car batteries: present analysis and future predictions. Environ Eng Res 24(4):699–710. https://doi.org/10.4491/eer.2018.383
2. Larouche F, Tedjar F, Amouzegar K, Houlachi G, Bouchard P, Demopoulos GP, Zaghib K (2020) Progress and status of hydrometallurgical and direct recycling of Li-Ion batteries and beyond. Materials 13(3):801. https://doi.org/10.3390/ma13030801
3. Meng F, Liu Q, Kim R, Wang J, Liu G, Ghahreman A (2020) Selective recovery of valuable metals from industrial waste lithium-ion batteries using citric acid under reductive conditions: leaching optimization and kinetic analysis. Hydrometallurgy 191:105160. https://doi.org/10.1016/j.hydromet.2019.105160
4. Munir H, Srivastava RR, Kim H, Ilyas S, Khosa MK, Yameen B (2020) Leaching of exhausted LNCM cathode batteries in ascorbic acid lixiviant: a green recycling approach, reaction kinetics and process mechanism. J Chem Technol Biotechnol 95(8):2286–2294. https://doi.org/10.1002/jctb.6418
5. Meng Q, Zhang Y, Dong P (2017) Use of glucose as reductant to recover Co from spent lithium ions batteries. Waste Manag 64:214–218. https://doi.org/10.1016/j.wasman.2017.03.017
6. Ghassa S, Farzanegan A, Gharabaghi M, Abdollahi H (2020) The reductive leaching of waste lithium ion batteries in presence of iron ions: process optimization and kinetics modelling. J Clean Prod 262:121312. https://doi.org/10.1016/j.jclepro.2020.121312
7. Prasetyo E, Muryanta WA, Anggraini AG, Sudibyo S, Amin M, Al Muttaqii M (2022) Tannic acid as a novel and green leaching reagent for cobalt and lithium recycling from spent lithium-ion batteries. J Mater Cycles Waste Manag 24:927–938. https://doi.org/10.1007/s10163-022-01368-y
8. Chen G, Jiang Ch, Liu R, Xie Zh, Liu Z, Cen S, Tao Sh, Guo Sh (2021) Leaching kinetics of manganese from pyrolusite using pyrite as a reductant under microwave heating. Sep Purif Technol 277:119472. https://doi.org/10.1016/j.seppur.2021.119472
9. Peng C, Hamuyuni J, Wilson BP, Lundström M (2018) Selective reductive leaching of cobalt and lithium from industrially crushed waste Li-ion batteries in sulfuric acid system. Waste Manag 76:582–590. https://doi.org/10.1016/j.wasman.2018.02.052
10. Tran TT, Moon HS, Lee MS (2022) Co, Ni, Cu, Fe, and Mn integrated recovery process via sulfuric acid leaching from spent lithium-ion batteries smelted reduction metallic alloys. Miner Process Extr Metall 43(8):954–968. https://doi.org/10.1080/08827508.2021.1979541
11. Meshram P, Abhilash PBD, Mankhand TR, Deveci H (2016) Comparision of different reductants in leaching of spent lithium ion batteries. JOM 68:2613–2623. https://doi.org/10.1007/s11837-016-2032-9
12. Chen WS, Ho HJ (2018) Recovery of valuable metals from Lithium-Ion batteries NMC cathode waste materials by hydrometallurgical methods. Metals 8(5):321. https://doi.org/10.3390/met8050321
13. Sohn JS, Shin SM, Yang DH, Kim SK, Lee CK (2006) Comparison of two acidic leaching processes for selecting the effective recycle process of spent lithium ion battery. Geosystem Eng. 9(1):1–6. https://doi.org/10.1080/12269328.2006.10541246
14. Zhu SG, He WZ, Li GM, Zhou X, Zhang XJ, Huang JW (2012) Recovery of Co and Li from spent lithium-ion batteries by combination method of acid leaching and chemical precipitation. Trans Nonferrous Met Soc China 22(9):2274–2281. https://doi.org/10.1016/S1003-6326(11)61460-X
15. Aaltonen M, Peng C, Wilson BP, Lundström M (2017) Leaching of metals from spent lithium-ion batteries. Recycling 2(4):20. https://doi.org/10.3390/recycling2040020

Recyclability of Proton Exchange Membrane Electrolysers for Green Hydrogen Production

Nawshad Haque, Sarb Giddey, Sejuti Saha, and Paul Sernia

Abstract Renewed interest in green hydrogen energy due to its versatility and ability to decarbonise numerous economic sectors prompted research on the evaluation of sustainability of associated technologies. Proton Exchange Membrane (PEM) water electrolysis is a promising technology to produce hydrogen gas from water electrolysis using renewable power. However, PEM electrolysers use rare noble metals and other expensive materials. Furthermore, the availability and supply risks are additional concerns for the critical metals. Hence, this paper explores the review of the recycling process of end-of-life PEM electrolysers from the point of collection to the final material recovery and the potential reuse in the manufacturing process. Several studies have highlighted existing and novel recycling technologies for the different materials used in electrolyser components. Some of these methods include hydrometallurgy, pyrometallurgy, transient electrochemical dissolution, selective electrochemical dissolution, and acidic process. Overview of these processes and implication of recycling are presented here.

Keywords Renewable · Hydrogen energy · PEM electrolyser · Recycling · Sustainability

Introduction

The global impacts of climate change have created an urgent need to transition to low-carbon economies. Driven by the Paris Agreement, many nations have placed active efforts to decarbonise communities and meet their emission reduction targets. For example, under the recent Climate Change Bill 2022, Australian government sets Australia's greenhouse gas emissions reduction targets to a 43% reduction from 2005

N. Haque (✉) · S. Giddey · S. Saha
CSIRO Energy, Private Bag 10, Clayton South, VIC 3169, Australia
e-mail: Nawshad.Haque@csiro.au

P. Sernia
Endua, 104 Boniface St, Archerfield, QLD 4108, Australia

R. G. Reddy et al. (eds.), *New Directions in Mineral Processing, Extractive Metallurgy, Recycling and Waste Minimization*, The Minerals, Metals & Materials Series,
https://doi.org/10.1007/978-3-031-22765-3_14

levels by 2030 and net zero emissions by 2050 [1]. Green hydrogen as a future energy system has received a considerable attention in the last decade. It is a high energy density fuel (on mass basis) and generated via water electrolysis using RE. Hydrogen energy has been a popular topic amongst researchers, industries, and governments due to its versatility [2]. By using excess RE for its production, hydrogen fuel can act as a storage system for intermittent renewable energy generators, improve grid stabilities, and enable the use of RE when generation is not possible, i.e., during night times when there is no solar power. Secondly, hydrogen fuel can decarbonise the transport sector through simultaneous scalable adoption of the fuel cell electric vehicle (FCEV) technologies in the industry. It can also penetrate and reduce emissions in the heavy industries.

Despite its potential to revolutionise the energy sector, the commercialisation of hydrogen energy systems has been hindered by high production costs and uncertainty about the technology's scalability. These production costs stem from current electricity prices and the capital costs of electrolysers. The latter is particularly the case for Proton Exchange Membrane (PEM) electrolysers, which are expensive due to the use of rare noble metals (platinum, iridium, and gold), titanium metal, and polymer membranes to achieve high efficiencies. While the price of renewable electricity is expected to decline with greater penetration, the cost of precious metals could further increase due to their scarcity [3]. Additionally, research suggests that the current platinum and iridium reserves are insufficient to support the expansion of growing hydrogen production via PEM electrolysers to gigawatt scales. These challenges have resulted in substantial research to reduce the metal requirements in electrolysers, find alternative lower cost materials, and investigate sustainable recycling or even upcycling (reusing of materials with required specifications in PEM) options. Secondary production of metals will be necessary to reduce the cost of electrolysers and recovery of materials and ensure that sufficient material is available for the growth of the nascent green hydrogen sector. This paper provides an overview of the PEM electrolysers and their material requirements. This review also highlights potentially available recycling technologies presented in published studies, identifies gaps in the current literature, and recommends future directions.

Methodology

Various publicly available literatures were reviewed to find the current state of the art of the PEM electrolysers or fuel cell recycling processes. Studies on the generic recycling of metals were also consulted. Various research groups are proposing to improve the current recycling rates (RR) for various metals and generating a database. Sometimes these RR values are displayed on a periodic table for each metal. These technologies for electronic waste recycling can potentially be customised or applied for recycling of the end-of-life PEM electrolysers.

Results and Discussion

PEM Cell and Stack Description

The PEM water electrolysis cell structure comprises a solid polymer electrolyte/membrane made from perfluorosulfonic acid (PFSA). Acidic electrolytes are necessary to enhance the ionisation of water molecules that have low self-ionisation capacities [4]. DuPont's Nafion membranes are most common materials in electrolysers and fuel cells. The Nafion separates two electrodes and prevents the crossover of product gases from oxygen evolution reaction side to hydrogen evolution reaction. The membrane is usually coated directly with electrocatalysts on either side (catalyst-coated membrane (CCM)), which serves as the electrodes. The anode comprises an iridium oxide (IrO_2) coating, while the cathode catalyst is carbon-supported platinum (Pt/C) [5]. The CCM is sandwiched between porous transport layers (PTL), which facilitate the transport of water, electrons, and product gases to and from the electrodes. Carbon paper is used for the cathodic PTL, while sintered titanium is used for the anodic PTL. The titanium is sometimes coated with gold to improve its corrosion resistance. Surrounding the active cell area are gasket sealings that prevent gas crossover between electrodes and maintain contact between the different components [6]. The cells assembled within an electrolyser stack are separated by titanium bipolar plates. Current collectors are made of copper. The bolts, nuts, flat springs, and endplates maintain the structural integrity of the stack. Individual stacks in the current market have reached capacities of 2.5 MW, while the multi-stack modules have exceeded the 20 MW system size [7–9].

Material Inventory of PEM Electrolysers: Material Utilisation in Stacks

The material composition of PEM electrolyser stacks depends on the component thicknesses and catalyst loadings required for efficient performance. Research is focusing to explore the feasibility of lower catalyst loadings, thinner component, and the use of cheaper, alternative materials without compromising the performance. A 1-MW PEM stack comprised 745 cells, each with an active area of 500 cm^2 [9]. This stack weighs 670 kg, and approximately 504 kg is attributed to the titanium bipolar plates. The iridium and platinum catalyst coatings only account for 0.1% and 0.04% of the total weight, respectively. According to an analysis, the CCM, which comprises expensive noble metal catalysts and the Nafion membrane, accounts for 87% of the cost, followed by the bipolar plates and PTLs. The total cost of the end plates and connectors is negligible despite making up 15% of the stack mass. This is attributed to the low unit cost of stainless steel compared to the other materials used. When manufacturing processes are considered, the cost breakdown is changed. A techno-economic assessment (TEA) reported bipolar plates can account for 51% of the total stack cost. On the other hand, the catalysts and the membrane collectively make up 23% of the costs, while PTLs contribute 17% towards the total stack cost. The TEA also provides a cost breakdown of the complete system, where the PEM stack makes up 60% of the total cost.

Several studies have investigated the properties of used Nafion membranes and their performance in new fuel cells or electrolyser cells [10–12]. The results showed increased ohmic losses, and this was consistent between the different literary works. The performance drops were attributed to membrane contamination [10], sulfonic group deactivation [10], and swelling [10–12]. Hence, reprocessing and recasting the Nafion ionomers into membranes is recommended to maintain performance.

Moghaddam et al. [12] experimented with different recycling methods and identified the optimum process based on the recast membranes' swelling behaviour, ion exchange, and proton conductivity. The separated Nafion membrane obtained after delamination was first cleaned using sulphuric acid (H_2SO_4). This was followed by the dissolution of the membrane into its ionomers in an ethanol–water solution at high temperature and pressure. This process was conducted in an autoclave reactor, and conditions specified in the process flow diagram were necessary to dissolve the PFSA (Nafion) membrane in the given solution. After dissolution, the ethanol–water solution was vaporised while the concentrated PFSA solution was mixed with dimethylacetamide (DMAc). Finally, the membrane was cast and dried.

Material Recyclability

The common conventional metal, such as steel, aluminium, copper, and lead recycling industry, is a relatively well-established system although the recycling rate (RR) can vary between specific metal sectors. For instance, while the overall RR for platinum is between 60 and 70%, its RR in industrial applications (catalysts and electrochemical processes) is within the range of 80–90%. Similarly, iridium (Ir) and ruthenium (Ru) report RR of 40–50% in industrial processes [10].

Figure 1 provides a process flow diagram for the overall recycling of PEM stacks, which can be broken down into three main stages: collection, dismantling, and material recovery. The collection logistics for material recycling can often be challenging and costly [13, 14]. Detailed analysis of the logistical chains and appropriate legislations would be needed to ensure efficient collection processes [14].

The second step is the dismantling process. The difficulty of this stage depends on the assembly method of the stack components. Usually, PEM stack components are bolted or screwed together, and the dismantling process would use manual separation methods [15]. Dismantling instructions are ideally provided by the manufacturers to minimise safety hazards [13]. Once dismantled, different material recovery pathways are adopted for the stack components based on the recycling technology's maturity, efficiency, safety, toxicity, and environmental and economic costs. The hydrometallurgical (HMT) and pyrometallurgical (PMT) processes are existing metal recycling methods and are used in numerous facilities. On the other hand, acid process (AP),

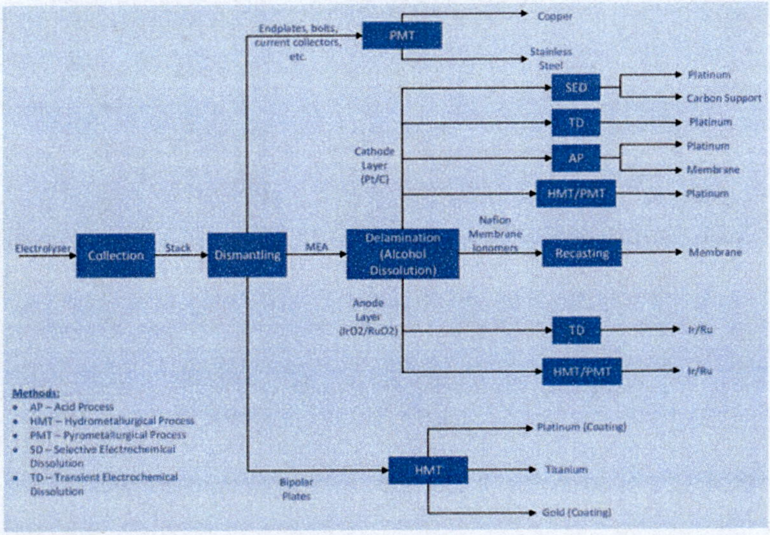

Fig. 1 Overall recycling process of PEM electrolysers and the various material recovery pathways [16]

transient electrochemical dissolution (TD), and selective electrochemical dissolution (SED) are more novel methods that are currently under investigation [16]. The membrane electrode assembly (MEA) may also require an additional delamination stage to separate the catalyst layers from the Nafion membrane.

Recycling Membrane Electrode Assembly

Delamination of the Membrane Electrode Assembly

MEA delamination is considered a pre-treatment step to separate and recover the catalyst metals from the PFSA membrane and the PTLs before further processing. Although there is limited literature on the process considers delamination a necessary step to improve the efficiency of subsequent recovery processes. This is particularly relevant if HMT methods are adopted, as the Nafion substrate and the ionomers present in the catalyst inks can reduce the leaching efficiencies of the precious metals unless these are separated [17]. Furthermore, delamination eliminates or minimises the formation of hydrogen fluoride gases when PMT processes are used to extract the noble metals [17]. Figure 2 shows process flow diagram from MEA delamination via alcohol dissolution.

Figure 2 shows an alcohol dissolution (AD) method developed by [10] to separate the catalyst metals from the Nafion substrate. In this process, the MEA was first

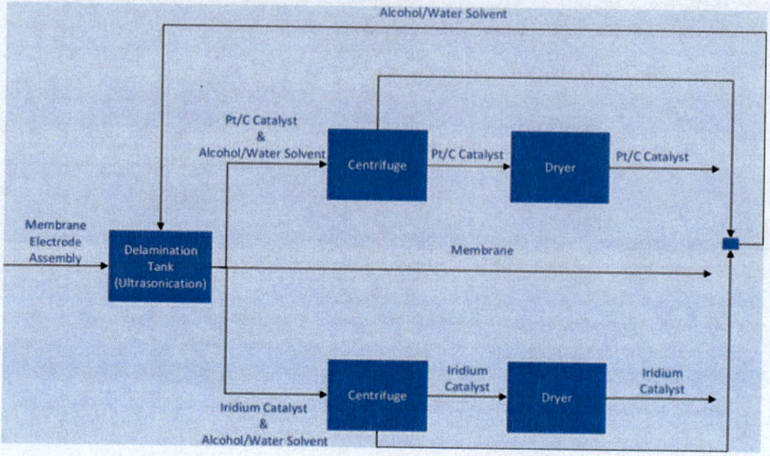

Fig. 2 Process flow diagram from MEA delamination via alcohol dissolution [10]

fixed in a delamination tank. A deionised water-alcohol mixture was circulated on either side of the membrane, and the catalyst layers were agitated via ultrasonication. Isopropanol solution was used in this experiment [17] suggesting that it allows for more efficient delamination than ethanol solutions or pure alcohol. The dissolved metals were collected in separate tanks, and the Nafion membrane was removed from the tank. Finally, the concentrated metal solutions were centrifuged, and the solids were obtained via further drying. The separated alcohol solution can also be recycled to save costs [18]. The experiment was conducted in mild conditions (safe pH, temperature, and voltage) and recorded metal recovery rates of more than 98% [10]. The process was also completed in 30 min. There are variations to the AD process described above. In [18], which focussed on MEAs from fuel cells, both the catalyst layers and PFSA ionomers were dissolved in the alcohol solution followed by separation using filter press and ultra-filtration. The process was also conducted at higher temperatures to facilitate ionomer dissolution. In contrast, [19] conducted the dissolution process in an autoclave reactor. AD provides promising results as a pre-treatment stage [10, 17]. However, the process can be complicated and may increase the costs of the complete recovery process compared to direct HMT and PMT methods [16]. Furthermore, some catalyst particles may diffuse into the membranes and be lost in the separation processes [13]. As a result, further analysis should be conducted to assess the technology's feasibility.

Recycling Electrode Catalysts

Recycling electrode catalysts of PEM cells, predominantly IrO_2 and sometimes RuO_2 of the anode and Pt/C of the cathode, is of importance due to the rich metal concentration of the layers, natural reserve scarcity, and increased demand for these metals. A number of methods which are used for platinum recovery are also relevant for iridium and ruthenium [16].

Hydrometallurgical Technology

Hydrometallurgy is a branch of metallurgy that focuses on metal extraction through selective dissolution. It is an established method in the metal mining industry and can also be applied to recover and recycle materials. The recovery of iridium metal via hydrometallurgy involves two main parts, the leaching stage and the separation and purification stage. During leaching, a mineral acid-oxidant mixture (lixiviant) dissolves the iridium oxide as a metal-ion complex [20]. The acid acts as the complexing agent, while the oxidising agent oxidises the metal to the appropriate ionic state [17]. The most commonly used lixiviant is aqua regia, a mixture of hydrochloric acid (HCl) and nitric acid (HNO3) in a 3:1 mol ratio. However, some studies suggest that stronger oxidants like perchlorate anions are needed for leaching iridium due to the metal's high corrosion resistance [21].

The leaching process for ruthenium follows a similar pathway, and experiments have reported recovery rates of 88% using 12 molars (M) HCl with 10% HNO3 (volume %) at 75 °C [17]. This is lower than the recovery rates of platinum due to the former metal's higher resistance to mineral acids. Recovery rates of iridium have not been specified in the available literature but can be expected to be similar to ruthenium's recovery rates due to their similar properties. Once the metal-ion complexes are formed and dissolved, they are separated from the leaching solution. This separation can be carried out via liquid–liquid extraction, which has a higher separation efficiency than other methods like ion exchange or precipitation [22]. The suitable extractants for Ir (IV), the oxidation state of iridium after leaching, are tributyl phosphate (TBP) or Aliquat 336 [22, 23]. On the other hand, an Alamine 336-LIX54 or Aliquat 336-LIX54 mixture is suggested for the effective extraction of ruthenium from chloride solutions [24]. The metal-ion complexes in the dissolved extractant are stripped off through precipitation using ammonium hydroxides or chlorides [23–25], the precipitates are ignited, and finally, high-purity metal is obtained via reduction [25].

The HMT recovery of platinum metal is almost identical to iridium and ruthenium recycling. The only difference is the higher recovery rate for platinum due to more efficient leaching using aqua regia. The higher efficiency is a result of the metal's lower resistance to mineral acids. [17] reported platinum leaching efficiencies of 97% with 12 M HCl (10% HNO_3 - volume basis) at 75 °C. Additionally, studies have explored the use of milder oxidants like hydrogen peroxide (H_2O_2), and comparable if not better results were obtained [17, 25]. However, H_2O_2 has the potential to decompose into water and oxygen at the operating temperatures due to its thermal instability. Furthermore, the use of chloride salts to reduce the HCl concentrations required has also been under investigation [17, 25]. The process flow diagram for platinum recycling shows two possible separation methods. Separation via ion exchange using resin was also experimented by [26]. However, solvent–solvent extraction reported higher overall recovery rates (76% compared to 72%). Hydrometallurgy is an attractive metal recovery method as it does not release toxic HF gases, consumes less energy, and can accommodate chemical recycling. However, the process produces toxic wastewater due to the use of concentrated acids (required for effective leaching of corrosion-resistant PGMs) [27]. Additionally, the complex separation and purification stages can reduce the overall metal recovery rate [16]. A strategy to improve the yield rates is CCM delamination (discussed previously), which can increase leaching efficiency by up to 25% [17]. Pre-treatment grinding may also be necessary to increase surface area and exposure to the lixiviants [15].

Pyro-Hydrometallurgical Technology

PMT utilises heat and high temperatures for metal extraction and recovery. Unless delaminated, one of the main drawbacks of this technique for recovering electrode catalysts is the formation of HF gases due to the destruction of the membrane ionomers. The PFSA ionomers in the catalyst ink could also produce smaller amounts of HF gases. Pyrometallurgy is also an energy-intensive process [15]. On the other hand, its benefit includes higher metal recovery rates than hydrometallurgical methods. The first stage of this technique comprised drying and calcination to remove carbon and other volatile compounds. In the second stage, the residues from the calcination process were dissolved in HNO_3 at 65 °C. This step was followed by the addition of HCl and heating the solution to 110 °C to decompose HNO3 into nitrous oxide (NO) and nitrogen dioxide (NO_2) [28]. The solids obtained from this step were $H_2(PtCl_6)$, and these were dissolved in deionised water and reduced to platinum metal using methanoic acid. Lastly, the solids were purified through vacuum filtration [15, 28]. The performance of the refabricated PEMFC was lower than that of pristine PEMFCs, but the performance was comparable [28]. Another problem with this technique was the formation of NO, a highly potent greenhouse gas emission, and therefore, this method may not be the most environmentally sustainable.

Transient Electrochemical Dissolution

TED is a novel metal recovery technique that applies cycles of low and high electric potentials (voltage) to drive cathodic and anodic dissolution of the metal within a selected electrolyte [29]. The electrolyte is typically a mineral acid, like HCl or perchloric acid ($HClO_4$), where the chloride ions accelerate the dissolution rates [27, 29]. As the voltage is increased from the lower potential to the upper potential, an oxide layer gradually forms on the metal surface. Before the oxide layer forms, the metal surface is momentarily exposed to these high potentials, which results in the oxidation of the metal into an ionic state, allowing it to dissolve in the electrolyte. Cycling back to the lower potential triggers the cathodic reaction, and this reduces the metal oxides. The removal of the oxide layer, once again, exposes the metal surface to the electrolyte, and further dissolution occurs. The cycle is repeated numerous times to increase dissolution and maximise the metal recovery rate. A few studies have investigated the electrochemical dissolution of Ir and Ru-based catalysts from a recycling perspective. Much of the published work focuses on the metals' dissolution rates from a corrosion aspect due to the high potentials applied at the anode for OER. Additionally, TED studies that focus on recycling have predominantly been conducted for platinum metal. However, the knowledge gathered from both areas of study can be utilised to develop potential recycling pathways for these anode catalysts [16, 30]. Several studies [30–33] have explored the electrochemical dissolution of Ir and Ru metals under the OER conditions of PEM cells and have noted slow but considerable dissolution of the catalyst, which is not desirable for long-term performance. As a result, more stable metal oxides, IrO_2 and RuO_2, with dissolution rates 2–3 order of magnitudes lower, are used [30, 32]. While this is beneficial for the operating lifetime of electrolysers, the improved stability creates challenges for recycling. Additional research needs to be conducted to identify an optimised TED pathway for IrO_2 and RuO_2 and assess whether this method is technically and economically feasible. The studies need to explore changes in the dissolution rate by adjusting the voltage range, the number of cycles applied, and the electrolyte concentrations. The voltages required for anodic and cathodic dissolution can be provided via an external circuit or through the use of appropriate oxidising and reducing agents [27].

Variations of TED with different potential ranges exist in the available literature. [29] explored the recycling of Pt/C from PEMFC catalysts by cycling between 0.05 V and 1.6 V, while [34] evaluated platinum dissolution using lower and upper potentials of 0.4 V and 1.6 V, respectively. Hodnik et al. [27] presents an interesting pathway for TED, where oxidising and reducing agents were used to induce the surface potentials that facilitate anodic and cathodic dissolution (voltage from an was not applied). The reducing agent was carbon monoxide (CO), while ozone (O_3) was added as the oxidant. These agents promote platinum dissolution and sweep the cycle between 1.3 V and 0.8 V. Furthermore, an HCl-NaCl mixture was used as the electrolyte. Using chloride salts reduced the acid concentrations required for efficient recovery. A maximum yield of 98.6% was attained when 20 cycles of 5-min O3/5-min CO were applied [27].

The method explored in [27] should be investigated for its applicability for cathode metal recycling in PEM electrolysers as the study was conducted on platinum black and not Pt/C catalyst. The presence of the carbon supports could potentially affect (improve or reduce) the effectiveness of the CO and O_3 used to induce the surface potentials. If successful, TED using oxidants and reductants presents a promising and low-energy technique for recycling catalyst metals. The process is also conducted in mild conditions in terms of pH and temperature [27]. However, the overall recovery rate is unlikely to be as high as 98.6% as additional steps would be required to extract pure platinum from the chloride solution.

Selective Electrochemical Dissolution

The upper potentials applied in the TED studies for platinum would result in the decomposition of the carbon supports through oxidation, especially if the scan rate (sweeping rate) is high [29]. Although these high potentials are favoured for maximising platinum dissolution, the anodic potential needs to be below 1.2 V to preserve the carbon supports. A study [35] presents a selective electrochemical dissolution (SED) process that applies a potential cycle of 0.5 V–1.1 V at a scan rate of 50 mV/s using an external power source. To compensate for the low potentials applied, the experiment was conducted in an oxygen-saturated electrolyte (HClO4-HCl mixture). Dissolution rates of 22 $\mu g/cm^2$ were recorded, which is considered favourable for the mild conditions applied. On the other hand, the carbon supports showed minimal corrosion with some changes in surface structure. As a result, reconditioning may be required before the carbon is used in new cells [16].

Recycling Bipolar Plates

The recycling of PEM electrolysers is an emerging and underdeveloped area of study. As a result, literature papers specific to bipolar plates recycling are not readily available. The recycling method adopted for titanium, like other metals, depends on the material stream comprising the titanium. In most cases, titanium metal or its alloys can be extracted via physical methods [16] or pyrometallurgy [36]. Additional steps may be required if alloying elements are to be removed. With bipolar plates, the recovery is more complex due to the platinum and gold coatings. Consequently, hydrometallurgical processes need to be adopted to separate and recover the metals. The precious metal coatings can be separated through selective leaching using aqua regia. The titanium base does not dissolve in aqua regia due to the formation of a thin layer of titanium dioxide (passivation) [37]. After leaching, base metal can be separated and purified by smelting. On the other hand, pure platinum and gold can be extracted via solvent–solvent extraction using TBP and dibutyl carbitol, respectively [21]. However, since the coatings are thin, the overall process may not be economical. If this is the case, platinum and gold recovery could be omitted.

Recycling Current Collectors

The current collectors are usually thin copper or aluminium blocks with an approximate thickness of 50 mm. Since the collectors are not coated with other materials, conventional pyrometallurgical processes are sufficient to recycle the metals. However, the recycling pathway adopted depends on the quality and metal grade of the obsolete collectors, as potential contamination could have occurred during their use. If the metal remains pure, melting and recasting are sufficient to recycle the material. Alternatively, the materials are smelted and refined to remove impurities [38]. The recycling process of copper and aluminium via pyrometallurgy is well-established. In central facilities, the metals are first separated based on material type and grade [39]. The separation techniques used depend on the other materials present in the stream but can include magnetic, specific gravity, or eddy-current separation [38]. The metals are then crushed and shredded to increase the surface area and improve the efficiency of the smelting process. If metal oxides are present in the scrap, the smelter is operated in reducing conditions (carbon or iron with fluxing agents facilitates this) [39]. Finally, the metals are refined and recast. The most common purification method for copper is electrorefining [38, 39]. On the other hand, the refining techniques for aluminium can include electrorefining, degassing, or floatation to remove specific impurities [40].

Recycling End Plates and Fixtures

Steel recycling is the most developed process amongst all metal recycling due to its high demand and importance in the construction industry. The recycling technique is based on pyrometallurgy and follows a similar path as copper and aluminium recycling. Literature recommends recycling steel-based and copper-based materials separately as the latter can contaminate and affect the steel quality. The separation techniques used include magnetic separation, sieving, and air separation. Additional methods can comprise hand sorting, density, and eddy-current separation. Since the end plates, bolts, screws, and springs in PEMWE stacks are manufactured from stainless steel, an additional step is required compared to ordinary-steel recycling. Stainless steel is predominantly made from steel scrap (new or old) melted in an electric arc furnace. However, since the volumes of stainless steel scraps generated are usually lower than the production demands, ordinary-steel scrap (with low chromium and nickel concentration) is mixed. As a result, additional ferrochromium and ferronickel are added. Following this process, excess carbon is removed as carbon monoxide in an argon oxygen decarburization converter, and the molten stainless steel is recast [41].

Conclusion

PEM electrolysers for green hydrogen production have received considerable focus in recent years due to their superior electrochemical performances. However, current PEMWEs are expensive due to the use of noble metal catalysts for electrodes, costly polymer membranes, and titanium-based bipolar plates and PTLs. These materials are necessary to withstand the corrosive conditions of the oxidation and reduction reactions. The scarcity of the noble metals also creates supply risks. Therefore, recycling electrolysers and using recycled materials during manufacturing may be crucial to reduce costs and ensure a steady material supply. The value chain of electrolyser recycling can be complex, and it will not take off until sufficient EOL materials are available for processing. Numerous recycling techniques exist in literature papers for platinum group metals. Some of the methods highlighted in the studies include hydrometallurgy, pyrometallurgy, and transient electrochemical dissolution. Most of these methods have been studied extensively for platinum metal due to its wider application. Several gaps in research have been identified, which should be addressed in future studies. Firstly, iridium recycling needs to be investigated more thoroughly. The overall efficiencies of the recycling pathways highlighted should also be investigated. Thirdly, the performance of electrolysers manufactured from recycled materials should be studied. Lastly, the scalability of the recycling techniques and the logistics of the complete value chain need to be assessed.

References

1. Parliament of Australia (2022) ParlInfo - Climate Change Bill 2022. https://parlinfo.aph.gov.au/parlInfo/search/display/display.w3p;query%3DId%3A%22legislation%2Fbillhome%2Fr6885%22. Accessed 5 Aug 2022
2. Bruce S, Temminghoff M, Hayward J, Schmidt E, Munnings C, Palfreyman D, Hartley P (2018) National hydrogen roadmap. CSIRO, Australia
3. Hughes, AE, Haque, N, Northey, SA, Giddey, S (2021) Platinum group metals: a review of resources, production and usage with a focus on catalysts. Resources 10(93)
4. Bessarabov D, Wang H, Li H, Zhao N (2016) PEM electrolysis for hydrogen production: principles and applications. CRC Press
5. Mergel J, Fritz DL, Carmo M (2016) Stack technology for PEM electrolysis. Hydrogen science and engineering: materials, processes, systems and technology 2016. Wiley-VCH Verlag GmbH & Co. KGaA, Weinheim, pp 331–357
6. Soriano RM, Rojas N, Nieto E, de Guadalupe González-Huerta R, Sandoval-Pineda JM (2021) Influence of the gasket materials on the clamping pressure distribution in a PEM water electrolyser: bolt torques and operation mode in pre-conditioning. Int J Hydrog Energy 46(51):25944–25953
7. Thomas, D (2019) Large-scale PEM electrolysis: technology status and upscaling strategies, Brussels
8. Cummins (2021) HyLYZER® water electrolysers. https://mart.cummins.com/imagelibrary/data/assetfiles/0070328.pdf. Accessed 21 Dec 2021
9. Bareiß K, de La Rua C, Möckl M, Hamacher T (2019) Life cycle assessment of hydrogen from proton exchange membrane water electrolysis in future energy system. Appl Energy 237:862–872

10. Carmo M, Keeley GP, Holtz D, Grube T, Robinius M, Muller M, Stolten D. PEM water electrolysis: innovative approaches towards catalyst separation, recovery and recycling. Int J Hydrog Energy 44(7):3450–3455
11. Sreeraj, P, Vedarajan, R, Rajalakshmi, Ramadesigan V (2021) Screening of recycled membrane with crystallinity as a fundamental property. Int J Hydrog Energy 46(24):13020–13028
12. Moghaddam JA, Parnian MA, Rowshanzamir S (2018) Preparation, characterisation, and electrochemical properties investigation of recycled proton exchange membrane for fuel cell applications. Energy 161:699–709
13. Wittstock R, Pehlken A, Wark M (2016) Challenges in automotive fuel cells recycling. Recycle 1(3):343–364
14. Reck BK, Graedel TE (2012) Challenges in metal recycling. Science 337(6095):690–695
15. HyTechCycling (2018) Existing end-of-life technologies applicable to FCH products
16. HyTechCycling (2018) New end-of-life technologies applicable to FCH products
17. Sandig-Predzymirska L, Barreiros TV, Thiere A, Weigelt D, Stelter VM, Charitos M (2021) Recycling strategy for the extraction of PGMs from spent PEM electrodes. In: EMC 2021. Freiberg
18. Shore, L (2012) Process for recycling components PEM fuel cell membrane electrode assembly. Washington, DC Patent 8,124,261
19. Grot S, Grot W (2007) Recycling of used perfluorosulfonic acid membranes. U.S.A Patent 7255798, 14 August 2007
20. Ohriner EK (2008) Processing of iridium and iridium alloys. Platin Met Rev 52(3):186
21. Serpe A (2008) Green chemistry for precious metals recovery from WEEE. In: Waste electrical and electronic equipment recycling 2008. Woodhead Publishing, pp 271–332
22. Nguyen TH, Sonu TH, Lee MS (2016) Separation of Pt(IV), Pd(II), Rh(III) and Ir(IV) from concentrated hydrochloric acid solutions by solvent extraction. Hydrometallurgy 164:71–77
23. Ueda T, Ichiishi S, Okuda A, Matsutani K (2016) Refining and recycling technologies for precious metals. Metal sustainability: global challenges, consequences and prospects 2016. John Wiley & Sons, West Sussex, pp 333–360
24. Kedari S, Coll MT, Fortuny A, Goralska E, Sastre AM (2004) Liquid-liquid extraction of Ir, Ru, and Rh from chloride solutions and their separation using different commercially available solvent extraction reagents. Sep Sci Technol 40(9):1927–1946
25. Jha MK, Lee JC, Kim MS, Jeong J, Kim BS, Kumar V (2013) Hydrometallurgical recovery/recycling of platinum by the leaching of spent catalysts: a review. Hydrometallurgy 133:23–32
26. Duclos L, Svecova L, Laforest V, Mandil G, Thivel PX (2016) Process development and optimisation for platinum recovery from PEM fuel cell catalyst. Hydrometallurgy 160:79–89
27. Hodnik N, Baldizzone C, Polymeros G, Geiger S, Grote JP, Cherevko S, Mingers A, Zeradjanin A, Mayrhofer KJ (2016) Platinum recycling going green via induced surface potential alteration enabling fast and efficient dissolution. Nat Commun 7(13164)
28. Zhao J, He X, Tian J, Wan C, Jiang C (2007) Reclaim/recycle of Pt/C catalysts for PEMFC. Energy Convers Manag 48(2):450–453
29. Pavlisic A, Jovanovic P, Selih VS, Sala M, Hodnik N, Gaverscek M (2018) Platinum dissolution and redeposition from Pt/C fuel cell electrocatalyst at potential cycling. J Electrochem Soc 165(6):F3161–F3165
30. Jovanovic P, Hodnik N, Ruiz-Zepeda F, Arcon I, Jozinovic B, Zorko M, Bele M, Sala M, Selih SV, Hocevar S, Gaberscek M (2017) Electrochemical dissolution of iridium and iridium oxide particles in acidic media: transmission electron microscopy, electrochemical flow cell coupled to inductively coupled plasma mass spectrometry and x-ray absorption spectroscopy study. J Am Chem Soc 139(36):12837–12846
31. Cherevko S, Geiger S, Kasian O, Kulyk N, Grote JP, Savan, Shrestha BR, Merzlikin S, Breitbach B, Ludwig A, Mayrhofer KJ (2016) Oxygen and hydrogen evolution reactions on Ru, RuO_2, Ir, and IrO_2 thin film electrodes in acidic and alkaline electrolytes: a comparative study on activity and stability. Catalysis Today 262:170–180

32. Cherevko S, Zeradjanin AR, Topalov AA, Kulyk N, Katsounaros I, Mayrhofer KJ (2014) Dissolution of noble metals during oxygen evolution in acidic media. ChemCatChem 6(8):2219–2223
33. Kasian O, Grote JP, Geiger S, Cherevko S, Mayrhofer KJ (2018) The common intermediates of oxygen evolution and dissolution reactions during water electrolysis on Iridium. Angew Chem Int Ed 57:2488–2491
34. Sharma R, Nielsen KR, Lund PB, Simonsen SB, Grahl-Madsen L, Anderson SM (2019) Sustainable platinum recycling through electrochemical dissolution of platinum nanoparticles from fuel cell electrodes. ChemElectroChem 6(17):4471–4482
35. Latsuzbaia R, Negro E, Koper GJM (2015) Environmentally friendly carbon-preserving recovery of noble metals from supported fuel cell catalysts. Chemsuschem 8(11):1926–2934
36. Takeda O, Okabe TH (2019) Current status of titanium recycling and related technologies. JOM 71:1981–1990
37. Wang D, Li Q, Xu M, Jiang G, Zhang Y, He G (2017) A novel approach to fabrication of three-dimensional porous titanium with controllable structure. Mater Sci Eng 71(1):1046–1051
38. Schlesinger ME, King MJ, Sole KC, Davenport WG (2011) Collection and processing of recycled copper. In: Extractive metallurgy of copper (Fifth Edition), Elsevier, pp 373–387
39. Samuelsson C, Bjorkman B (2014) Copper recycling. In: Handbook of recycling: state-of-the-art for practitioners, analysts, and scientists 2014. Newnes, pp 85–94
40. Schlesinger ME (2006) Common impurities in molten aluminium. In: Aluminium recycling 2006. CRC Press, Boca Raton, pp 171–192
41. Bjorkman B, Samuelsson C (2014) Recycling of steel. In: Handbook of recycling: state-of-the-art for practioners, analysts, and scientists 2014. Newnes, pp 65–83

Rare Earth and Critical Base Metals Electrodeposition Using Urea-Choline Chloride Ionic Liquids

Rajyashree Lenka and Ramana G. Reddy

Abstract The electrodeposition of critical base metals and rare earth metals using a eutectic mixture of urea and choline chloride (IL) was discussed. The maximum current efficiencies obtained for Co, Zn, Pb, and Cu were 95%, 91.3%, 96.4%, and 97%. The deposition of alloys of zinc like Zn-Sn, Zn-Cu, Zn-Ni, and Zn-Mn and alloys of nickel like Ni-Mn, and Fe-Co–Ni was reported. The co-deposition of rare earth elements like Co-Nd and Sm-Co alloy was also carried out, because rare earth metals cannot be deposited alone in urea melt due to their high negative reduction potentials of the rare earth ions. The effect of Urea and choline chloride or EMIC or BMIC on the electrodeposition of critical base metals and rare earth elements was discussed. The critical parameters in the electrodeposition processes were identified, such as current density and efficiency in metal and alloy production.

Keywords Base metal · Rare earth · Ionic liquid · Electrodeposition

Introduction

Ionic liquids (ILs) are defined as salts which comprise cations and anions with melting point at or below 100 °C [1]. Ionic liquids provide various benefits which include a broad potential window, high electrical conductivity compared to non-aqueous solvent, thermal stability, low volatility, structural designability, high polarity, and non-flammability. Abott et al. [2] reported that there is a deep melting point depression when eutectic composition is formed between hydrogen bond donors and hydrogen bond acceptors. Compared to the classical ILs, Deep Eutectic Solvents (DESs) are easily formed by mixing two components where a eutectic mixture forms through hydrogen bond interaction Choline chloride is a promising candidate for DESs due to its cheap price, availability and biodegradability, and easy storage due to its non-volatility and non-flammability. Choline chloride mixed with urea in the ratio

R. Lenka · R. G. Reddy (✉)
Department of Metallurgical and Materials Engineering, The University of Alabama, Tuscaloosa, AL 35487, USA
e-mail: Rreddy@eng.ua.edu

R. G. Reddy et al. (eds.), *New Directions in Mineral Processing, Extractive Metallurgy, Recycling and Waste Minimization*, The Minerals, Metals & Materials Series,
https://doi.org/10.1007/978-3-031-22765-3_15

1:2 forms eutectic that is liquid at ambient temperature and has exceptional solvent properties [2]. This mixture leads to a major depression of freezing point to 12 °C which is lower than the choline chloride (melting point-302 °C) and Urea (melting point-133 °C). All DES have a similar density which is close to water (1.00 g/cm^3). Oxides such as ZnO, CuO, and PbO_2 show a good solubility in ChCl/Urea.

The most common ionic liquid used for the electrodeposition of cobalt, lead, aluminum, bismuth, selenium, copper, tin, nickel, zinc, and manganese is a eutectic mixture of urea and choline chloride in the molar ratio 2:1. Abbot et al. [3] investigated that zinc oxide shows an appreciable amount of solubility in Urea-choline chloride deep eutectic solvent which is eco-friendly and highly efficient. Yang et al. [4] showed that the maximum solubility of ZnO was 1.23 M at 373 K and reported a current efficiency of 88.9% at temperature of 363 K and applied potential of 3.3 V. Further, a high current efficiency of 92.6% was reported when a suitable additive (2 mg/ml of [BMIM] HSO_4) was added [5]. Kim et al. [6] reported cobalt electrodeposition with a highest current efficiency of 95% at 323 K and 2.7 V while Cao et al. [7] studied that electrodeposition potential was the main driving force for deposition of cobalt recording a higher current efficiency of 96.4%. Liao et al. [8] reported that lead compounds show acceptable solubility in Urea-ChCl, although they are most likely non-soluble in water. Ru et al. [9] reported that lead deposition shows a highest efficiency of 96.4% at a current density of 130 Am^{-2} and 343 K. The electrodeposition of copper and its composites from its metal chloride salts in ChCl-Urea/ethylene glycol with current efficiency of around 100% was studied by Abbot et al. [10]. Ali et al. [11] reported that copper deposition shows a cathodic efficiency of 97% at a current density of 4.0 Am^{-2} and 303 K. The electrodeposition of metals like Al, Bi, Se, and Sn in Urea-ChCl deep eutectic solvent was confirmed by XRD and SEM analysis. This eutectic mixture is also used for the deposition of alloys of zinc like Zn-Sn, Zn-Cu, Zn-Ni, Zn-Mn, and alloys of nickel like Ni-Mn, and Fe-Co–Ni. Manganese and its based alloy are commonly used to improve corrosion resistance. Guo et al. [12] confirmed the co-deposition of Ni-Mn films where the Mn content increased from 0 to 9.4% as the current density increased from 1 to 9 mA/cm^2 at 343 K. Lui et al. [13] studied that at temperature of 373 K and potential of −3.5 V, Co-Nd films can be deposited with cobalt content of 96.26% and Nd of 3.74% while Gomez et al. [14] reported that Sm-Co can give samarium deposit of more than 70 wt.% at 343 K and −1.6 V. EMIC has been used as electrolyte in lead storage batteries due to its cost-effectiveness, high capacity, and safety. Zn, Pb, and Cu were electrodeposited using Urea-EMIC. He et al. [15] studied the electrodeposition of zinc from ZnO in Urea-EMIC as an alternative to zinc electroplating reporting the current efficiency of 84.20% at −1.25 V and 353 K which is lower when compared to the Urea-ChCl system. He et al. [16] investigated the deposition of Pb from PbO at 353 K in Urea-EMIC and also studied the reduction of Cu_2O to metallic copper which follows a 3D progressive nucleation and diffusion growth model [17]. An ideal alternative to EMIC can be Urea-BMIC which is more economical and easily synthesized. Lui et al. [18] reported that the current efficiency for Zn increases to 99.81% on using Urea-BMIC at 383 K and potential of −1.2 V while for Pb, the current efficiency was 96.17% at 373 K and potential of −0.4 V [19]. At 383 K and -0.8 V, the co-deposition of Cu–Zn can be done on a steel electrode in eutectic Urea-BMIC containing 0.05 mol/L Cu_2O and 0.5 mol/L ZnO as investigated by Liu et al. [20]. So, for zinc electrodeposition, the optimized current efficiency (99.81%) [18] in Urea-BMIC is higher than Urea-EMIC (84.2%) [15] and Urea-ChCl (91.3%) [5] ionic liquids. In the current study, the electrodeposition of zinc is carried out in a eutectic mixture of urea/ChCl. It and important areas in electrodeposition which include current density, cathodic current efficiency, and energy consumption are investigated during the process.

Experimental Procedure

Preparation of Ionic Liquid and Materials

The eutectic mixture of choline chloride and Urea in the ratio 2:1 was synthesized in the laboratory. Urea (NH_2CONH_2, 98 + %) and choline chloride ($HOCH_2CH_2N(CH_3)_3Cl$, 98 + %) were purchased from Alfa Aesar and zinc oxide from Fisher Scientific (certified ACS, ≥99.0%). All chemicals were dried for 3 h in vacuum to remove the residual moisture. The eutectic mixture of choline chloride and Urea in the ratio 2:1 was synthesized by maintaining the correct molar ratio, and the salts were mixed in a beaker. The solution was stirred constantly for 12 h at 90 °C till a clear homogenous solution was obtained. The required concentration of ZnO (0.82 M) was mixed with 2:1 Urea/Choline chloride ionic liquid at a specified temperature The electrolyte was mixed for more than 24 h until all the ZnO powders were dissolved in the solution.

Electrodeposition Set-Up

All the experiments were conducted in a 50 ml glass beaker and were carried out in a Kepco programmable power supply. The current was recorded using a multimeter (Keithley 2000 Multimeter), and lab view software was used. A two-electrode cell was used where copper plate (0.25 mm in thickness) was used as a working electrode and platinum wire (0.5 mm in diameter) was used as counter electrode, respectively. Prior to the electrochemical measurements, the electrodes were polished SiC paper, washed with deionized water, and then dried. A thermometer was inserted in the electrolyte solution to control the temperature, and a constant potential difference was applied between cathode and anode to facilitate the electrodeposition experiment. After the electrodeposition experiment, the electrode surface was washed thoroughly in deionized water until the particles and ionic liquids are removed and then dried in vacuum for 24 h before further measurements.

Characterization of the Deposits on the Substrate

The zinc deposits on the copper substrate were characterized using X-ray diffraction pattern obtained from the Phillips MPD XRD, and morphology and elemental analysis of the deposits were carried out using scanning electron microscopy, the APREO model. Cathodic current efficiency (η) was calculated using Eq. (1)

$$\eta = \frac{\Delta \mathrm{W}}{\mathrm{Wt}} X100\% \tag{1}$$

where Δw is the experimental weight gain of Zinc deposits and w_t is the theoretical weight of zinc deposits (g). Energy consumption in kwh/kg can be calculated using the equation.

$$E = \frac{VQ}{\Delta W} \tag{2}$$

where V is the applied potential in volts and Q is the total area of the current versus time plot.

Results and Discussion

Effect of Temperature

Temperature is an important variable that has significant effect on current density, current efficiency, and morphology of the deposits. The variation of cathodic current density with time was recorded in Fig. 1. It can be noted that current increases slightly with increase in temperature. This can be attributed to enhanced diffusion rate at higher temperature which further increases the transportation rate. At a fixed ZnO concentration (0.82 M), current efficiency was calculated to be 75.24% at 3.0 V and 373 K and the energy consumption was 3.26 kWh/kg. The morphology of zinc deposited on the cathode at different temperatures is shown in Fig. 2. At 363 K, the zinc deposits show a hexagonal crystalline structure with average grain size of 1.7 μm. The overall morphology appears the same at higher temperature such as 373 K but with a larger grain size of 2.1 μm showing that temperature largely influences the grain size and surface morphology of zinc particle. EDS analysis shows that peaks correspond to metallic zinc and copper. The composition of deposits is Zn (96.9 wt.pct) and Cu (3.1 wt.pct). XRD analysis shown in Fig. 5a. confirmed the presence of zinc peaks, and at 373 K, the peaks at $2\theta = 43.3°$ become more distinct which means thicker zinc deposition on the cathode substrate. The distinct peaks at $2\theta = 50.4°$ and 74.1° is coming from copper substrate according to ICDD File No.01–089-0831.

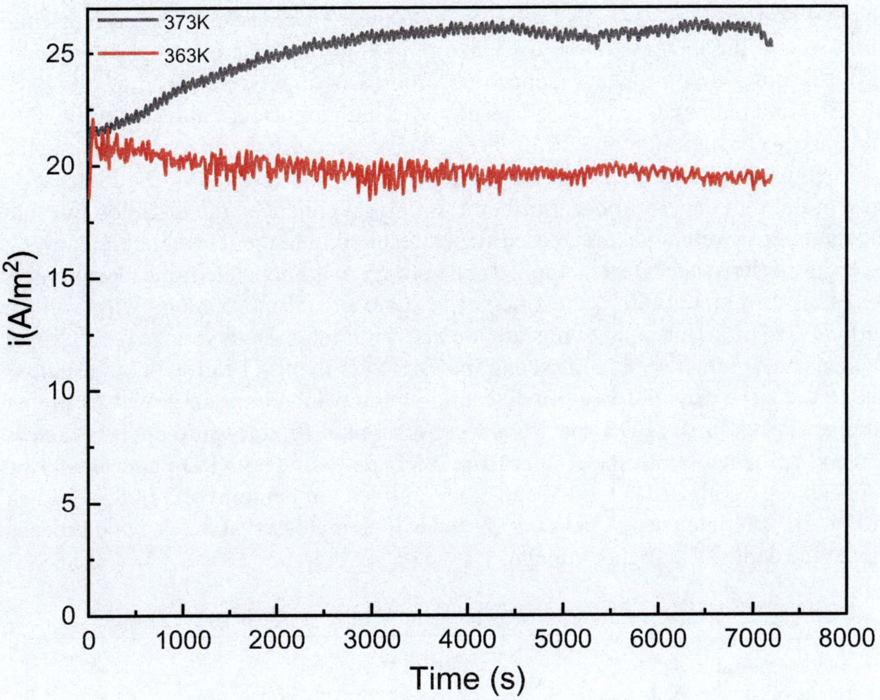

Fig. 1 Variation of current density with time at 3.0 V at different temperature, **a** 373 K, **b** 363 K

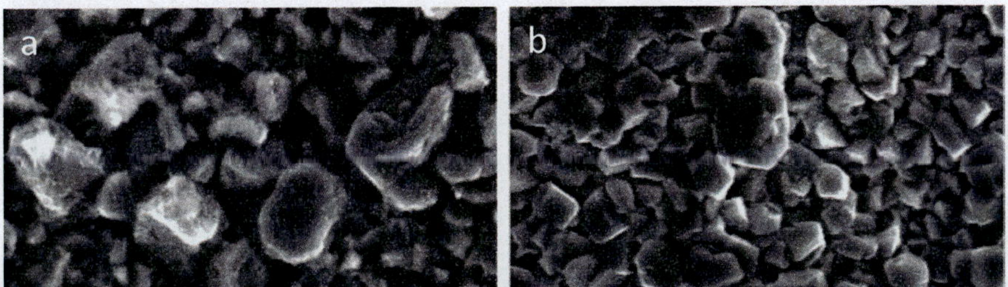

Fig. 2 SEM images of deposited Zn at 3.0 V and different temperatures, **a** 373 K, **b** 363 K

Effect of Applied Voltage

The effect of applied cell voltage on current density, current efficiency, energy consumption, and morphology of the zinc deposits is investigated at a fixed temperature of 373 K. The variation of current density with time is shown in Fig. 3, and the summarization of the current efficiency is included in Table 1. As the voltage was stepped up from 3.0 V to 3.3 V, the current density increased in the current–time curve due to the increase in the total charge pass through the electrochemical cell. Table 1 shows that the electrodeposition process is largely enhanced at a voltage of 3.3 V and temperature of 373 K resulting in higher current efficiency of 96.85% and energy consumption of 2.79 Kwh/Kg. This high current efficiency is due to the high temperature and high applied cell potential which resulted in the increase in current through the electrodes. Further, this current enhanced the diffusion rate and the theoretical weight gain increased which resulted in higher current efficiency.

Figure 4 shows the effect of applied cell voltage on the deposited zinc morphology. It can be seen that at 3.0 V, the zinc shows a hexagonal morphology with average particle size of 2.1 μm and grains are thicker. With increase in voltage to 3.3 V, the deposits are brighter with increased grain size to 2.53 μm. The growth of particle is due to the large driving force of nucleation which results in grain growth at higher voltage. Peaks in the EDS spectrum were attributed to zinc and copper which is in good agreement with the results from XRD pattern. The XRD patterns of zinc deposits were compared at 3.0 V and 3.3 V at fixed temperature of 373 K as shown in Fig. 5b. The intensity of peaks at $2\theta = 36.1°$ is increased at 3.3 V which means more zinc deposits at higher voltage.

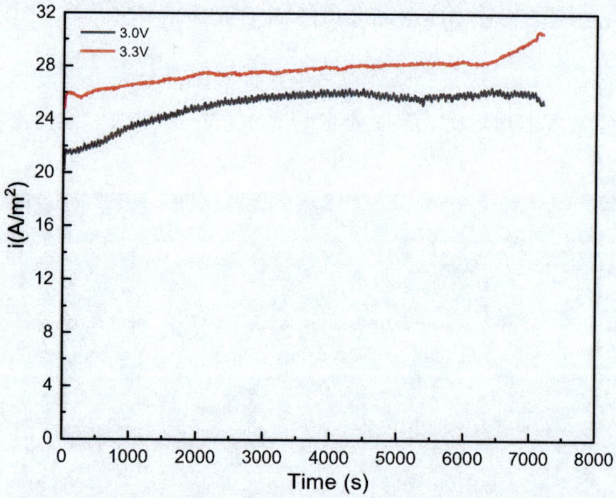

Fig. 3 Variation of current density with time at 373 K at different applied voltage, **a** 3.0 V, **b** 3.3 V

Table 1 Comparison of current efficiency of zinc in various ILs

System	Temperature(K)	Potential (V)	C.E (%)	References
ZnO in Urea-ChCl	373	3.0	89.70	Yang et al. [5]
ZnO in Urea-ChCl	363	3.3	88.90	Yang et al. [21]
ZnO in Urea-BMIC	383	−1.1	99.24	Liu et al. [18]
ZnO in Urea-EMIC	353	−1.25	84.20	He et al. [15]
ZnO in Urea-ChCl	373	3.3	96.85	This work

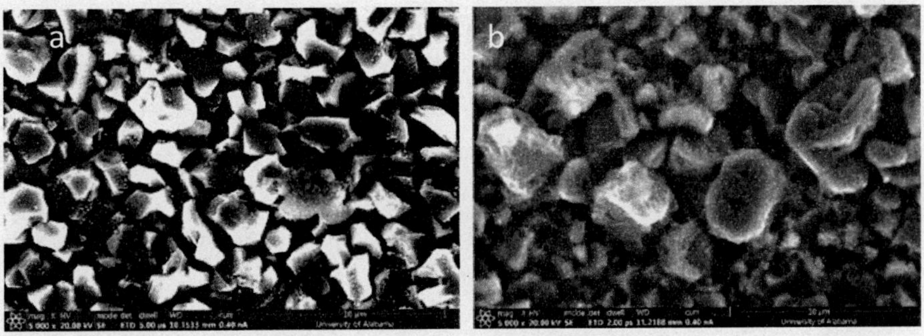

Fig. 4 SEM images of deposited Zn at 373 K and different cell voltages, **a** 3.3 V, **b** 3.0 V

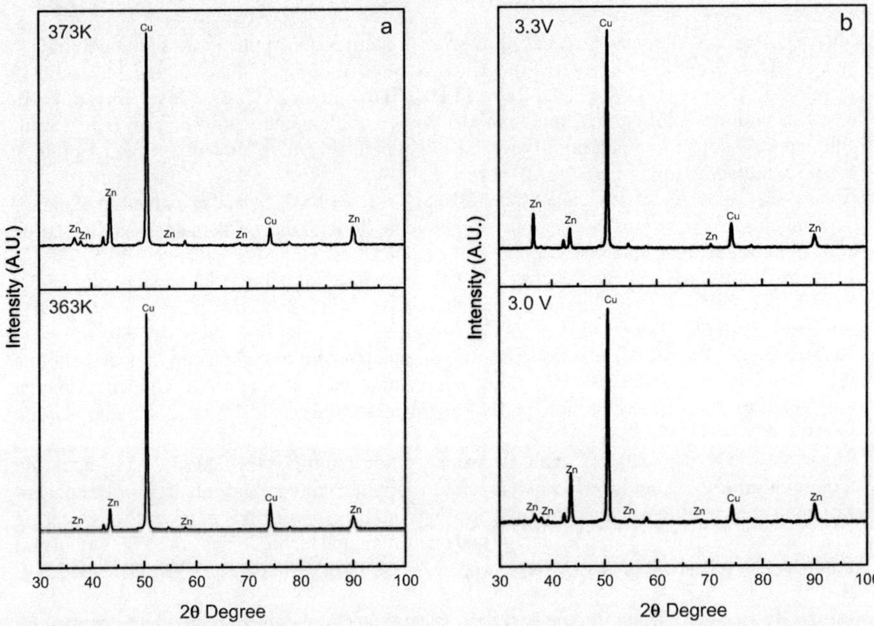

Fig. 5 XRD pattern for deposited zinc, **a** as a function of temperature at 3.0 V, **b** as a function of applied voltage at 373 K

Conclusions

Zinc oxide was successfully electrodeposited from ZnO precursor using 2:1 urea/ChCl ionic liquid, and the effect of experimental variables on the zinc deposition was investigated. At a fixed potential of 3.0 V, a higher current density was recorded at 373 K which is due to the enhanced diffusion rate at elevated temperatures. At a fixed concentration of 0.82 M of ZnO, temperature of 373 K and an applied potential of 3.3 V, the current efficiency can reach up to 96.85% and the energy consumption as low as 2.79 kWh/kg. An increased in grain size was observed at high potential and high temperature. XRD and EDS analysis confirmed the presence of zinc layer of high purity on the copper substrate.

Acknowledgements The authors acknowledge the financial support received from the Department of Energy (DOE), Office of Fossil Energy, National Energy Technology Laboratory, DE-FE0032119, and the ACIPCO for this research project. We thank the Department of Metallurgical and Materials Engineering (MTE) and The University of Alabama for providing experimental and analytical facilities.

References

1. Shah FU, An R, Muhammad N (2020) Editorial: properties and applications of ionic liquids in energy and environmental science. Front Chem 8. https://doi.org/10.3389/fchem.2020.627213
2. Abbott AP, Capper G, Davies DL, Rasheed RK, Tambyrajah V (2002) Novel solvent properties of choline chloride/urea mixtures electronic supplementary information (ESI) available: spectroscopic data see http://www.rsc.org/suppdata/cc/b2/b210714g/. Chem Commun 1:70–71. https://doi.org/10.1039/b210714g
3. Abbott AP, Capper G, Davies DL, Rasheed RK, Shikotra P (2005) Selective extraction of metals from mixed oxide matrixes using choline-based ionic liquids. Inorg Chem 44(19):6497–6499. https://doi.org/10.1021/ic0505450
4. Yang H, Reddy RG (2014) Electrochemical deposition of zinc from zinc oxide in 2:1 urea/choline chloride ionic liquid. Electrochim Acta 147:513–519. https://doi.org/10.1016/j.electacta.2014.09.137
5. Yang H, Reddy RG (2014) Electrodeposition of zinc from zinc oxide using urea and choline chloride mixture: effect of (BMIM)HSO_4, temperature, voltage on current efficiency, energy consumption, and surface morphology. In: Rare metal technology 2014, pp. 21–26. https://doi.org/10.1002/9781118888551.ch5
6. Kim AR, Reddy RG (2017) Cobalt electrodeposition from cobalt chloride using urea and choline chloride ionic liquid: effect of temperature, applied voltage, and cobalt chloride concentration on current efficiency and energy consumption. In: Applications of process engineering principles in materials processing, energy and environmental technologies: an EPD symposium in honor of professor Ramana G. Reddy, pp. 97–114. https://doi.org/10.1007/978-3-319-51091-0_9
7. Cao X, Xu L, Shi Y, Wang Y, Xue X (2019) Electrochemical behavior and electrodeposition of cobalt from choline chloride-urea deep eutectic solvent. Electrochim Acta 295:550–557. https://doi.org/10.1016/j.electacta.2018.10.163
8. Liao Y-S, Chen P-Y, Sun I-W (2016) Electrochemical study and recovery of PB using 1:2 choline chloride/urea deep eutectic solvent: a variety of PB species pbso4, PbO_2, and PbO exhibits the analogous thermodynamic behavior. Electrochim Acta 214:265–275. https://doi.org/10.1016/j.electacta.2016.08.053
9. Ru J, Hua Y, Xu C, Li J, Li Y, Wang D, Gong K, Zhou Z (2015) Preparation of sub-micrometer lead wires from PbO by electrodeposition in choline chloride-urea deep eutectic solvent. Adv Powder Technol 26(1):91–97. https://doi.org/10.1016/j.apt.2014.08.008
10. Abbott AP, El Ttaib K, Frisch G, McKenzie KJ, Ryder KS (2009) Electrodeposition of copper composites from deep eutectic solvents based on choline chloride. Phys Chem Chem Phys 11(21):4269. https://doi.org/10.1039/b817881j
11. Electrodeposition of copper from a choline chloride based ionic liquid. J Electrochem (2014, April 17). Retrieved September 13, 2022, from http://electrochem.xmu.edu.cn/EN/Y2014/V20/I2/139

12. Guo J, Guo X, Wang S, Zhang Z, Dong J, Peng L, Ding W (2016) Effects of glycine and current density on the mechanism of electrodeposition, composition and properties of Ni–Mn films prepared in Ionic Liquid. Appl Surf Sci 365:31–37. https://doi.org/10.1016/j.apsusc.2015.12.248
13. Liu A, Shi Z, Reddy RG (2020) Electrochemical synthesis of Co-Nd films in urea and choline chloride deep eutectic solvents. Metall Mater Trans B 51(3):1162–1168. https://doi.org/10.1007/s11663-020-01826-8
14. Gómez E, Cojocaru P, Magagnin L, Valles E (2011) Electrodeposition of Co, Sm and SmCo from a deep eutectic solvent. J Electroanal Chem 658(1–2):18–24. https://doi.org/10.1016/j.jelechem.2011.04.015
15. He W, Shen L, Shi Z, Gao B, Hu X, Xu J, Wang Z (2016) Zinc electrodeposition from zinc oxide in the urea/1-ethyl-3-methylimidazolium chloride at 353 K. Electrochemistry 84(11):872–877. https://doi.org/10.5796/electrochemistry.84.872
16. He W, Liu A, Guan J, Shi Z, Gao B, Hu X, Wang Z (2017) Pb electrodeposition from PBO in the urea/1-ethyl-3-methylimidazolium chloride at room temperature. RSC Adv 7(12):6902–6910. https://doi.org/10.1039/c6ra27383a
17. He W, Shi Z, Liu A, Guan J, Yang S (2021) Electro-reduction of Cu_2O to Cu in urea/1-ethyl-3-methylimidazolium chloride. J Appl Electrochem. https://doi.org/10.1007/s10800-021-01565-x
18. Liu A, Shi Z, Reddy RG (2017). Electrodeposition of zinc from zinc oxide in 2:1 urea/1-butyl-3-methylimidazolium chloride ionic liquid. J Electrochem Soc 164(9). https://doi.org/10.1149/2.1501709jes
19. Liu A, Shi Z, Reddy RG (2017) Electrodeposition of Pb from PbO in urea and 1-butyl-3-methylimidazolium chloride deep eutectic solutions. Electrochim Acta 251:176–186. https://doi.org/10.1016/j.electacta.2017.08.011
20. Liu A, Shi Z, Reddy RG (2020) Mechanism study of Cu-Zn alloys electrodeposition in deep eutectic solvents. Ionics 26(6):3161–3172. https://doi.org/10.1007/s11581-019-03418-2
21. Yang H, Reddy RG (2015) Electrochemical kinetics of reduction of zinc oxide to zinc using 2:1 urea/ChCl Ionic liquid. Electrochim Acta 178:617–623. https://doi.org/10.1016/j.electacta.2015.08.050

Estimated End-of-Life Lithium-Ion Battery Resources for Potential Recycling in Bangladesh

Md. Rakibul Qadir, Miao Chen, Nawshad Haque, and Warren Bruckard

Abstract This study analyzes open access data on the input and generation of end-of-life lithium-ion battery waste supply for a potential commercial battery recycling industry in Bangladesh. Four main sources were identified in the battery waste pool: mobile phones, laptop and tablet PCs, small handheld devices, and hybrid electric vehicles. Their predicted generation rate and volume by type were forecasted to the year 2041 based on the available historic data and assumptions. Such batteries contain commercially recoverable quantity of base metals like Co, Ni, and Mn as well as other common metals like Li, Cu, Al, and Fe. These metals are in high demand for Bangladesh where urban mining and 'informal recycling' currently present challenges to public health, ecological safety, and resource efficiency. A precise estimate of secondary resource inventory, cumulative growth, and economic forecasting of battery waste can pave the way for 'formal recycling' which will attract national and international investments.

Keywords Lithium-ion battery · e-waste generation · Bangladesh

Introduction

The use of secondary lithium-ion batteries (LIB) is widespread from small wearable electronics, like smart watches or wireless earbuds, to battery operated electric vehicles (EV). The most significant quantity of LIBs is found in small to medium

Md. R. Qadir (✉) · M. Chen · N. Haque · W. Bruckard
CSIRO Mineral Resources/CSIRO Energy, Private Bag 10, Clayton South, VIC 3169, Australia
e-mail: rakibul.qadir@csiro.au

Md. R. Qadir · M. Chen
School of Science, RMIT University, GPO Box 2476, Melbourne, VIC 3001, Australia

Md. R. Qadir
PP&PDC, Bangladesh Council of Scientific and Industrial Research, Dr. Qudrat-I-Khuda Road, Dhaka 1205, Bangladesh

R. G. Reddy et al. (eds.), *New Directions in Mineral Processing, Extractive Metallurgy, Recycling and Waste Minimization*, The Minerals, Metals & Materials Series,
https://doi.org/10.1007/978-3-031-22765-3_16

devices related to information technology (IT) and portable electronic devices. This segment holds great significance in terms of numbers, short service life, need for subsequent replacement, lack of 'second life' options or incentives, and a mixed mode of formal-informal recycling practices. Reliable data on battery recycling is scarce. Heelan [1] estimated a minimum 5,569 tonnes of LIB recycling in North America by the six major LIB recyclers (Retriev, Umicore, Recupyl, Xstrata Nickel, Batrec, and Accurec). Notably, they identified 3 frequent sources of LIB waste: mobile phones, laptop, and tablet PCs, while commenting that the recycling industry was driven predominantly by laptop batteries. The recycling rate is believed to have significantly improved in the US recently, especially in the EV battery sector [2].

Warnken [3] examined retailer surveys and customs statistics based on the tariff code and classified the battery stream into; handheld (<1 kg), automotive starting, lighting and ignition (SLI), and large and heavy batteries (>1 kg). He reported that about 350 million batteries per annum are consumed in Australia, with 98% of them (about 16,000 tonnes) being portable electronic device batteries. O'Farrell [4] considered handheld (<5 kg) batteries in Australia and reported that about 50% of the annual sales were rechargeable batteries and that LIBs were about 24% by weight of all batteries. Original equipment manufacturers (OEM), wholesalers, and retailers were surveyed for sales data, and the literature and customs data were reviewed for forecasting. This data revealed that for the year 2012–13 about 98% (1,720 tonnes) of LIB waste was illegally dumped in Australian landfills. Randell [5] reviewed data for handheld (<5 kg) devices, EVs, and photovoltaic panel energy storage (PVS) and concluded that it is 'most likely' that the Australian combined LIB waste would be about 137,000 tonnes by the year 2036 with an annual growth rate of 20%. It was estimated that about AU$813 million—3.09 billion total recoverable metal values would be contained in waste LIBs by 2034 [6].

A European Union (EU) study [7] analysed the LIB put on market (POM) by different databases across the EU and correlated it with the collected LIB waste designated in three different categories: portable (including IT devices), traction (EVs and e-bikes), and emergency supply and storage (ESS). The report considered the effect of informal storage (hoarding effect) in determining the average lifespan of LIBs. It was estimated to be more than 3 years. It was also recognized that the lifespan prediction of LIBs involved high level of uncertainty. A complete historical inventory report of the EU, Switzerland, and Norway was constructed by modelling corroborated survey results on batteries, waste electronics, and vehicles on 2017 considering supply placed on market, in-use stock, and waste generation [8]. According to the database, about 2.7 million tonnes of batteries were placed on the EU28 + 2 market in 2017 alone–120,000 tonnes of which were LIBs. The results suggested about 85–90% of the Al, Cu, Li, and Co from the LIB waste is unaccounted-for in the model. The European Compliance Organizations for Batteries considers the end-of-life (EoL) battery quantities to be about 80% of the batteries' POM, though speculates at the same time that the amount of batteries available for collection varies greatly due to the increasing lifespan of the batteries

[9]. It was specified that about 19% of EoL batteries are used as second life outside of the EU or get embedded with the appliance when exported because some appliances need qualified persons with designated tools to remove the battery—which are unavailable to the recycler. The average lifespan of the newer chemistry LIBs used in cordless power tools, portable computers, and cellular phones remains about 2–5 years [10].

Although electronic vehicle (EV) batteries are enormous in volume, the current EV market is nascent. Fortunately, large batteries are relatively easy to track, collect, and have a clear future as 'second life' applications or have battery metals recovered using formal recycling. It is estimated that about 1.6 billion electric drive vehicles (including hybrid, plug-in, and pure electric) would have been manufactured globally by 2050 [11]. Furthermore, EV battery chemistry is still evolving to meet the growing demand for faster charging, deep discharge capacity, and safety. As a result, the type of EV batteries may vary widely in different regional economies. Historically, many of the HEV manufacturers used nickel-metal hydride (NiMH) type batteries; however, the next generation of hybrids are available with LIB variants.

Bangladesh Socio-Economic Factors

This country has very recently drafted electronic waste legislations [12]. Bangladesh was reported to produce about 126 kilo tonnes of e-waste (0.8 kg per capita) in 2014 [13] and 199 kilo tonnes (1.2 kg per capita) in 2019 [14]. As an emerging economy, the country is increasing infrastructure development for IT and digitalization over the past two decades. People are encouraged to access many of the government services online and embrace digitalization at and outside of work. The country now has about 184 million registered mobile phone subscribers [15] and about 126 million Internet subscribers [16]. There was an initiative to manufacture laptop computers locally but this project failed after two years in 2013 [17]. From 2022, about 19 million mobile handsets have been manufactured domestically [18], though initiatives to export locally produced electronics have been in the planning stage[19]. The country has a practical fourth generation (4G) network and is currently evolving towards the 5G standard. The growing demand for online application-based services (i.e., ride sharing, delivery) and over-the-top media entertainment has increased the demand for smart phones, and laptops as well as affiliated devices like power banks, wireless audio systems, and many other portable devices. In addition, the COVID-19 pandemic transformed many existing services towards online models (e.g., grocery shopping, telemedicine, home grooming) and helped conceive new services like-commerce of countless commodities and online primary education). Furthermore, the introduction of electric mobility is gaining fast popularity in urban regions as global fuel prices increase. Although the category is currently limited to private passenger cars, electric bikes

(E-bikes) and lead acid battery (LAB) powered three-wheelers, and the sector is expected to grow and widen [20].

There is no established lithium-ion battery manufacturer or recycler in Bangladesh. LIB recycling requires implementing advanced technology, making infrastructure investment, and building a legally certifiable supply chain. As a result, growing economies suffer the deprivation of large international investment due to the lack of a significant enough volume of waste to feed an industrial scale recycling plant. The objective of this work is to examine the LIB generation potential of Bangladesh not only to forecast the future waste volumes but also to examine the environmental burden to the country's ecosystem and the latent metal resources that will be lost.

Determination of Sources of Lithium-Ion Batteries in Bangladesh

Considering the socio-economic aspects of Bangladesh four categories of possible LIB waste generators were identified—(i) mobile phones, (ii) portable computers (laptops and tablet), (iii) small portable appliances ('electric shavers and hair clippers' and 'television and digital camera'), and iv) electric vehicles including two-wheelers. The source descriptions are described in Table 1 beside the relevant harmonized system (HS) code. In addition, the import of stand-alone batteries and accumulators was considered. Regarding the electric vehicle segment, pure electric vehicle data (HS 870,380) is intentionally left blank. This segment also includes LAB powered 3-wheelers import statistics. Several other sources were also overlooked, e.g., rechargeable toys, drones, video game consoles, smart watches, wireless audio devices and peripherals (MP3 players, headphones, microphones, voice recorders, etc.), and small medical devices (oximeter, glucometer, etc.) due to the low potential volumes of the resources. Unfortunately, there is minimal statistical data available on the specific category of import and manufacture of these devices. In addition, many of them use a mixed mode of power, e.g., AA/AAA dry cell/other primary batteries. Furthermore, active users of such electronics make up a small percentage of the overall. Also noted that the LIB batteries used in most of them are significantly smaller in size.

Methodology

The estimation of LIB waste was calculated following the procedure of Forti et al. [22]. Because Bangladesh does not record detailed category-wise sales data and has no formal e-waste collection infrastructure, the 'apparent consumption method' descriptor was adopted. The e-waste generated in an evaluation year n is the

Table 1 Considered sources of LIB waste generation [21]

Category	HS code	Description
Cells and Accumulators	850,650	Cells and batteries; primary, lithium
	850,760	Electric accumulators; lithium-ion, including separators
Phones	851,712	Telephones for cellular networks/for other wireless networks, other than line telephone sets with cordless handsets
Portable PC	847,130	Portable automatic data processing machines, not over 10 kg, consisting of at least a central processing unit, keyboard, and display
Handheld devices	852,580	Television cameras, digital cameras, and video camera recorders
	851,010 and 851,020	Hair clippers and shavers, with self-contained electric motors
Automobiles	870,350 and 870,360	Vehicles: with both compression-ignition internal combustion piston engines (diesel or semi-diesel) and electric motors for propulsion, incapable of being charged by plugging to external sources of electric power; with both spark-ignition internal combustion reciprocating piston engines and electric motors for propulsion, capable of being charged by plugging to external sources of electric power
	871,160	Motorcycles (including mopeds) and cycles; fitted with auxiliary motor, with electric motor for propulsion, with or without sidecars

product of sales or placed on market (POM) from the initial year t_0 to any historical year t prior to n, and the discard-based lifetime profile L^p (t, n) for the batch of products sold in year t (see Eqs. (1) and (2))

$$e\text{-}waste\ generated\,(n) = \sum_{t=t_0}^{n} POM\,(t) * L^{(p)}\,(t,n) \tag{1}$$

where

$$POM\,(t) = Domestic\ production\,(t) + Imports\,(t) - Exports\,(t) \tag{2}$$

The domestic production of several electronic and electrical devices (laptop, mobile phones, etc.) by a handful of companies depend on LIB imports to locally assemble the appliances because there are no LIB manufacturers in the country. Accordingly, the POM was calculated based on imports and subtracting exports (if any) statistics in the primary data. For cases other than 'cells, batteries and accumulators' which is represented in kilogram (kg) data, most trade info is reported as number of units in primary data sources. In such cases, the average representative battery mass was assumed based on the available literature summarized in Table 2.

Table 2 Weibull functions for LIB lifetime estimation

Type of cell/ appliance	Weibull factors		Life estimated on	Source
	Shape parameter, α	Scale parameter, β		
Mobile phone	1.52	5.62	Years	[22]
EV all chemistry (2005–2019)	3.10017	7.93695	Years	[26]
LCO (Pouch)	5.22	516.88	Number of Cycles	[24]
LCO (Pouch)	1.5	250	Number of Cycles	[25]
LFP	11.17	926.78	Number of Cycles	[27]
LMO	5.75	144.20	Number of Cycles	

The lifetime profile was calculated using the Weibull probability distribution function [22–25] as shown in Eq. (3).

$$L^{(p)}(t,n) \equiv f(t) = \frac{\alpha}{\beta}\left(\frac{t}{\beta}\right)^{\alpha-1}.e^{-\left(\frac{t}{\beta}\right)^{\alpha}} \quad (3)$$

The probability function *f(t)* is the lifetime distribution function (also known as density function and failure rate indicator), α is the shape parameter, and β is the scale parameter of the distribution profile. The factors can vary widely depending on the chemistry and shape (see Table 2). While such lifetime estimations are more correct when fitted to 'cycle life' data, the actual lifetime of an appliance in terms of years could significantly vary depending on the user's habit or the service conditions. Correspondingly, the current scope of work requires a much more generalized approach. Hence, the Weibull factors for 'year' data reported by Forti [22] for mobile phones were used. It was assumed that among the variety of electrical and electronic appliances the lifetime of the mobile phones was relatively closer to that of the embedded LIB unit. Finally, the historical data of aggregate POM of LIBs was matched with a third-degree polynomial curve and projected out to the year 2041.

Data Sources

The current work utilizes the reported trade data from the United Nations Comtrade database website according to the HS code of commodities using LIB. The data reports by countries all over the world to Bangladesh from the year 2002 to 2021 were used. For trade direction, export, import, re-export, and re-imports (if any) were considered. In comparison, formal import values declared by the Bangladesh Bureau of Statistics (BBS) under the Ministry of Planning, Government of Bangladesh were also used. Import volume of identified commodities was only accessible for the past four consecutive fiscal years. Relevant previous literature, open articles, media coverages, and press releases were also reviewed.

There are several issues with the data itself. Firstly, the sources only report formal imports which were declared to certain categories. There are other sources of undeclared import/input. For example, batteries brought by luggage of international travellers or expatriates, by salvage/scrap vehicles or ships, or intentional or unintentional miscategorised import (i.e., 'informal'). Secondly, there could be discrepancies in the data itself. For example, some transactions were made in different units or quantities, per kg and in some cases, neither the units nor volumes were recorded. This could either be the continuation of transactions of the previous fiscal term or a typographical error. Nonetheless, such entries were ignored, where it was applicable, the volumes were projected to a value from historical reports from the same country. In addition, the presence of data discrepancies for the combined EVs and E-bikes for the FY 2017–18 and FY 2020–21 of the BBS data were noted and that information was not used for calculations. Evidently, the number of imports reported for the EV categories for those two years is erroneous when compared to other sources [28] or to the number of private passenger vehicles and motorcycles actually registered in those years [29]. Finally, it is recognised that the reported export, re-export, and re-import values could have been avoided, but was included anyway to keep the integrity of the data source.

Previous Surveys and Modelling for Bangladesh e-waste Scenarios

In published work, Environment and Social Development Organization-ESDO conducted surveys among importers, retailers, mechanics, and consumers of electronic and electrical equipment (EEE) in the capital city Dhaka during 2009–10 [30]. The study focused on 6 types of electronic waste (television, computer, mobile, CFL bulb, and medical and dental equipment). Some of these contain EoL LIBs, but unfortunately, there was no precise estimation of total LIB waste generated. The authors estimated about 2.7 million tonnes e-waste generated annually in the country, and ship breaking yards accounted for most (2.5 million tonnes). In a later work, it was estimated that 10.02 million tonnes of e-waste per year in 2011–12 [31]. Another report [32] from the same organization emphasised the impact of ship breaking yards as well as the informal trade of e-waste from developed countries into developing ones–making a reference to the UNEP report [33]. Yousuf and Reza [34] estimated the number of computers and mobile phones, following the work of Sinha [35], by average lifetime versus the weighted distribution lifetime for 2010–2012 and reported significant growth rates in both cases, but EoL LIB volumes were not determined. The Centre for Environmental and Resource Management conducted a study [36] on e-waste in Bangladesh reviewing existing literature and surveying primary data from recycling shop owners, dealers, and scrap collectors. Material Flow Analysis method was adopted according to the various United States Environmental Protection Agency (USEPA) models to forecast e-waste generation from eight EEE items (TV, AC, Computer, Mobile phone, IT equipment, CFL Bulb, Fridge, and Electric fan) based on the weighted End of Life quantity. In their study, the authors deferred to the claims of ESDO that the actual amount of wastes generated would be much less.

Average Battery Weight and Lifespan

One of the advantages of LIBs is the customizable size, shape, and charge density–depending on chemistry. Unfortunately, that renders it difficult to precisely estimate the total volume of LIB waste. Historically, button cells of primary lithium battery were available, and secondary rechargeable varieties are available still. For cylindrical batteries, there are several common geometries used individually as well as a building block (series–parallel) for larger batteries. In addition, the type 18,650 (18 mm diameter, 65 mm length) was the most popular among the battery and electronics manufacturers for past several years. The prismatic and pouch type of LIBs are fundamentally similar, except that the prismatic cell is larger and consists metallic (commonly Al) casing compared to the soft polymeric bag of pouch cell. The pouch type LIBs are made for smaller applications and are mostly embedded inside the appliances (i.e., cellular phones).

Table 3 summarizes several types of LIB average mass and average lifespan reported by various sources. Due to increasing demands of appliances, battery lifespan has decreased, and in many cases, the appliance embedded may be discarded with functional battery. Regardless, a conservative approach was taken while selecting average mass and lifespan of different LIB types in this study to keep resemblance to the available literature as well as to minimize the risk of over-estimation. The few categories not considered in this study include ‘Cordless power tools’, ‘PV energy storage (PVS)’, and ‘Emergency power storage (EPS)’ because of less popularity (power tools) and the use of lead acid alternatives (PVS and EPS) in the Bangladesh scenario.

Results and Discussion

According to the primary data sources, the LIB POM for historical years 2002–2021 is displayed in Fig. 1. The BSB data is depicted in the inset. The largest contribution was by the mobile phone batteries until 2017. Portable PC and handheld devices have had meaningful contributions to the EoL battery supply for

Table 3 Average weight and lifespan reported of LIBs used for various appliances

Type of LIB source		Type of cell	Average mass	Average lifespan (years)	Considered in this study Avg. mass (g)
Mobile phone	Feature phone	Pouch	22 g [1] 21.2 g [3] 32 g [4]	6.5 [4] 6.4 [7] 4.8 [37]	20 g
	Smart phone	Pouch	36–45 g [38]	2–3 [39]	40 g
Small-medium IT	Laptop PC	Cylindrical	274 g [1] 563.3 g [3] 325 g [4]	6 [3] 6.5 [4] 6.4 [7] 4.1 [37]	325 g
	Tablet PC	Pouch	50 g [1] 125 g [4]	6.5 [4] 6.4 [7] 2–3 [40]	
	Digital devices	Pouch	21.3 g [3] 14 g [41]	2–3 (~500 cycles) [41]	20 g
	Power banks	Prismatic	140–640 g [42]	2–3 (~500 cycles) [41]	250 g
Handheld devices (<5 kg)	Cordless power tools	Pouch/ cylindrical/ prismatic	546.6 g [3] 635.8 g [4]	5 [3] 3.4 [4]	Not considered
	Portable devices	Cylindrical (type-1865)	44–49.5 g [41]	2–3 (~500 cycles) [41] 6.4 [7]	50 g
Electric vehicles (EV)	4 wheels	Cylindrical/ Prismatic	~ 300 kg [5] 0.415–1.288 kg (per cell) [41]	10 [5] 5–15 [7]	300 kg
	E-bikes	Cylindrical/ Prismatic	2.5–17.4 kg [43]	2–5 (500–2000 cycles) [43]	10 kg
Emergency supply and storage (ESS)	PV energy storage (PVS)	Pouch/ prismatic	Variable	10 [5]	Not considered
	Emergency power storage (EPS)	Prismatic	Variable	5.8 [4] 15–20 [7]	

many years. Beginning in 2017, the EoL electric automobile (EV) batteries had significant volume—which it will increase in the future. The largest volume was cells and accumulators' segment. The first reason is the domestic manufacture of mobile phones and other portable devices might have triggered bulk import of LIBs. In addition, EV batteries reach EoL around the same time which might have accompanied some EV battery as replacement spare parts. However, the BSB data is incomplete or erroneous due to the reasons mentioned earlier.

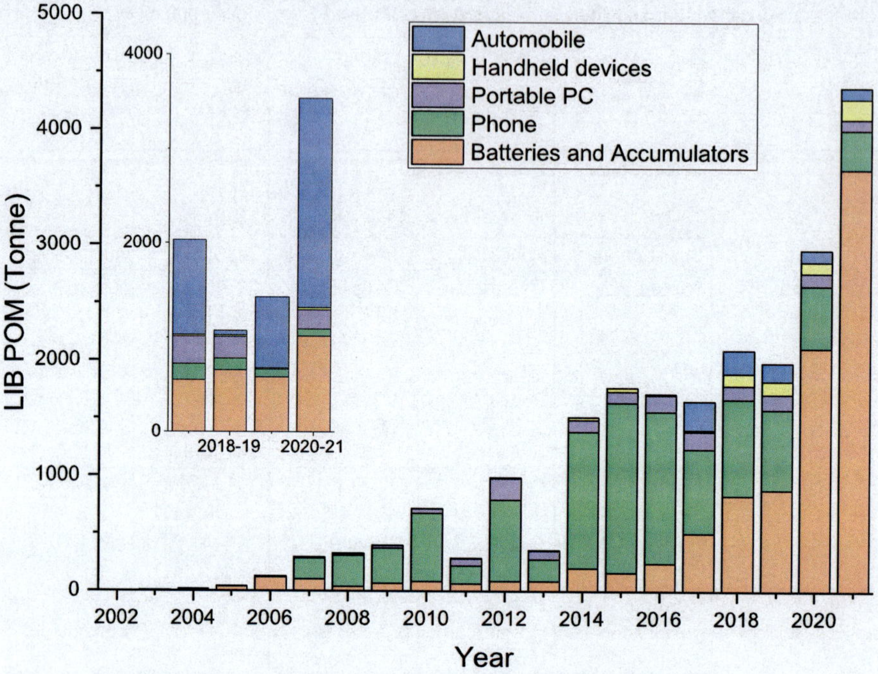

Fig. 1 LIB put on market from Comtrade data and BBS data (inset)

The LIB POM projections are depicted in Fig. 2. It is expected that by 2023 the national LIB POM will reach 5,000 tonnes and will nearly double in the next 5 years. At this rate, the total POM will probably surpass 40,000 tonnes by 2041.

Based on the historical data, the yearly LIB waste generation is shown in Fig. 3. The Weibull density function is given on the inset. The distribution was taken as impactful up to 15 years, considering the variety of types and chemistries of LIBs expected including EVs. Beginning with 2022, about 2,500 tonnes of LIB wastes are generated each year—which is going to grow significantly in the future. The volume of this waste is significant considering zero formal recycling facilities exist for lithium-ion batteries. Accordingly, it is assumed zero battery waste is being dumped into landfills. Commercial LIB plants all around the world have annual capacities (operating and planned) ranging from 100–100,000 metric tonnes [44]. Since several of them operating at present have an average capacity of 3,000–6,000 tonnes per annum, the raw material supply in Bangladesh for formal recycling ventures could be considered comparatively promising.

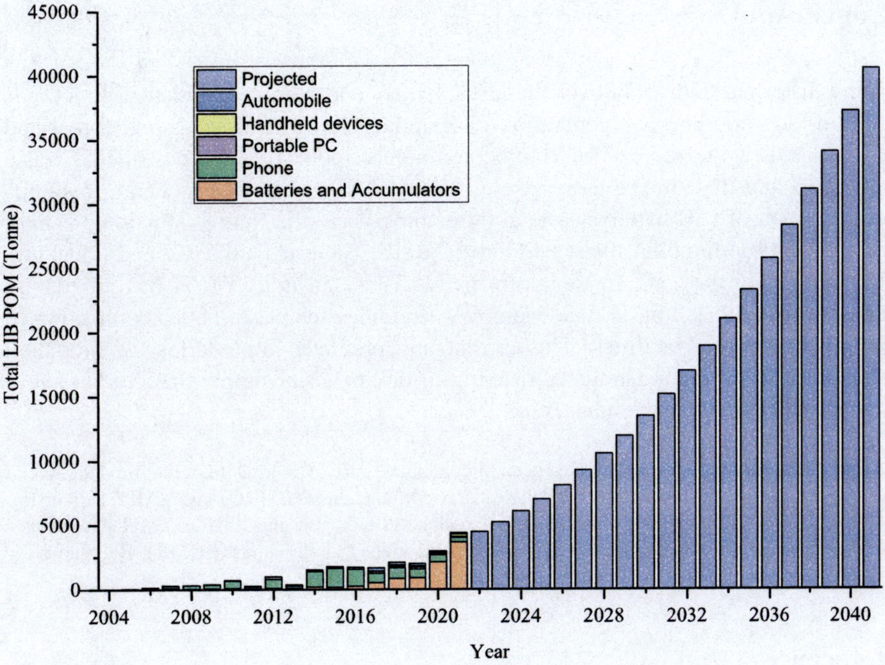

Fig. 2 LIB put on market, based on Comtrade data (projected from year 2022)

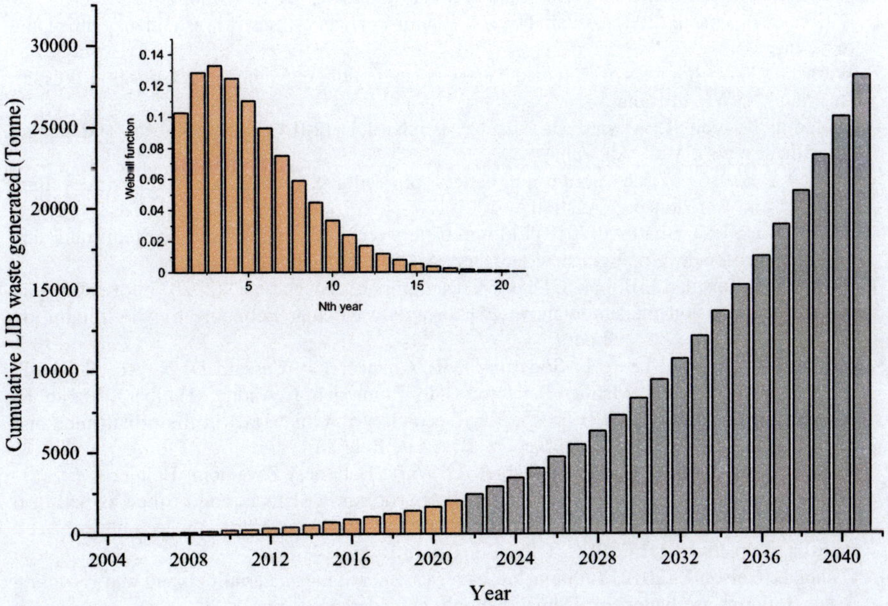

Fig. 3 Estimation of LIB waste generated (forecasted from year 2021) and Weibull distribution (inset)

Conclusion

Using historical trade data over the last 20 years and common statistical functions, lithium-ion battery waste generation in Bangladesh was estimated and forecasted for the next two decades. The volume and a description of the types of LIB waste generated and their future trends were displayed. Bangladesh is generating significant amount of LIB battery waste at this moment and the future. This issue cannot be resolved by dumping the waste into landfills. Batteries are toxic and constitute fire hazards if they are treated informally. These insights could be utilized by decision makers as a basis for developing strategies for waste battery collection to establish recycling facilities. The current and predicted time series for predicted waste generation demonstrates a meaningful raw material supply that could lead to formal battery recycling industry.

Acknowledgements This paper has been prepared as part of a PhD research funded by the Commonwealth Scientific and Industrial Research Organization (CSIRO) and RMIT University, Australia, and facilitated under the trilateral collaboration between CSIRO, RMIT University, Australia, and Bangladesh Council of Scientific and Industrial Research (BCSIR), Bangladesh.

References

1. Heelan J, Gratz E, Zheng Z, Wang Q, Chen M, Apelian D, Wang Y (2016) Current and prospective Li-Ion battery recycling and recovery processes. JOM 68:2632–2638
2. O'Connor P, Wise P (2021) An analysis of lithium-ion battery fires in waste management and recycling. US
3. Warnken (2012) Analysis of lead acid battery consumption, recycling and disposal in Western Australia. NSW, Australia
4. O 'farrell K, Veit R, A 'vard D, Allan P, Perchard D (2014) Trend analysis and market assessment report. VIC, Australia
5. Randell P (2016) Waste lithium-ion battery projections: Lithium-ion forums: recycling, transport and warehousing. Australia
6. King S, Boxall NJ, Bhatt AI (2018) Lithium battery recycling in Australia - current status and opportunities for developing a new industry, A CSIRO Report. Australia
7. Di Persio F, Huisman J, Bobba S, Dias PA, Blengini GA, Blagoeva D (2020) Information gap analysis for decision makers to move EU towards a Circular Economy for the lithium-ion battery value chain. Luxembourg
8. Huisman J, Leroy P, Tertre F, Söderman ML, Chancerel P, Cassard D, N. AL, Wäger P, Kushnir D, Rotter VS, Mählitz P, Herreras L, Emmerich J, Anders, Hallberg, Habib H, Wagner M, Downes S (2017) Prospecting Secondary raw materials in the urban mine and mining wastes (ProSUM) - Final Report. Brussels, Belgium
9. Eucobat Position Paper Collection Target for Waste Batteries. Zaventem, Belgium
10. EUR-Lex (2006) Directive 2006/66/EC of the European Parliament and of the Council of 6 September 2006 on batteries and accumulators and waste batteries and accumulators and repealing Directive 91/157/EEC. Off J Eur Union L 266:1–14
11. Gaines L, Nelson P (2012) Lithium-ion batteries: examining material demand and recycling issues. Lithium use batter. In: Demand supply considerations, pp 27–39
12. Kengo Y (2021) Bangladesh publishes E-waste Management Rule – Enviliance ASIA.

13. Baldé CP, Wang F, Kuehr R, Huisman J (2015) The global e-waste monitor 2014 Quantities, flows and resources. IAS-SCYCLE, United Nations University, Bonn, Germany
14. Forti V, Baldé CP, Kuehr R, Bel G (2020) The global E-waste monitor 2020: quantities, flows and the circular economy potential. United Nations University (UNU)/United Nations Institute for Training and Research (UNITAR) – co-hosted SCYCLE Programme, International Telecommunication Union (ITU) & International Solid Waste Association (ISWA), Bonn/Geneva/Rotterdam
15. BTRC Bangladesh telecommunication regulatory commission-government of the People's Republic of Bangladesh. http://www.btrc.gov.bd/site/page/0ae188ae-146e-465c-8ed8-d76b7947b5dd/-. Accessed 15 Aug 2022
16. BTRC Internet Subscriber, Bangladesh telecommunication regulatory commission-government of the people's Republic of Bangladesh. http://www.btrc.gov.bd/site/page/347df7fe-409f-451e-a415-65b109a207f5/-. Accessed 15 Aug 2022
17. Bhattacharjee J Why Doel Laptop Failed? - Business inspection BD. https://businessinspection.com.bd/why-doel-laptop-failed/. Accessed 30 Aug 2022
18. BTRC Mobile phone handset Info, Bangladesh telecommunication regulatory commission-government of the People's Republic of Bangladesh. http://www.btrc.gov.bd/site/page/cf3588dd-31c9-45b2-b9d5-10f9f2990526/. Accessed 15 Aug 2022
19. Ovi IH Walton to export 'Made in Bangladesh' tagged smartphones to US | Dhaka Tribune. https://archive.dhakatribune.com/business/2020/02/18/walton-to-export-made-in-bangladesh-tagged-smartphones-to-the-us. Accessed 30 Aug 2022
20. UKPACT (2016) Road to a green Bangladesh
21. UN Download trade data | UN Comtrade: international trade statistics. https://comtrade.un.org/data/. Accessed 29 Aug 2022
22. Forti V, Balde C, Kuehr R (2018) E-waste statistics, guidelines on classification, reporting and indicators, Second edi. Bonn, Germany
23. Islam MK, Haque N, Somerville M, Pownceby MI, Bhargava S, Tardio J (2022) Estimation of the generation and value recovery from E-waste printed circuit boards: Bangladesh case study. Miner Met Mater Ser 91–102. https://doi.org/10.1007/978-3-030-92563-5_11/TABLES/3
24. Mouais T, Kittaneh OA, Majid MA (2021) Choosing the best lifetime model for commercial lithium-ion batteries. J Energy Storage 41:102827. https://doi.org/10.1016/j.est.2021.102827
25. Harris SJ, Harris DJ, Li C (2017) Failure statistics for commercial lithium ion batteries: a study of 24 pouch cells. J Power Sources 342:589–597. https://doi.org/10.1016/j.jpowsour.2016.12.083
26. Xu C, Dai Q, Gaines L, Hu M, Tukker A, Steubing B (2020) Future material demand for automotive lithium-based batteries. Commun Mater 1. https://doi.org/10.1038/s43246-020-00095-x
27. Ganjeizadeh F, Tapananon T, Lei H (2014) Predicting reliability of lithium ion batteries. Int J Eng Res Technol 3. https://doi.org/10.17577/IJERTV3IS080846
28. Light castle is Bangladesh ready for the adoption of electric vehicles? - LightCastle partners. https://www.lightcastlebd.com/insights/2021/06/the-current-shape-of-ev-and-hybrid-vehicle-ecosystem-in-bangladesh/. Accessed 12 Aug 2022
29. BRTA Number of registered Vehicles in Whole BD. http://brta.gov.bd/site/page/. Accessed 12 Aug 2022
30. Hossain S, Sulatan S, Shahnaz F et al (2010) E-waste: Bangladesh situation. Dhaka, Bangladesh
31. ESDO (2014) Magnitude of the flow of E-waste in Bangladesh. Dhaka, Bangladesh
32. Hossain S (2011) Illegal import and trade off of e-waste in Bangladesh. Dhaka, Bangladesh
33. Schluep M, Hagelueken C, Kuehr R, Ma Galini F, Maurer C, Meskers C, Mueller E, Wang F (2009) Sustainable innovation and technology transfer industrial sector studies. Recycling – from E- waste to resources

34. Yousuf TB, Reza A (2011) E-waste management in Bangladesh: present trend and future implication. In: World congress of international solid waste association. Korea society of waste management, Daegu, Korea
35. Sinha S, Wankhade K, Sinha-Khetriwal D (2007) Mumbai: choking on e-waste, a study on the status of e-waste in Mumbai. New Delhi, India
36. CERM (2018) Assessment of generation of e-waste, its impacts on environment and resource recovery potential in Bangladesh. Dhaka, Bangladesh
37. Bakker C, Wang F, Huisman J, Den Hollander M (2014) Products that go round: exploring product life extension through design. J Clean Prod 69:10–16. https://doi.org/10.1016/j.jclepro.2014.01.028
38. Mejame PPM, Jung DY, Lee H, Lee DS, Lim SR (2020) Effect of technological developments for smartphone lithium battery on metal-derived resource depletion and toxicity potentials. Resour Conserv Recycl 158:1–19. https://doi.org/10.1016/j.resconrec.2020.104797
39. Cordella M, Alfieri F, Clemm C, Berwald A (2021) Durability of smartphones: a technical analysis of reliability and repairability aspects. J Clean Prod 286:125388. https://doi.org/10.1016/j.jclepro.2020.125388
40. Clemm C, Sinai C, Ferkinghoff C, Dethlefs N, Nissen NF, Lang KD (2017) Durability and cycle frequency of smartphone and tablet lithium-ion batteries in the field. In: 2016 Electronics goes green 2016+, EGG 2016. Fraunhofer IZM, Berlin
41. Diao W, Kulkarni C, Pecht M (2021) Development of an informative lithium-ion battery datasheet. Energies 14:1–19. https://doi.org/10.3390/en14175434
42. Cygnett Care What are the dimensions and weight of this power bank? – Cygnett Care. https://support.cygnett.com/hc/en-us/articles/360023716114-What-are-the-dimensions-and-weight-of-this-power-bank-. Accessed 12 Aug 2022
43. KOK Ebike batteries, electric scooter batteries, lithium ebike batteries. https://www.kokpower.com/electric-bike-scooter-batteries_c2. Accessed 12 Aug 2022
44. Baum ZJ, Bird RE, Yu X, Ma J (2022) Lithium-ion battery recycling—overview of techniques and trends. ACS Energy Lett 7:712–719. https://doi.org/10.1021/acsenergylett.1c02602

Recycling of Spent Lithium-Ion Batteries at Swerim

Xianfeng Hu, Elsayed Mousa, Ludvig Ånnhagen, and Guozhu Ye

Abstract For a smooth transition to a CO_2-neutral society, spent lithium-ion batteries (LIBs) must be recycled. Swerim has made significant advances in this field. This paper will summarize the major projects and highlight their major results. In the pyrometallurgical study, an optimized pyrometallurgical recycling process was developed and demonstrated in a pilot electric arc furnace (8-tonne scale). The developed process demonstrated a high recovery of cobalt and nickel of >98%, and most of the lithium was recovered into a Li-rich dust fraction which could be upgraded into lithium carbonate by carbonated water leaching. In the shredding and thermal treatment study, a homemade simulator was constructed to simulate the combined mechanical and thermal treatment of spent LIBs cells. Studies were conducted to examine the possibility of recovering electrolyte components as well as to understand the mechanisms behind the formation of gases generated during shredding.

Keywords Recycling · Spent lithium-ion batteries · Pilot testing · Simulator · Hazardous gases · Shredding

Introduction

The rapid growth in the market for electric vehicles using lithium-ion batteries (LIBs) in the recent and coming years will lead to a vast accumulative burden of spent LIBs in the future [1–4]. Recycling of spent LIBs and recovery of valuable elements from spent LIBs is imperative due to the following reasons. Firstly, spent LIBs contain toxic and flammable components (such as nickel, cobalt, and electrolytes); improper disposal of the spent LIBs could incur explosion and contamination of the soil and groundwater [1, 5–7]. Secondly, spent LIBs contain valuable elements such as cobalt, nickel, copper, and lithium. On the one hand, the recovery of these elements is critical to sustaining the ever-increasing production and market needs for LIBs [8, 9]. On the

X. Hu · E. Mousa · L. Ånnhagen · G. Ye (✉)
Swerim (Swedish Research Institute for Mining Metallurgy and Materials), Box 812, 97125 Luleå, Sweden
e-mail: guozhu.ye@swerim.se

R. G. Reddy et al. (eds.), *New Directions in Mineral Processing, Extractive Metallurgy, Recycling and Waste Minimization*, The Minerals, Metals & Materials Series,
https://doi.org/10.1007/978-3-031-22765-3_17

other hand, producing these valuable elements from secondary resources consumes less energy and releases less greenhouse gas than that from virgin resources [10].

The recycling processes for spent LIBs can be categorized as hydrometallurgical-dominant processes and pyrometallurgical-dominant processes. Regardless of the applied recycling processes, pretreatment of spent LIBs, including dismantling, crushing, thermal treatment, sieving, etc., is generally needed to remove the toxic electrolytes and concentrate the valuable fractions. In a typical hydrometallurgical-dominant process, the electrode materials obtained after pretreatment of spent LIBs were subjected to acid leaching, solvent extraction, and precipitation to extract Co, Ni, and Mn from electrode materials as corresponding sulfates, which can be further purified and used as the precursors for producing new LIBs [1, 10]. In a typical pyrometallurgical-dominant process, the electrode materials obtained after pretreatment of spent LIBs were subjected to smelting reduction, for example, in a shaft or electric furnace. Oxides of Co, Ni, and Mn in the electrode materials could be reduced to form a metal alloy, which can be used as a master alloy for the production of special steels or be further processed to extract Co, Ni, and Mn for producing new LIBs. Meanwhile, lithium in the electrode materials is often lost in the slag phase. Generally speaking, hydrometallurgical processing might have the advantage of recovering most of the metals including lithium. However, the processes are time intensive and sensitive to changes in the battery chemistry and impurities levels. The pyrometallurgical-dominant processes still have several advantages such as high throughput efficiency, fewer unit operations, and readiness for scaling up [1, 10]. Furthermore, chemistries of LIBs are changing in a rather rapid phase, and cell manufacturers are switching to cheaper cathode materials, such as lithium manganese oxide and lithium iron phosphate, to lower the price of batteries [2]. In such cases, the pyrometallurgical-dominant processes are to be preferred due to their high flexibility in battery types and chemistries [6]. The major disadvantage is the poor or no Li-recovery in the existing pyrometallurgical processes.

Another important issue for a sustainable recycling industry for spent LIBs is the safety and secure working environment during battery production and recycling. In this regard, a shredding simulator was built and installed at Swerim to investigate the possible hazards that might be formed and to understand the gas formation mechanisms under various shredding and thermal treatment conditions.

This paper is a compilation of our major accomplishments (also published elsewhere [11–13]) from two research projects conducted at Swerim: one on developing an optimized pyrometallurgical process to efficiently recover Co, Ni, Mn, and Li from Spent LIBs and another one on the safety and working environment management during crushing/shredding of LIBs.

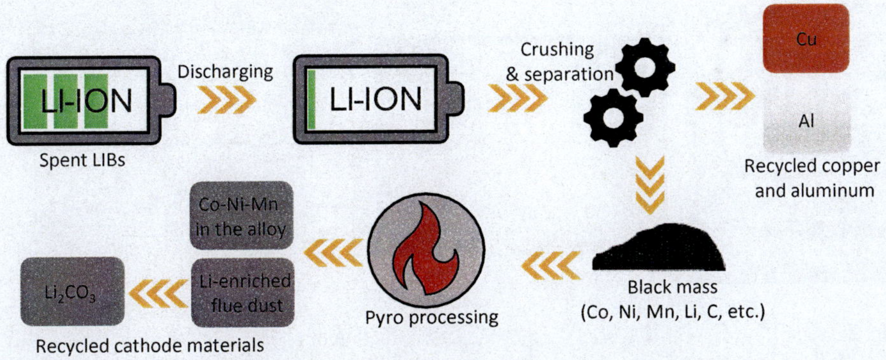

Fig. 1 Illustration of the ReLion concept on the recycling of spent LIBs

Recovery of Valuables from Spent LIBs—The ReLion Concept

The ReLion Concept

The ReLion concept has been developed by detailed thermodynamic considerations. It is indicated that it should be possible to control the process so that Ni-, Co-, and Mn-oxides can be reduced to a metal alloy phase while lithium can be reduced to a gas phase and recovered as a Li-compound in the off-gas, as illustrated in Fig. 1.

Laboratory Tests [11]

To validate the developed ReLion concept, laboratory-scale tests were carried out using a Tamman furnace (Fig. 2a). The set-up of the Tamman furnace and testing procedure is given elsewhere [14]. The furnace has a graphite heating element, which is protected by Ar gas (6 l/min). The testing materials were prepared by well mixing the 'de-coked' black mass with needed carbon for the reduction process without the addition of any slag-forming materials. MgO type crucible (OD: 64 mm; H: 240 mm) with each testing sample of 120 g was placed in the even temperature zone of the furnace. The furnace sample was heated up to 1600 °C at a rate of 15 °C/min and then held at 1600 °C for 180 min. After that, the sample was cooled down to room temperature in the furnace. During the smelting process, the flue dust was collected by a cone-shaped stainless-steel cover placed at the outlet of the furnace. After the experiment, the dust sample was taken from the cover and analyzed by XRD method. The metal and slag samples (if any) were taken from the furnace, manually separated, and, respectively, analyzed by Optical Emission Spectrometry (OES) and XRD methods.

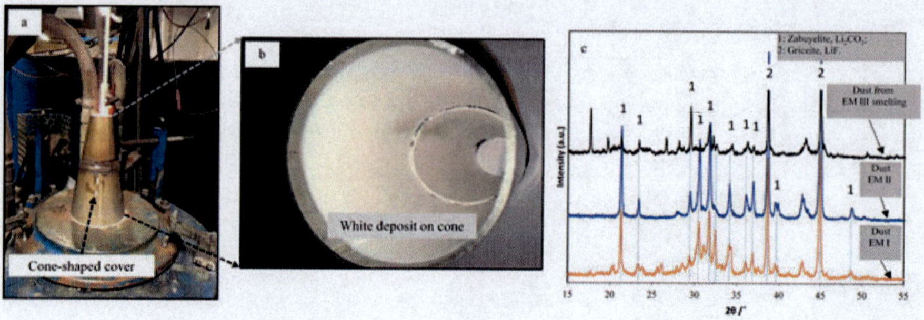

Fig. 2 **a** The Tamman furnace used for the smelting reduction of black mass materials from spent LIBs; **b** the white deposit formed on the cone-shaped stainless-steel cover; **c** XRD analysis of the white deposits [11]

Figure 2b shows the dust collection cover cone at the top of the furnace. When the Li-bearing furnace gas flowed through the cone, the evaporated Li reacted with the exit CO-gas and air and formed Li_2CO_3 which was evenly deposited as a white layer on the inner side of the cone. XRD analysis of the collected white powder (Fig. 2c) shows that it mainly consists of Li_2CO_3 and LiF for all three different battery materials tested. In one of the tests, there was no slag formed meaning that all Li could be recovered to the dust fraction. Co- and Ni-recovery yield was as high as 100%.

Trials in Pilot Scale [12]

Based on the promising laboratory testing results, a pilot-scale testing campaign was carried out at Swerim in 2018. The pilot testing set-up is shown in Fig. 3. Three major types of battery wastes were tested and processed to demonstrate the ReLion concept. These include production waste of (i) cathode material only (no graphite, Type A), (ii) mixture of cathode and anode material (Type B), and iii) battery cells (Type C). Cells were charged during smelting, whereas the dust fraction was added via injection and −4 mm fraction via top feeding as shown in the figure.

Two pilot smelting trials, namely Trial I and Trial II, were conducted consecutively in two days with one trial heat each day in a pilot EAF (with a capacity of 10 tonnes and 5 MVA). The LIBs materials tested in the Trials I and II are, respectively, 5.3 and 4.3 tonnes.

The testing procedure is shortly described as follows:

(i) A start charge was prepared and melted in the EAF aiming to start a metal bath of around 4 tons. The recovered metals from battery materials should be reported there. A certain amount of $CaO-Al_2O_3$ slag was also added in the start charge with the purpose to absorb the formed Al_2O_3 as Al in the battery waste will be oxidized and needed to be treated. The other start-charge materials

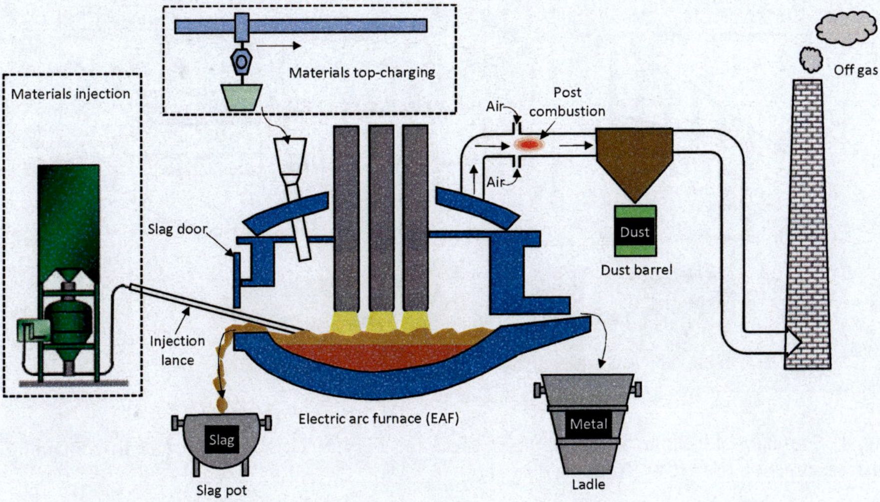

Fig. 3 The set-up for the pilot testing trials [12]

include about 300 kg of A + 4 mm material in Trial I and 560 kg of spent LIBs materials (C materials) in Trial II.

(ii) Top charging of the LIBs materials (A −4 mm and B −4 mm fractions). The top charging of the materials was implemented by the overhead crane and the top charging bin equipped at the EAF (referring to Fig. 3). The -4 mm fraction materials were mixed with the anthracite and slag formers. The charging speed of the material mix was controlled at 1000–1300 kg per hour.

(iii) Injection of the LIBs materials (A dust and B dust). Due to the fineness of the A dust and B dust, they were charged into the EAF through a submerged injection lance (referring to Fig. 3). The injection rate was controlled at around 45 kg per minute, and the operating temperature was controlled at > 1500 °C. During the injection, the anthracite (only applied for A dust), lime, and dolomite were loaded in the furnace by top charging.

(iv) EAF dust collection. During the entire EAF smelting, the off-gas was exhausted with a flow rate of around 6000 m^3/h. The off-gas was subjected to post-combustion at the off-gas outlet of the EAF and subsequentially cleaned by a bag filter, where the EAF dust was collected in the dust barrel.

(v) Tapping of slag and metal. The tapping of the slag was carried out during and at the end of the trials, while tapping of the metal was carried out only at the end of the trials.

The tapped slag and metal as well as collected dust samples were weighted and analyzed. Based on that, an overall heat and mass can be made. The recovery yield of Co, Ni, and Mn could be calculated, and Li distribution in the dust and slag could also be calculated.

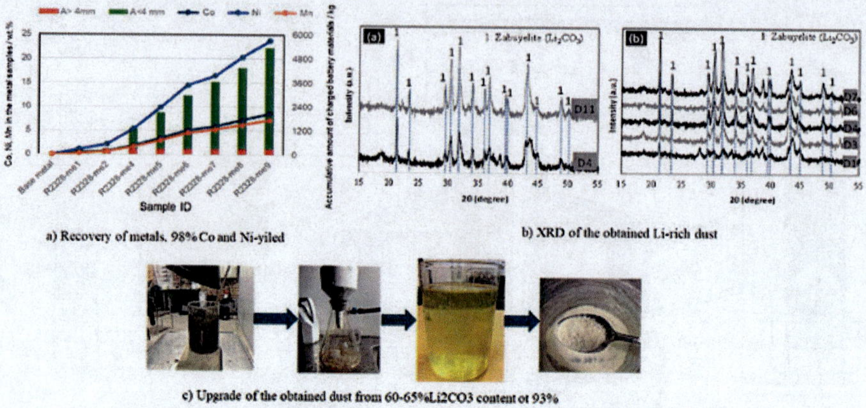

Fig. 4 Summary of the pilot testing results: **a** Metal recovery; **b** XRD analysis of the Li-containing dust; **c** extraction of Li_2CO_3 from the dust [12]

The main testing results achieved are summarized below and shown in Fig. 4.

- A metal recovery of Co and Ni of 98%, and around 94% for Mn
- About 75% of Li was recovered as Li_2CO_3 in the off-gas dust containing about 60–65% Li_2CO_3
- The obtained Li-rich dust was further upgraded to Li_2CO_3 (purity > 93%) by carbonated water leaching and crystallization

Summary of ReLion Testing Results

The pilot testing results validated of the ReLion concept. It is also demonstrated in the pilot scale that most of the Li in the spent LIBs or black mass can be recovered without having to compromise the high recovery yield of Co and Ni. To further increase the Li-recovery, the Al content needs to be reduced to a minimum level as it will easily end up in the slag. Some of Li-oxide will dissolve in the slag. A possible approach for LIBs recycling based on the ReLion testing results is illustrated in Fig. 5.

LIBs Shredding Simulator at Swerim [13]

Lithium-ion batteries (LIBS) contain volatile and reactive chemicals possibly generating toxic and/or flammable gases during the related recycling. The studies on gas formation during the recycling of spent LIBS are scarce. In order to understand what happens during shredding of spent LIBs or pyrolysis of LIBs black mass, a reactor for simulating shredding conditions and for pretreatment of LIBS cells or

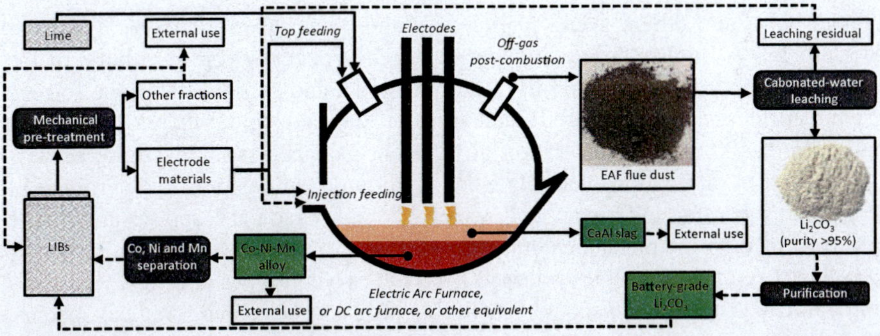

Fig. 5 A pyrometallurgical approach (based on the results from ReLion project) for efficient recovery of valuable elements from LIBs [12]

black mass is installed at Swerim in 2020. During the trials, a multi-component mass spectrometer is normally employed to investigate the formation of gas species (especially those that are flammable and toxic) during combined mechanical and thermal treatments of the spent LIBS cells. This makes it possible to qualitatively and quantitatively identify the gas species formed during mechanical and thermal treatments and thereby providing important knowledges needed for risk management and for a non-hazardous working environment. The set-up of the simulator is shown in Figure 6.

In the followings, one of the studies will be used to illustrate how the simulator can be used for various purposes. In the study, two types of spent LIBS cells were subjected to combined mechanical and thermal treatments at the constant temperatures of 20 °C, 120 °C, 200 °C, and 400 °C under a nitrogen atmosphere. A total of 46 gaseous species, including electrolyte components, oxygenated hydrocarbons,

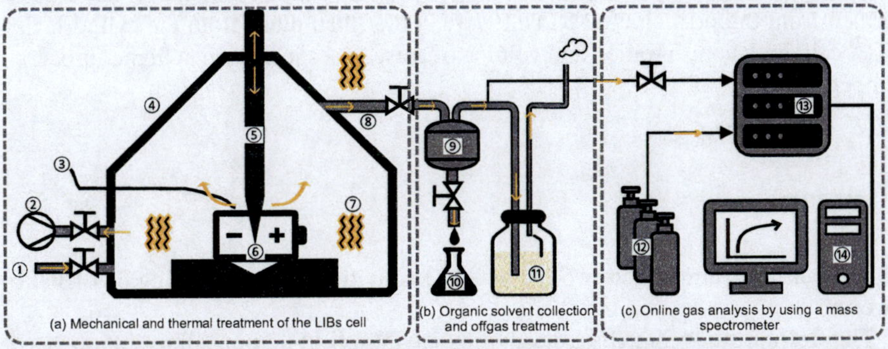

Fig. 6 Experimental set-up used in this study: (a) mechanical and thermal treatments system; (b) the electrolyte components condensation system; (c) the in-situ gas analysis system. ① nitrogen inlet; ② vacuum pump; ③ thermocouple; ④ chamber; ⑤ stainless-steel nail; ⑥ LIBS cell; ⑦ heating elements; ⑧ thermal gas duct; ⑨ ice-water-cooled capsule; ⑩ electrolyte components; ⑪NaOH solution; ⑫calibration gases; ⑬ mass spectrometer; ⑭computer [13]

hydrocarbons, and others, were qualitatively and quantitatively analyzed by mass spectrometry. At higher process temperatures, the concentration or volume of the formed gases increased accordingly. Additionally, at and below 120 °C, the formed gaseous species slightly differed depending on the cell type, whereas they were analogous at 400 °C. The formation of different gas species involved the activity of electrolyte volatilization, electrolyte degradation/decomposition, and pyrolysis of the organic separator and binder, followed by complex radical reactions among the species formed by the physicochemical reactions.

Table 1 lists the gas species released from the two cell types at different process temperatures. A total of 46 species, including 43 gas species and 3 radical species, were identified and divided into four groups: electrolyte-component species (α1–α5), oxygenated hydrocarbon (OCH) species (β1–β6), hydrocarbon (HC) species (γ1–γ18), and other gas species (δ1–δ17). Among the identified gas species, the release of several gas species (such as DMC, EC, CH_2O, CH_4, C_6H_6, C_8H_8, and CO_2) was independent from cell type and process temperature. This indicates that some gas species could already be present in the cells before the mechanical and/or thermally treatments. As an example, DMC and EC could be the components of the electrolyte; CH_4 and CO_2 could have formed inside the cells during their cyclic charging-discharging process [15–18]. The difference in the gas species released from the two cell types was notable at T $\leq$120 °C. As an example, at 20 °C, C_2H_2 and HCN were found in the prismatic-NMC cell but absent in the cylindrical-NCA cell. Furthermore, the number of gas species formed generally increased at higher process temperatures, and some gas species were only detected at high temperatures, e.g., C9–C16 (γ11–γ18) at 400 °C. The presence of free radicals indicated the residual reactivity of the gas that did not reach a stable state.

Figure 7 shows the formation of $(CN)_2$ and POF_3 in the two cell types. The formed $(CN)_2$ in the gas reached 30 ppm and 26 ppm, corresponding to 0.56 ml and 0.90 ml from the cylindrical-NCA and prismatic-NMC cells in the 60-minute measurement time, respectively, at 400 °C. The concentration and volume of the $(CN)_2$ and POF_3 formed from the prismatic-NMC cell were higher than those from the cylindrical-NCA cell; both concentration and volume of these gases increased at higher process temperatures.

Summary

This paper has summarized two major R&D activities on the recycling of LIBs at Swerim.

The ReLion study aimed for developing an optimized pyrometallurgical process with a focus on Li-recovery. It has demonstrated in a large pilot scale that it is possible to recover most of the Li from the spent batteries/black mass without compromising a high yield of recovery of Co, Ni, and Mn, by using pyrometallurgical approaches. It has also shown that the best way to further improve the Li-recovery degree is to

Table 1 Species detected by mass spectrometry during the combined mechanical and thermal treatments of the cylindrical-NCA and prismatic-NMC spent LIBs cells at different process temperatures in the nitrogen atmosphere [13]

Cell type	Temperature / °C	ID and chemical formula of the identified species in the gas																						
		α1	α2	α3	α4	α5	β1	β2	β3	β4	β5	β6	γ1	γ2	γ3	γ4	γ5	γ6	γ7	γ8	γ9	γ10	γ11	γ12
		$C_3H_6O_3$ (DMC)	$C_3H_4O_3$ (EC)	$C_4H_8O_3$ (EMC)	$C_5H_{10}O_3$ (DEC)	$C_3H_2O_3$ (VC)	CH_2O	•CHO	•CH_3O_3	C_2H_4O	C_3H_4O	C_6H_5OH	CH_4	C_2H_2	C_2H_4	C_2H_6	C_3H_6	C_4H_6	C_6H_6	C_7H_8	C_8H_8	C_8H_{10}	C_9H_{10}	$C_{10}H_8$
A	20	√	√	×	×	×	√	√	√	×	×	×	√	×	×	√	×	×	√	×	√	×	×	×
B		√	√	√	√	×	√	√	√	×	×	×	√	√	×	√	×	×	√	×	√	×	×	×
A	120	√	√	×	×	√	√	√	√	×	×	×	√	√	×	√	×	×	√	×	√	×	×	×
B		√	√	√	√	√	√	√	√	√	√	×	√	√	√	√	√	×	√	√	√	√	×	×
A	200	/	√	√	√	√	√	√	√	√	√	×	√	√	√	/	×	×	√	√	√	√	×	×
B		√	√	√	√	√	√	√	√	√	√	×	√	√	√	√	√	√	√	√	√	√	×	×
A	400	√	√	√	√	√	√	√	√	√	√	√	√	/	√	/	√	√	√	√	√	√	√	√
B		√	√	√	√	√	√	√	√	√	√	√	√	√	√	/	√	√	√	√	√	√	√	√

Cell type	Temperature / °C	ID and chemical formula of the identified gas species																						
		γ13	γ14	γ15	γ16	γ17	γ18	δ1	δ2	δ3	δ4	δ5	δ6	δ7	δ8	δ9	δ10	δ11	δ12	δ13	δ14	δ15	δ16	δ17
		$C_{11}H_{10}$	$C_{12}H_{10}$	$C_{13}H_{10}$	$C_{14}H_{10}$	$C_{16}H_{26}$	$C_{16}H_{34}$	HCl	HBr	•CH_2F	CH_3F	CHF	CF_4	COF_2	PF_5	POF_3	SiF_4	$(CN)_2$	HCN	NO	NH_3	O_2	H_2O	CO_2
A	20	×	×	×	×	×	×	×	×	×	×	×	×	×	×	×	×	×	×	×	×	√	×	√
B		×	×	×	×	×	×	×	×	×	×	×	×	×	×	√	√	×	√	×	×	√	×	√
A	120	×	×	×	×	×	×	×	√	×	×	×	×	×	×	√	√	√	√	×	×	√	×	√
B		×	×	×	×	×	×	×	√	×	√	×	√	×	×	√	√	√	√	√	×	/	×	√
A	200	×	×	×	×	×	×	×	√	×	√	×	√	×	×	√	√	√	√	×	×	/	√	√
B		×	×	×	×	×	×	√	√	×	√	×	√	√		√	√	√	√	×	×	/	√	√
A	400	√	√	×	×	×	×	√	/	√	√	√	√	√	√	√	√	√	√	√	√	/	√	√
B		√	√	√	√	√	√	√	√	√	√	√	√	√	√	√	√	√	√	√	√	/	√	/

The three radical species are labeled with a dot (˙), α1–α5: electrolyte-component gas species; β1–β6: oxygenated hydrocarbon; γ1–γ18: hydrocarbon gas species; δ1–δ17: other gas species. '√' and ' × ' indicate the presence and absence of the species, respectively; '/' indicates no data available. A and B refer to the cylindrical-NCA and prismatic-NMC cells, respectively.

minimize the slag amount. For this reason, a feed material with a lower Al content is essential.

Safety and working environment are essential for sustainable battery production and recycling processes. The initial studies using the shredding simulator at Swerim have shown that hazardous gases like cyanides and F-bearing gases like HF and POF_3 can be formed during mechanical and thermal treatment. It has also shown that the organic constituents of the electrolyte used in the LIBs can be recovered during mechanic and thermal treatment.

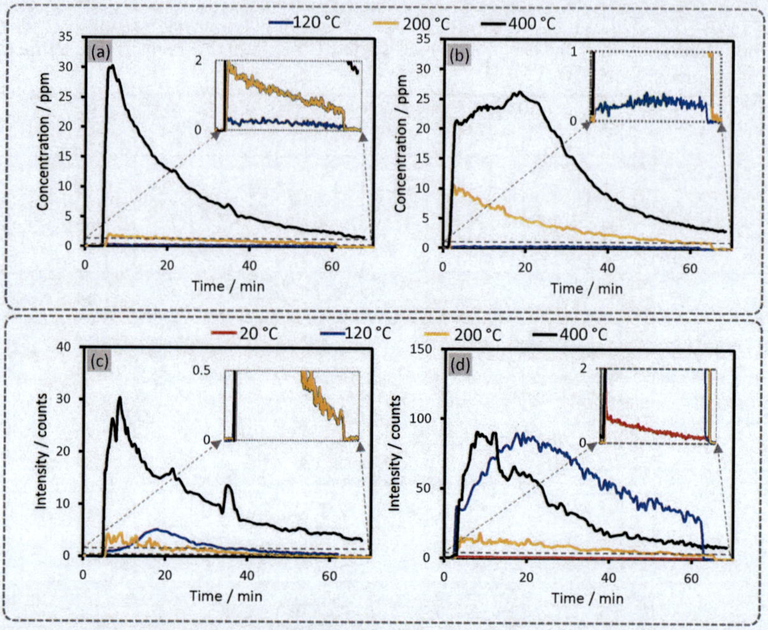

Fig. 7 Quantitative identification of $(CN)_2$ and qualitative detection of POF_3 formed during the combined mechanical and thermal treatments of the cylindrical-NCA and prismatic-NMC cells at different process temperatures in the nitrogen atmosphere. **a** and **b** show $(CN)_2$ formed from the cylindrical-NCA cells and prismatic-NMC cells, respectively; **c** and **d** show POF_3 formed from the cylindrical-NCA and prismatic-NMC cells, respectively [13]

Acknowledgements The studies presented in this paper were supported by the 'Batterifondsprogrammet' of the Swedish Energy Agency (Energimyndigheten) for the ReLion project (dnr: 2016–006027) and the SCOPE-LIBs project (Projectnr: 48208-1). Preparation of this manuscript was supported by CAMM—Centre of Advanced Mining and Metallurgy at the Luleå University of Technology, Sweden.

References

1. Harper G, Sommerville R, Kendrick E, Driscoll L, Slater P, Stolkin R, Walton A, Christensen P, Heidrich O, Lambert S, Abbott A, Ryder K, Gaines L, Anderson P (2019) Recycling lithium-ion batteries from electric vehicles. Nature 575:75–86. https://doi.org/10.1038/s41586-019-1682-5
2. Sonoc A, Jeswiet J, Soo VK (2015) Opportunities to improve recycling of automotive lithium ion batteries. Procedia CIRP 29:752–757. https://doi.org/10.1016/j.procir.2015.02.039
3. Liu C, Lin J, Cao H, Zhang Y, Sun Z (2019) Recycling of spent lithium-ion batteries in view of lithium recovery: a critical review. J Clean Prod 228:801–813. https://doi.org/10.1016/j.jclepro.2019.04.304
4. Huang B, Pan Z, Su X, An L (2018) Recycling of lithium-ion batteries: recent advances and perspectives. J Power Sources 399:274–286. https://doi.org/10.1016/j.jpowsour.2018.07.116

5. Zheng X, Zhu Z, Lin X, Zhang Y, He Y, Cao H, Sun Z (2018) A mini-review on metal recycling from spent lithium ion batteries. Engineering 4:361–370. https://doi.org/10.1016/j.eng.2018.05.018
6. Knights BDH, Saloojee F (2015) Lithium battery recycling-keeping the future fully charged. Department of environmental affairs, greenfund and development Bank of South Africa, Midrand
7. Lebedeva NP, Di Persio F, Kosmidou T, Dams D, Pfrang A, Kersys A, Boon-Brett L (2019) Amount of free liquid electrolyte in commercial large format prismatic li-ion battery cells. J Electrochem Soc 166:A779–A786. https://doi.org/10.1149/2.1151904jes
8. Zeng X, Li J, Singh N (2014) Recycling of spent lithium-ion battery: a critical review. Crit Rev Environ Sci Technol 10:1129–1165. https://doi.org/10.1080/10643389.2013.763578
9. Pagliaro M, Meneguzzo F (2019) Lithium battery reusing and recycling: a circular economy insight. Heliyon 5(6):e01866. https://doi.org/10.1016/j.heliyon.2019.e01866
10. Zhang X, Li L, Fan E, Xue Q, Bian Y, Wu F, Chen R (2018) Toward sustainable and systematic recycling of spent rechargeable batteries. Chem Soc Rev 47:7239–7302. https://doi.org/10.1039/C8CS00297E
11. Hu X, Mousa E, Tian Y, Ye G (2021) Recovery of Co, Ni, Mn, and Li from Li-ion batteries by smelting reduction - part I: a laboratory-scale study. J Power Sources 483:228936. https://doi.org/10.1016/j.jpowsour.2020.228936
12. Hu X, Mousa E, Ye G (2021) Recovery of Co, Ni, Mn, and Li from Li-ion batteries by smelting reduction - part II: a pilot-scale demonstration. J Power Sources 483:229089. https://doi.org/10.1016/j.jpowsour.2020.229089
13. Hu X, Mousa E, Ånnhagen L, Musavi Z, Alemrajabi M, Hall B, Ye G (2022) Complex gas formation during combined mechanical and thermal treatments of spent lithium-ion-battery cells. J Hazard Mater 431:128541. https://doi.org/10.1016/j.jhazmat.2022.128541
14. Lindvall M, Rutqvist E, Ye G, Björkvall J, Sichen D (2021) Possibility of selective oxidation of vanadium from iron and phosphorus in Fe-V-P Melt. Steel Res Int 81:105–111. https://doi.org/10.1002/srin.200900123
15. Kumai K, Miyashiro H, Kobayashi Y, Takei K, Ishikawa R (1999) Gas generation mechanism due to electrolyte decomposition in commercial lithium-ion cell. J Power Sources 81–82:715–719. https://doi.org/10.1016/S0378-7753(98)00234-1
16. Zhang SS (2014) Insight into the gassing problem of Li-ion battery. Front Energy Res 2:59. https://doi.org/10.3389/fenrg.2014.00059
17. Metzger M, Strehle B, Solchenbach S, Gasteiger HA (2016) Origin of H2 evolution in LIBs: H2O reduction vs. Electrolyte Oxid J Electrochem Soc 163:A798–A809. https://doi.org/10.1149/2.1151605jes
18. Galushkin NE, Yazvinskaya NN, Galushkin DN (2019) Mechanism of gases generation during lithium-ion batteries cycling. J Electrochem Soc 166:A897–A908. https://doi.org/10.1149/2.0041906jes

Comparison of Hydrogen Reduction of Different Lead-Bearing Materials for Lead Recovery

A. Rukini, M. A. Rhamdhani, G. A. Brooks, and A. Van den Bulck

Abstract Understanding the fundamental knowledge of lead reduction from lead-bearing materials using hydrogen is essential for developing a carbon neutral lead production process. Recycling of lead from secondary resources using hydrogen as a reducing agent is industrially important to promote circular economy, resource efficiency, and reduction of carbon footprint. The current study evaluates and compares the general reduction mechanism of PbO with a more complex lead-containing compound, PbO-SiO_2 glass/slag. Isothermal reductions were carried out on pelletized PbO and PbO-SiO_2 (70wt% PbO and 30wt% SiO_2) using hydrogen at 350 to 800 °C (for PbO) and 300 to 700 °C (for PbO-SiO_2). The results from microstructural observation showed that uniform, globular, non-wetting lead droplets formed on the surface of PbO pellets. The lead droplets covered the whole surface, which appeared to inhibit further reduction. In the case of PbO-SiO_2, the viscous nature of the glass/slag appeared to significantly hinder diffusion of hydrogen within the pellet, resulting in a low reduction rate.

Keywords PbO · Lead silicate · Hydrogen reduction · Reduction mechanism · Decarbonization

A. Rukini (✉) · M. A. Rhamdhani · G. A. Brooks
Fluid and Process Dynamics (FPD) Group, Department of Mechanical and Product Design Engineering, Swinburne University of Technology, Melbourne, VIC 3122, Australia
e-mail: wrukini@swin.edu.au

M. A. Rhamdhani
e-mail: arhamdhani@swin.edu.au

A. Van den Bulck
Umicore Group Research and Development, B-2250 Olen, Belgium

R. G. Reddy et al. (eds.), *New Directions in Mineral Processing, Extractive Metallurgy, Recycling and Waste Minimization*, The Minerals, Metals & Materials Series,
https://doi.org/10.1007/978-3-031-22765-3_18

Introduction

Current global lead production dominantly relies on secondary resources which include three major sources: spent lead-acid batteries; degraded metallic lead (alloy); and by-products from metal industries such as dross and slags. It is worth to note that some metal plants (e.g. copper smelting plant) generate slag with substantial amount of lead in which economically attractive to be recycled. The recovery of lead from complex slag (typically contains more than three types of oxide compounds) is commonly carried out through reduction by coke or carbon-based reductant at high temperature. Following the reduction, lead is then separated from the molten slag. A number of studies on lead-rich slag reduction using carbon-based reductant have been published [1–5]. Amidst the decarbonization effort globally, hydrogen utilization is one of the most sought alternatives to carbon. Fundamental understanding on the reduction of lead oxide and its multicomponent slag system by hydrogen is essential for optimizing lead extraction and recovery in industrial practice. However, no existing work in an open literature was found on the reduction of complex lead bearing slag using hydrogen gas.

Slags from industrial metal productions are quite complex and consist of various components that are mainly the result of reactions during the process. An approach in studying the reduction from the simplest system of PbO to an increasingly complex system will help in understanding of the broader behavior of the industrial system. The current study is part of broader study that intends to investigate the mechanism and the kinetics of the reduction of industrial lead-bearing oxides step by step from the simplest compound of PbO to a more complex slag system to get a clear understanding of Pb recovery. From the perspective of thermodynamics, based on the Gibbs free energy formation of PbO and hydrogen oxide on Ellingham diagram, PbO is thermodynamically reducible by molecular, atomic, and plasma hydrogen gas. There are few studies reported on how PbO behave under hydrogen atmosphere in which the previous studies encompassed an observation of kinetics behavior of different PbO crystal structure [6] and kinetics of a packed bed of PbO powder under hydrogen atmosphere [7]. There are still research gaps that need to be answered such as the effect of water vapor on the reduction behavior, the reduction of bulk PbO-containing solids, and how the liquid Pb affect the solid-state reduction. No hydrogen reduction studies were found in the open literature on more complex lead-bearing slags with more than two components except for lead silicate ($PbO.SiO_2$) glass. Hydrogen treatment is one of the common steps for microchannel plate production; thus, hydrogen reduction on a lead silicate glass type has been relatively explored. However, these studies mainly focused on its semiconductor properties that are essential for microchannel plate applications [8, 9]. The current work aims to study from the point of view of metal extraction from the Pb-containing resources in support of decarbonization effort in non-ferrous industries. In the current work, reduction of PbO and $PbO\text{-}SiO_2$ pellets under hydrogen containing atmosphere was carried out at temperatures from 350 °C to 800 °C with a focus on the microstructure evolution during the process.

Experimental Procedures

Pellet Making

The starting material for the PbO reduction was a high purity PbO powder (>99.9%) sourced from Sigma Aldrich. During pellet making, 5 g of PbO was pressed into a pellet in a steel mould, resulted in a pellet with an average diameter of 29.5 mm and thickness of 4.2 mm. The PbO powder was hydraulically pressed under 10 tons of weight for 30 min. A lead silicate slag/glass composition selected for the study was 70wt% PbO—30wt% SiO_2. The SiO_2 powder was sourced from Sigma Aldrich (with purity of 99% +). The mixture was mixed for 24 h in a ball mill before melted twice in an alumina crucible at 800 °C for 30 min. The cast slag then crushed and sieved into 45 μm powder before pelletized. Approximately 2.5 g of $PbO.SiO_2$ powder was pressed in a hydraulic pressing machine under 5 tons weight for 20 min, which resulted in a pellet with the same dimension as PbO pellets.

Hydrogen Reduction Set-Up

In this study, solid-state PbO and $PbO.SiO_2$ reductions in a mixture of 15% hydrogen-85% nitrogen (H_2/N_2) atmosphere were carried out. An isothermal reduction was performed in a vertical tube furnace (as shown in Fig. 1) at temperatures of 350 °C to 800 °C. The furnace was preheated to the designated temperature prior to inserting pellet into hot zone. Temperature in the hot zone was calibrated with a standard reference thermocouple, and it was found that the temperature measured in hot zone has ±1 °C accuracy. Argon gas was first flowed for 5 min into the furnace in order to flush out any air inside the furnace chamber. Then, the H_2/N_2 gas mixture was flowed into the furnace chamber with constant gas flowrate of 500 mL/min. The pedestal, which supported a crucible containing the pellet, was then positioned in the hot zone to let the reaction occur (with reaction times from 30 min to 4 h). In anticipation of having liquid Pb metal produced during the reduction, the crucible holder was equipped with nickel chromium wire net as shown in Fig. 1, to let any possible lead liquid to drip down through it. For stopping the reaction, the furnace chamber was flushed again with Ar gas to let the hydrogen mixture gas flushed out of the system. The sample was then lowered into a water cooling zone at the bottom of the furnace and cooled down for 30 min to minimize direct re-oxidation when exposed to air.

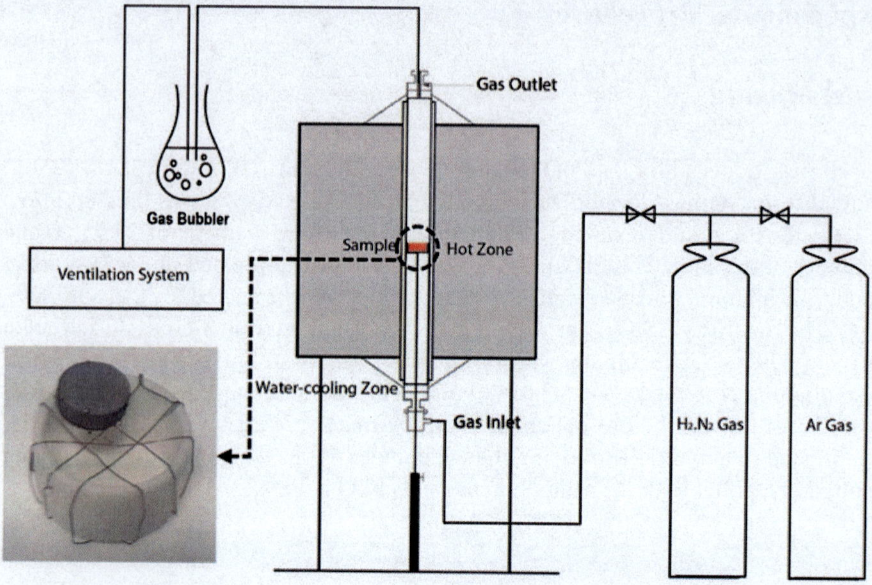

Fig. 1 A schematic of experimental apparatus consisting of a vertical resistance tube furnace; with sample holder, gas, and ventilation systems; (inset) a crucible holder system equipped consisting a shallow alumina crucible with Ni–Cr wire net to hold the pellet sample and also functioning as a drain for liquid Pb

Results and Discussion

A summary of comparison between hydrogen reduction of PbO and $PbO.SiO_2$ pellets is presented in Table 1. In general, samples from both experiments experienced a physical alteration such as color change during the early stage of reactions. In the case of the PbO reduction, lead droplets were observed visually at temperatures 500 °C to 800 °C. In the case of $PbO.SiO_2$ reduction, softening and significant change in the shape of the samples were observed, especially at high temperatures (beyond the softening point). The chemical reaction and mechanism appeared to be different between the PbO and $PbO.SiO_2$. In general, the reduction of both PbO and $PbO.SiO_2$ appeared to be controlled by the diffusion process although the hindering factor for diffusion was different for each case. In the case of reduction of PbO, a dense liquid Pb product covered the whole surface and this appeared to hindered diffusion of hydrogen to allow further reduction. In the case of $PbO.SiO_2$, a viscous melt was formed, and bubble formation was observed. These phenomena appeared to provide a hindrance for hydrogen reduction; hence, the reaction was also halted.

Table 1 Comparison of PbO and $PbO.SiO_2$ reduction process under hydrogen atmosphere

Key observations	PbO reduction	$PbO.SiO_2$ reduction
Physical alteration	350–500 °C: Color change from reddish yellow to greyish pellet 500–800 °C: Formation of visible lead droplets	300–500 °C: Color change of white pellet into brownish and black pellet 600–700 °C: Softening and drooping of pellets
Chemical reaction	$PbO + H_2 \rightarrow Pb + H_2O$	$(PbO) + H_2 \rightarrow Pb + H_2O$
Reduction Mechanism	Diffusion controlled process by dense metal product	Diffusion controlled process by viscous silicate glass
Misc. Phenomena	Involving a polymorphic transition of PbO and grey oxide formation At temperature range 700–800 °C, a heavily liquid lead was formed	Viscous silicate bubble formation was observed Reduction reaction occurred involving PbO embedded

Macro- and Micro-Observation

Physical observations at macroscopic and microscopic level were carried out on the reduced PbO and $PbO.SiO_2$ samples. As mentioned in Table 1, there was a color change for both samples due to reduction process at the early stage of the reaction. There was also shape alteration observed at later stage of reduction, especially on the lead silicate samples which was not the effect of hydrogen reduction but highly related to glass softening.

Figure 2 shows the macrograph of the surface and cross section of PbO pellets reduction by hydrogen at different temperatures, taken at 3 h of reaction. It can be seen from Fig. 2 that a reddish yellow PbO pellet turned into greyish color as the reduction proceed. Few Pb droplets were observed at 400 °C and 500 °C while significant numbers of Pb droplets were observed at 600 °C and 700 °C. At these high temperatures, the lead droplets covered all the surface and that some evidence of droplets coalescence could be observed. Phase analyses have been carried out by the authors in other work [12], indicated that the greyish color was a result of partial reduction of PbO which generated a mixture of Pb with unreduced PbO known as 'grey PbO' or 'leady oxide'. Meanwhile, a slight color change from reddish yellow of PbO pellet into bright yellow observed on the cross section of the pellet was due to the polymorphic transition of α-PbO to β-PbO that has been observed in previous studies to occur in the temperature range 489 to 530 °C [6, 10, 11]. The cross section of PbO samples showed a characteristic of a diffusion-controlled process where reduction was not fully occurred at the center of the pellet especially for samples that were uniformly covered by lead droplets. A comprehensive kinetics study has been carried out by the authors [12] which demonstrated that the hydrogen reduction of PbO pellet was diffusion-controlled.

Selected SEM images showing the surface morphology of PbO pellets after reduction at higher magnification are given in Fig. 3. Fine Pb seeds with average size of 0.5 μm were observed on sample reduced at 400 °C (Fig. 3a). These Pb seeds grew

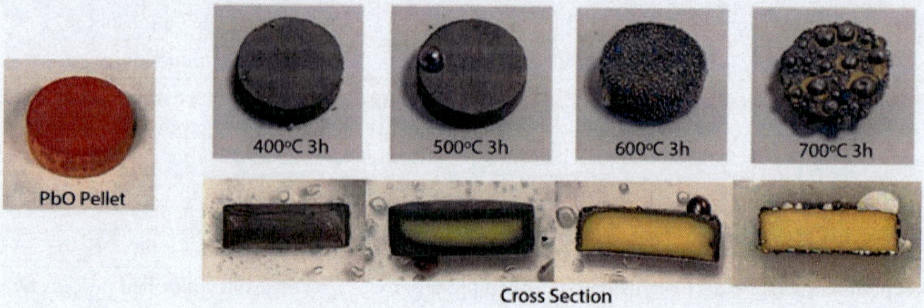

Fig. 2 Macrograph of PbO pellets reduced by hydrogen for 3 h at different temperatures

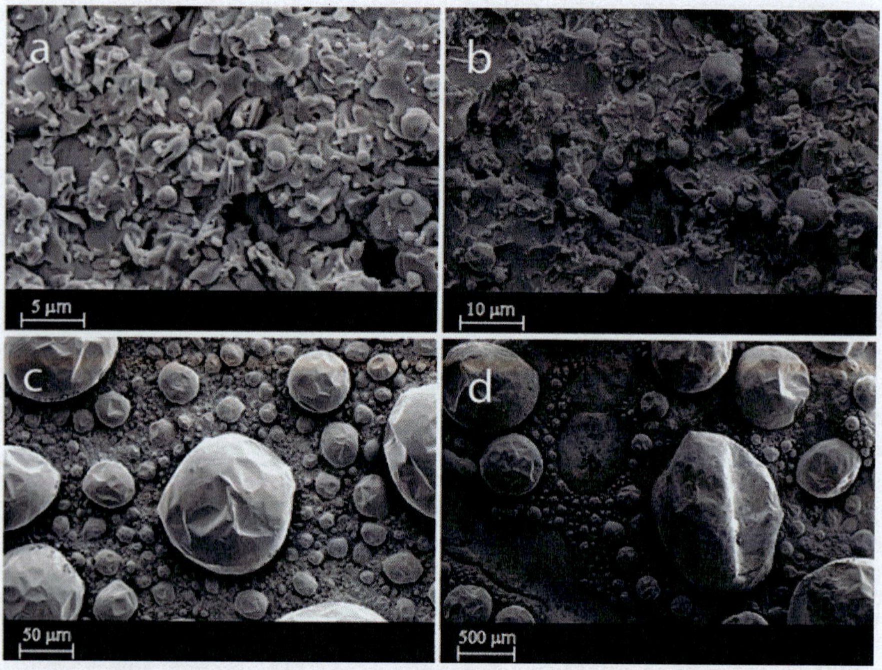

Fig. 3 SEM observation of PbO pellet's surface reduced in 15%H_2-N_2 for 3 h at: **a** 400 °C, **b** 500 °C, **c** 600 °C, and **d** 700 °C

bigger in diameter as temperature increase and eventually form globular lead droplets (Fig. 3c and 3d). It can be seen from Fig. 3c and 3d that lead droplets were uniformly formed on the surface of the pellets, which from the cross section observation (shiny metallic layer in Fig. 2) it formed a seclusion layer on the pellets. This seclusion was the main factor that hinder further reduction of pellet as it provided barrier for hydrogen diffusion. The globular shape of lead droplets was a sign of non-wetting of liquid Pb on PbO. It can also be seen from Fig. 3 that some droplets underwent coalescence to form bigger droplets and eventually dripping down by gravity force

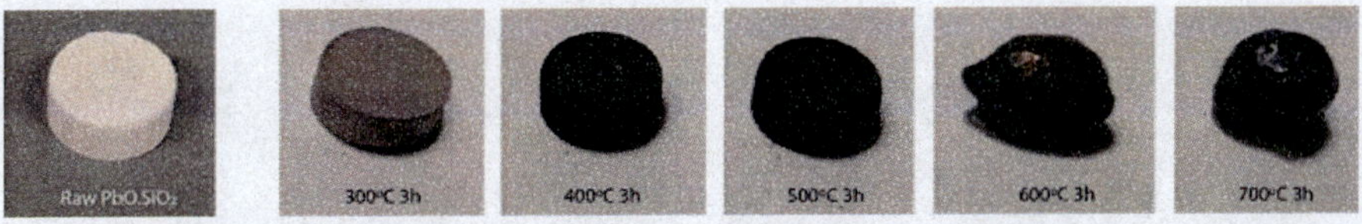

Fig. 4 Macrograph of $PbO.SiO_2$ pellets reduced by hydrogen for 3 h at different temperatures

from the pellets to the crucible underneath. It can also be seen from Fig. 3d that the coalescence of the droplets created a new bare PbO surface which allowed the hydrogen to meet and react with PbO. Therefore, it appeared that the overall reduction at high temperatures will be affected by the rate of coalescence and dripping of the droplets as they directly affected the availability of PbO surface for further reaction.

Figure 4 shows the macrograph of surface of $PbO.SiO_2$ pellets reduced in hydrogen for 3 h at different temperatures. It can be seen from Fig. 4 that a white $PbO.SiO_2$ pellet gradually turned into black as the reduction proceed, followed by shape alteration at high temperatures. The color changing from white to brownish and then blackish happened due to the formation of Pb metal in glass structure, in which a similar observation was also reported by Blodgett [8]. Shape alteration on the lead silicate pellets occurred at high temperatures was due to glass softening nature of the materials. Initially, some part of the pellet started to become viscous, and this resulted in that part to be drooping down into the crucible. This softening however, not ensued the viscous part to be separated from the main pellet body. As the softening occurred more uniformly within the body, the whole pellet started to form a more rounded shape but bloated. This might indicate that the highly viscous melt has a high surface tension and it tended to shape into a single big droplet to reduce its overall surface energy.

Another phenomenon observed in the case of reduction of $PbO.SiO_2$ pellet was that many bubbles were observed to form in the interior of the pellet which contributed to its bloated shape. Figure 5 (bottom) shows the appearance of the cross section of the bloated samples, showing the presence of bubbles of different size. The coalesced big bubbles on the top of the sample created a dome shape and appeared to be stuck inside the sample and could not break up the thin top layer of the sample. Figure 5 (top) shows the SEM images of microstructure of the cross section as well as the inside surface of the bubble/dome on the top. At higher magnification (Fig. 5 top left), micro-lead droplets were observed under the bubbly dome, near the sample's top surface. At the center of the samples (Fig. 5 top right), there appears to be unreduced melt. This might indicate that for samples reduced above softening point, the reduction mainly happened near the surface of samples and further reduction in the interior/center of the sample was halted due to the presence of these bubbles. When softening phenomena happened and silicate network become viscous, formation of bubbles has been reported in glass making literature which is attributed to the presence of impurities and entrapment of air. It may be that a similar mechanism occurred in the current sample. Some of the gas (15%H_2 and N_2) was initially entrapped in the

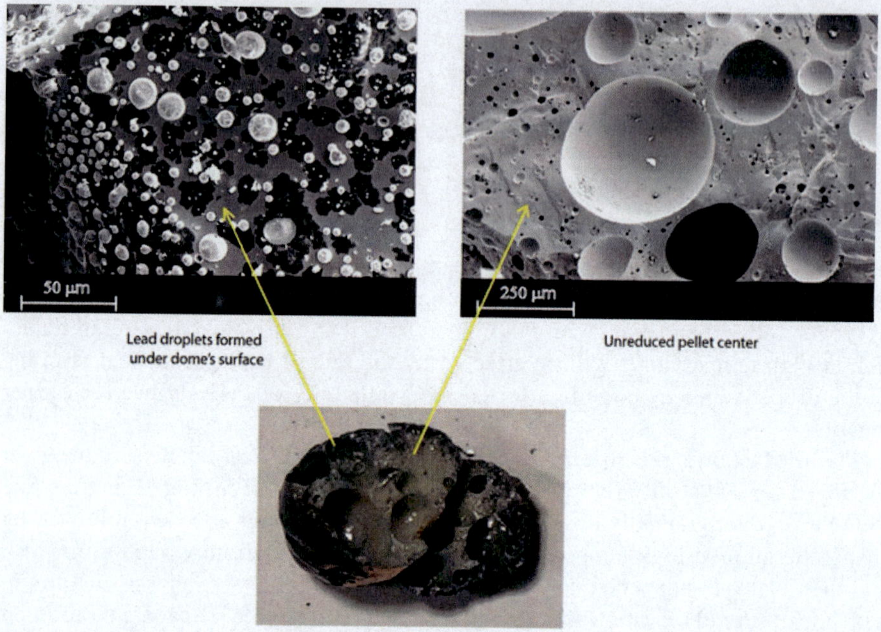

Fig. 5 Detailed cross section observation under SEM of $PbO.SiO_2$ sample reduced at 700 °C for 2 h

sample. At the beginning, this entrapped gas still allowed for reduction to occur in the interior of the sample. However, once the hydrogen was consumed and H_2O was generated, the entrapped gas/bubble was no longer reducing. Also, the trapped bubbles provided a barrier for hydrogen to diffuse into the center of the sample and resulted in unreduced $PbO.SiO_2$ in the center.

General Chemical Reaction and Mechanism

Lead reduction from PbO follows the reaction shown in Eq. 1, where lead monoxide is reduced into lead (solid or liquid) and also results in water vapor. There is no intermediate product expected as reported by previous study. It was previously postulated that PbO reduction resulted in a Pb_2O intermediate. However, this was confirmed later to be just a mixture of Pb metal with unreduced PbO [7] which is known as 'grey PbO' or 'leady oxide' [13].

$$PbO_{(s)} + H_{2(g)} \rightarrow Pb_{(s,l)} + H_2O_{(g)} \tag{1}$$

$$PbO_{(s,l)} + H_{2(g)} \rightarrow Pb_{(s,l)} + H_2O_{(g)} \tag{2}$$

Equation 2 shows a generic reaction of $PbO.SiO_2$ reduction by hydrogen and in this case (PbO) represents the PbO in the solution slag/glass. The presence of PbO in the glass/slag is characterized based on the bonding it makes with silicate tetrahedra. The role and effect of PbO on the silicate structure depend on their concentration in the melt. It has been reported by Kannunikova et al. [14] and Kacem et al. [15] that PbO acts as network modifier of silicate network. The current $PbO.SiO_2$ mixture contained 70wt% PbO which is equal to ~40 mol% PbO. It was expected that the PbO would act as network modifier in the sample. The reduction in $PbO.SiO_2$ would involve a bonding alteration between PbO and silicate network and might follow a sequence presented in Eq. 3.

$$(Pb_{modifier} - O) \rightarrow (Pb - O)_{ion} \rightarrow Pb^0 \tag{3}$$

The generic reduction mechanism that happened to PbO pellet during reduction is presented in Fig. 6. PbO samples reduced at lower temperature involved a grey lead oxide formation. The existence of unreduced PbO in the grey lead oxide gave the pellet a 'diffusible' structure that allows the hydrogen to further reduce the PbO in the interior of the pellet. Meanwhile, PbO samples that were reduced at higher temperatures experienced a rapid reduction on the surface which created dense lead droplets covering the surface which eventually hindered the diffusion of hydrogen. Further reduction would depend on the rate of droplets coalescence and rate of lead dripping which can create a fresh PbO surface.

Figure 7 shows a generic mechanism of $PbO.SiO_2$ reduction by hydrogen. There are two distinct reduction based on the lead silicate glass softening point. Samples that were reduced below the softening point undergone a hydrogen reduction from the surface through into the center of pellet. The reduction resulted in the Pb formation within glass which eventually turned the raw white pellet into black pellet. In this case, no change in shape was observed. The samples that were reduced above the glass softening point undergone a softening phenomenon which resulted in a shape change and formation of viscous dome structure with many bubbles entrapped inside the sample. This viscous silicate glass trapped initially trapped the gas within the furnace chamber as well as the generated H_2O from the reaction in the interior of the sample.

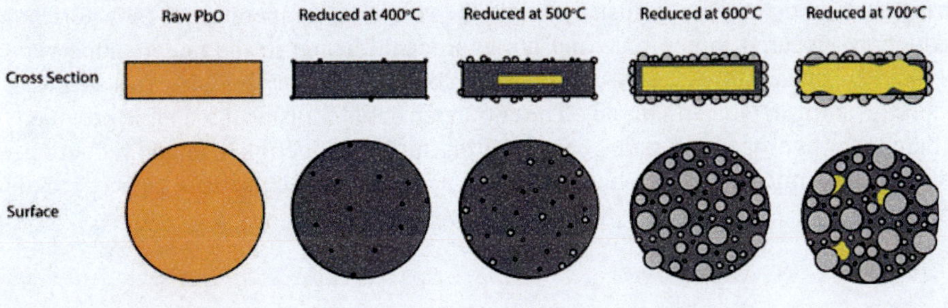

Fig. 6 An illustration of the cross section and surface appearance of PbO pellet after reduced with hydrogen

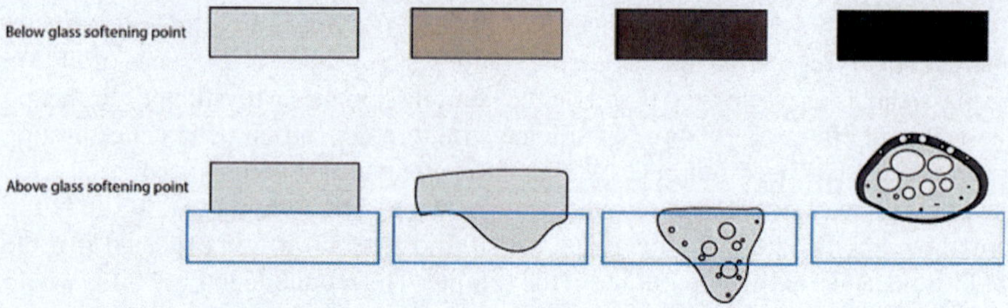

Fig. 7 An illustration of $PbO.SiO_2$ pellet reduction mechanism by hydrogen

The accumulation of bubbles within softened glassy sample provided a buoyancy effect. Meanwhile, a hydrogen reduction happened on the top surface of viscous sample that resulting on a blackish color. The reduction process was supported by the presence of lead micro-droplets under the blackish dome in the interior as shown in Fig. 5. The bloated structure, however, hindered hydrogen diffusion that causing the center of the sample unreduced.

Conclusions

A study of reduction of PbO and $PbO.SiO_2$ pellets by hydrogen at temperatures from 350 °C to 800 °C has been carried out. The focus of the study was on the observation of macrostructure and microstructure evolution during the reduction process. In both cases, the reduction appeared to be limited by diffusion process. In the case of experiment with PbO pellets, at low temperatures (< 500 °C) the microstructure evolution allowed the hydrogen to diffuse through to the interior of the sample, hence further reduction could occur. At high temperatures (> 500 °C), a formation of dense lead droplets covering the whole surface was observed. This provided a barrier for hydrogen diffusion and appeared to significantly affect the reduction process. Further reduction would be dependent on the rate of lead droplets coalescence and dripping which provides fresh PbO surface. The reduction of $PbO.SiO_2$ pellet also involved two distinct mechanisms. Below the softening temperature (~600 °C), the reduction appeared to continue and no significant change in the pellet shape was observed. Above 600 °C, a viscous sample containing many bubbles was formed, hence resulting in bloated samples. The entrapped bubbles inside the sample provided a barrier for hydrogen diffusion, hence halting further reduction. Limited reduction, evidenced by micro-lead droplets formation, appeared to occur on the top surface of the sample.

Acknowledgements This work is part of a PhD project of Mrs Asywendi Rukini funded under Joint Umicore-Swinburne SUPRA (Swinburne University Postgraduate Research Award) scholarship.

Conflict of Interest The authors declare that they have no conflict of interest.

References

1. Kinaev NN, Jak E, Hayes P (2005) Scand J Metal 34:150–157
2. Zhao B, Errington B, Jak E, Hayes P (2008) Gaseous reduction of ISASMELT lead slag and lead blast furnace sinters. In: The Southern African institute of mining and metallurgy. Lead and Zinc, pp 133–146
3. Hou X, Chou KC, Zhao B (2013) J Min Metall Sect B-Metall 49(2):201–206
4. Wang C, Li K, Yang H, Li C (2017) ISIJ Int 57(6):996–1003
5. Vanparys R, Brooks G, Rhamdhani MA, Crivits T (2020) Reduction of lead-rich slags with coke in the lead blast furnace. In: PbZn 2020: 9th international symposium on lead and zinc processing, pp 173–185
6. Ivanov II, Shelmet VM, Ulyanov VV, Teplyakov YA (2015) Kinet Catal 56:304–307
7. Culver RV, Matthew IG, Spooner ERC (1962) Aust J Chem 15:40–55
8. Blodgett KB (1951) J Am Ceram Soc 34:14–27
9. Huang Y, Zhang Y, Liu H, Gu Z (2011) XPS study on microporous surface composition of microchannel plates. In: Proceedings of SPIE 8194, international symposium on photoelectronic detection and imaging 2011: advances in imaging detectors and applications, 81941Q (18 August 2011). https://doi.org/10.1117/12.900283
10. Oka K, Unoki H, Sakudo T (1979) J Cryst Growth 47:568–572
11. Risold D, Nagata JI, Suzuki RO (1998) J Ph Equilib 19(3):213–233
12. Rukini A, Rhamdhani MA, Brooks GA, Van den Bulck A (2022) submitted to Metall Mater Trans B
13. Pavlov D (2017) Leady oxide, a handbook of lead-acid battery technology and its influence on the product. In: Lead-acid batteries: science and technology, 2nd edn, pp 245–273
14. Kannunikova OM, Gilmutdinov FZ, Shakov AA (2002) Int J Hydrogen Energy 27:783–791
15. Kacem IB, Gautron L, Coillot D, Neuville DR (2017) Chem Geol 461:104–114

Selective Separation and Recovery of Titanium from Titanium Alloy Grinding Scrap Via the Electrostatic Separation-Oxidative Roasting-Acid Leaching Process

Xianglin Cheng, Shangfeng Xu, Zijian Su, and Yuanbo Zhang

Abstract Titanium alloy grinding scraps (TAGS) have a production of 30 thousand tons annually. They contain large titanium resources as by-product from the titanium alloy machining process. TAGS cannot be sent back directly to the vacuum smelting process due to its high-content of impurities such as the grinding wheel ash and titanium oxides. In this study, a process of Electrostatic Separation-Oxidative Roasting-Acid Leaching was applied to separate and recover titanium from TAGS. The results indicate that 76.61% titanium could be recovered during electrostatic separation process under the optimal conditions. The final product of TAGS containing 91.44% titanium dioxide was obtained after optimized physical separation, oxidative roasting followed by hydrochloric acid leaching process. The product meets the requirements for feed raw materials required to prepare the industrial titanium dioxide. It was found that aluminum was the major impurity in the titanium-based product in the form of corundum.

Keywords Titanium · Grinding scrap · Electrostatic separation · Selective leaching

Introduction

Titanium alloys are widely used at the cutting edge of national defense and chemical industry due to material properties such as high strength, superior biocompatibility, and chemical corrosion resistance. In China, the output of titanium alloy reached only about 136 thousand tons in 2021. The reason for limitations on its large-scale applications is the high production costs due to high scraping rates during casting and difficulties machining the metal. The loss during the casting process was in the range of 40–60%, [1] which creates 140 thousand tons of the titanium alloy scrap annually.

X. Cheng · S. Xu · Z. Su · Y. Zhang (✉)
School of Minerals Processing & Bioengineering, Central South University, Changsha 410083, Hunan, China
e-mail: sintering@csu.edu.cn

R. G. Reddy et al. (eds.), *New Directions in Mineral Processing, Extractive Metallurgy, Recycling and Waste Minimization*, The Minerals, Metals & Materials Series,
https://doi.org/10.1007/978-3-031-22765-3_19

The grinding process uses the grinding wheel to remove the surface defects and oxide layers of the titanium alloy, which was a type of machining, which was distinguished from other types of machining which changed the shape of the alloy. Titanium alloy grinding scraps (TAGS) with annual output more than 30 thousand tons were a typical by-product from the titanium alloy grinding process.

Currently, superior titanium alloy scraps are sent back to the vacuum smelting after simple cleaning and grading, whereas inferior titanium alloy scraps are utilized as an additive in steel industry [2]. Distinguished from other scraps, TAGS were regarded as solid waste and disposed as the garbage due to its high-content impurities such as the grinding wheel ash and oxide layer. However, TAGS contains a significant titanium resource and plenty of grinding wheel ash which is considered as a high-quality secondary resource.

Researchers focus on the development of the molten salt electrolysis of the direct reduction of high-quality titanium alloy scraps [3], as well as the influence of the grinding wheel on the surface of titanium alloy materials [4]. However, little research is conducted on the grinding process of titanium alloys and there are few reports on the reutilization of the generated TAGS.

The discrepancy in magnetism and density among the TAGS components was not obvious, but huge in electrical properties and acid resistance according to the research of the characteristics of grinding scraps. Based on the study of the recovery of copper from the wasted printed wiring boards [5] and the high temperature oxidation characteristics of titanium alloy [6], a process route was proposed.

In this study, representative TAGS was served as raw materials, and the process of Electrostatic Separation-Oxidative Roasting-Hydrochloric Acid Leaching was applied to separate and recover titanium, which provides new ideas for the resource recovery and utilization of the TAGS.

Experimental

Materials

TAGS used in this study was derived from Baoji, Shaanxi Province of China. After sieving, TAGS accounted for 19% in the particle size range of 400–850 microns, about 36% in the 150–400 microns, 95–150 microns accounted for 29%, and less than 95 microns accounted for 16%. The ICP-AES assay showed that the content of Ti, Al, Fe, Ca, Mg, and Si was 18.36%, 14.63%, 8.83%, 6.04%, 5.03%, and 3.12%, respectively. The electrostatic separator comes from Changsha research institute of mining and metallurgy (model: YD 32,190-23 N), which is mainly used for the sorting of rutile and zircon placer minerals. The processing capacity of the electrostatic separator is 3–5 tons per hour, and the power of the machine is 30 kilowatts. The hydrochloric acid was analytical reagent (AR) grade. The analytical assay solutions were prepared using ultrapure water for the experiments.

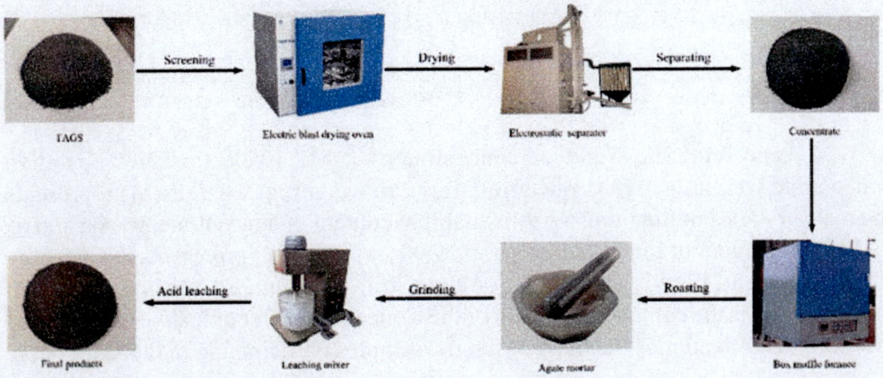

Fig. 1 TAGS experimental procedure

Methods

Experimental Procedure

The TAGS should be processed through a series of operation in order to get the final products. The experimental flowsheet is shown in Fig. 1.

First, the TAGS was screened to suitable size and dried to proper moisture. Then, TAGS was separated by the electrostatic separator. After that, the electrostatic separation concentrate was subjected to oxidative roasting. Finally, the roasted concentrate was grinded and sent to the leaching process.

Characterization

The phase of the sample was analyzed by X-ray diffraction (XRD, D/Max 2500, RIGAKU, Japan) under the conditions of radiation: Cu Kα, tube current and voltage: 250 mA, 40 kV, scanning range: 10–80°/2θ, step size: 0.02°/2θ, and scanning speed: 5°/min. Element content of solid material dissolved in solution and tested by using a plasma emission spectrometer (ICP-AES, Icap7400 Radial, Thermo Fisher Scientific, USA). Topographic analysis was examined by scanning electron microscope (SEM–EDS, JSM-6490LV, JEOL, Japan).

Assessment indexes

Yields and recovery rate of concentrate were used to evaluate the electrostatic separation process. Yield of concentrate was calculated according to Eq. (1). Recovery rate of concentrate was calculated according to Eq. (2). Leaching ratios of Ti were used to evaluate the leaching process. It is calculated according to Eq. (3).

$$Y(M) = (M_1 / M_0) \times 100\% \quad (1)$$

$$R(M) = (\alpha_1/\alpha_2) \times Y(M) \tag{2}$$

$$L(M) = (c \times V)/(\alpha_3 \times m) \times 100\% \tag{3}$$

Y(M) represents the Yield of concentrate, %; M1 is electrostatic separation concentrate weight, g; M0 is raw grinding scrap weight, g; where R(M) represents the recovery rates of titanium, %; α_1 is titanium content of concentrate, wt. %; and α_2 is titanium content of raw grinding scrap, wt.%; where L(M) represents the leaching ratio of element, %; c is the content of element in lixivium, g/L; V is the lixivium volume, L; α_3 is the content of element of the Concentrate after oxidative roasting and grinding, wt.%; and m is the mass of treated samples participating in the reaction, g.

Results and Discussion

Effect of Electrostatic Separation Parameters on the Separation and Recovery of TAGS

Through preliminary experimental exploration, the rough experimental conditions were determined. The following will explore the influence in detail of various electrostatic separation parameters upon the enrichment of titanium from TAGS. The yield, titanium recovery rate, and titanium content of the concentrate have positive or negative correlation, which was all taken as the criteria for evaluating the separation effect.

Figure 2a showed that with the increase of electrostatic separation voltage from 15 to 21 kV, the titanium content increased sharply, the yield decreased gradually, and the ti recovery rate raised slightly. In the case of a certain recovery rate, the higher the concentrated titanium content, the lower the yield. When the electrostatic separation voltage reached 21 kV, the titanium content of the concentrate reached the maximum value of 25.41 wt.% and the yield reached the minimum of 62.1%. When the voltage increased, the yield rose but other metrics decreased significantly instead. This increase in the electrification voltage led to increase in non-conductive substances, which report to decreasing the purity of the concentrate. Therefore, the optimal electrostatic separation voltage was selected as 21 kV.

It could be observed from Fig. 2b that the feed granularity of TAGS had great influence on the electrostatic separation effect. The titanium purity increased as the particle size decreased, but the yield declined. The recovery rate increased slowly while granularity increased from 150 to 400 microns, with the highest recovery rate reaching 98.87%. Under the conditions, titanium content had been improved to 25.28%, which meant that electrostatic separation was most suitable for TAGS in this granularity. Then, the recovery rate decreased rapidly but the titanium purity increased, which reason was large particles of electrically conductive substances

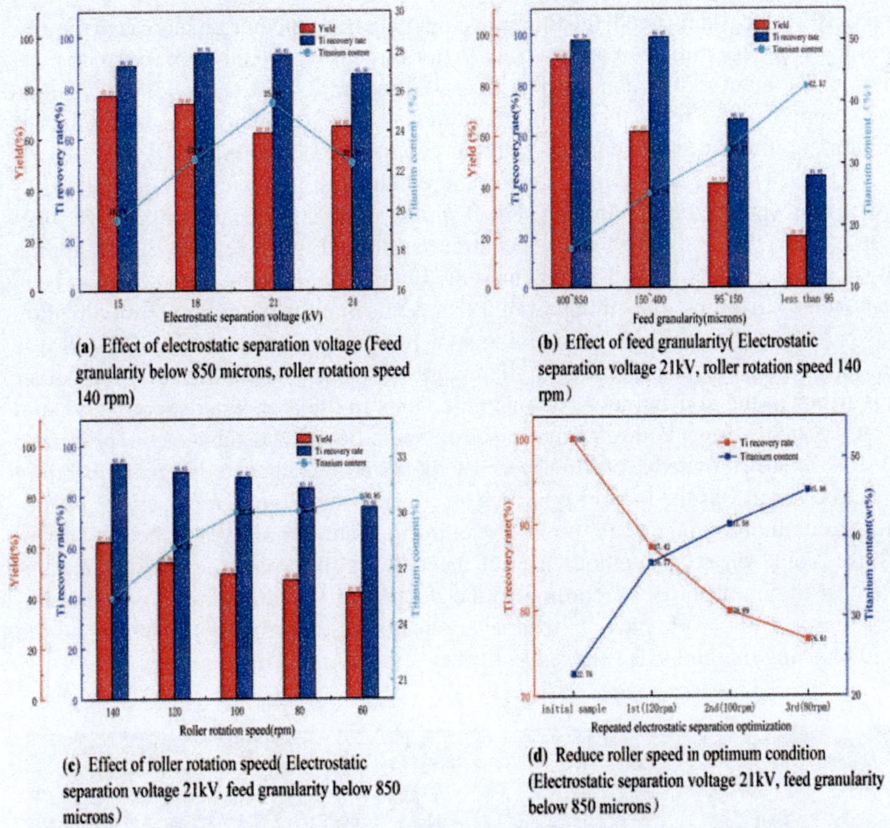

(a) Effect of electrostatic separation voltage (Feed granularity below 850 microns, roller rotation speed 140 rpm)

(b) Effect of feed granularity(Electrostatic separation voltage 21kV, roller rotation speed 140 rpm)

(c) Effect of roller rotation speed(Electrostatic separation voltage 21kV, feed granularity below 850 microns)

(d) Reduce roller speed in optimum condition (Electrostatic separation voltage 21kV, feed granularity below 850 microns)

Fig. 2 Effect of electrostatic separation parameters on enrichment of titanium from TAGS

were selected into concentrate while smaller particles were difficult to be sorted out. Combining the results of titanium purity and yield, granularity was the important selection parameter from process optimization. The optimal feed granularity was controlled between 150 and 400 microns. Considering the original granularity and resource recovery, the experiment feed granularity was under 850 microns.

It is shown in Fig. 2c that the titanium content of the concentrate increased when the roller rotation speed was decreased. Generally speaking, the faster the roller speed, the more the TAGS processed. When the optimal rotation speed is reached, the recovery rate and the titanium content were basically constant which owing to the conductivity of the conductor substances was basically unchanged in the electric field of the electrostatic separation. When the rotation speed dropped to 60 rpm, the recovery rate significantly decreased and the titanium purity did not change significantly. Optimum condition was the rotation speed between 80 and 120 rpm and using the previous concentrate to repeat the electrostatic separation.

Figure 2d showed the repeated electrostatic separation experiment with the above optimal conditions. With the increase of the number of electric selection and the decrease of the roller speed, the titanium content in the concentrate increased rapidly, and the recovery rate decreased slowly. After three times electrical separations, the titanium content of the concentrate increased from 22.78wt.% to 45.86wt.%, and the comprehensive recovery rate of titanium was 76.6%, which had a great effect of separating impurities and enriched titanium elements for TAGS.

Figure 3a showed the typical backscattered electron image of electrostatic separation concentrate. Figure 3b represented spot scan of X-ray energy dispersive spectrometer (EDS) analysis of the concentrate surface. Figure 3c was EDS-mapping of specific elements. Spot 1 represented the titanium-aluminum alloy that had been removed by grinding machining. Spot 2 represents that the titanium-aluminum alloy has been partially oxidized. Combining spot 1 and spot 2, the surface of TAGS was smooth, which was conspicuous in the image. Obviously, there are still many other impurities in the concentrate. The main elements in the area where spot 3 and spot 4 are located were calcium magnesium iron and silicon. The substance represented by this area originated from the wear of the grinding wheel during the grinding process, which was the major impurities on concentrate. Point 6 represented magnesium–aluminum spinel while point 5 was iron-chromium alloy attached to spinel. Conventional physical methods cannot meet the requirements for further enrichment of titanium purity. According to the document [7], titanium alloys and high valence titanium oxide have the characteristics of acid corrosion resistance, so the acid leaching method was selected to further remove impurities.

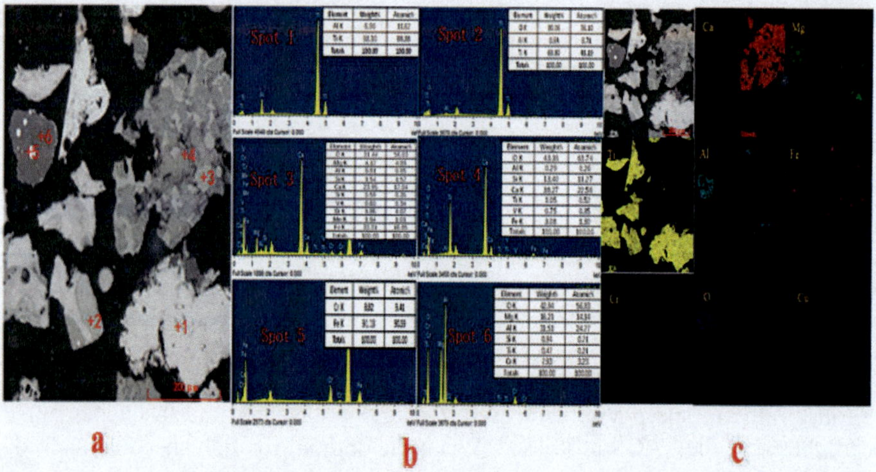

Fig. 3 EDS image of electrostatic separation concentrate

Analysis of the Acid Leaching of Concentrate and Pretreated Concentrate

The pretreatment process before leaching was to roast the concentrate at 950 °C for 4 h under air atmosphere on the box muffle furnace and then use the agate grinder to grind the sample by hand for 30 min. The leaching conditions were as follows: the molar concentration of hydrochloric acid was 6 mol/L, the liquid–solid ratio was 10 mL/g, the leaching temperature was 60 °C, and the time was 4 h.

Figure 4 showed that the main phase composition of the concentrate after oxidative roasting is the rutile and corundum. According to Fig. 3, there were plenty of impurities in the concentrate. The reason why the phase diffraction peaks of other impurity elements could not be found may be that the high temperature oxidized the surface of titanium-aluminum alloy into dense rutile and corundum structures, which caused a volume expansion and encapsulated other impurities. The other reason the main diffraction peak intensity of the rutile and corundum phase was large enough to mask the weak diffraction peaks of other impurities. Acid leaching increased the diffraction peak intensity of the main phase of the concentrate after oxidative roasting.

In order to more intuitively analyzing the leaching process and clearly comparing the leaching effects of concentrate and pretreated concentrate in hydrochloric acid system, the leaching rate of main element after leaching is shown in Table 1. The EDS-mapping of specific elements of the concentrate after acid leaching and the pretreatment concentrate are shown in Fig. 5.

It can be observed from Table 1 that the hydrochloric acid leaching decreased calcium and iron impurities in the concentrate. This also had affected the magnesium–aluminum spinel phase. However, about 24.8% of the titanium element was brought into the solution system when leaching impurities, which removes valuable titanium resources. Combined with Figs. 3c. and 5a, it can be found that part of the surface

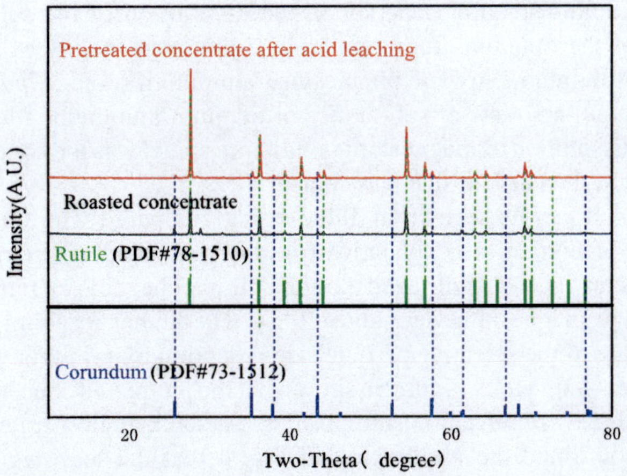

Fig. 4 XRD pattern of roasted concentrate and pretreated concentrate after acid leaching

Table 1 The leaching rate of main element in two samples

Main element	Leaching rate of concentrate (%)	Leaching rate of pretreated concentrate (%)
Titanium	24.86	2.66
Iron	86.44	99.43
Aluminum	31.75	19.41
Magnesium	66.31	98.38
Calcium	99.39	99.87

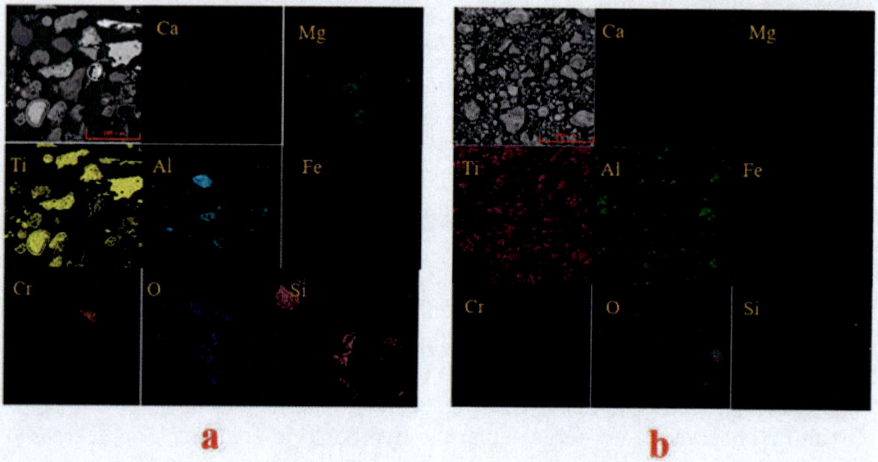

Fig. 5 **a** EDS image of concentrate after acid leaching. **b** EDS image of pretreated concentrate after acid leaching

of the ground titanium alloy had disappeared which indicated that the leaching of titanium mainly depends on the dissolution of the suboxide layer on the surface of the titanium aluminum alloy. Aluminum mainly exists in the form of titanium-aluminum alloy and magnesium–aluminum spinel. Under the acid leaching system, small amounts of sub-oxides of titanium-aluminum alloy dissolve. The incomplete dissolution of magnesium–aluminum spinel was the concentrate had not been ground and the particle size was coarser.

It can be seen from Table 1 that compared with the untreated concentrate, the concentrate after oxidative roasting and grinding pretreatment increased impurity removal and maximized titanium purity. The removal rate of impurities iron, calcium, and magnesium were above 98%. The higher leaching rate of magnesium may be due to the selective particle size reduction by grinding which increased magnesium leaching yield. Figure 5b shows that the pretreated concentrate after acid leaching had already removed most impurities. The bulk agglomerate structure formed in Fig. 5b confirmed the assumption in Fig. 4 that the leaching rate of titanium decreased because the low-valent superficial titanium oxides oxidize to dense rutile due to the high temperature, which lead to superior acid corrosion resistance during leaching. After chemical analysis, it was found that the final products contained 91.44% titanium (IV) dioxide after oxidative roasting followed by hydrochloric acid leaching process. This product meets the requirements of a raw material to produce industrial titanium dioxide. The aluminum is mainly in the form of corundum which combines with titanium to decrease leachability.

Conclusions

1. The optimal conditions of the electrostatic separation were obtained as follows: The electrostatic separation voltage was selected as 21 kV. Feed granularity was controlled between 150 and 400 microns. The rotation speed was held between 80 and 120 rpm. Considering the original granularity and resource recovery, the experiment feed granularity was under 850 microns and using the previous concentrate to repeat the electrostatic separation. After three electrical separations, the titanium content of the concentrate increased from 22.78wt.% to 45.86wt.%, with 76.61% comprehensive recovery rate. Through the analysis of process mineralogy of concentrate, it was found that titanium and aluminum were closely combined in the form of titanium-aluminum alloy. The main impurities in the concentrate were the grinding wheel ash and magnesia-aluminum spinel.
2. The acid leaching results showed that the selective impurity leaching rates iron, calcium, and magnesium on the pretreated concentrate were all more than 98%. The leaching loss of titanium was only 2.66% which was less than untreated concentrate. The products obtained by chemical analysis contained 91.44% titanium dioxide, which met the requirements of raw materials to prepare the industrial titanium dioxide. The aluminum as corundum had resistance to acid—which caused the impurity leaching rates. Future work is to prevent the oxidation of aluminum into corundum by selective oxidation, then further leaching out the aluminum to produce a raw material with a higher titanium content.

References

1. Reitz J, Lochbichler C, Friedrich B (2011) Recycling of gamma titanium aluminide scrap from investment casting operations. Intermetallics 19(6):762–768. https://doi.org/10.1016/j.intermet.2010.11.015
2. Liu M, Shigang LU, Kan S, Guoxun LI (2007) Effect of electrolysis voltage on electrochemical reduction of titanium oxide to titanium in molten calcium chloride. Rare Met 26(6):5. https://doi.org/10.1016/S1001-0521(08)60005-4
3. Jiao H, Song WL, Chen H, Wang M, Fang D (2020) Sustainable recycling of titanium scraps and purity titanium production via molten salt electrolysis. J Clean Prod 261(1):121314. https://doi.org/10.1016/j.jclepro.2020.121314

4. Xu X, Yu Y, Hui H (2003) Mechanisms of abrasive wear in the grinding of titanium (tc4) and nickel (k417) alloys. Wear 255(7–12):1421–1426. https://doi.org/10.1016/S0043-1648(03)00163-7
5. Ma JW, Wang ZZ, Li JH (2006) Study on recovery of copper from waste printed wiring boards by electrostatic separation. Chinese J Anal Chem 27(9):1895–1900. https://doi.org/10.1016/S1872-2040(06)60041-8
6. Narayana PL, Kim JH, Yun DW, Kim SE, Reddy NS, Yeom JT (2022) High temperature isothermal oxidation behavior of electron beam melted multi-phase γ-tial alloy. Intermetallics 141:107424. https://doi.org/10.1016/j.intermet.2021.107424
7. Long M, Rack HJ (1998) Titanium alloys in total joint replacement—a materials science perspective. Biomaterials 19(18):1621–1639. https://doi.org/10.1016/S0142-9612(97)00146-4

Investigation of Heavy Metal Levels in Tin Mine Wastes and the Implication to Mine Closure Plan: A Case Study of Rutongo Mine, Rwanda

Jean Claude Ishimwe, Abubakary Salama, and Kenneth Sichone

Abstract Rwanda has not experienced any mine closure since mining operations started in the 1930s. To date, there is no information on good practice mine closure procedures despite the presence of old mining operations. Therefore, this study is the first of its kind to scrutinise the closure plan for Rutongo, a tin mine that has existed for over 80 years and is approaching its mine life. The study seeks to quantify heavy metal levels in mine wastes and suggests a closure plan for waste rock dumps (WRDs) and tailings storage facility (TSF) so as to reduce contamination effects during and after cessation of operations. Geochemical characterization using X-Ray Fluorescence (XRF) and Inductive-Coupled Plasma Mass Spectroscopy (ICP-MS) was conducted. Results indicated the presence of pyrite and arsenopyrite, the deleterious elemental concentration ranging from 120 mg/kg to 710 mg/kg As, 10.32 mg/kg to 22.35 mg/kg Co, and 8.6 mg/kg to 128 mg/kg Cr. When these results are benchmarked against various soil quality standards, it is evident that the mine wastes are contaminated by As and a critical concentration of Cr and Co. A mine closure plan with remediation and containment strategies for bio-accumulation of As, Cr, and Co and use of covers that have a capillary barrier to minimise water percolation, and a top soil that supports vegetation backed by hyper-accumulators was recommended.

Keywords Mine closure · Plan · Heavy metals · Mine wastes · Waste rocks · Tailings · Tin

J. C. Ishimwe · A. Salama
School of Mines and Geosciences, Department of Mining and Mineral Processing, University of Dar es Salaam, P. O. Box 35131, Dar es Salaam, Tanzania

K. Sichone (✉)
School of Science & Technology, Harvest University, P.O. Box 37866, Lusaka, Zambia
e-mail: bmkspure@gmail.com

R. G. Reddy et al. (eds.), *New Directions in Mineral Processing, Extractive Metallurgy, Recycling and Waste Minimization*, The Minerals, Metals & Materials Series,
https://doi.org/10.1007/978-3-031-22765-3_20

Introduction

The problem of deleterious elements, particularly heavy metals in the earth's ecosystems, is a global concern [16]. Heavy metals have been studied for their toxicity, mobilisation, and impacts on people's lives. Investigations into heavy metals in soil facilitate quantification of their occurrence and interaction with the environment [6]. Heavy metals occur naturally in the earth's crust; however, at high concentrations they can be intolerable for ecosystems [27].

The deleterious effect of heavy metals is their inherent toxicity, bio-accumulation, and persistence in the environment [2]. Use of heavy metal-contaminated waters has been associated with raised distress and death rates worldwide [34]. Elevated heavy metal levels have hindered usage of some agricultural practices and products; rice, tomatoes, and potatoes have been classified unsuitable for human consumption [7, 14].

Areas geologically disturbed by mineral exploitation activities such as mining and processing are the drivers of heavy metal contamination [21] due to the huge quantities associated effluents and their high susceptibility to mobilisation of heavy metals [13]. For instance, arsenic from abandoned mine workings and waste rock dumps is the foremost source of water contamination in South Africa [1, 22]. In the UK, heavy metals emanating from abandoned mines have contaminated surface waters and it is the leading pollution threat [12]. The highly contaminated mining region of Shangba village and villages around Dabaoshan mine in China are widely known to have elevated cancer rates [14].

In Rwanda, mining activities are the main source water-borne pollution [23]. Heavy metal levels exceeding acceptable standards in parts of Nyabarongo river and nearby swamps [24]. A number of the tributaries of Nyabarongo river traverse Rutongo mine workings and waste rock dumps. Surface dumping of mine wastes can elevate heavy metal levels in the environment [20]. Since mining activities continue contributing harmful concentrations of heavy metals even after mine life, restoring the contaminated mining area should be prioritised by regulators, the community, and mining firms. To mitigate the adverse heavy metal contamination of the environment, mine closure plans should incorporate studies of heavy metal deportment in the environment.

Mitigation and minimisation of heavy metals should be guided by physical and geochemical characterisation; therefore, investigation into the deportment of heavy metals is an essential component in mine closure planning, particularly when dealing with mine wastes generated by metal mines. Common practice in heavy metals investigation have concentrated on abandoned mine sites and rarely on operating mines where much opportunity and flexibility in sustainable management of mine wastes and closure planning might exist. Most research has been limited to the fate of heavy metals in soils, and their mobilisation rather than the sources of heavy metals primary including waste rock dumps, tailings storage facilities, underground openings, and pits [5, 8, 17, 30].

Moreno-Jiménez et al. [17] have reported Cd, Cu, and Zn concentrations above the permissible limits in the soil and plants from a mining valley in Spain. Corales et al. [8] reported Pb, Ni, and Cu concentrations above permissible limits in the water near lateritic nickel ore stockpiles on Maniacani Island, and [30] reported that heavy metals concentrations in tailings of small-scale abandoned metallic mines in South China had adversely impacted the quality of surface water and soils downstream the mining district.

The status of heavy metals in operating mines is rarely made public, due to confidentiality policies and increased public sensitivity to environmental issues. Despite the legacy of old workings, mine wastes dumps, and degraded landscapes from the colonial and immediate post-independence period adjacent to or near current mining operations, there is no information on good practice mine closure procedures in Rwanda. Rwanda has never experienced any systematic mine closure. Therefore, this study aims to quantify heavy metal levels in mine wastes and suggest a closure plan for waste rock dumps (WRDs) and tailings storage facilities (TSF) to reduce the contamination effects during and after mine life.

Study Area Description

Location

Rutongo Mine is located 10 km North of Kigali, the capital city of Rwanda. The mine is within the vicinity of a large and mainly subsistence farming population. It is dominated by underground hydrothermal quartz veins and some alluvial deposits. The mining area comprises waste rock dumps and tailings in relatively small dumps that occupy approximately 72.5 hectares in more than 13 different places (Rutongo mines 2015).

The area is situated in a background with distinctive steep slopes originating from the consecutive folding and erosion of the earth strata. The overall average gradient is 25% but in some areas the average gradient exceeds 75%. There are several deep and narrow well drained valleys that bound this array of mountains. The elevation in the area ranges from 1391 to 2133 m (Twagiramungu et al. 2015).

Climate and Topography

Rutongo has a humid tropical climate with alternating rains and drought. The dry season running from June to August is followed by a short rainy season from September to November and a dry season during the months of December, January, and February before another short rain season. Due to current global climate change, the four seasons can probably be assumed as temporal (Twagiramungu et al. 2015).

The temperature ranges from 17.5 °C to 24 °C while the precipitation ranges from 1061 to 1140 mm annually.

Mine Operations Overview

Rutongo Mine is the largest producer of tin concentrates in Rwanda, with an average production of 100 tonnes per month of 71% Sn grade concentrates. The estimated mine life in 2015 was 20 to 25 years [33].

The mine selectively recovers high-grade cassiterite ore manually using hand tools; blasting is rarely used. Manual ore handling enables careful dilution control required in the shrinkage stoping method by selectively extracting the ore while leaving the waste rock in place. While hand-sorting tools may be time-consuming, it reduces the amount of waste exposed outside the mine and dilution. The remaining lower-grade ore is beneficiated using a gravity concentration processing plant at the site.

Mine sites far from the mineral processing plant produce hand-sorted high-grade cassiterite that does not need further beneficiation. However, host rocks, sands, and fines with the mineral of interest are processed on-site using alternative traditional methods, including sluicing and panning. The wastes at these sites tend to be enormous, and both tailings and waste rocks are deposited in the same place. The waste rocks are sourced from underground openings and dumped together with sluicing wastes.

Since the mine is in a traditionally agricultural region and in alignment with the government's development goals and Maslow's hierarchy of human needs for a developing country as applied to land use drivers [10], it is anticipated that the post-mining land use for the Rutongo mining area would be agriculture and forestry. The closure objectives of Rutongo should return waste rock dumps and tailings storage facilities into areas that can support vegetation.

Materials and Methods

The study materials were waste rock samples taken from Nyamyumba, Gasambya, and Sanzari mine sites and tailings samples taken from Gasambya process plant tailings storage facility (Fig. 1). Ten 40 kg-lots were collected from waste rock dumps and another three 40 kg-samples from the tailings storage facility using a pick. The samples were kept in clean cotton bags and transported to Rutongo mine office for further preparation and analysis.

Before analysis, the samples were left to dry naturally for 3 days, to facilitate sieving. The 40 kg of waste rocks or tailings were coned and quartered to obtain three samples of 1000 g and two samples of 250 g from each 40 kg lot for sieve analysis and chemical analysis, respectively.

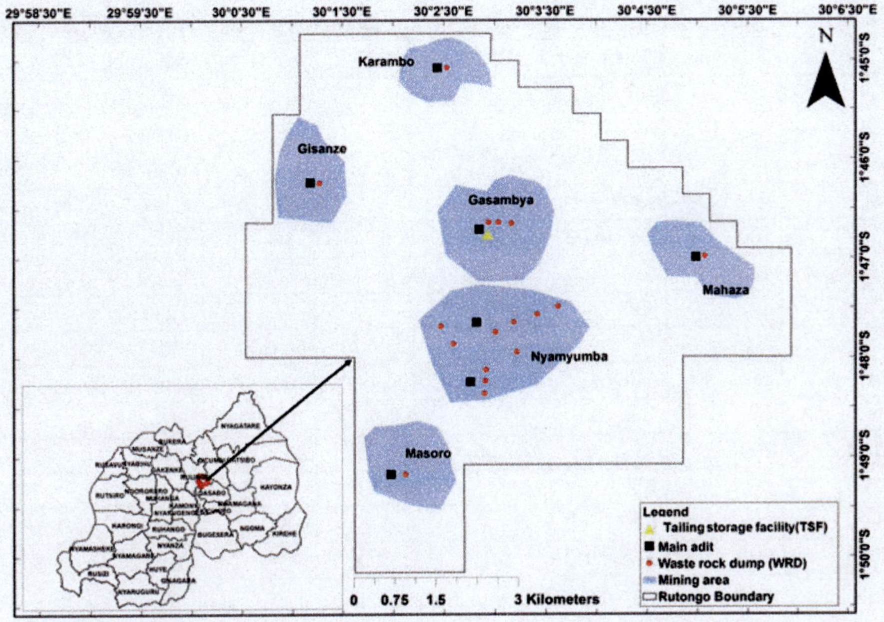

Fig. 1 Main locations of mine wastes at Rutongo mine, Rwanda

To assess the potential for heavy metals mobilisation and reactivity of the materials studied, their hydraulic conductivity was determined based on particle size analysis. Hydraulic conductivities were computed using the Terzaghi model HydrogeoSieveXL [9]. The concentration of heavy metals in waste rocks and tailings was done using XRF and ICP-MS. Samples to be analysed using XRF were pulverised into powder using a miller machine, model: RS 200, manufactured by Retsch.

Six confirmatory samples were also analysed at another laboratory where the samples were air-dried and screened on a 10 mesh (<2 mm), quartered, and pulverised before being passed through an 80 mesh (<180 µm) sieve. To determine the low concentrations of heavy metals, 6 ml of HCl and 2 ml of HNO_3 were added to 0.5 g of the sample. The mixture was heated to 150 °C while being stirred for 1 h. The solution was diluted with 10 ml of distilled water before being introduced into ICP-MS for heavy metals analysis.

The analysis results were compared with maximum acceptable heavy metals concentration in soil quality standards of different countries (Table 1).

Table 1 Comparison of the regulatory standard of heavy metals in soil in different countries

Elements	Tanzania	South Africa	WHO	India	China	Australia
Sn	–	–	–	–	–	34*
Fe	–	–	–	–	–	20,000–50,000**
Ti	–	–	–	–	–	4400***
V	–	150	–	–	–	–
Cr	100	6.5–46,000	100	–	150–300	50
Mn	1800	740		–	–	
Co	–	300	–	–	–	0.2–0.5****
Cu	200	16	30	135–270	50–200	100
Zn	150	240	300	300–600	200–300	200
As	1	5.8	–	–	20–40	20
Pb	200	20	100	250–500	80	300
Cd	1	7.5	3	3–6	0.3–0.6	3
Hg	2	1	–	–	0.3–1	1
Source	[3]	S. A, 2013	[4]	[25]	[3]	[3]

1. Tanzania: Maximum permissible limits of contaminants in habitat and agricultural soils
2. South Africa: Average soil quality standards that are defensive of both human health and eco-toxicological risk for multi- exposure pathways, inclusive of pollutant relocation to the water reserve
3. WHO: The maximum threshold heavy metal concentration in soil
4. India: Guideline for safe limits of heavy metals, Indian Standard
5. China: Regulatory standards of heavy metals in agricultural soils
6. Australia: Regulatory standards of heavy metals in agricultural soils

*Adapted from [28] as maximal permissible addition of heavy metals and metalloids by the data of Dutch and Ecologists

** Adapted from [26] as general concentration of Iron in soils

*** Global average concentration of Titanium in soils adapted from [31]

**** Tolerable *limit of Co in soil for the growth of plant* [15]

Results and Discussion

Particle Size Distribution Curves of Waste Rocks and Tailings

The particle size distribution shows that both waste rocks and tailings are dominated by gravels with an average cumulative percentage retained on a 4.75 mm sieve of 58.99 $\pm$ 7.96, for new waste rocks, 70.96 $\pm$ 8.39 for old waste rocks, and 52.26 $\pm$ 9.17 for tailings. Based on the Terzaghi model, the hydraulic conductivity of waste rocks ranged from 1.5×10^{-5} cm/s to 9.33×10^{-1} cm/s as against 9×10^{-5} cm/s to 3.6×10^{-4} cm/s for tailings. As both waste rocks and tailings samples contain a large proportion of gravel (>50%), and fewer fines, they would tend to have higher hydraulic conductivities. The high hydraulic conductivity characteristic of gravel-dominated wastes is less likely to attain natural restoration as no vegetation would be sustained by pebbles-dominated waste rocks despite extended years without disturbance.

High hydraulic conductivities which are associated with higher gravel content (>50%) and less silt are expected to reduce the water holding capacity and nutrient content of the areas occupied by the waste materials. Adding a soil layer with lesser quantities of gravel on top of waste rocks and tailings can enhance the potential for waste rocks and tailings areas to restore and regain the previous soil properties. Moreover, the higher hydraulic conductivity of waste rocks and tailings can enhance heavy metal migration and percolation to the surrounding areas. Heavy metals mobilisation chemical reactions from waste rocks and tailings include oxidation reactions of pyrite and arsenopyrite. Sulphide minerals are prone to substitution reactions involving different trace metals [29]. The overall oxidation reaction of pyrite is as follows:

$$FeS_2 + \frac{15}{4}O_2 + \frac{7}{2}H_2O = Fe(OH)_3 + 2H_2SO_4 \tag{1}$$

High hydraulic conductivities facilitate the sulphides oxidation reactions by allowing water movement within the waste rocks and tailings while also exposing more waste rocks and tailings to oxygen. Heavy metals are highly mobile in the acidic environment created by the 2 mol released by sulphuric acid in Eq. 1. A closure plan for WRDs and TSF that deals with higher hydraulic conductivity and gravel-dominated materials needs greater effort to cover design to limit exposure to the atmosphere, provide desirable substrate layer, and containment barriers to prohibit further metal-leaching.

Elemental Composition Using XRF

The results show variations in elemental concentration among the waste rocks from different WRDs. This could be attributed to lithological variations arising from geological profile and depth of operations during mining (Fig. 2). The lithological variation is also evident within tailings samples taken from the same TSF (Fig. 3).

The average Sn in new waste rocks, old waste rocks, and tailings were 266 ± 161.0 mg/kg, 162 ± 77.6 mg/kg, and 66.6 ± 11.5 mg/kg, respectively. The Sn concentration in tailings generated by the mineral processing plant was lower compared to waste rocks and sluiced materials, implying that high recoveries enhance economies of the mine and reduce the environmental burden. Waste rocks from current operations are high in tin; therefore, containment of the wastes and a closure plan that mitigates the risks of mobilisation is necessary.

Fe content in old waste rocks was seven to eleven times higher than current (Table 2), arising probably due to lithological variations. Old waste rocks came from

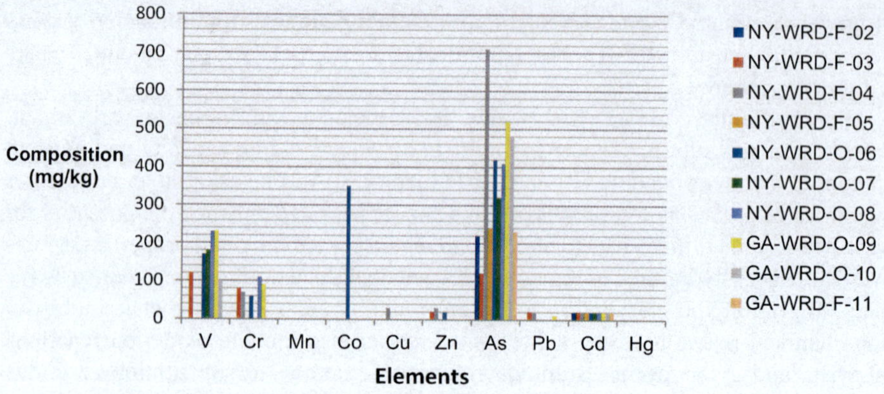

Fig. 2 Elemental composition of sample of waste rocks measured by XRF

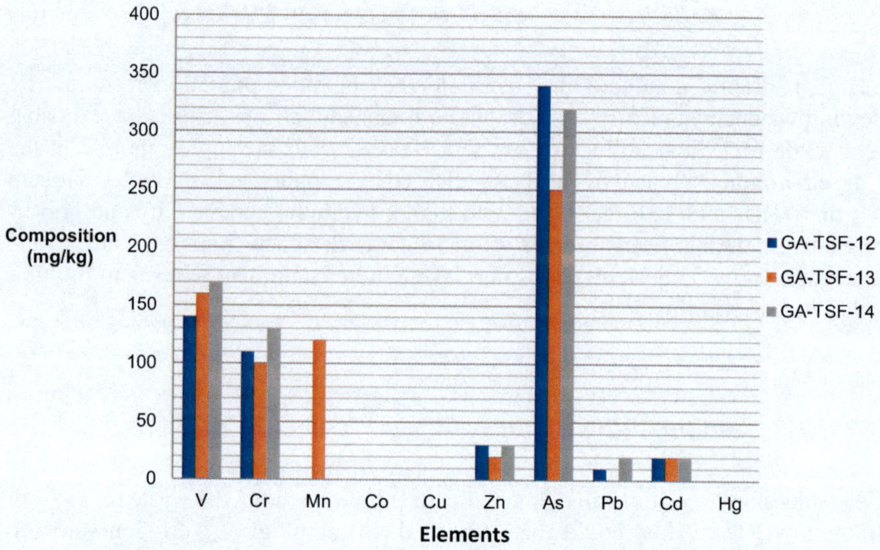

Fig. 3 Elemental composition of sample of tailings measured by XRF

the surface and shallow depth while current waste rocks originated from deeper zones. The higher concentrations of iron in tailings are probably due to the presence of pyrite and arsenopyrite in the cassiterite ore, alluvial from surrounding hills, and eroded waste rocks. The measured concentrations of Fe in all samples were within the characteristic concentration range of Fe (0.2% to 55%) in soils. Therefore, Fe is not a major concern in mine closure provided oxidation reactions of Fe are contained.

Table 2 Elemental composition of mine wastes measured by XRF in mg/kg

Sample	Sn	Fe	Ti	V	Cr	Zn	As	Pb
NY-WRD-F-02	530	12,540	1,430	< LOD	< LOD	< LOD	220	< LOD
NY-WRD-F-03	180	12,180	4,860	120	80	20	120	20
NY-WRD-F-04	160	21,070	4,650	< LOD	70	30	710	20
NY-WRD-F-05	150	17,000	2,450	< LOD	< LOD	< LOD	240	< LOD
NY-WRD-O-06	300	88,590	2,770	170	60	20	420	< LOD
NY-WRD-O-07	120	72,610	4,040	180	< LOD	< LOD	320	< LOD
NY-WRD-O-08	120	145,010	2,790	230	11	< LOD	410	< LOD
GA-WRD-O-09	130	114,380	2,550	230	90	< LOD	520	10
GA-WRD-O-10	140	91,200	2,580	100	< LOD	< LOD	480	< LOD
GA-WRD-F-11	310	9,190	570	< LOD	< LOD	< LOD	230	< LOD
GA-TSF-12	80	91,200	2,850	140	110	30	340	10
GA-TSF-13	60	78,320	4,450	160	100	20	250	< LOD
GA-TSF-14	60	134,250	3,070	170	130	30	320	20

Fe from pyrite and arsenopyrite can trigger acid mine drainage through oxidation reactions in aqueous media. Based on the typical mineralogical composition of tailings [18], the arsenopyrite is less than 1%. Considering the mineralogical composition, the net acid producing potential estimator (ABATES v.1.4), the maximum net acid produced would be 6.02 kg H_2SO_4 per tonne material. The mine closure plan has to attain to the intended land use. Mine drainage preventive measures are the best available techniques to be considered first during mine closure planning, particularly where the future land use is agriculture. Therefore, the acid producing and non-acid producing waste rocks and tailings should be identified and managed accordingly; appropriate covers normally limit oxidation by limiting water/oxygen ingress into the facilities. Thick covers are normally applied when high risks of acid mine drainage are likely.

Arsenic concentration of the tailings and waste rocks varied between 120 and 710 mg/kg, which exceeds allowable maximum concentration by Tanzania, South Africa, China, and Australia standards. Arsenic in the ore exists as pyrite and arsenopyrite. The minerals are assumed to have originated in the reducing conditions of the earth. When oxidised, the sulphide minerals produce sulphuric acid which can potentially release arsenic and other heavy metals into the surroundings thereby polluting soil, surface, and groundwater resources [11]. An effective mine closure plan needs to evaluate physical and chemical methods of removing as from waste rocks and tailings. Deemed as the most socially acceptable method and currently industry best heavy metal remediation practice is phytoremediation [19, 32].

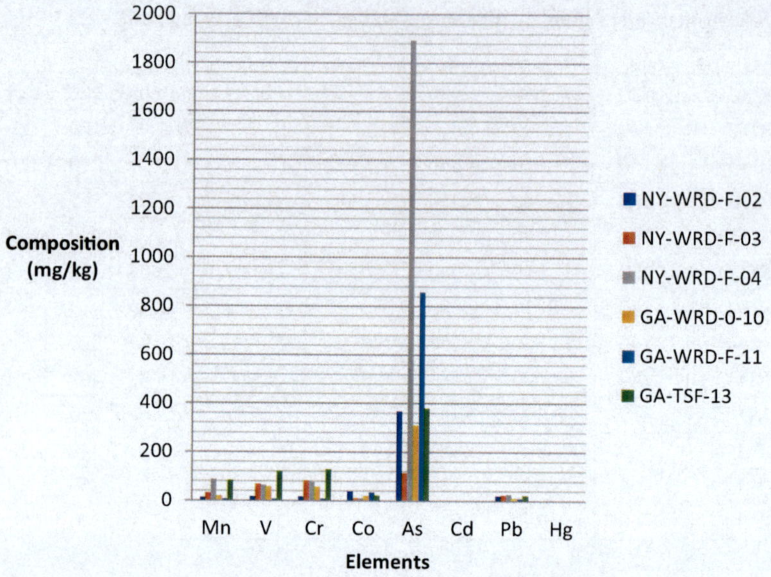

Fig. 4 Elemental composition of sample of tailings measured by ICP-MS

Elemental Composition Using ICP-MS

Combining XRF and ICP-MS analyses enabled measurement of deleterious metals at low levels. Figure 4 shows the concentrations of Mn, V, Cr, Co, As, Cd, Pb, and Hg in 6 samples.

Cr of tailings varied from 8.6 to 128 mg/kg which was below maximum permissible values by Tanzania and WHO soil standards. The GA-TSF-13 sample was slightly higher than the maximum allowable concentration limit. All measured concentrations were below the maximum permissible values stipulated in soil standards of China. The materials studied contained critical concentrations of Cr with potential to migrate and contaminate humans, plants, animals, and water bodies. The containment and remediation strategies of Cr should be incorporated into the mine closure plan to mitigate the risk of contaminating the environment.

The average Co concentration of all 6 samples tested using ICP-MS was 22.35 mg/kg ± 12.03 mg/kg which was below the maximum limit of South Africa Standards. However, the tolerable limit of Co in soil for the growth of plants ranges from 0.2 mg/kg to 0.5 mg/kg [15]. The closure of WRDs and TSF should therefore ensure the containment and remediation of Co in waste rocks and tailings.

Conclusions

The main objective of this research was to determine the closure plan for Rutongo mine wastes in relation to heavy metals in waste rock dumps (WRDs) and tailings storage facilities (TSFs), to reduce the contamination effects during and after mine life. The following were the main findings of the research:

- The physical characterisation showed that waste rocks and tailings are dominated by gravels with an average cumulative percentage retained materials at 4.75 mm aperture sieve of 58.99% for fresh waste rocks, 70.96% for old waste rocks, and 52.26 for tailings. Therefore, a cover layer is required to provide a substrate capable of supporting vegetation and growth after mine closure.
- The higher hydraulic conductivity of waste rocks and tailings predicted using the Terzaghi model indicate that mine wastes are unsuitable for use as a future sealing material and are unlikely to support vegetation; hence, an additional soil cover would be required.
- Due to the typically high levels of As, Co, and Cr (ranging from 120 mg/kg to 710 mg/kg, 10.32 mg/kg to 22.35 mg/kg, and 8.6 mg/kg to 128 mg/kg, respectively) when benchmarked against various soil quality standards. The mine wastes would need to be profiled, and contained with a cover that is able to limit their migration.
- The revegetation might occur after providing soil cover.

The waste rocks and tailings are contaminated with As, Cr, and Co; hence, a comprehensive mine closure plan to attain to the future land use should include remediation and effective containment strategies to mitigate for bio-accumulation of the three heavy metals. Mn, V, Pb, Cd, Hg, and Zn concentrations have minimal impact on mine closure planning as they were within acceptable limits of soil quality standards. Future WRDs and TSF should be located in areas with minimal risks of erosion and avoiding waste rocks-dumping near water sources and channels.

Mine closure has to ensure physical, biological, and chemical stability of the mine wastes. Heavy metal levels in excess of the acceptable limits and/ or background values indicate a failure to accomplish acceptable standards of heavy metals in soils and consequently pose a risk to agricultural soils. Elevated heavy metal levels can result in additional work during mine closure such as thickening of covers, addition of barricades around the facilities, and additional drainage requirements to meet agreed land use. The study recommends using covers backed by hyper-accumulators as a method of closing WRDs and TSF. The covers should have a capillary barrier that minimises water percolation and a top soil that supports vegetation. The recommended technique can potentially contain and remove heavy metals in WRDs and TSF and enhance revegetation after closure.

Acknowledgements The authors gratefully acknowledge TINCO Investment Ltd, Rutongo mines branch for facilitating sample collection within their mine sites, providing access to archived documents and their mine operations; Rwanda Mines, Petroleum and Gas Board for permitting the exportation of samples; and Mr J Iranzi at the Korea Basic Science laboratory for ICP-MS sample analyses. The work is a part of master's dissertation entitled "Establishment of Mine Closure Plan of Underground Openings, Waste Rock Dumps and Tailings Storage Facility for Rutongo Mine in Rwanda" that was submitted to the University of Dar es Salaam.

References

1. Akcil A, Koldas S (2005) Acid mine drainage (AMD): causes, treatment and case studies 14:1139–1145. https://doi.org/10.1016/j.jclepro.2004.09.006
2. Ali H, Khan E, Ilahi I (2019) Environmental chemistry and ecotoxicology of hazardous heavy metals : environmental persistence, toxicity, and bioaccumulation. J Chem Hindawi 201:15. https://doi.org/10.1155/2019/6730305
3. Bakshi S, Banik C, He Z (2018) The impact of heavy metal contamination on soil health. In: Reicosky D (ed) Managing soil health for sustainable agriculture, volume 2: monitoring and management (Issue August, pp 63–95). Burleigh Dodds Science Publishing. https://doi.org/10.19103/as.2017.0033.20
4. Bello S, Zakari YI, Ibeanu IGE, Muhammad BG (2016) Characterization and assessment of heavy metal pollution levels in soils of Dana steel limited dumpsite, Katsina state , Nigeria using geo-accumulation. Am J Eng Res (AJER) 5 (1):49–61. www.ajer.org
5. Bu Q, Li Q, Zhang, H, Cao H, Gong W, Zhang X, Ling K, Cao Y (2020) Concentrations, spatial distributions, and sources of heavy metals in surface soils of the coal mining city Wuhai, China. J Chem. https://doi.org/10.1155/2020/4705954
6. Chileshe MN, Syampungani S, Festin ES, Tigabu M, Daneshvar A, Odén PC (2020) Physico-chemical characteristics and heavy metal concentrations of copper mine wastes in Zambia: implications for pollution risk and restoration. J For Res 31(4):1283–1293. https://doi.org/10.1007/s11676-019-00921-0
7. Cobb GP, Sands K, Waters M, Wixson BG, Dorward-King E (2000) Accumulation of heavy metals by vegetables grown in mine wastes. Environ Toxicol Chem 19(3):600–607. https://doi.org/10.1002/etc.5620190311
8. Corales-Ultra OG, Peja RP, Casas EV (2019) Baseline study on the levels of heavy metals in seawater and macroalgae near an abandoned mine in Manicani, Guiuan, Eastern Samar, Philippines. Mar Pollut Bull 149:110549. https://doi.org/10.1016/j.marpolbul.2019.110549
9. Devlin JF (2015) HydrogeoSieveXL: an excel-based tool to estimate hydraulic conductivity from grain-size analysis. Hydrageology J. https://doi.org/10.1007/s10040-015-1255-0
10. Hattingh R, Williams D, Corder G (2019) Applying a regional land use approach to mine closure : opportunities for restoring and regenerating mine-disturbed regional landscapes. Aust Centre Geomechanics 18. https://doi.org/10.36487/ACG
11. Jamieson HE, Walker SR, Parsons MB (2015) Mineralogical characterization of mine waste. Elsevier 57:85–105. https://doi.org/10.1016/j.apgeochem.2014.12.014
12. Johnston D, Potter H, Jones C, Rolley S, Watson I, Pritchard J (2008) Abandoned mines and the water environment. Sci Rep (Issue SC030136/SR41\rSCHO0508BNZS-E-P). www.environment-agency.gov.uk
13. Ledin M, Pedersen K (1996) The environmental impact of mine wastes - roles of microorganisms and their significance in treatment of mine wastes. Earth Sci Rev 41(1–2):67–108. https://doi.org/10.1016/0012-8252(96)00016-5
14. Liao J, Wen Z, Ru X, Chen J, Wu H, Wei C (2015) Distribution and migration of heavy metals in soil and crops affected by acid mine drainage: public health implications in Guangdong Province, China. Ecotoxicol Environ Saf 124:460–469. https://doi.org/10.1016/j.ecoenv.2015.11.023
15. Mahey S, Kumar R, Sharma M, Kumar V, Bhardwaj R (2020) A critical review on toxicity of cobalt and its bioremediation strategies. SN Appl Sci 2(7):1–12. https://doi.org/10.1007/s42452-020-3020-9

16. Miller M, Gosar M (2009). Application of SEM / EDS to environmental geochemistry of heavy metals Application of SEM/EDS to environmental geochemistry of heavy metals Uporaba SEM/EDS v okoljski geokemiji težkih kovin. GEOLOGIJA, June 2009. https://doi.org/10.5474/geologija.2009.008
17. Moreno-Jiménez E, Peñalosa JM, Manzano R, Carpena-Ruiz RO, Gamarra R, Esteban E (2009) Heavy metals distribution in soils surrounding an abandoned mine in NW Madrid (Spain) and their transference to wild flora. J Hazard Mater 162(2):854–859. https://doi.org/10.1016/j.jhazmat.2008.05.109
18. Nambaje C, Mweene L, Subramanian S, Krishnan S, Santosh M (2020) Xanthan gum based investigations into the surface chemistry of cassiterite and beneficiation of cassiterite tailings. Miner Process Extr Metall Rev 00(00):1–15. https://doi.org/10.1080/08827508.2020.1825957
19. Nedjimi B (2021) Phytoremediation: a sustainable environmental technology for heavy metals decontamination. SN Appl Sci 3(3):1–19. https://doi.org/10.1007/s42452-021-04301-4
20. Nieder R, Weber TKD, Paulmann I, Muwanga A, Owor M, Naramabuye F, Gakwerere F, Biryabarema M, Biester H, Pohl W (2014) The geochemical signature of rare-metal pegmatites in the Central Africa Region : Soils, plants, water and stream sediments in the Gatumba tin – tantalum mining district, Rwanda. J Geochem Explor 1–13. https://doi.org/10.1016/j.gexplo.2014.01.025
21. Rashed MN (2010) Monitoring of contaminated toxic and heavy metals, from mine tailings through age accumulation, in soil and some wild plants at Southeast Egypt. J Hazard Mater 178(1–3):739–746. https://doi.org/10.1016/j.jhazmat.2010.01.147
22. Robson TC, Braungardt CB, Keith-roach MJ, Rieuwerts JS, Worsfold PJ (2012) Impact of arsenopyrite contamination on agricultural soils and crops. J Geochem Explor 8. https://doi.org/10.1016/j.gexplo.2012.11.013
23. Sekomo CB, Birori MM, Nkurunziza E (2019) Water quality monitoring in Rwanda final report. In: Report
24. Sekomo CB, Nkuranga E, Rousseau DPL, Lens PNL (2011) Fate of heavy metals in an urban natural wetland: the Nyabugogo Swamp (Rwanda). Water Air Soil Pollut 214(1):321–333. https://doi.org/10.1007/s11270-010-0426-9
25. Singh A, Sharma RK, Agrawal M, Marshall FM (2010) Risk assessment of heavy metal toxicity through contaminated vegetables from waste water irrigated area of Varanasi, India. Tropical Ecol 51(2 SUPPL.):375–387
26. USEPA (2003) Ecological soil screening level for iron (Issue November)
27. Vhahangwele M, Khathutshelo ML (2018) Environmental contamination by heavy metals. In: Hosam E-DMS, Refaat FA (eds) IntechOpen, p 13. https://doi.org/10.1007/s11270-010-0426-9
28. Vodyanitskii YN (2016) Standards for the contents of heavy metals in soils of some states. Ann Agrar Sci 14(3):257–263. https://doi.org/10.1016/j.aasci.2016.08.011
29. Vriens B, Plante B, Seigneur N, Jamieson H (2020) Mine waste rock: Insights for sustainable hydrogeochemical management. MDPI 10(9):1–38. https://doi.org/10.3390/min10090728
30. Wang P, Sun Z, Hu Y, Cheng H (2019) Leaching of heavy metals from abandoned mine tailings brought by precipitation and the associated environmental impact. Sci Total Environ 695:133893. https://doi.org/10.1016/j.scitotenv.2019.133893
31. WHO (1982) Enviromental health criteria 24: Titanium
32. Wuana RA, Okieimen FE (2011) Heavy metals in contaminated soils : a review of sources, chemistry, risks and best available strategies for remediation. Int Sch Res Netw. https://doi.org/10.5402/2011/402647
33. Yager BTR (2019) 2015 minerals yearbook the mineral industry of Rwanda (Issue May)
34. Zhushan F, Shuhua X (2019) The effects of heavy metals on human metabolism. Toxicol Mech Methods 34. https://doi.org/10.1080/15376516.2019.1701594

Processing of Luanshya Copper Smelting Slag

Yaki Chiyokoma Namiluko, Yotamu Rainford Stephen Hara, Agabu Shane, Makwenda Thelma Ngomba, Ireen Musukwa, Alexander Old, Ronald Hara, and Rainford Hara

Abstract Copper smelting slag from Luanshya of the Copperbelt province of Zambia with about 2 weight% copper, 0.2 weight % cobalt, 0.8 weight % chromium, 6 weight% sulphur, and 18 weight% iron, the remainder being silica, aluminium, calcium, and magnesium oxides, was processed. Sulphur is mostly present in elemental form. The presence of sulphur and chromium makes this slag an environmental hazard. Mineralogical examination showed that chromium and cobalt are predominantly present in chromite and complex fayalite phases, respectively. However, copper is mainly present in the sulphide or matte form. Copper was upgraded to more than 25 weight % via a combination of screening, magnetic separation, and froth flotation. Sulphur was recovered as a concentrate at a grade of 73 weight % via pre-flotation. Reductive roasting followed by magnetic separation or acidic leaching was carried out. Chromium was upgraded by a factor of 8 in the weakly magnetic fraction. Recovery of 55–70% for cobalt was obtained after acidic leaching of reductively roasted samples.

Keywords Chromium · Slag · Zambia · Hazard · Sulphur · Cobalt · Magnetic separation

Introduction

Copper smelting slags are good source of copper as they have significant grade of copper (0.6–2 weight %) and cobalt (0.1–0.8 weight %) [1–4]. A majority of copper is present in the sulphide or matte form as copper was lost out, mainly due to mechanical entrainment [1]. On the other hand, cobalt was lost to the slag phase as result of oxidation. In general, copper smelting slags are inert. However, copper smelting slag in Luanshya is unique as it has high levels of chromium and elemental sulphur. Chromium is a toxic heavy metal found in different oxidation states in the

Y. C. Namiluko · Y. R. S. Hara (✉) · A. Shane · M. T. Ngomba · I. Musukwa · A. Old · R. Hara · R. Hara
The Copperbelt University, Kitwe, Zambia
e-mail: yotamuhara@gmail.com

R. G. Reddy et al. (eds.), *New Directions in Mineral Processing, Extractive Metallurgy, Recycling and Waste Minimization*, The Minerals, Metals & Materials Series,
https://doi.org/10.1007/978-3-031-22765-3_21

environment, ranging from –2 to + 6 with the most stable states being the trivalent (chromium III) and hexavalent (chromium VI) [5]. High levels of chromium pose a serious environmental threat due to its leachability, mobility, and toxicity of the higher valence chromium (VI) [6].

Recovery of Copper from Slag

Owing to the fact that copper in the copper smelting slag is mainly present as matte or oxide (due to oxidation), froth flotation is widely used for recovering copper. Copper is recovered as a high grade concentrate (more than 25 weight % copper). The advantages of froth flotation are low capital and operating costs as compared to hydrometallurgical processing [4]. The copper concentrates may either be roasted and then leached or, smelted to recover copper. The extraction efficiency for recovering copper matte or sulphide from slag is very high (more than 90%). Even though froth flotation is widely used for processing copper smelting slag, no attention has been paid to study the behaviour of chromium.

Recovery of Cobalt from Slag

There is no established process that efficiently extracts cobalt from copper smelting slag. This is because cobalt is chemically dissolved in calcium-iron-silicate matrix. Reductive smelting has been tried but the low grade content of cobalt in the slag does not justify this process and this is why Chambishi Metals PLC in Zambia abandoned this process after commercialising it.

Recent studies have indicated that cobalt might be recovered via low temperature reduction (700–900 °C) followed by acidic media leaching [3]. This method looks promising as low processing temperatures are involved compared to reductive smelting which requires a temperature of more than 1500 °C [7, 8]. Since Luanshya copper smelting slag has significant amount of chromium, it is necessary to study the effect of chromium on low reduction—acidic leaching.

Aims of the Study

Considering a significant amount of sulphur and chromium in Luanshya copper smelting slag, this work focuses on efficient flotation of Luanshya copper smelting slag. In addition, low temperature reductive roasting—acidic leaching of Luanshya copper smelting slag is also investigated.

Experimental

Materials

Copper smelting slag was obtained from Luanshya district of the Copperbelt province of Zambia. The material was collected from several different locations of the slag dump site in order to have a good representative sample. Chemical composition of the as-received material is shown in Table 1. Chemical analysis results in Table 1 clearly show that the slag has significant grade of sulphur and chromium which are both hazards elements [5, 9]. The main constituents are Fe and SiO_2, and this is typical for any copper smelting slag [1].

Unless stated, all test works were carried out at particle size of 100% passing through the 106 microns sieve size.

Froth Flotation Test Works

Froth flotation test works were carried out for the purpose of recovering copper from slag. Froth flotation was carried out at a pulp density of 33%. Since the material has some copper oxide, a sulphidiser (sodium hydrogen sulphide) was used during froth flotation. Pine oil which is widely used as a frother for copper smelting flotation in Zambia was used.

Reductive Roasting

For reductive roasting, the as-received ground slag material was mixed with charcoal and transferred into a salamander crucible before loading into the furnace. The material was heated isothermally for 1.5 h at the predetermined temperature.

Leaching of Reductive Roasted Sample

The reductive roasted sample was leached for 2 h at pH 1.8 and 33% solids. The leached sample was filtered to separate out the leach liquor from residue (solids).

Table 1 Chemical analysis in weight % of the as-received Luanshya slag

TCu	TCo	Fe	S	Cr	SiO_2	Al_2O_3	CaO	MgO	K_2O	MnO
2.11	0.17	18.13	6.26	0.86	40.69	8.19	9.76	0.10	2.36	0.18

Chemical Analysis

Copper, cobalt, iron, and chromium were all analysed by atomic absorption spectrometer (Perkin Elmer Analysis 300) technique while the rest of the compounds were analysed by gravimetric technique.

Results and Discussion

The as-received material was characterised via scanning electron microscope, and the images under backscattered electron imaging are shown in Fig. 1. The major phases that were identified in the material were fayalite, quartz, complex calcium silicates, bornite, chalcocite, malachite, etc. Chromium was present as Fe–Cr-Mg-Al_O, and this is due to substitution between chromite ($FeCr_2O_4$) with magnesiochromite ($MgCr_2O_4$) and ($FeAl_2O_4$). Point analysis using energy dispersive X-ray point analysis showed that cobalt is predominantly contained in complex iron—silicate phase (fayalite). The bright phase belonging to copper matte can be observed in Fig. 1b.

Froth Flotation

Since Luanshya copper smelting has high levels of chromium and sulphur, the as-received material was magnetically separated prior to froth flotation. The high, weakly, and non-magnetic fractions were floated individually, and the results are given in Table 2. A copper grade of 41.43 weight % copper was obtained in the concentrate after flotation of magnetic fraction. It is worth noting the grade of concentrate obtained after flotation of the magnetic fraction is higher than what is obtained

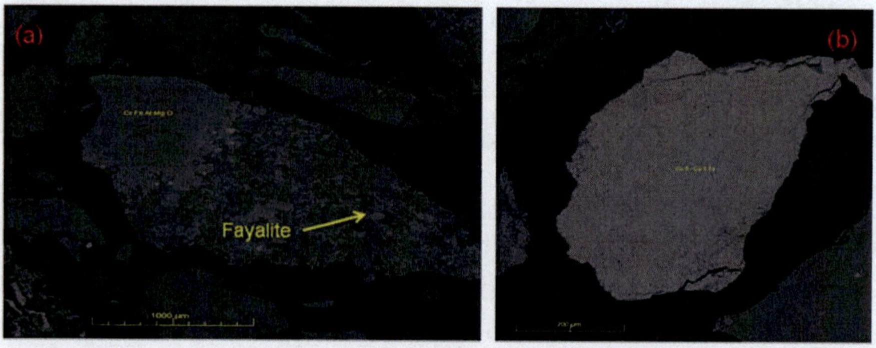

Fig. 1 Scanning electron microscopy (SEM) Images of the as-received samples under backscattered electron imaging

Table 2 Flotation results of the magnetic, weakly, and non-magnetic fractions

Magnetic separation fraction	% Grade in feed				% Grade in concentrate				% Grade in tailings				% Cum. recovery		
	Cu	Cr	Co	S	Cu	Cr	Co	S	Cu	Cr	Co	S	Cu	Cr	Co
Non-mag	3.07	0.52	0.11	3.42	8.21	0.13	0.05	23.54	1.47	0.62	0.11	1.11	89.33	73.53	66.40
Weakly mag	2.01	0.64	0.17	1.96	13.56	0.22	0.09	27.18	0.97	0.67	0.10	0.77	65.02	32.36	37.30
Mag	2.06	0.87	0.21	1.60	41.43	0.54	0.17	7.96	0.62	0.77	0.15	0.96	73.41	16.67	16.16

when bulk material is floated. Commercial flotation of Luanshya copper slag yields a concentrate with less than 20 weight % copper owing to higher content of sulphur.

On the other hand, the grade of concentrate was low (8.21 weight %) when the non-magnetic fraction was floated and this is because of the high content of sulphur as seen from the results.

Visual inspection showed that sulphur particles are generally finer than most slag particles. As a result, the as-received slag material was screened on a 2.36 mm sieve size. The oversize (plus 2.36 mm) fraction was milled to less than 106 microns and subjected to froth flotation, and the results are shown in Table 3. Sulphur was upgraded to 73.65 weight % in the pre-float concentrate. The results in Table 3 further show that copper was upgraded to 27.98% weight in the final concentrate while the final tailings stream has 0.49 weight % copper, thereby representing a cumulative % recovery of copper of 84.36%. The final tailings have less than 0.01 weight % sulphur. Cobalt and iron were slight upgraded in the re-cleaner tails.

Due to high presence of sulphur, the undersize (minus 2.36 mm) fraction was subjected to magnetic separation to obtain magnetic and non-magnetic fractions. The magnetic fraction was 30 weight % of the undersize fraction. The magnetic fraction was subjected to froth flotation, and the results are presented in Table 4 from which it can be observed that a concentrate grade of 25.98 weight % copper was obtained. Chromium was slight concentrated in the scavenger 3 stream.

Table 3 Results after flotation of the oversize (+2.36 mm)

Stream	Weight (g)	% Weight	Weight %					% Recovery			
			Cu	Fe	Cr	Co	S	Cu	Cr	S	Co
Pre-conc	12.25	4.08	3.77	13.58	0.197	0.079	73.65	9.19	2.79	85.2	2.52
Concentrate	6.38	2.13	27.98	17.67	0.20	0.09	6.28	33.01	1.46	3.8	1.56
Re- Cleaner Tails	3.4	1.13	8.39	24.82	0.24	0.15	3.42	5.01	0.94	1.1	1.30
Cleaner Tails 2	12.23	4.08	2.43	22.88	0.28	0.13	2.72	5.92	3.91	3.1	4.04
Cleaner Tails	43.59	14.53	1.30	22.14	0.28	0.14	1.62	11.26	14.13	6.7	15.87
Scavenger 1	39.75	13.25	1.27	21.86	0.28	0.12	0.00	10.09	12.93	0.0	12.71
Scavenger 2	18.56	6.19	1.66	21.56	0.32	0.12	0.00	6.14	6.96	0.0	5.60
Scavenger 3	18.16	6.05	1.04	22.85	0.36	0.13	0.00	3.75	7.65	0.0	6.04
Final Tails	145.64	48.55	0.49	21.15	0.29	0.13	0.00	15.64	49.23	0.0	50.36
Total	299.96	100									

Table 4 Results after flotation of the undersize (−2.36 mm) magnetic fraction

Stream	Weight (g)	% Weight	Weight %					% Recovery			
			Cu	Fe	Cr	Co	S	Cu	Cr	S	Co
Concentrate	2.92	2.02	25.98	22.72	0.13	0.12	6.81	32.15	0.82	86.12	1.53
Re-cleaner Tails	1.16	0.80	9.31	29.16	0.19	0.17	2.53	4.58	0.47	12.71	0.88
Cleaner Tails 2	7.52	5.21	3.53	28.00	0.25	0.20	0.00	11.26	3.90	0.00	6.85
Cleaner Tails	15.44	10.70	1.63	28.87	0.36	0.13	0.00	10.69	11.54	0.00	8.84
Scavenger 1	17.25	11.95	1.33	26.62	0.29	0.18	0.00	9.74	10.46	0.00	13.85
Scavenger 2	12.87	8.92	0.88	26.55	0.35	0.17	0.00	4.79	9.40	0.00	9.63
Scavenger 3	17.50	12.13	0.79	27.89	0.46	0.17	0.00	5.82	16.79	0.00	13.18
Final Tails	69.65	48.26	0.71	27.17	0.32	0.14	0.00	20.95	46.63	0.00	45.23
Total	144.31	100.00									

Reductive Roast—Acid Leaching

The results for the samples that were roasted at various temperatures and leached in sulphuric acid media are shown in Fig. 2. The following observations can be made from Fig. 2.

- The recoveries of Co / Fe and Cu / Cr are similar, and this is because cobalt is hoisted in the iron silicate matrix.
- The recoveries of Cu and Cr are lower compared to Co and Fe. The maximum recoveries of Cu and Cr are less than 15%.
- The highest recoveries were obtained at 900 °C.

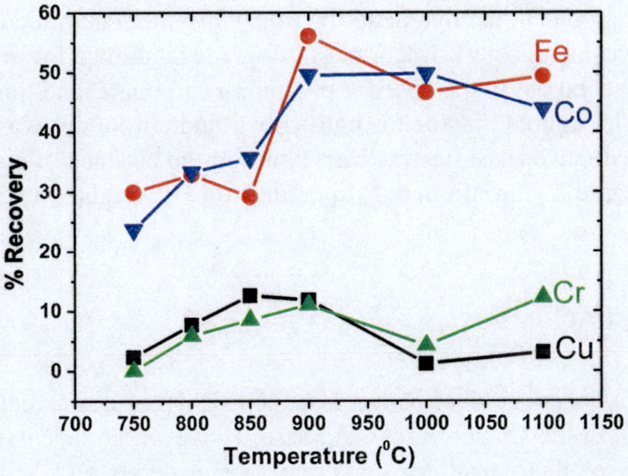

Fig. 2 Plot of % recovery against calcinations temperature

Table 5 Leach test results for the samples that were roasted with 25 weight % charcoal and catalysis (sodium carbonate or borax) at 900 °C and leached at pH of 1.8

Additive	Initial weight (g)	Final weight (g)	% Grade				% Recoveries			
			Cu	Fe	Cr	Co	Cu	Fe	Cr	Co
10%Borax	200	118.22	1.04	8.82	0.82	0.14	67.5	70.8	13.8	49.8
None	200	132.4	0.84	9.08	0.72	0.11	70.8	66.4	14.7	55.9
10 Na_2CO_3	200	126.2	1.33	9.85	0.71	0.09	55.5	65.3	19.9	64.7
10% Na_2CO_3	200	122.5	1.40	10.69	0.83	0.11	54.7	63.4	8.8	59.3

As earlier stated, most of the copper is present in the sulphide or matte form and hence reductive roasting did not metallise it. Chromium oxide is highly refractory, and this is why it was less reduced to metallic state. It is worth noting that the recovery of cobalt and copper is lower than what it was obtained on a copper slag material from Democratic Republic of Congo [3].

Since copper is mostly present in the sulphide form and with the aim of improving the recovery, the material was roasted in air at 700 °C before reductive roasting in the presence of borax or sodium carbonate. The summary of results is given in Table 5 from which it can be observed that the recovery of copper increased to 70.8%. However, there was minimal increase in the recovery of cobalt as the maximum recovery was 64.7%.

Reductive Roasting—Magnetic Separation

Magnetic separation of the reductive roasted sample was carried, and the results are shown in Table 6. It can be observed from Table 6 that cobalt was slight upgraded in the magnetic fraction with a recovery of 97.77%. The low upgrading factor of cobalt is because cobalt is present in the complex iron silicate matrix which reduces its grade in the magnetic fraction. Chromium upgraded by a factor of about 5 in the weakly magnetic fraction. The upgrade factor was lower when reductive roasting was carried out in the absence of sodium carbonate, and this is due to low metallisation. The upgrade factor for both cobalt and chromium was lower for the sample when reductive roasting was carried out in the absence of sodium carbonate as shown in Table 7. Chromium only upgraded to 1.01 weight % in the weakly magnetic fraction.

Conclusions

- Scanning electron microscope (SEM) studies revealed that chromium is present in the Cr-Fe–Mg-Al-O phase. However, copper is mostly present in the matte, oxide (malachite), and complex silicate phases.

Table 6 Magnetic separation results for the sample after reductive roasting in the presence of 10% sodium carbonate, at 900 °C

Stream	Wt. (g)	Wt (%)	% Grade				% Recovery			
			Cu	Fe	Cr	Co	Cu	Fe	Cr	Co
Magnetic	32.29	75.13	1.84	21.92	0.24	0.38	78.23	87.76	28.19	97.77
Weakly Magnetic	4.34	10.10	1.72	14.09	4.34	0.05	10.24	7.58	68.52	1.73
Non-magnetic	6.35	14.77	1.35	5.92	0.14	0.01	11.53	4.66	3.29	0.51
Total	42.98	100.00								

Table 7 Magnetic separation results for the sample, after reductive roasting at 900 °C

Stream	Wt. (g)	Wt. (%)	% Grade				% Recovery			
			Cu	Fe	Cr	Co	Cu	Fe	Cr	Co
Magnetic	63.89	47.36	2.28	24.52	0.36	0.28	65.00	68.17	33.26	74.72
Weakly Magnetic	5.94	4.40	2.04	20.96	1.01	0.17	5.41	5.42	8.58	4.33
Non-magnetic	65.06	48.23	1.02	9.33	0.62	0.08	29.61	26.42	58.17	20.95
Total	134.89									

- Efficient froth flotation of Luanshya slag requires screening the as-received material on a 2.36 mm sieve size, subjecting the undersize to magnetic fraction, and floating the oversize and under size magnetic fraction.
- Sulphur may be recovered via pre-flotation as a grade of about 73 weight % sulphur was obtained after pre-flotation.
- Low temperature reductive roasting—acid leaching of the material was carried out. However, the recovery of cobalt was 55–70% while that of copper and chromium was less than 15%.
- Chromium upgrades by a factor of about 8 in the weakly magnetic fraction when the reductive roasted sample is subjected to magnetic separation. By comparison, cobalt was only upgraded by a factor of 2 in the magnetic fraction.

References

1. Davenport WGL et al (2002) Extractive metallurgy of copper, 4th edn. Elsevier
2. Hara YSR, Jha A (2014) Kinetic and thermodynamic analysis of the reduction of oxides of Cu and Co in a SiO2-CaO-(Al, Fe)2O3 slag. In: Celebrating the megascale. Wiley, pp 553–562
3. Hara YRS, Musowoya SCDM, Kaluba G, Machona J, Chishimba P, Tina Chanda Phiri, Parirenyatwa S (2017) Leaching of copper–cobalt tailings from Democratic Republic of Congo. In: characterization of minerals, metals, and materials 2017. The minerals, metals & materials series. Springer

4. Yotamu RS, Hara AJ (2016) Carbothermic processing of copper–cobalt mineral sulphide concentrates and slag waste for the extraction of metallic values. CRC Press
5. Shekhawat K, Chatterjee S, Joshi B (2015) Chromium toxicity and its health hazards. Int J Adv Res 3(7):167–172
6. Matinde E, Simate G, Ndlovu S (2018) Mining and metallurgical wastes: a review of recycling and re-use practices. J South Afr Inst Min Metall 118(8):825–844
7. Jones RT et al (2002) Recovery of cobalt from slag in a DC arc furnace at Chambishi, *Zambia*. J South Afr Inst Min Metall 102(Compendex): 5–9
8. Jones RT, Deneys AC (1998) Using a direct-current arc furnace to recover cobalt from slags. JOM 50(10):53–56
9. Cui Y et al (2004) Effect of elemental sulphur on solubility of soil heavy metals and their uptake by maize. Environ Int 30(3):323–328

Separation of Li and Co from $LiCoO_2$ Cathode Material Through Aluminothermic Reduction Using Different Aluminum Sources: Chemical Grade, Swarf, and Dross

D. C. Nababan, R. Mukhlis, Y. Durandet, L. Prentice, and M. A. Rhamdhani

Abstract The worldwide ever-increasing vehicle uptake escalates the number of end-of-life lithium-ion batteries that need to be managed. Aluminothermic reduction is considered as an alternative approach for separating Li and Co from $LiCoO_2$, the most common material used as batteries' cathode. The current study evaluates the potential use of three different aluminum sources (chemical grade, waste swarf, and waste dross) as reductants for the aluminothermic reduction process. Systematic thermodynamic analyses have been carried out using the FactSage thermochemical package to identify the optimum conditions and the reaction products at various temperatures. Selected experiments were carried out to demonstrate the process. In the case of reduction with chemical grade aluminum, Co could be extracted as Co metal or Co–Al alloy while Li is distributed to $Li_{(g)}$ and slag. The impurities in the waste swarf and Al dross were found to affect the Co and Li products and their separation mechanism.

Keywords Lithium-ion-batteries recycling · Aluminothermic reduction · Thermite reaction

Introduction

The rapid growth of lithium-ion batteries (LIBs) usage due to the increased global uptake of electric vehicles (EVs) has provided challenges in supplying the demand of the raw materials as well as management of large quantities of end-of-life/spent

D. C. Nababan (✉) · R. Mukhlis · Y. Durandet · M. A. Rhamdhani
Fluid and Process Dynamics (FPD) Group, Department Mechanical and Product Design Engineering, Swinburne University of Technology, Melbourne, VIC 3122, Australia
e-mail: dnababan@swin.edu.au

M. A. Rhamdhani
e-mail: arhamdhani@swin.edu.au

L. Prentice
CSIRO Manufacturing, Clayton, Melbourne, VIC 3169, Australia

R. G. Reddy et al. (eds.), *New Directions in Mineral Processing, Extractive Metallurgy, Recycling and Waste Minimization*, The Minerals, Metals & Materials Series,
https://doi.org/10.1007/978-3-031-22765-3_22

LIBs [1]. The valuable metals contained in the spent LIBs may potentially be recovered through hydrometallurgy and pyrometallurgy processes. The pyrometallurgy approach commonly utilizes elevated temperatures for the extraction and separation processes. Reduction roasting using carbon as a reducing agent is the most common approach for extracting the valuable metals [2]. Nevertheless, metal such as aluminum has also been used as a reductant for the recycling of cathode LIBs in the recent years [3–6].

Aluminum is widely known to be a reducing agent for many metal oxides due to its high reduction potential. Apart from the chemical grade from aluminum producers, metallic Al could be contained in wastes, e.g. swarf and dross. Aluminum swarf or chip are wastes generated from machining during manufacturing process, and the amount can be quite considerable, i.e. 3–5% of the casting weight [7]. The Al swarf used in this study is a waste swarf from Al7075 manufacturing process. Aluminum dross is an industrial waste generated from aluminum smelter plants containing metallic aluminum entrapped in aluminum oxide along with some salt flux. The dross is classified into different categories based on its aluminum content. The dross evaluated in this work contains 10 wt.% Al which is classified as dross salt cakes [8].

This study evaluates the potential use of aluminum from three different sources, e.g. chemical grade (99.98% Al), waste swarf (90% Al), and waste dross (10% Al), as a reductant to separate Li and Co from $LiCoO_2$, the most common material used in batteries' cathode. Thermodynamic assessment and selected experimental works were carried out in the current study to demonstrate the potential of using the different Al sources mentioned. This study provides some alternative route for which can be beneficial for developing future strategy of waste cathode LIBs recycling, especially to recover the Li and Co.

Methodology

Sample Preparation

A commercial reagent-grade $LiCoO_2$ powder (~10 μm, 99.8% purity) supplied from Sigma-Aldrich was used in this study. In order to evaluate the use of Al-only as the reductant, a reagent-grade Al powder (~125 μm, 99.5% purity) supplied by Sigma-Aldrich was used. The waste Al swarf used in the study was collected from machining process at a local workshop, with a typical chemical composition shown in Table 1. It can be seen from Table 1 that the main element in the swarf was Al (~90 wt.%), followed by Zn (~5 wt.%).

Figure 1a shows the physical appearance of the Al swarf with various sizes (around 1 cm in length). The secondary electron (SE) image from a scanning electron microscope (SEM) of one of the Al swarf is provided in Fig. 1b including the mapping of the elements.

Table 1 Chemical composition of typical waste Al swarf (wt%)

Item	Element											
	Al	Si	Cu	Fe	Mg	Zn	Cr	Mn	Ni	Ti	Sr	Zr
Al swarf	90.65	0.15	1.41	0.23	2.1	5.1	0.19	0.09	0.01	0.05	0.001	0.005

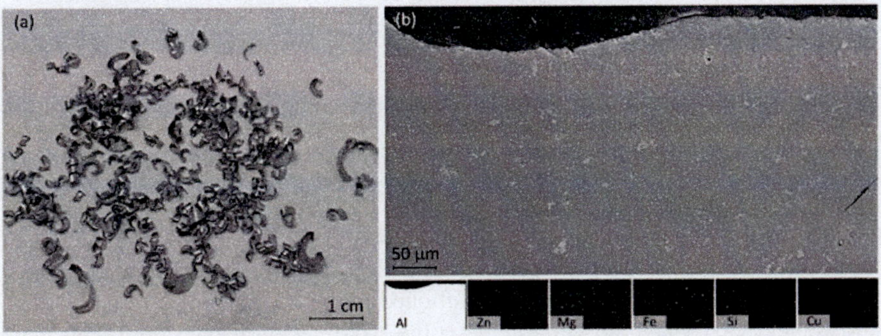

Fig. 1 **a** The typical appearance of Al swarf with the SE SEM image including the mapping of the elements (**b**)

Table 2 Chemical composition of typical waste Al dross (wt%)

Item	Component						
	Al_2O_3	Na_3AlF_6	Al	AlN	KCl	Mg $(AlO_2)_2$	CaF_2
Al dross	64.5	9.25	10	7.5	2.5	3.75	2.5

The Al dross used in this study was obtained from a local aluminum smelter factory where its typical composition is shown in Table 2. The dross mainly contained up to 64 wt% Al_2O_3, while the amount of metallic Al was relatively small, i.e. 10 wt.%. The rest of the composition was dominated by salts.

Thermodynamics Assessment

The thermodynamic assessment was carried out using the thermochemical package FactSage™ 8.2 (Thermfact/CRCT-Ecole Polytechnique de Montreal, QC, Canada). The effects of different parameters including initial reaction temperature and the addition of Al, swarf, and dross were investigated. The details about the FactSage can be seen in Bale et al. [9, 10]. Five databases in FactSage were used for this assessment: FactPS (for pure substances), FToxid (for solid and liquid oxides compound and solution), FTlite (for Al alloys), FTsalt (for chloride salts), and FThall (for cryolite). A private thermochemical database was developed for $LiCoO_2$ compound by considering existing thermodynamics data from literatures. The molar heat capacity

(Cp) data of $LiCoO_2$ was taken from Jankovsky et al. [11], while the enthalpy (H) and entropy (S) data were taken from Chang [12]. The Equilib module was used to predict the final products of the $LiCoO_2$–Al/Al swarf/Al dross system at equilibrium conditions.

Experimental Techniques

Selected experiments for the reactions of $LiCoO_2$–Al and $LiCoO_2$–Al swarf were conducted in a horizontal electrical resistant furnace. Mixtures of the reactants with compositions of 72 wt.% $LiCoO_2$ – 28 wt.% pure Al and 72 wt.% $LiCoO_2$-28 wt.% Al swarf were compacted with a pressure of 10 kPa to form pellets. The furnace was equipped with a finger condenser at the outlet end of the tube, built from a copper tube, and cooled by circulated water to allow the condensation of volatile phases (such as Li phases). The sample was positioned in the hot zone of a furnace filled with argon gas to initiate the reaction. After the end of reaction, the sample was then retrieved for further characterizations.

Results and Discussion

Thermodynamics Assessment Results

It has been observed in the other unpublished studies by the authors that reactions between Al and $LiCoO_2$ can exhibit a thermite reaction/ignition which causes a significant increase in the system's temperature from its initial reaction temperatures of 800 and 1000 °C. To investigate the possible maximum temperature of the reacting system, a set of thermodynamic calculations were carried out. For all the assessments, the ΔH on the final equilibrium condition was set to be 0 in order to simulate the adiabatic condition where the ΔH is converted to increase the temperature of the system. This condition is to illustrate a thermite reaction that occurs with an ignition which leads to an increase in the temperature.

The discussion and comparison of the results of calculations of reactions between $LiCoO_2$ and Al, Al swarf, and Al dross were carried out for a fixed metallic Al to $LiCoO_2$ ratio of 1.39 (in mol) and initial reaction temperature of 1000 °C to allow meaningful comparison. This ratio is represented by the dotted vertical lines in Figs. 2, 3, and 4.

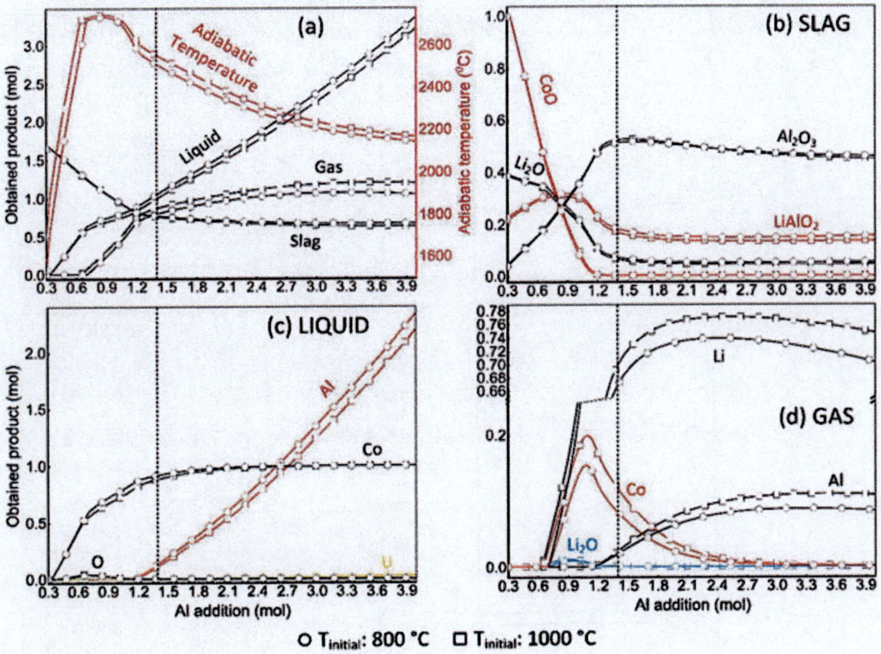

Fig. 2 Equilibrium calculation results of $LiCoO_2$–Al at different initial temperatures, i.e. 800 and 1000 °C at 1 atm with the **ΔH on the final condition set to 0 kJ; a** the red y-axis shows the adiabatic temperature while the rest show the amount of the obtained products in mol. **b** Shows the distribution within the slag, **c** liquid, and **d** gas

Thermodynamics Assessment of Aluminothermic $LiCoO_2$ Using Pure Aluminum

Figure 2 shows the results of the thermodynamic assessment of the $LiCoO_2$–Al system. The adiabatic temperature generated by reaction with 1.39 mol Al was predicted to be 2566 °C. At this temperature, the slag was in liquid state and consisted of CoO–Al_2O_3–Li_2O–$LiAlO_2$. The liquid metal was predicted to be dominated by Co with a small amount of Al. In the gas product, approximately 0.7 mol of Li were predicted with a small amount of Co and Al.

Thermodynamics Assessment of Aluminothermic $LiCoO_2$ Using Waste Aluminum Swarf

Figure 3 shows the results of the thermodynamic assessment of the $LiCoO_2$–Al swarf system. At 0.413 mol of Al swarf (equivalent to 1.39 mol metallic Al) with initial reaction temperature of 1000 °C, the adiabatic temperature generated was predicted to reach 2519 °C, similar to the case of reaction with pure Al. The slag produced was

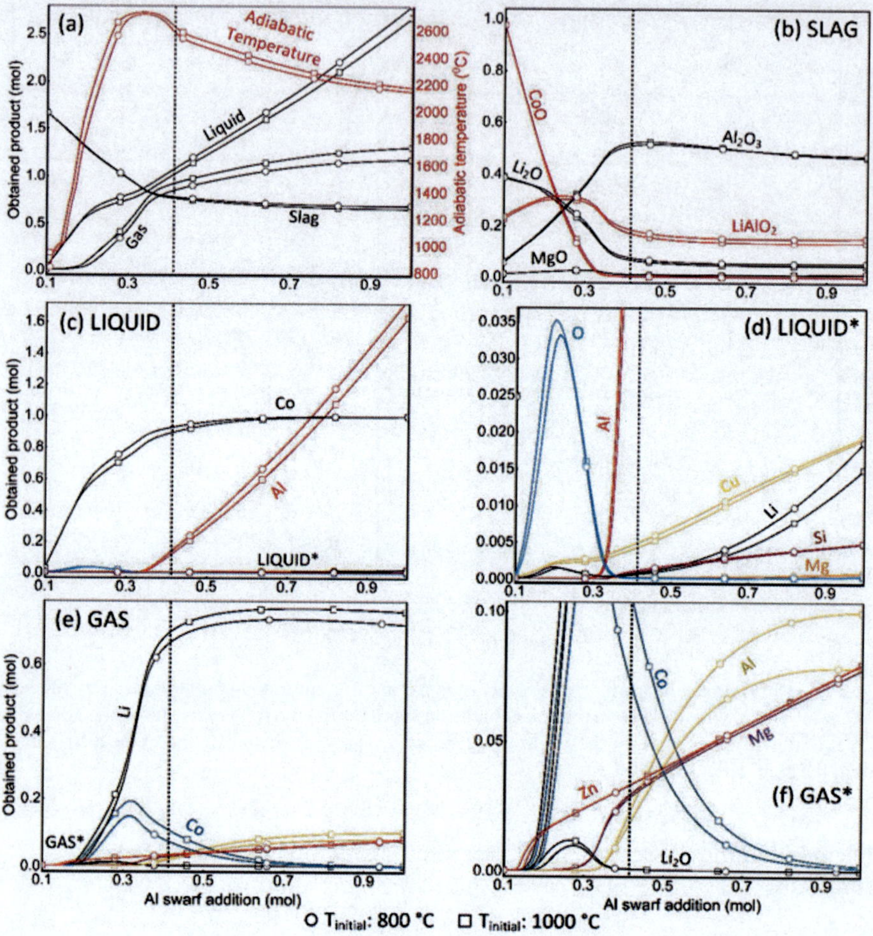

Fig. 3 Equilibrium calculation results on $LiCoO_2$–Al swarf at different initial temperatures, i.e. 800 and 1000 °C at 1 atm with the ΔH **on the final condition set to 0 kJ; a** the red y-axis shows the adiabatic temperature while the rest show the amount of the obtained products in mol. **b** Shows the distribution within the slag, **c** liquid, and **d** gas

predicted to be dominated by Al_2O_3, while the liquid metal contained majority of Co with a small amount of Al. Different from the gas produced from reaction with pure Al, the gas from the reaction with Al swarf was predicted to contain Zn and Mg.

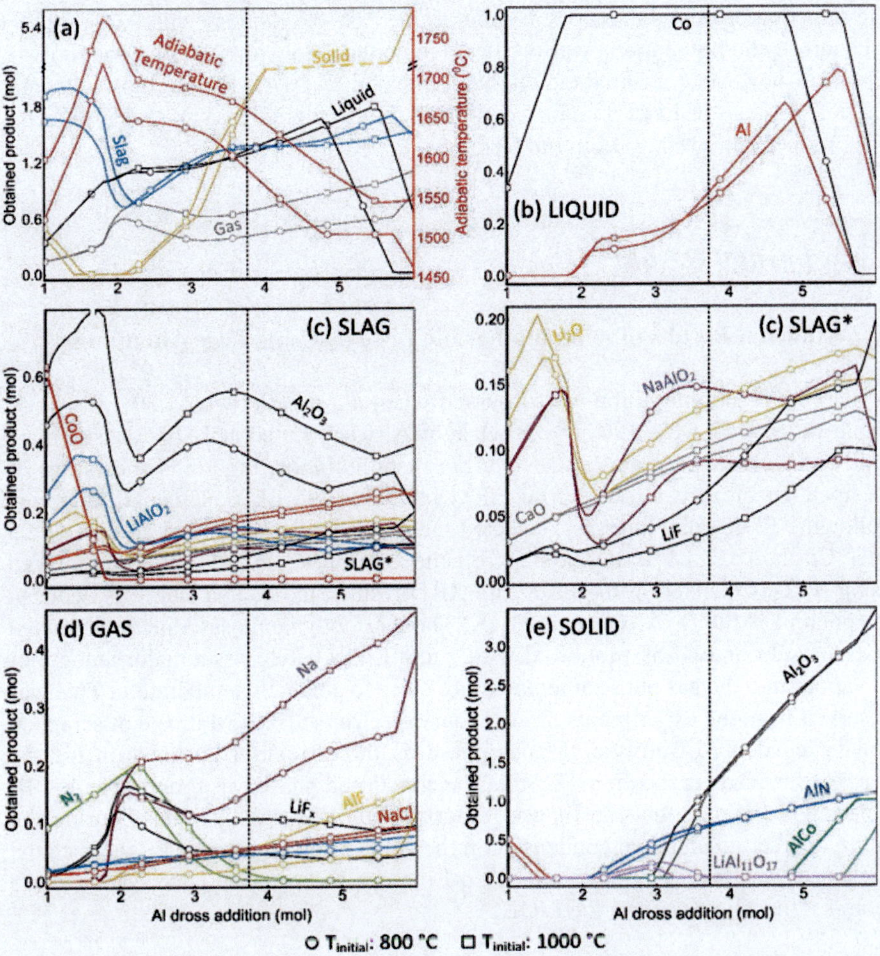

Fig. 4 Equilibrium calculation results on $LiCoO_2$–Al dross at different initial temperatures, i.e. 800 °C and 1000 °C at 1 atm with the **ΔH on the final condition set to 0 kJ**; **a** the red y-axis shows the adiabatic temperature while the rest show the amount of the obtained products in mol. **b** Shows the distribution within the liquid, **c** slag, **d** gas, and **e** solid

Thermodynamics Assessment of Aluminothermic $LiCoO_2$ Using Waste Aluminum Dross

Figure 3 shows the results of the thermodynamic assessment of the $LiCoO_2$–Al dross system. With initial reaction temperature of 1000 °C and at 3.744 mol of Al dross (equivalent to 1.39 mol metallic Al), the adiabatic temperature was predicted to be 1664 °C. This temperature was much lower compared to the case of reactions with pure Al and Al swarf. This was mainly because the amount of Al in dross is quite small (~10 wt%) and much of the heat generated was also used to heat up the rest of the salt components in the dross. Similar to the case of reactions with the other two Al sources, the liquid metal was predicted to contain Co with a small amount of Al. The slag, however, was predicted to be dominated by Al_2O_3 followed by AlF_3 and $LiAlO_2$. Instead of Li as a major component, Na was predicted to dominate in the gas product, followed by Li in the form of LiF.

Experimental Results

Experimental Results of Aluminothermic $LiCoO_2$ Using Pure Aluminum

A thermite reaction/ignition was observed when a 3 g-pellet with 1.39 Al/$LiCoO_2$ molar ratio or 72 wt.% $LiCoO_2$-28 wt.% of Al was reacted at 1020 °C. The reaction was violent and produced several big metallic blobs. Figure 5a shows one of the metallic blobs captured by the SEM including the EDX elemental mapping. Following EDX point analyses, selected metallic blob was found to contain 89.29 wt%. Co and 10.71 wt.% Al. The slag from the reaction was ground and then analyzed using XRD. Considering the EDX and XRD results, the cooled slag was deduced to contain $LiAlO_2$, $LiAl_5O_8$, and Al_2O_3. The thermodynamic assessment predicted that the addition of 1.39 mol of Al with 1 mol $LiCoO_2$ releases a major amount of Li vapour into the gas phase at temperature close to adiabatic temperature. This was observed from the experiments. A secondary electron (SE) SEM image of scrapped powder condensed from the gaseous phase on the condenser is shown in Fig. 5c whereas the XRD is shown in Fig. 5d. The condensed phase was found to be LiOH. This might indicate that the Li was vaporized into the gaseous phase during the thermite reaction and then condensed on the condenser's surface as Li(s). Once the condenser was taken out, the condensed Li interacted with the water vapor in the room's atmosphere to form the LiOH.

Experimental Results of Aluminothermic of $LiCoO_2$ Using Waste Aluminum Swarf

Similar to the case of $LiCoO_2$–Al experiment, the reaction of a 3 g pellet containing 72 wt.% $LiCoO_2$-28 wt.% Al swarf at 1020 °C showed an ignition/thermite reaction and also resulted in metallic and slag phases; as well as condensed phase on the condenser. Figure 6a shows one of the metallic blobs captured by the SEM including the EDX elemental mapping. It shows that the obtained metal contains highly pure Co whereas the slag still contains a relatively small amount of Co. The different outcome with the $LiCoO_2$–Al reaction might be caused by the amount of Al in the pellet being lesser; hence, some of the Co still was in its oxide form in the slag.

Figure 6c shows an SE SEM image of the scrapped powder from the condenser including its XRD result. Based on the EDX analysis, the powder was found to be dominated by Zn with a composition of 87 wt% Zn, 6.5 wt% O, 5 wt% Al, and

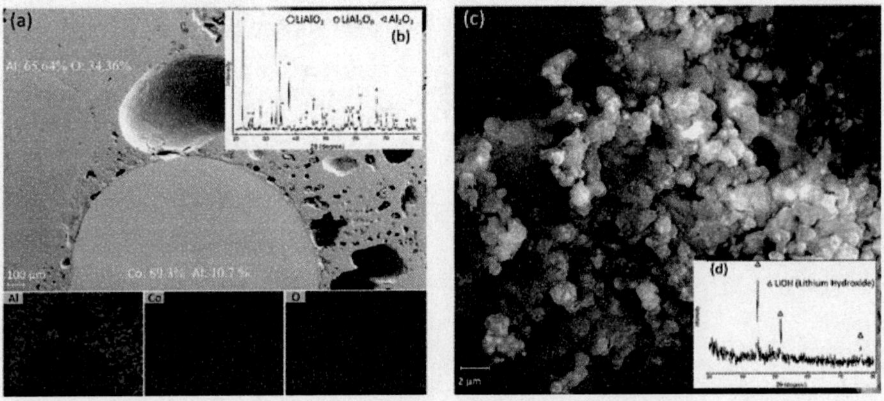

Fig. 5 Products obtained from heating 3 g-pellet containing 72 wt.% $LiCoO_2$ and 28 wt.% pure Al powder: **a** SE SEM of the metal-slag image including the EDX elements mapping with. **b** XRD analysis of the slag; **c** SE SEM of the condensed powder including, **d** its XRD analysis

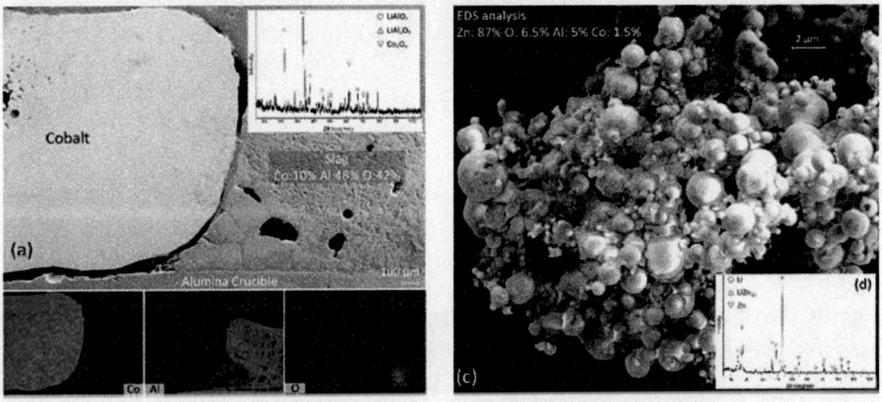

Fig. 6 Products obtained from heating 3 g-pellet containing 72 wt.% $LiCoO_2$ and 28 wt.% pure waste Al swarf: **a** SE SEM of the metal-slag image including the EDX elements mapping with. **b** XRD analysis of the slag; **c** SE SEM of the condensed powder including, **d** its XRD analysis

1.5 wt% Co. However, the EDS could not detect Li, although the XRD shows some peaks of Zn, Li, and $LiZn_{13}$. This might suggest that there were two main elements, Li and Zn, vaporized during the thermite reaction.

As the thermodynamic assessment shows, when the adiabatic condition was achieved, the temperature of the system was increased significantly and resulted in the formation of gas, liquid slag, and liquid metal/alloy. Selected experimental results show that the ignition of thermite reaction occurred leading to smelting of the pellet at high temperature and producing the same products, i.e. gas, liquid slag, and liquid metal/alloy. In order to outline the simplified mechanism of the Li and

Co separation from $LiCoO_2$ using Al as the reductant, the reactions presented in Eqs. 1–4 below were proposed:

$$LiCoO_2 \rightarrow 1/2\,Li_2O + CoO + 1/4\,O_2 \tag{1}$$

$$LiCoO_2 + 1/3\,Al \rightarrow 1/2\,Li_2O + CoO + 1/6\,Al_2O_3 \tag{2}$$

$$3/2\,CoO + Al \rightarrow 3/2\,Co + 1/2\,Al_2O_3 \tag{3}$$

$$3/2\,Li_2O + Al \rightarrow 3Li + 1/2\,Al_2O_3 \tag{4}$$

The thermal treatment of $LiCoO_2$ could lead to its decomposition into Li_2O and CoO as suggested from previous studies [13–15], following Eq. 1. The presence of Al, following Eq. 2, might facilitate further the decomposition of $LiCoO_2$. The highly exothermic reaction between CoO and Al (Eq. 3) might increase the temperature of the system which allows the vaporization of Li (Eq. 4). In the case of using Al swarf as the reductant, the Zn contained in the Al swarf might be vaporized accordingly due to the high temperature of the system.

Conclusions

The thermodynamic assessments and experimental works have demonstrated the use of Al in a form of pure Al powder and waste Al swarf as reductant for separating Li and Co from $LiCoO_2$ into different forms. Li was distributed into the gaseous phase and slag phase while the Co could be recovered as Co metal or Co–Al alloy depending on the amount of Al added in the sample. This separation mechanism was enhanced by the occurrence of ignition from the thermite reaction. The minor element, Zn, in the Al swarf was found to vaporize and interact with the Li.

Acknowledgements This work was conducted under the Swinburne-Automotive Engineering Graduate Program (AEGP) scholarship, funded by Australian Government through the Department of Industry, Science, Energy and Resources (previously known as the Department of Industry, Innovation, and Science (DIIS)). The work was also co-funded by the Commonwealth Scientific and Industrial Research Organisation (CSIRO), Australia.

Conflict of Interest The authors declare that they have no conflict of interest.

References

1. Miao Y et al (2022) An overview of global power lithium-ion batteries and associated critical metal recycling. J Hazard Mater 425:127900
2. Fan E et al (2020) Sustainable recycling technology for Li-ion batteries and beyond: challenges and future prospects. Chem Rev 120(14):7020–7063
3. Wang W et al (2019) A simplified process for recovery of Li and Co from spent LiCoO2 cathode using Al foil as the in situ reductant. ACS Sustain Chem Eng 7(14):12222–12230
4. Wang W et al (2020) Cleaner recycling of cathode material by in-situ thermite reduction. J Clean Prod 249:119340
5. Huang Z et al (2021) In situ recombination of elements in spent lithium-ion batteries to recover high-value γ-LiAlO2 and LiAl5O8. Environ Sci Technol 55(11):7643–7653
6. Dang S et al (2021) In Situ aluminothermic reduction induced by mechanochemical activation enhances the ability of the spent LiCoO2 Cathode to activate peroxymonosulfate. ACS Sustain Chem Eng 9(46):15375–15385
7. Puga H et al (2009) Recycling of aluminium swarf by direct incorporation in aluminium melts. J Mater Process Technol 209(11):5195–5203
8. Meshram A, Jha R, Varghese S (2021) Towards recycling: understanding the modern approach to recover waste aluminium dross. Mater Today: Proc 46:1487–1491
9. Bale C et al (2009) FactSage thermochemical software and databases—Recent developments. Calphad 33(2):295–311
10. Bale CW et al (2002) FactSage thermochemical software and databases. Calphad 26(2):189–228
11. Jankovský O et al (2016) Thermodynamic properties of stoichiometric lithium cobaltite LiCoO2. Thermochim Acta 634:26–30
12. Chang K, Seifert HJ, Phase equilibria, thermodynamic and electrochemical properties of cathodes in lithium ion batteries based on the Li-(Co, Ni)–O system. In: 2013: Hochschulbibliothek der Rheinisch-Westfälischen Technischen Hochschule Aachen
13. Li J, Wang G, Xu Z (2016) Environmentally-friendly oxygen-free roasting/wet magnetic separation technology for in situ recycling cobalt, lithium carbonate and graphite from spent LiCoO2/graphite lithium batteries. J Hazard Mater 302:97–104
14. Mao J, Li J, Xu Z (2018) Coupling reactions and collapsing model in the roasting process of recycling metals from LiCoO2 batteries. J Clean Prod 205:923–929
15. Li J, et al (2020) Pyrolysis kinetics and reaction mechanism of the electrode materials during the spent LiCoO2 batteries recovery process. J Hazard Mater 122955

Towards Framework Development for Benchmarking Energy Efficiency in Foundation Industries: A Case Study of Granulation Process

Shoaib Sarfraz, Ziyad Sherif, Mark Jolly, and Konstantinos Salonitis

Abstract The manufacturing sector depends mainly on the foundation industries which have a major contribution to the country's economy. These foundation industries include glass, metals, ceramics, cement, paper, and chemical sectors that support the demands of our modern lives. On the other hand, these industries are energy intensive. Energy requirements in foundation industries can be improved by benchmarking and comparing the actual energy consumption of individual processes with their theoretical minimum value. In this study, an energy taxonomy approach has been used to identify the elements responsible for energy consumption in the granulation process. Bond's law has been used to calculate the theoretical minimum energy consumption of one such individual process, i.e., granulation—a common process among the foundation industries which accounts for an average of 50% of the total energy consumption. A framework has also been developed that can be used by foundation industries to benchmark their energy efficiency and that provides an insight into the practical and theoretical potential for reducing their energy requirements.

Keywords Benchmarking · Energy efficiency · Foundation industries · Manufacturing · Theoretical minimum

S. Sarfraz (✉) · Z. Sherif · M. Jolly · K. Salonitis
Sustainable Manufacturing Systems Centre, School of Aerospace, Transport and Manufacturing, Cranfield University, Cranfield, Bedfordshire MK43 0AL, UK
e-mail: shoaib.sarfraz@cranfield.ac.uk

Z. Sherif
e-mail: z.sherif@cranfield.ac.uk

M. Jolly
e-mail: m.r.jolly@cranfield.ac.uk

K. Salonitis
e-mail: k.salonitis@cranfield.ac.uk

R. G. Reddy et al. (eds.), *New Directions in Mineral Processing, Extractive Metallurgy, Recycling and Waste Minimization*, The Minerals, Metals & Materials Series,
https://doi.org/10.1007/978-3-031-22765-3_23

Introduction

Manufacturing sector is the backbone of the economic output of a country. On the other hand, it is responsible for 38% of global energy consumption [1]. The manufacturing sector can be split into energy-intensive industries and non-energy-intensive industries based on their energy usage [2]. Food and electronics industries are examples of non-energy-intensive industries. On the other hand, energy-intensive industries include glass, metals, ceramics, cement, paper, and chemical industries, also known as foundation industries. These industries provide essential commodities as well as a major contribution to the overall economy of a country. For instance, in the UK economy around £52 billion is contributed by the foundation industries [3]. In the last few decades, a growing trend has been observed in the manufacturing industry production in response to consumption per capita. Along with labour and raw material costs, energy is a major component of production costs in these "foundation" industries, driving a shift toward improved energy efficiency. Moreover, global environmental concerns raise the focus on energy efficiency. According to IEA [4], energy efficiency is the most contributing factor towards reducing CO_2 emissions by up to 53%. Rogelij et al. [5] stated energy efficiency as one of the keys to retain the potential of keeping global warming to below 1.5 °C.

Benchmarking is a traditional approach for identifying and prioritising particular opportunities to improve energy efficiency in the industry. It has been used in steel and iron [6], construction [7], automotive [8], glass [9], and majority of other processing industries [10], acknowledging the benefits of benchmarking in identifying energy efficiency improvements. Hagenbruch and Zeumer [11] provided a proven benchmarking approach for the steel industry. They also explained how benchmarking can produce useful data and provide performance transparency that creates sound discussions, which can result in new concepts and improved operating procedures. All of this contributes to creating an industrial culture that is focused on performance and improvement. The study concluded that while benchmarking is a widely used tool, many production businesses are sceptical of its practical application due to the difficulty in identifying reliable reference point for the evaluation of their performance versus a specified "best" standard.

Mahamud et al. [12] and Fresner and Krenn [13] recommended using the approach of theoretical minimum consumption calculation to support benchmarking as it can provide an unbiased reference for the specific manufacturing industry. The studies also suggested using the mentioned approach as a base point for looking for ways to improve energy efficiency. The majority of industrial processes consume at least 50% more energy than the theoretical minimum defined by the fundamental rules of thermodynamics [4]. The theoretical minimum energy use is being approached by many countries, particularly in the chemical [14], cement [15], and metal [16] sectors. Although theoretical minimum energy consumption has been investigated as a target for energy reduction measures at the industry level, it has not been generally employed to direct energy reduction efforts at the process level, especially for energy-intensive processes, such as granulation.

Granulation could refer to a size enlargement process such as agglomeration, or to a size reduction such as crushing and grinding. This paper focuses on the size reduction aspect as it is one of the most important processes in the foundation industries. The energy efficiency of the granulation process is usually 5–10% only, with the rest being transformed to heat [4]. It is also an energy-intensive process; for instance, in a typical cement industry, 62% of overall power is consumed by the granulation process [17]. Determining and understanding the energy use of the manufacturing process is important to improve energy efficiency.

The majority of the published research offers benchmarking at the industry, factory, or machine level, which is unable to precisely predict the energy performance at the process level. Literature review reveals that theoretical energy consumption calculations can assist manufacturing industries in benchmarking their energy efficiency, a step toward energy efficiency improvements. From the ongoing studies, it is observed that little or no studies have been found reporting the theoretical minimum energy calculations for the granulation process. Moreover, there is no framework available to benchmark the energy efficiency of granulation process in the foundation industries. Therefore, this study aims to theoretically determine the minimum required energy for the granulation process based on Bond's law. An effort has also been made to develop a high-level framework to benchmark energy efficiency in foundation industries, focusing on the granulation process.

Energy Taxonomy

Energy is one of the most fundamental resources for manufacturing industries for future economic growth and prosperity. Its consumption is expected to continue to increase over the coming decades, with world energy demand estimated to be 45% higher in 2030 than today's levels [18]. The foundation industries were responsible for 59% of the total energy used by the manufacturing sector of the United States in 2018, as portrayed in Fig. 1. The data is obtained from the U.S Energy Information Administration (EIA) [19]. This highlights the need for energy management resources aiming at the foundation industries and their operations.

In order to reduce the need for energy, the first step would be identifying and characterising the energy consumption of the process at hand. This includes any element that can influence the required energy of the process. According to Seow and Rahimifard [20], the energy used by various processes required to manufacture a product is defined as direct energy. This energy consists of two components: (i) minimum energy which refers to the lowest possible energy (bound by the physics of the process) and (ii) auxiliary energy which can be defined as the energy required by the supporting activities and auxiliary equipment for the process. Therefore, the parameters affecting energy consumption can be recognised based on their influence on both aspects of direct energy.

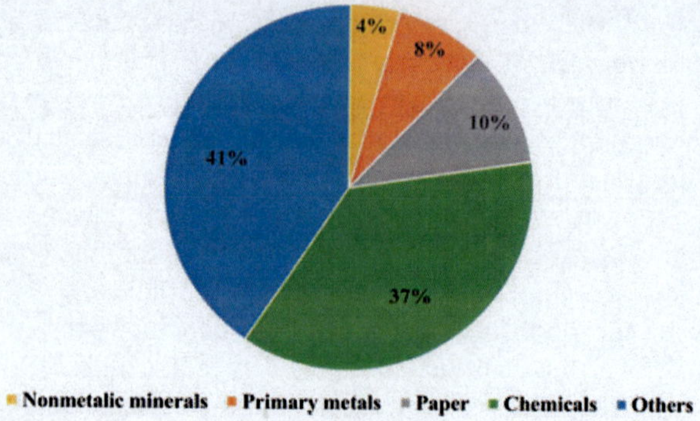

Fig. 1 Manufacturing industries' energy consumption by sector in the US (2018); Non-metallic minerals: cement, glass, gypsum, and limestone; Primary metals: iron and steel; Paper, Chemicals; Others: non-foundation industries. (Total = 19,436 Btu)

In the case of granulation, the minimum energy consists of the energy required for the size reduction of a material and therefore will depend on material characteristics. Meanwhile, auxiliary energy incorporates the energy consumed by machine tools and production related equipment and will depend on its type and efficiency [21, 22]. In reality, the direct energy supplied is mostly converted into heat (losses) whereas only a minor portion is consumed by the main size reducing equipment and auxiliary operations [23]. Hence, the actual energy consumed in the granulation process consists of direct energy and unwanted losses. Figure 2 depicts how the energy supplied for the granulation process is allocated along with the elements responsible for the consumption.

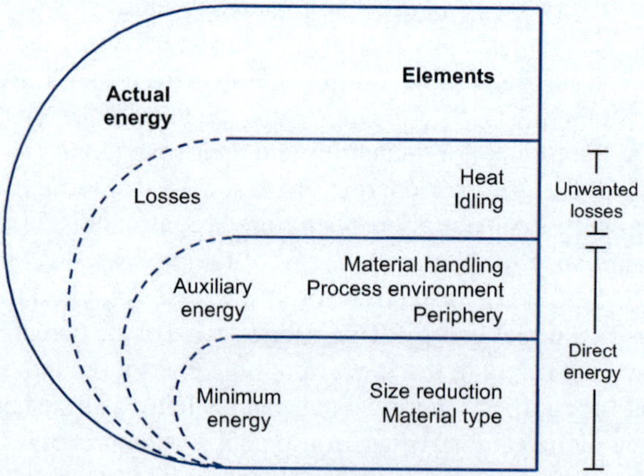

Fig. 2 Energy consumption taxonomy of granulation process (not to scale)

Framework Development

A basic framework has been developed in this study for the purpose of benchmarking, where the theoretical minimum energy acts as a reference point. This will allow for the identification of potential room for improvements once compared against actual consumption by recognising means to minimise the gap between the minimum energy required for the process and the actual consumption. It is essential to have an accurate measure of the actual energy consumed during a process and cannot solely rely on theoretical modelling for estimating energy consumption [24].

According to Klein et al. [25], the energy efficiency of the granulation process can be defined as the energy required for size reduction divided by the energy used by the mechanical system. Based on that, and the work done by Nadolski et al. [26], the benchmarking of granulation energy efficiency will involve the use of the following proposed parameters:

- B = Baseline efficiency of existing scenario = theoretical minimum energy (TME)/Baseline actual energy (AE_b)
- B_n = Improved efficiency of potential scenario (where n represents the number of potential scenarios) = theoretical minimum energy (TME)/Actual energy after improvements (AE_i)
- Efficiency improvement = $B_n - B$

The theoretical minimum energy (TME) represents the lowest energy theoretically required to reduce the size of a material, where its calculation will be examined in the following section. The actual energy (AE) is the value of the energy consumed during real time operations of a granulation system. As seen in Fig. 3, the benchmarking is done by computing the efficiency parameters B and B_n, which represent the baseline efficiency and the improved efficiency, respectively. Possible improvements could target the process and the way in which it is being conducted, or the equipment being utilised.

It would be challenging to identify an inclusive list of material properties that affect the energy required to reduce the size of a particular material. For instance, the hardness of a material, which is one of the most important factors in the amount of energy consumed by the milling unit, has a minor impact on the amount of energy consumed by jaw crushers [21]. To solve this problem, a material property, known as Bond's work index [27] which expresses the resistance of the material to crushing

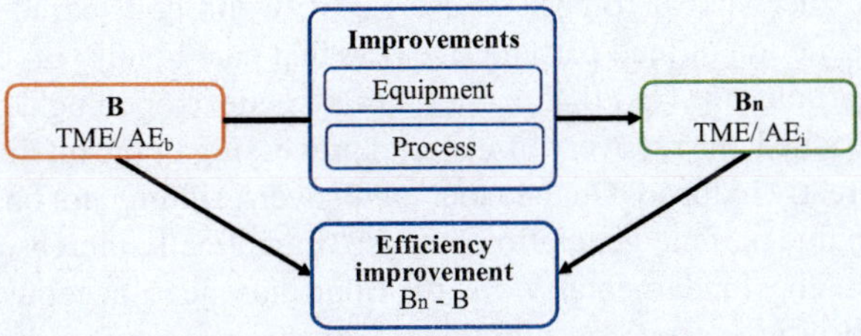

Fig. 3 Basic benchmarking framework overview

and grinding, was introduced. Crushing and grinding are the two industrial means of conducting the granulation process. The following section describes the theoretical minimum calculations using the work index.

Theoretical Minimum Energy Calculation

Although it is impossible to estimate accurately the amount of energy required to reduce the size of a given material, a number of empirical laws have been proposed. Earlier two laws were introduced by Kick [28] and Rittinger [29] in 1885 and 1867, respectively, and then, the third law by Bond [27] was proposed in 1952. These three laws may all be derived from the basic differential equation [30]:

$$\frac{dE}{dL} = -KL^n \tag{1}$$

which states that the energy *dE* required to result in a change in the size *dL* of a unit mass of material is a simple power function of the size (L) and the crushing strength of the material (K) [31]. It has been found that Rittinger's expression, $n = -2$, is more applicable for coarse size reduction while that of Kick, $n = -1$, is more appropriate for finer size reductions in the region where large surface areas of particles are exposed as in the case of grinding operations. Bond has suggested a law intermediate between Rittinger's and Kick's laws which covers almost the entire range of particles, by equating $n = -3/2$ leads to:

$$E = -10\,W_i\left[\frac{1}{\sqrt{L_1}} - \frac{1}{\sqrt{L_2}}\right](\text{kWh/t}) \tag{2}$$

where E expresses the minimum amount of specific energy required to reduce the size of a particle from L_1 to L_2. W_i presents the Bond's work index which is defined as the amount of energy required to reduce the unit mass of material from infinite particle size to a size of 100 μm, which is usually computed experimentally with standardised tests.

Since the main consequence of granulation is the change in size dL, this parameter is considered crucial for determining the minimum required energy E for the process. Similarly, it is a process in which the dispersion state of solids changes with the action of mechanical forces. To use Eq. 2, Bond considered the energy required for the reduction of feed particle size (L_1) that passed 80% of a particular sieve to a product particle size (L_2) that passed 80% of a sieve opening. In case this certain size is not reached, the recovery of material processing or the quality of the final product will be greatly reduced. On the other hand, over-grinding not only reduces the efficiency due to fine particle generation but also dramatically increases the granulation costs [32].

The fundamental work by Bond has been accepted universally to predict the specific energy consumption of crushing machines during material size reduction

[33, 34]. The limitation to bond's law is its dependence on the work index of the material which is obtained experimentally. However, since the development of the concept many decades ago, the work index W_i of several materials has been computed and documented regularly. This led to an abundance of data for each material based on its type and composition. The final version of the TME equation for a particular size change in mm is given as:

$$\text{TME} = -0.3162\, W_i \left[\frac{1}{\sqrt{L_1}} - \frac{1}{\sqrt{L_2}} \right] (\text{kWh/t}) \tag{3}$$

Case Study of Granulation Process

As mentioned before, granulation could refer to the grinding process which consumes the most electricity in the cement industry. It is responsible for about 50% of the whole electrical energy utilization [35–37]. The consumption is strongly influenced by the production method, compositions of raw materials, and grinding fineness [38]. Clinker grinding consumes the largest amount of electrical energy in a cement plant [39]. This is attributed to various types of equipment incorporated in a circuit revolved commonly around a mill. The minimum power required is consumed by the main drive systems of mills, system fans, and separators. However, the process also includes the consumption of auxiliary equipment such as material transport systems, weigh-feeders, and pneumatic and hydraulic units. The input energy is lost in the form of heat produced from the frictional forces encountered by the mill as it rotates [40]. To demonstrate the use of the framework, the benchmarking process was implemented on possible improvements on equipment changeover bases.

Accordingly, a cement ball mill was used as a baseline case while the roller press and ring roller mill (Horomill) were taken as the potential equipment for the improved cases. The work index of Limestone, the main ingredient of clinker, was determined by Heiskari [41] as 10.21 kWh/t. The size of the feed and product together with the actual energy consumption values are presented in Table 1 as obtained from [42] and TransFIRe industry partner. The values represent an estimation of the general figures across the grinding stage of the cement industry.

Accordingly, using the formulas presented in Sect. "Framework Development", the benchmarking was performed for the base case and the subsequent possible improved scenarios. The values, presented in Table 1, indicate that the ball mill system is highly inefficient due to the large difference between its actual consumption (40 kWh/t) and the theoretically calculated minimum energy (18.88 kWh/t) required for such size reduction. On the other hand, the Horomill machine had the highest efficiency when compared to the other scenarios as seen in Fig. 4. Hence, a 27% increase in efficiency is possible just by replacing the grinding equipment.

While it is not always possible to completely change the equipment used without incurring large capital costs. However, by altering just the process conditions, the

Table 1 Grinding parameters and energy consumption values for case study

L_1 (mm)	30		
L_2 (mm)	0.045–0.01		
Limestone Wi (kWh/t)	10.21		
	Ball mill	**Roller press mill**	**Horomill**
AE (kWh/t) specified	40	27.50	21
TME (kWh/t) calculated	18.88	14.63	15.55
B	47.20	–	–
B1	–	53.20	–
B2	–	–	74.06

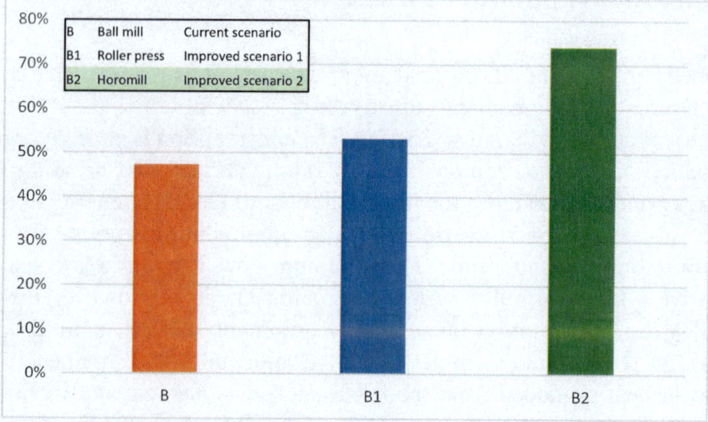

Fig. 4 Efficiency values for current and potential improved scenarios

overall energy required could be reduced. For instance, modifying the charge amount and pattern of the ball mill to increase the average surface area and the effective contact surface of balls can enhance the production rate and, therefore, reduce electricity consumption [43]. Moreover, implementing closed-circuit grinding enhances the overall efficiency of the mills [44]. Altun [45] discovered that altering the flow sheet of the circuit can result in up to 4% in energy savings. Additionally, if both coarse particles and very fine particles are minimized overall and the particle size distribution is made more uniform, the energy consumption would be more stable [40]. Advancements made in Waste Heat Recovery (WHR) systems make it a viable option to improve overall efficiency beyond reducing the input energy [46]. The currently available WHR systems are established in dealing with the heat produced by the kilns [47]; however, systems targeting grinding circuits are lacking. According to Napier [48], much can be done now to reduce granulation energy consumption by up to 30%, and energy savings of over 50% will be attainable in the next 10 years with technological advances currently under development.

Conclusions

Enhancing energy efficiency is the most effective and environmentally benign approach to meet the world's energy demands. Reduction in energy supply investment is also a benefit of energy efficiency improvements. Many industrial processes are energy inefficient, and therefore, the overall energy consumption is significantly higher than what the best technology would allow.

This study is a step towards developing a framework for benchmarking energy efficiency of foundation industries in relation to the minimum energy required to execute the process. The framework incorporates benchmarking parameters that have been established to benchmark energy efficiency for the granulation process. This is due to its importance in the foundation industries and the fact that it is one of the most energy-intensive processes. The energy taxonomy approach reveals the elements accountable for actual energy use in the process. In this paper, a case study has been presented to demonstrate how the proposed benchmarking concept can be applied to a granulation process for cement production. The efficiencies of three types of grinding machines have been computed relative to their theoretically obtained minimum energy required. This allowed for the recognition of the most efficient equipment as well as the energy that is not used directly for production. However, optimising the process conditions could also enhance the efficiency without having to change the equipment.

This is an ongoing project and will be explored further towards an advanced level of the framework by introducing practical minimum and best practice values to identify further energy saving and energy recovery opportunities. Moreover, some case studies will be performed to evaluate the framework with the industrial partners of TransFIRe. Further research is in progress to identify and benchmark the resource efficiency of the common processes across the foundation industries.

Acknowledgements The authors would like to acknowledge the Transforming Foundation Industries Research and Innovation hub (TransFIRe) for the support of this work. TransFIRe (EP/V054627/1)) is funded by UKRI (https://gow.epsrc.ukri.org/NGBOViewGrant.aspx?GrantRef=EP/V054627/1).

References

1. IEA (2021) Tracking industry 2021. https://www.iea.org/reports/tracking-industry-2021. Accessed 28 Apr 2022
2. Worrell E (2018) Industrial energy use, status and trends. Encycl. Anthr. 1:421–430. https://doi.org/10.1016/B978-0-12-809665-9.09045-5
3. Jolly M, Velenturf APM, Salonitis K, Paddea S (2022) The UK transforming the foundation industries research and innovation hub (TransFIRe). In: Developing tomorrow's technical cycles, vol I, pp 341–353. https://doi.org/10.1007/978-3-030-92563-5_35
4. IEA (2006) Energy technology perspectives 2006: scenarios and strategies to 2050. OECD Publishing, Paris. https://doi.org/10.1787/9789264109834-en

5. Rogelj J et al (2015) Energy system transformations for limiting end-of-century warming to below 1.5 °C. Nat Clim Chang 5(6):519–527. https://doi.org/10.1038/nclimate2572
6. Haider S, Mishra PP (2019) Benchmarking energy use of iron and steel industry: a data envelopment analysis. Benchmark Int. J. 26(4):1314–1335. https://doi.org/10.1108/BIJ-02-2018-0027
7. Costa DB, Formoso CT, Kagioglou M, Alarcón LF, Caldas CH (2006) Benchmarking initiatives in the construction industry: lessons learned and improvement opportunities. J Manag Eng 22(4):158–167. https://doi.org/10.1061/(ASCE)0742-597X(2006)22:4(158)
8. Oh S-C, Hildreth A (2014) Estimating the technical improvement of energy efficiency in the automotive industry—Stochastic and deterministic frontier benchmarking approaches. Energies 7(9):6196–6222. https://doi.org/10.3390/en7096196
9. Ross CP (1997) Energy benchmarking: a tool for continuing process improvement for the glass industry. In: 57th conference on glass problems: ceramic engineering and science proceedings, pp 95–108. https://doi.org/10.1002/9780470294406.ch9
10. Chan DY-L, Huang C-F, Lin W-C, Hong G-B (2014) Energy efficiency benchmarking of energy-intensive industries in Taiwan. Energy Convers Manag 77:216–220. https://doi.org/10.1016/j.enconman.2013.09.027
11. Hagenbruch T, Zeumer B (2017) How benchmarking can improve cost competitiveness in steel. https://www.mckinsey.com/industries/metals-and-mining/our-insights/how-benchmarking-can-improve-cost-competitiveness-in-steel?cid=other-eml-alt-mip-mck-oth-1708&hlkid=e49fb3ad97f2405b9e49069d7eda7b13&hctky=1298066&hdpid=79b0d5b8-b6ff-4b41-9db9. Accessed 16 Aug 2022
12. Mahamud R, Li W, Kara S (2017) Energy characterisation and benchmarking of factories. CIRP Ann 66(1):457–460. https://doi.org/10.1016/j.cirp.2017.04.010
13. Fresner J, Krenn C (2018) Theoretical minimum consumption calculation as starting point for cleaner production option identification as a new approach to benchmarking. J Clean Prod 172:1946–1956. https://doi.org/10.1016/j.jclepro.2017.11.233
14. Worrell E, Phylipsen D, Einstein D, Martin N (2000) Energy use and energy intensity of the U.S. chemical industry. Berkeley, CA. https://doi.org/10.2172/773773
15. IEA (2018) Technology roadmap: low-carbon transition in the cement industry. https://www.iea.org/reports/technology-roadmap-low-carbon-transition-in-the-cement-industry. Accessed 17 Aug 2022
16. Fruehan RJ, Fortini O, Paxton HW, Brindle R (2000) Theoretical minimum energies to produce steel for selected conditions. U.S. Department of Energy, Washington DC. https://doi.org/10.2172/769470
17. Xu D, Cui Y, Li H, Yang K, Xu W, Chen Y (2015) On the future of Chinese cement industry. Cem Concr Res 78:2–13. https://doi.org/10.1016/j.cemconres.2015.06.012
18. IEA (2021) World energy outlook 2021. https://iea.blob.core.windows.net/assets/4ed140c1-c3f3-4fd9-acae-789a4e14a23c/WorldEnergyOutlook2021.pdf. Accessed 01 Sep 2022
19. EIA (2018) Manufacturing energy consumption survey (MECS). https://www.eia.gov/consumption/manufacturing/data/2018/?src=‹ConsumptionManufacturingEnergyConsumptionSurvey(MECS)-f1. Accessed 01 Sep 2022
20. Seow Y, Rahimifard S (2011) A framework for modelling energy consumption within manufacturing systems. CIRP J Manuf Sci Technol 4(3):258–264. https://doi.org/10.1016/j.cirpj.2011.03.007
21. Kujundžić T, Bedeković G, Kuhinek D, Korman T (2008) Impact of rock hardness on fragmentation by hydraulic hammer and crushing in jaw crusher. Rud Zb 20(1):83–90
22. Li Y, He Y, Wang Y, Yan P, Liu X (2014) A framework for characterising energy consumption of machining manufacturing systems. Int J Prod Res 52(2):314–325. https://doi.org/10.1080/00207543.2013.813983
23. Balasubramanian S, Rajkumar R, Singh KK (2017) Determination of grinding parameters of fenugreek seed. J Spices Aromat Crop 26(1):16–26. https://doi.org/10.25081/josac.2017.v26.i1.802

24. Salonitis K, Ball P (2013) Energy efficient manufacturing from machine tools to manufacturing systems. Procedia CIRP 7:634–639. https://doi.org/10.1016/j.procir.2013.06.045
25. Klein B, Wang C, Nadolski S (2018) Energy-efficient comminution: best practices and future research needs. In: Energy efficiency in the minerals industry. Springer, Cham, pp 197–211. https://doi.org/10.1007/978-3-319-54199-0_11
26. Nadolski S, Klein B, Gong D, Davaanyam Z, Cooper A (2015) Development and application of an energy benchmarking model for mineral comminution
27. Bond FC (1952) Third theory of comminution. Min Eng 4:484–494
28. Kick F (1885) Das Gesetz der proportionalen Widerstände und seine Anwendungen: Nebst Versuchen über das Verhalten verschiedener Materialien bei gleichen Formänderungen sowohl unter der Presse als dem Schlagwerk. Arthur Felix, Leipzig, Germany
29. von Rittinger PR (1867) Lehrbuch der aufbereitungskunde in ihrer neuesten entwicklung und ausbildung systematisch dargestellt. Ernst & Korn, Berlin, Germany
30. Richardson JF, Harker JH, Backhurst JR (2002) Particle size reduction and enlargement. In: Chemical engineering series, chemical engineering, 5th ed. Elsevier, pp 95–145. https://doi.org/10.1016/B978-0-08-049064-9.50013-8
31. Gupta A, Yan D (2016) Size reduction and energy requirement. In: Mineral processing design and operations, 2nd ed. Elsevier, pp 71–121. https://doi.org/10.1016/B978-0-444-63589-1.00003-4
32. Salopek B, Sobota I, Bedeković G (2002) Construction features and efficiency of vertical impact crushers. Rud Zb 14(1):65–75
33. Refahi A, Rezai B, Aghazadeh Mohandesi J (2007) Use of rock mechanical properties to predict the bond crushing index. Miner Eng 20(7):662–669. https://doi.org/10.1016/j.mineng.2006.12.015
34. Lindqvist M (2008) Energy considerations in compressive and impact crushing of rock. Miner Eng 21(9):631–641. https://doi.org/10.1016/j.mineng.2007.11.013
35. Hosten C, Fidan B (2012) An industrial comparative study of cement clinker grinding systems regarding the specific energy consumption and cement properties. Powder Technol 221:183–188. https://doi.org/10.1016/j.powtec.2011.12.065
36. Njoku HO, Bafuwa OR, Mgbemene CA, Ekechukwu OV (2017) Benchmarking energy utilization in cement manufacturing processes in Nigeria and estimation of savings opportunities. Clean Technol Environ Policy 19(6):1639–1653. https://doi.org/10.1007/s10098-017-1353-x
37. Fellaou S, Bounahmidi T (2017) Evaluation of energy efficiency opportunities of a typical Moroccan cement plant: Part I. Energy analysis. Appl Therm Eng 115:1161–1172. https://doi.org/10.1016/j.applthermaleng.2017.01.010
38. Madlool NA, Saidur R, Rahim NA, Kamalisarvestani M (2013) An overview of energy savings measures for cement industries. Renew Sustain Energy Rev 19:18–29. https://doi.org/10.1016/j.rser.2012.10.046
39. Zhang W, Maleki A, Khajeh MG, Zhang Y, Mortazavi SM, Vasel-Be-Hagh A (2019) A novel framework for integrated energy optimization of a cement plant: an industrial case study. Sustain Energy Technol Assess 35:245–256. https://doi.org/10.1016/j.seta.2019.06.001
40. Ghalandari V, Iranmanesh A (2020) Energy and exergy analyses for a cement ball mill of a new generation cement plant and optimizing grinding process: a case study. Adv Powder Technol 31(5):1796–1810. https://doi.org/10.1016/j.apt.2020.02.013
41. Heiskari H, Kurki P, Luukkanen S, Gonzalez MS, Lehto H, Liipo J (2019) Development of a comminution test method for small ore samples. Miner Eng 130:5–11. https://doi.org/10.1016/j.mineng.2018.10.005
42. Genç Ö, Benzer AH (2016) Effect of high pressure grinding rolls (HPGR) pre-grinding and ball mill intermediate diaphragm grate design on grinding capacity of an industrial scale two-compartment cement ball mill classification circuit. Miner Eng 92:47–56. https://doi.org/10.1016/j.mineng.2016.02.009
43. Treiber K, Fleiger P (2019) Increase in energy efficiency and enhancement of cement properties by means of separate ultra-fine grinding. Cem Int 17(1):26–35

44. Altun O, Benzer H, Dundar H, Aydogan NA (2011) Comparison of open and closed circuit HPGR application on dry grinding circuit performance. Miner Eng 24(3–4):267–275. https://doi.org/10.1016/j.mineng.2010.08.024
45. Altun O (2018) Energy and cement quality optimization of a cement grinding circuit. Adv Powder Technol 29(7):1713–1723. https://doi.org/10.1016/j.apt.2018.04.006
46. Giles A, Williams C, Pugh D, Bowen P, Marsh R, O'Doherty T (2021) A waste heat recovery strategy and its deployment: an integrated steelworks case study. Proc Inst Civil Eng Waste Resour Manag 174(1):3–11. https://doi.org/10.1680/jwarm.20.00011
47. International CCS Knowledge Centre (2021) Powering amine regeneration with waste heat energy for CCS on cement
48. Napier-Munn T (2015) Is progress in energy-efficient comminution doomed? Miner Eng 73:1–6. https://doi.org/10.1016/j.mineng.2014.06.009

Alternative Fluxes for Lead Bullion Refining

Samuel Asante and Patrick Taylor

Abstract The Harris process for lead softening uses sodium hydroxide and sodium nitrate to remove arsenic, tin, and antimony from molten lead by oxidation. It is known to be better than oxygen softening; however, the process has a higher cost in terms of the reagents used. Sodium hydroxide is corrosive and expensive. In this research, calcium carbonate, calcium hydroxide, and magnesium hydroxide were used as alternatives for sodium hydroxide in the softening process. Based on the initial results, the research focused on using calcium carbonate for subsequent softening process. In this study, the activities of the impurities, arsenic, tin, and antimony, were calculated and compared with that obtained with sodium hydroxide as the melt. The softening experiments were conducted on lead bullion by design of experiment (DOE) method by varying temperature, time, and reagent quantity. The dross obtained was analyzed chemically and with X-ray diffraction (XRD) to identify the phases formed. Experimental and theoretical results are presented.

Keywords Lead bullion · Dross · Alternative fluxes · Lead softening

Introduction

Secondary lead produced by recycling process accounts for much of the world market for lead. Lead recycling is gaining prominence due to societal pressure to mitigate health and environmental hazards of lead. Lead is a heavy metal, soft, malleable, relatively low melting point, and high boiling point. It is silvery when freshly cut but oxidizes to form a dull gray coating when exposed to air. The coating makes lead resistant to corrosion as evidenced by its continuous use as water piping by ancient Romans [1]. It is used in roofing and as coverings for electric cables [2]. Lead is used

S. Asante · P. Taylor (✉)
Department of Metallurgical and Materials Engineering, Colorado School of Mines, Golden, CO 80401, USA
e-mail: prtaylor@mines.edu

S. Asante
e-mail: skasante@mines.edu

R. G. Reddy et al. (eds.), *New Directions in Mineral Processing, Extractive Metallurgy, Recycling and Waste Minimization*, The Minerals, Metals & Materials Series,
https://doi.org/10.1007/978-3-031-22765-3_24

as shield to protect nuclear reactors because it can effectively absorb electromagnetic radiation. The awareness of lead toxicity in recent times has restricted its usage in water pipes, paints, and cosmetic powder. Children are at high risk of lead poisoning and its effect on their mental development [3, 1]. Lead usage is recently dominated by the automotive lead-acid battery. Lead is the most recycled metal; about 60% of the world's supply of lead is produced from recycling [3]. Lead is estimated to be 13–20 ppm in the earth crust. Significant deposits of lead are in United States, Australia, Spain, Germany, Canada, Africa, and South America. In the ancient times, lead was mined and produced as co-product of silver [3]. Lead is mainly extracted from galena which contains lead sulfide (PbS) and silver. Other lead ores are cerussite ($PbCO_3$), anglesite ($PbSO_4$), and lead ochre (PbO) [3, 2, 1]. Primary lead is mainly produced from galena which is often mined with zinc sulfides. Demand for zinc metal has outweighed primary lead resulting in decline in mined lead [3]. Galena is concentrated by froth flotation; the concentrate contains impurities such as silver, copper, tin, bismuth, gold, cadmium, arsenic, antimony, gallium, mercury, and germanium. Lead refining process involves recovery of by-product such as silver and gold [3].

Lead production in the US mainly comes from secondary production. In January 2022, lead production from mines was 20,800 metric tons and 75,500 metric tons from the secondary production (mostly from lead/acid batteries) [4]. In secondary lead production, the alloying elements (in recycled acid/lead batteries) are removed by refining processes. The presence of arsenic, tin, and antimony causes hardness in lead, which is otherwise soft in its solid state. Softening process is used to remove arsenic, tin, and antimony from lead bullion. The pyrometallurgical method for lead softening process uses air (enriched air) or sodium hydroxide and sodium nitrate (Harris process). Studies have been conducted by several investigators on lead bullion to improve energy efficiency in electrorefining (Betts process), to optimize reagent usage, and to understand the thermodynamics and kinetics of lead softening. There is an ever-increasing interest in finding alternative reagents for the refining process.

The author has been investigating alternative chemistries and fluxes, which are less expensive and less corrosive for lead bullion softening/refining. The scope of the research is to find an alternative reagent for sodium hydroxide for both softening and final refining largely because the physical chemistry of lead softening has not been fully examined and there is a possibility to improve the process. The study examines the thermodynamics of using calcium carbonate and sodium nitrate for the softening process to remove arsenic, tin, and antimony.

Materials and Method

Materials

Lead (Pb 99.99%) and arsenic (As 99%) used for preparing crude lead were bought from Thermo Scientific, and antimony (Sb: 99.5%) and tin (Sn: 99.5%) were from

Sigma Aldrich. Calcium carbonate ($CaCO_3$: 98%), sodium nitrate ($NaNO_3$: 99%) from Alfa Aesar, sodium hydroxide (NaOH: 98%), and magnesium hydroxide ($Mg(OH)_2$: 95%) from Sigma Aldrich were reagents used in this study.

Method

The crude lead was prepared based on average impurities concentrations of As, Sn, and Sb from industrial partners as shown in Table 1. These are companies that funded this research through Center for Resource Recovery and Recycling (CR^3). Lead (Pb), arsenic (As), tin (Sn), and antimony (Sb) were weighed. The lead was split into two fractions, the first fraction of the lead was placed in the reactor vessel, and the impurities were added on top of the lead in the reactor vessel. The second fraction of the lead was added on top of the impurities. The reactor vessel is covered with a lid and sealed with graphite casket. Type-K thermocouple is used to measure the temperature, and at about 330 °C, the alloy (mixture) begins to melt. The stirrer/agitator is lowered into the crucible (reactor vessel), and stirring speed was maintained between 400 and 500 rpm. A sample is drawn from the molten lead (alloy) at time $t = 0$; prior to adding the reagents. The reagent is weighed and added to the molten alloy through the vortex created by the stirrer. The agitator/stirrer is turned off after the required reaction time, and the dross rises to the top of the molten lead and skimmed off. The stirrer is turned on, and $NaNO_3$ is added to the molten lead/alloy through the vortex. The stirrer is turned off, and the dross is skimmed off. A sample is drawn from the molten lead (softened lead), representing $t =$ 'end'.

The initial experiments were ran for 60 min using $CaCO_3$, NaOH, $Mg(OH)_2$, and $Ca(OH)_2$ as reagents. The quantity of reagent used in each experiment was 400 g. The quantities of crude lead and sodium nitrate used were 5000 g and 450 g, respectively. The experiments were run for 60 min, and samples were drawn at 15 min intervals. In subsequent experiments, design of experiment (DOE) with full factorial with one replicate was used to evaluate the effects and interactions of the following three controllable factors: temperature (°C), reaction time (minutes), and reagent ($CaCO_3$) quantity (grams). Each factor had three levels: high, mid-, and low points. The levels for temperature were selected based on the HSC equilibrium composition diagram. The levels for reaction time (i.e., $CaCO_3$ reaction time) were chosen based on the maximum time practiced by industrial partners. The levels for reagent quantity used

Table 1 Impurities concentrations prior to softening process

Company	Sb (ppm)	As (ppm)	Sn (ppm)
Company 1	2,400–2,600	9,000–11,000	900–1,100
Company 2	27,000–29,000	2,150–2,350	12,000–14,000
Company 3	34,000–36,000	7,400–7,600	7,400–7,600
Average concentration (ppm)	21,800	7,167	6,583

Table 2 Levels of factors for design of experiment for refining 5000 g of crude lead

Level	Factor 1: temperature (°C)	Factor 2: time (min)	Factor 3: reagent quantity (g)
Low	400	10	16
Mid	450	15	24
High	500	20	32

was 86% (86%) less than stoichiometric quantity of the proposed reaction products. This was to match amount used in industry. The quantity of sodium nitrate was reduced by 77% (77%). The levels of factors for the design of the experiment are summarized in Table 2.

Softening experiments were conducted for the $CaCO_3$ and NaOH processes based on the optimal parameters obtained from the DOE statistical analysis. In both experiments, sodium nitrate was used as the oxidizing agent. The results are shown in Table 3. The activities of arsenic, tin, and antimony in the softened lead for the $CaCO_3$ and NaOH processes were calculated using mole fractions in the refined lead and activity coefficients at the refining temperature. The calculated activities of the impurities for the two processes are shown in Table 4.

Nitric acid was used to digest the samples. Sn, As, and Sb content in the solution were analyzed by ICP-MS (inductively coupled plasma-mass spectrometry). The acid used for digestion was reagent grade. The drosses were characterized using XRD (X-ray Diffraction).

Results and Discussion

Softening Processes for Reaction Time of 60 Min

Initial experiments in the research were performed using $Mg(OH)_2$, $Ca(OH)_2$, $CaCO_3$, and NaOH as reagents. Sodium nitrate was used as the oxidizing agent HSC thermochemistry software was used to determine the feasibility of these reagents for the removal of arsenic, antimony, and tin from crude lead. The chemical analyses of these experiments are shown in Figs. 1, 2, and 3. Figure 1 shows the effect of time on arsenic content in crude lead for experiments with NaOH, $Mg(OH)_2$, $Ca(OH)_2$, and $CaCO_3$ as reagents. In all these experiments, sodium nitrate was used as the oxidizing agent. As shown in Fig. 1, there was sharp decrease in arsenic concentration from 0 to 15 min; concentrations decreased by 55, 46, 35, and 30% for NaOH, $CaCO_3$, $Mg(OH)_2$, and $Ca(OH)_2$, respectively. From 15 to 60 min, the arsenic concentration decreased gradually for all the reagents. The overall decreases in arsenic concentrations after one hour of the softening experiments were 80, 52.1, 52.2, and 52.7% for NaOH, $CaCO_3$, $Mg(OH)_2$, and $Ca(OH)_2$, respectively. Figure 2 shows the effect of time on tin concentration. As shown in Fig. 2, there is a general decrease in tin content for all the reagents. The decreases in tin concentration from 0 to 15 min were

Table 3 Chemical analysis of softening process for 5000 g of crude lead at 500 °C for 15 min

Process	Temp (°C)	Time (min)	Mass of reagent (g)	Pb (wt.%)	As (wt.%)		Sn (wt.%)		Sb (wt.%)	
				Product	Feed	Product	Feed	Product	Feed	Product
$CaCO_3$ process	500	15	32	98.70	0.551	0.239	0.368	0.144	1.728	0.917
NaOH process	500	15	32	99.04	0.528	0.150	0.340	0.028	1.859	0.782

Table 4 Comparison of activities of impurities for $CaCO_3$–$NaNO_3$ and NaOH–$NaNO_3$ processes

Reagent	Activity of arsenic		Activity of tin		Activity of antimony	
	Feed	Product	Feed	Product	Feed	Product
$CaCO_3$–$NaNO_3$	0.02476	0.00838	0.09831	0.02169	0.02844	0.01220
NaOH–$NaNO_3$	0.02476	0.00528	0.09831	0.00423	0.02844	0.01044

28.5, 19.8, and 13.7% for NaOH, $Ca(OH)_2$, and $CaCO_3$, respectively. Tin concentration decreased gradually from 15 to 60 min. The overall decreases in tin content in the crude lead during the softening processes were 71.8, 66.7, 68.0, and 68.2% for NaOH, $Mg(OH)_2$, $CaCO_3$, and $Ca(OH)_2$, respectively. Figure 3 shows antimony concentration over time using NaOH, $Mg(OH)_2$, $Ca(OH)_2$, and $CaCO_3$ as reagents; and sodium nitrate as the oxidizing agent. The figure shows a sharp decrease in antimony concentrations from 0 to 15 min for all reagents. The decreases in antimony concentration from 0 to 15 min were 61.1, 64.9, 39.6, and 35.1% for NaOH, $Mg(OH)_2$, $Ca(OH)_2$, and $CaCO_3$, respectively. Tin concentration decreased gradually from 15 to 60 min. The overall decrease in tin content in the crude lead during the softening processes were 84.9, 86.3, 82.5, and 74.1% for NaOH, $Mg(OH)_2$, $CaCO_3$, and $Ca(OH)_2$, respectively. Based on the results on the initial experiments, calcium carbonate was chosen as alternative for sodium hydroxide because it was the cheapest among the reagents used and its effectiveness in removing the impurities was relatively better than $Mg(OH)_2$ and $Ca(OH)_2$.

Fig. 1 Effect of time on arsenic content in the crude lead at 450 °C for 60 min

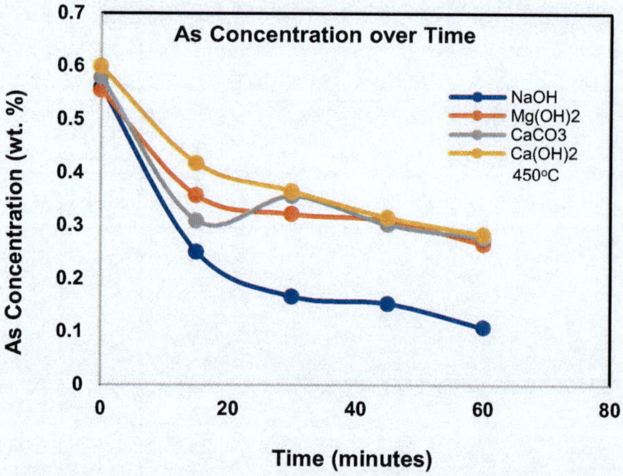

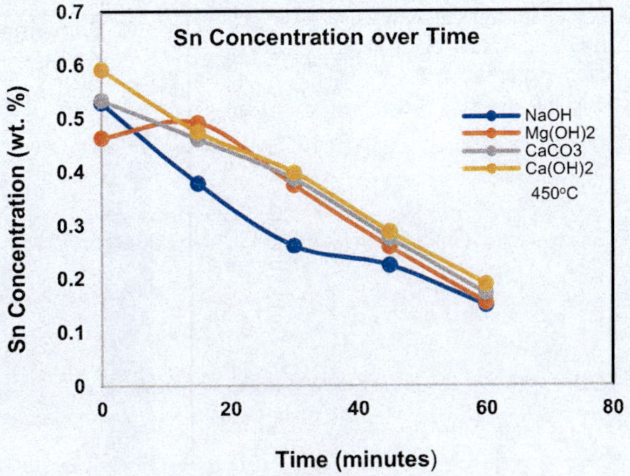

Fig. 2 Effect of time on Sn content in the crude lead at 450 °C for 60 min

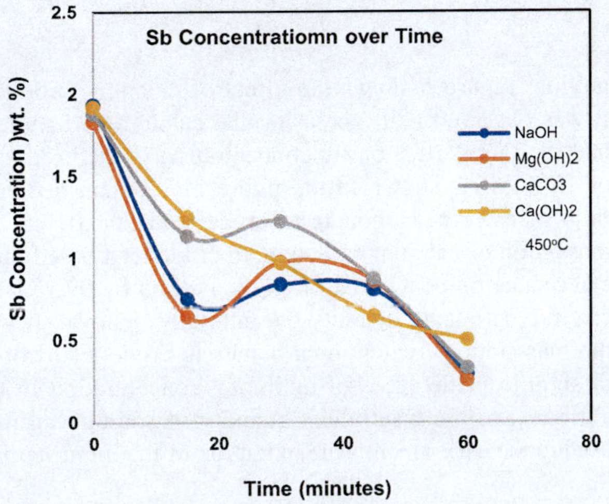

Fig. 3 Effect of time on antimony content in the crude lead at 450 °C for 60 min

Chemical Analyses of the Design of Experiment Based on Stoichiometric Quantities

Figures 4, 5, and 6 show the results of the design of experiment. The reaction time for the $CaCO_3$ was 15 min. Quantity of $CaCO_3$ used in these experiments was based on the stoichiometric equations from the proposed reactions Figs. 4, 5, and 6 derived from HSC thermochemistry software. Mass of crude lead used was 5178 g (with 96,44 wt.% Pb, 2.18 wt.% Sb, 0.72 wt.% As, and 0.66 wt.% Sn) and 124 g of sodium nitrate. Figure 4 shows the effect of calcium carbonate quantity on arsenic removal; as shown in the figure, increasing the calcium carbonate quantity from 98 to 243 g did not result in significant decrease of arsenic concentration in the product

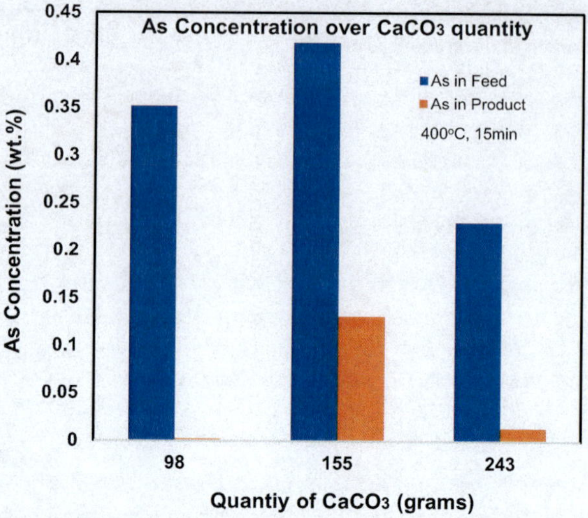

Fig. 4 Effect of calcium carbonate quantity on arsenic conten: experiment conducted at 400 °C for 15 min

bullion. Figure 5 shows the effect of calcium carbonate quantity on tin removal. It was observed that increasing the calcium carbonate quantity from 98 to 243 g had the same effect on tin concentration in the product bullion for a reaction time of 15 min. As shown in the figure, tin content decreased by 99.9% with a mass ratio of calcium carbonate to crude lead of 0.019:1. Hence, using the minimum mass ratio of calcium carbonate to crude lead based on the stoichiometric quantity can reduce tin concentration in the bullion by 99.9%. Figure 6 shows the effect of calcium carbonate quantity on antimony removal. It was observed that increasing the mass ratio of calcium carbonate to crude lead from 98 to 243 g did not result in significant decrease of antimony concentration in the product bullion. This is attributed to the insufficient quantity of sodium nitrate, the main oxidizing agent used in the experiment. The oxidation of the impurities decreases in the order: As > Sn > Sb.

Interpretation of Results of the Design of Experiment

The mass ratio of $CaCO_3$ to crude lead was reduced by 86% of the stoichiometric amount as shown in Table 2. This was to conform to current industrial practices where the mass ratio of NaOH to crude lead is approximately 0.0025:1. Pareto chart was used to visually depict factors and or interactions that are more significant in the softening process. Figures 7, 8, 9, and 10 are the Pareto charts for results of lead (Pb), arsenic (As), tin (Sn), and antimony (Sb), respectively. As shown in Fig. 7, the main factors, time and temperature, and the two-way interactions of 'temperature-time' and 'temperature-reagent quantity' impacted the lead concentration significantly. The main effect: 'reagent quantity' was not a significant factor. From Fig. 8, time,

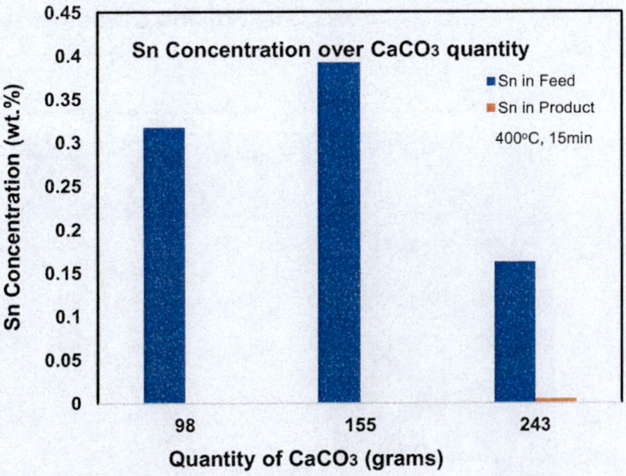

Fig. 5 Effect of calcium carbonate quantity on tin content; experiment conducted at 400 °C for 15 min

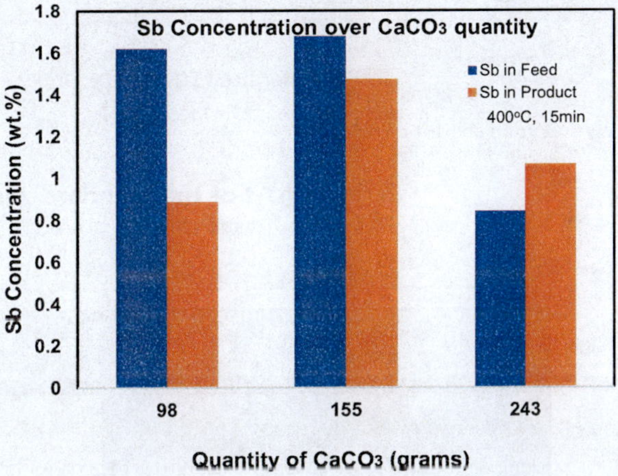

Fig. 6 Effect of calcium carbonate quantity on antimony content: experiment conducted at 400 °C for 15 min

temperature-time, and temperature-reagent quantity were the most significant factors in arsenic removal. Temperature and reagent quantity were not significant factors. According to Fig. 9, interactions of temperature-time and temperature-reagent quantity are significant in the tin removal. Time, temperature, and reagent quantity were not significant factors because they were below the reference line. For the antimony removal, time was the most significant factor as shown in Fig. 10.

Nature of Dross

Solid dross was formed during the refining process. The first dross is obtained after the calcium carbonate ($CaCO_3$), or sodium hydroxide (NaOH) is added to the molten

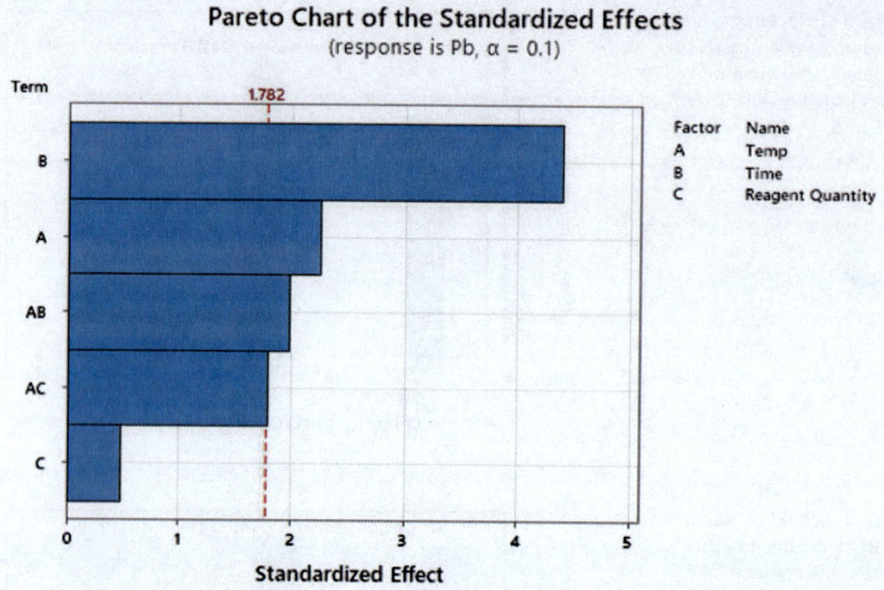

Fig. 7 Pareto chart for Pb results

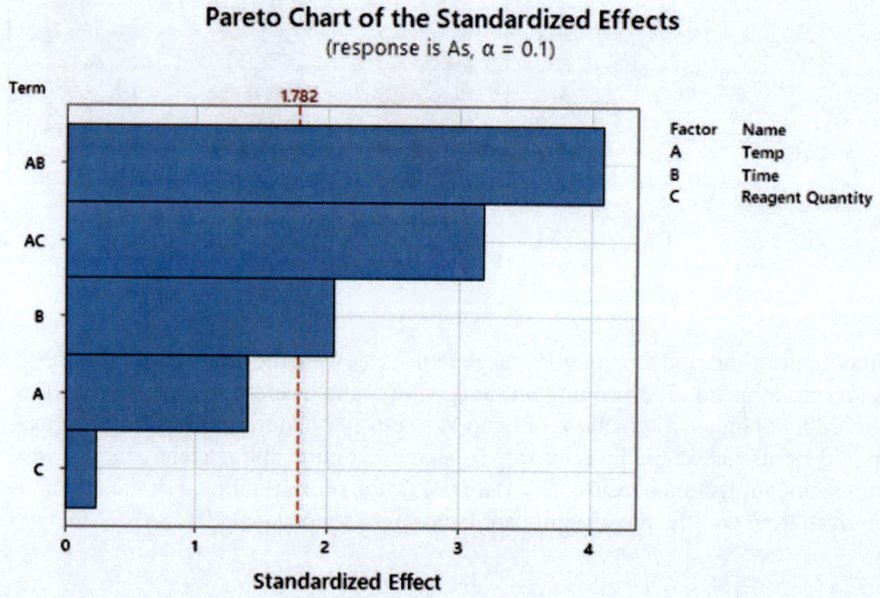

Fig. 8 Pareto chart for As results

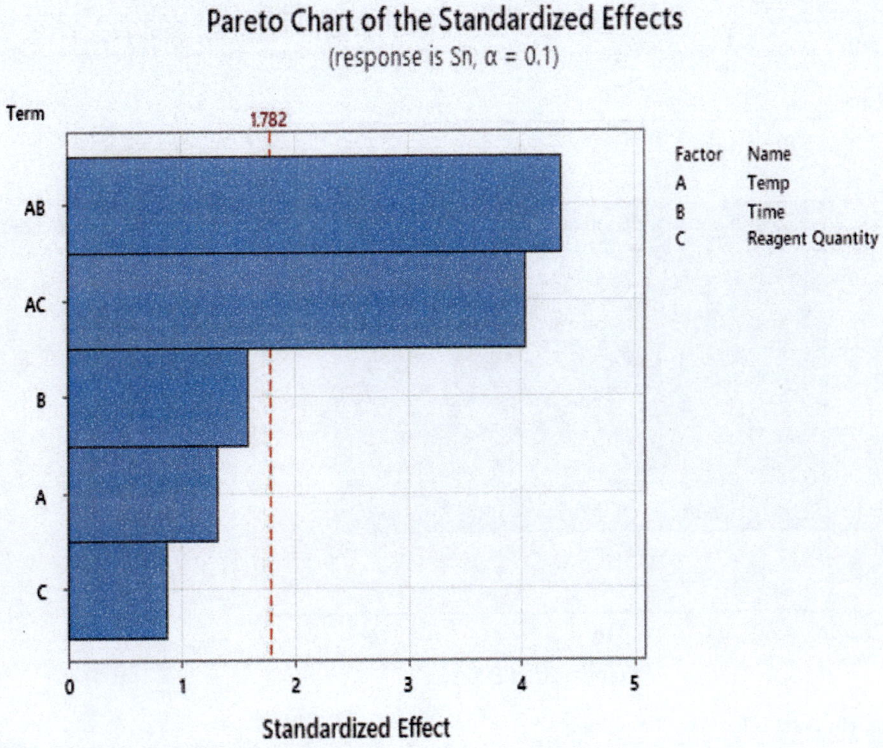

Fig. 9 Pareto chart for Sn results

lead and ran for the required reaction time. The second dross is formed when sodium nitrate ($NaNO_3$) is added to the molten lead. The calcium carbonate does not melt at the softening process temperature; a solid dross was formed for the entire reaction time. Initial dross formed by sodium hydroxide or sodium nitrate is molten; however, viscosity increases during the refining process which subsequently turn solid. The dross formed by calcium carbonate is relatively fine powder with little or no entrained lead as shown in Fig. 11. The calcium carbonate changes from white to yellow or brown. As shown in Fig. 12, the dross formed by NaOH was coarser with much entrained lead.

Characterization of Dross and XRD Plots

The $CaCO_3$ dross and NaOH dross were characterized using XRD. The XRD patterns show PbO, Sb_2O_3, As_4O_6, and SnO_2 as the major phases in both drosses as shown in Figs. 13 and 14. The phases found in the drosses were relatively similar as both sodium hydroxide and calcium carbonate cannot remove or oxidize the impurities

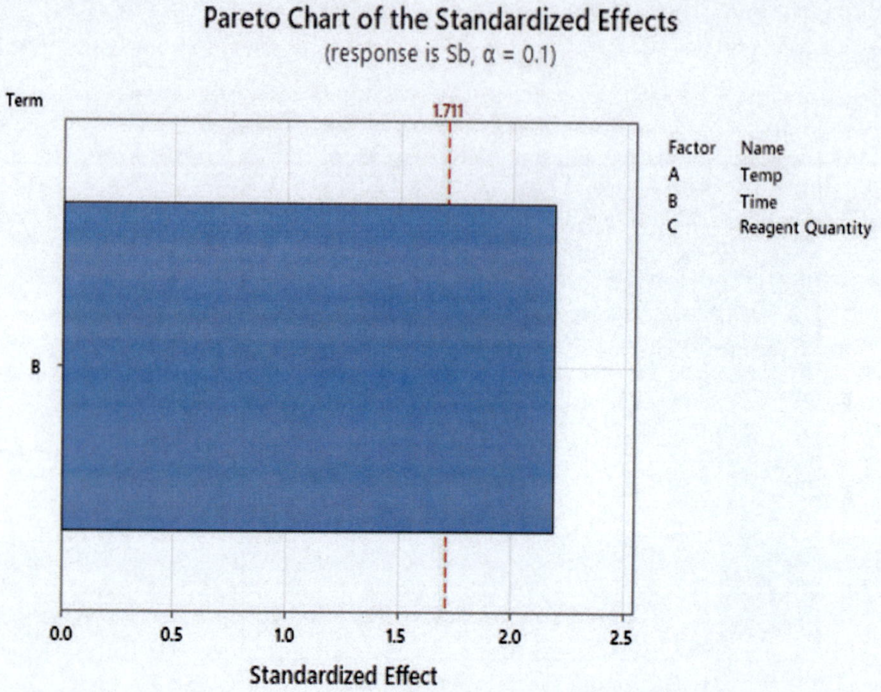

Fig. 10 Pareto chart for Sb results

Fig. 11 $CaCO_3$ dross

Fig. 12 NaOH dross

[5]. The sodium arsenate and sodium stannate, which are found in literature [6] as the sodium salts for Harris process, were undetected. Calcium arsenate and calcium antimonite from the proposed $CaCO_3$ process were not detected. This is attributed to the low concentrations of the calcium salts at 500 °C because calcium carbonate does not decompose at this temperature.

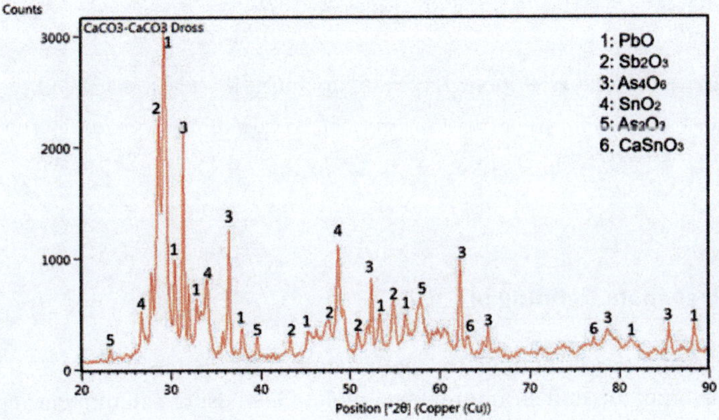

Fig. 13 XRD pattern of $CaCO_3$ dross

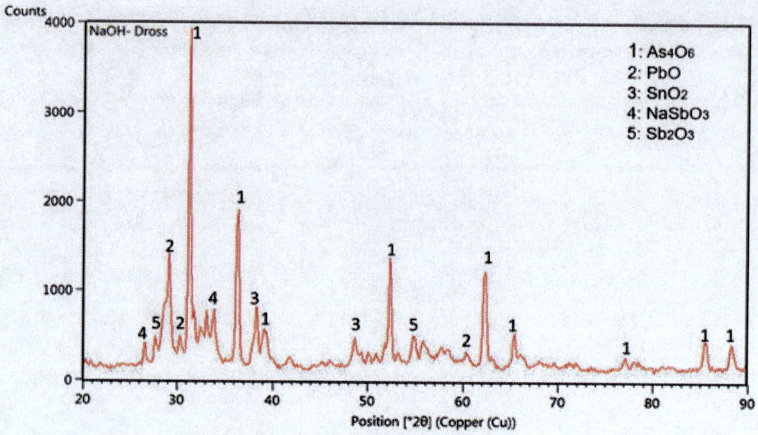

Fig. 14 XRD pattern of NaOH dross

Thermodynamic Consideration

Caustic Soda Refining of Lead

The Harris process of lead refining to remove arsenic, tin, and antimony from lead bullion using sodium hydroxide and sodium nitrate is made up of the following reactions [3]:

$$2As + 2NaNO_3 + 4NaOH = 2Na_3AsO_4 + N_{2(g)} + 2H_2O_{(g)} \quad (1)$$

$$5Sn + 4NaNO_3 + 6NaOH = 5Na_3SnO_3 + 2N_{2(g)} + 3H_2O_{(g)} \quad (2)$$

$$2Sb + 2NaNO_3 + 4NaOH = 2Na_3SbO_4 + N_{2(g)} + 2H_2O_{(g)} \quad (3)$$

Calcium Carbonate Refining of Lead

Similarly, the proposed reactions for the 'modified' Harris process for lead refining to remove arsenic, tin, and antimony from molten lead using calcium carbonate and sodium nitrate are:

$$2As + 2NaNO_3 + 3CaCO_3 = Ca_3(AsO_4)_2 + Na_2O + N_{2(g)} + 3CO_{2(g)} \quad (4)$$

$$5Sn + 4NaNO_3 + 5CaCO_3 = 5CaSnO_3 + 2Na_2O + 2N_{2(g)} + 5CO_{2(g)} \quad (5)$$

$$2Sb + 2NaNO_3 + 3CaCO_3 = Ca_3(SbO_4)_2 + Na_2O + N_{2(g)} + 3CO_{2(g)} \quad (6)$$

Arsenic Removal

It is stated in the literature that arsenic can be removed from lead with the addition of caustic soda without oxidizing agent [3]. However, the reaction does not correspond to reaction (7) because of the high positive Gibbs free energy at 450 °C (723 K) [6]. Thermodynamic calculations of possible reaction equations are:

$$2As + 6NaOH = 3Na_2O + As_2O_3 + 3H_2 \quad (7)$$

$$\Delta G_{723K} = 117.739 \text{ kcal}$$

$$2As + 6NaOH + O_2 = 3Na_2O + As_2O_3 + 3H_2O \quad (8)$$

$$\Delta G_{723K} = -31.154 \text{ kcal}$$

$$2As + O_2 = As_2O_3 \quad (9)$$

$$\Delta G_{723K} = -113.309 \text{ kcal}$$

The reaction (9) has the greatest probability of removing arsenic from lead bullion.

Similarly, thermodynamic data for possible reaction equations between calcium carbonate and arsenic are:

$$2As + 3CaCO_3 = As_2O_3 + 3CaO + 3CO \quad (10)$$

$$\Delta G_{723K} = 90.330 \text{ kcal}$$

$$2As + 3CaCO_3 + O_{2(g)} = Ca_3(AsO_4)_2 + 3CO_{2(g)} \quad (11)$$

$$\Delta G_{723K} = -205.922 \text{ kcal}$$

Reaction (11) has the greatest possibility since the Gibbs free energy is negative. This suggests that sodium hydroxide or calcium carbonate cannot remove arsenic from molten lead without the presence of an oxidizing agent. The removal of tin and antimony by NaOH and $CaCO_3$ is represented by reactions Eqs. (12–19). From reactions Eqs. (12, 13, 16, and 17), tin and antimony cannot be removed from molten lead without oxidizing agent.

$$Sn + 4NaOH = SnO_2 + 2Na_2O + 2H_{2(g)} \tag{12}$$

$$\Delta G_{723K} = 52.255 \text{ kcal}$$

$$Sn + 2CaCO_3 = SnO_2 + 2CaO + 2CO_{(g)} \tag{13}$$

$$\Delta G_{723K} = 33.969 \text{ kcal}$$

$$Sn + 2PbO = SnO_2 + 2Pb \tag{14}$$

$$\Delta G_{723K} = -31.415 \text{ kcal}$$

$$5Sn + 4NaNO_3 = 5SnO_2 + 2Na_2O + 2N_{2(g)} \tag{15}$$

$$\Delta G_{723K} = -440.874 \text{ kcal}$$

$$2Sb + 6NaOH = Sb_2O_3 + 3Na_2O + 3H_2O_{(g)} \tag{16}$$

$$\Delta G_{723K} = 107.376 \text{ kcal}$$

$$2Sb + 3CaCO_3 = Sb_2O_3 + CaO + 3CO_{(g)} \tag{17}$$

$$\Delta G_{723K} = 79.948 \text{ kcal}$$

$$2Sb + 2NaNO_3 = Sb_2O_5 + Na_2O + N_{2(g)} \tag{18}$$

$$\Delta G_{723K} = -114.461 \text{ kcal}$$

$$2Sb + 3PbO = Sb_2O_3 + 3Pb \tag{19}$$

$$\Delta G_{723K} = -18.328 \text{ kcal}$$

Conclusion

This study provided important information to understand the thermodynamics and mechanism of Harris process. It provided knowledge base understanding of industrial practice. Under the same conditions, the NaOH–$NaNO_3$ produced a refined lead of 99.04 wt.% and the $CaCO_3$–$NaNO_3$ produced refined lead of 98.70 wt.%. It was discovered that sodium hydroxide and calcium carbonate do not oxidize (remove) arsenic, tin, and antimony. The reactions are not thermodynamically feasible with high positive Gibbs free energies. Atmospheric oxidation of the impurities occurs prior to reaction with caustic soda or calcium carbonate. The impurities removal mainly depends on the quantity of sodium nitrate, the main oxidizing agent. The Pareto charts showed that calcium carbonate was not a significant factor for the removal of arsenic, tin, and antimony. Time and temperature were significant parameters for impurities removal. In using the stoichiometric quantities of calcium carbonate and sodium nitrate based on the average impurity concentrations from industrial partners and HSC reaction models, it was observed that 99.29% of tin was removed at 400 °C for $CaCO_3$ reaction time of 15 min. The average concentration of antimony removed by NaOH–$NaNO_3$ and $CaCO_3$–$NaNO_3$ processes were 57.9% and 47.9%, respectively. This is attributed to insufficient quantity of sodium nitrate used in the softening processes. The phases identified in both $CaCO_3$ dross and NaOH dross were As_4O_6, PbO, SnO_2, and Sb_2O_3. The residual calcium content in the softened lead was 0.006 wt.%, which is below the threshold of 0.1 wt. %. Calcium concentration of 0.1wt.% increases the rate of oxidation of lead (i.e., it corrodes lead) [7]. Oxidation of the impurities is by direct (i.e., $NaNO_3$ or air/atmospheric) and indirect via lead oxide.

References

1. Rosenquist T (2004) Principles of extractive metallurgy, 2nd edn
2. Morachevskii AG, Butukhanova TV (2011) On the limiting activity coefficient of arsenic in liquid lead. Russian J Appl Chem 84(4):602–606
3. Schlesinger ME, Lynch DC (1986) Activity of arsenic in molten lead. Metall Trans 17B:235–237
4. Moriyama J, et al (1964) Suiykaishi 15(4):191–194
5. McClincy RJ, Larson AH (1969) Trans Metall Soc AIME 245:173
6. Predel B, Emam AZ (1973) Metallkunde 64(9):647–652
7. Suleimanov DM et al (1974) Izv Vyssh Uchebn Zaved Khim Khim Tekhnol 17(3):365–367
8. Zaleska E (1974) Roczn Chem 48(2):195–200
9. Morachevskii AG, Kokhatskaya MS (2008) Applied chemical thermodynamics. Politekhn. Univ, St. Petersburg
11. Morachevskii AG, Butukhanovathe TV (2009) St. Petersburg State Polytechnical University, St. Petersburg, Russia
13. Onderka B, Wypartowicz J, Metallkunde Z (1980) 81(5):345–348
14. Redlich O, Kister A (1948) Ind Eng Chem 40(2):345–348
15. Hultgren R, Desai PD, Hawkins DT et al (1973) Selected values of the thermodynamic properties of binary alloys. Am Soc Metals

17. Krupkowski A (1974) Podstawowe zagadnienia teorii procesow metallurgicznich, Warszawa
18. Fitzner K (1981) Computer coupling phase diagram thermochemistry, vol 5, no 4, pp 239–253
19. Palmer JG (1989) Battery recycling-a solution in peril. In: Primary and secondary lead processing; proceedings of the international symposium, metallurgical society of the Canadian institute of mining and metallurgy, vol 15 pp 113–118
22. Kleppa OJ (1955) Phys CherrL 59:175–181
25. Karumb ET, The removal of arsenic and antimony from complex lead bullion via vacuum distillation. Ph.D. thesis, Colorado School of Mines, Golden, Colorado

Comprehensive Utilization of Copper Slag in a Pyro-Hydrometallurgical Process for Iron Phase Reconstruction and Valuable Metals Recovery

Changda Zhang, Bin Hu, Minge Yang, Guiqing Zhang, Shijie Zhao, Zhiqin Liao, Qiuxiang Liu, and Mingyu Wang

Abstract Copper slag is a major waste produced in pyrometallurgical plants during the smelting and converting operations and has caused heavy environmental contamination. A pyro-hydrometallurgical process for the treatment of copper slag has been developed to accomplish recovery of nickel, cobalt, and copper and reuse of iron source. It was proved that the slag phase was reconstructed by sulfation-sulfate decomposition roasting followed by water leaching, over 95% iron existing as hematite was transferred into residues which were subsequently to produce iron concentrates. Nickel-cobalt mixed hydroxide precipitate was obtained from liquors after the separations of copper, iron, and aluminum step by step. By adding concentrated HCl to re-dissolve the precipitate, then cobalt was extracted over nickel using extraction agent N235 and the raffinate was subsequently crystallized to produce nickel chloride by introducing HCl gas. On the basis of the optimum conditions from a laboratory-scale test, about 70% of copper, nickel, and cobalt were recovered as respective product, indicating the feasibility of the process.

Keywords Copper slag · Roasting · Leaching · Copper · Nickel · Cobalt · Hematite

Introduction

Copper slag, as a typical hazardous waste generated primarily during smelting and converting operations in pyrometallurgical plants, is an important secondary resource containing not merely Fe, Cu, Zn, Pb, in particular, Ni, and Co. It is estimated that 2.2–3 tons of copper slags are generated per ton of copper produced, but unfortunately, more than 80% of that, worldwide, is discarded directly without a suitable

C. Zhang · B. Hu · M. Yang · G. Zhang · S. Zhao · Z. Liao · M. Wang (✉)
School of Metallurgy and Environment, Central South University, Changsha 410083, China
e-mail: wmydxx@163.com

Q. Liu
Zhejiang Light Industrial Products Inspection and Research Institute, Hangzhou 310018, China

R. G. Reddy et al. (eds.), *New Directions in Mineral Processing, Extractive Metallurgy, Recycling and Waste Minimization*, The Minerals, Metals & Materials Series,
https://doi.org/10.1007/978-3-031-22765-3_25

treatment, which not only wastes the limited valuable resources, but occupies space and causes potential pollution [1]. In recent decades, attention towards considerable accumulation of slags and contamination of the environment has been paid, which stimulates researches for the purpose of recycling and cleaning these slags.

Currently, numerous literatures have reported that the copper slag can be applied as an additive to produce sand-blasting, abrasive tools, cement, and pavements [2]. However, the economic benefits of copper slag lie in the valuable and recyclable metals it contains. A pyrometallurgical modification process relying on reduction roasting, smelting reduction, or molten oxidation [3–5] followed by beneficiation methods such as flotation, magnetic separation, or gravity separation [6, 7] have been certified for metallurgical recycling of different types of copper slags. The pyrometallurgical process mainly focuses on the iron and copper content recovery as high-grade magnetite, metallic copper (or matte), or iron-enriched alloys. Hydrometallurgical treatment is becoming increasingly popular due to its lower energy consumption, easier operation, and recovery ability [8, 9]. However, by-products generated during the leaching process are discarded directly and cause a non-negligible environmental problem. Presently, a high-pressure oxidative acid leaching (HPOXAL) technology has gradually replaced the traditional processes of atmospheric acid leaching and oxidant-assisted acid leaching, as it avoids serious problems such as dissolution of iron, generation of silica gel, high acid consumption, and long leaching duration for the treatment of copper slag [10, 11]. Meanwhile, Li [12] found that fayalite phase in copper slag can be transformed into hematite and amorphous silica after HPOXAL process, which was subsequently used as iron enrichment material. To minimize the stacked time and grinding costs before leaching, molten copper slags were treated by quick water quenching instead of slowly natural cooling common in industry. Unfortunately, amorphous copper slags will not obtain an unsatisfied metal extraction efficiency while using HPOXAL [13]. Baghalha [14] compared the HPOXAL performance for different copper slags treated by water quenching and natural cooling methods and found that the leaching of Ni, Co, and Cu from quenched slags only reached 20–30%, whereas 75–85% of Ni, Co, and Cu could be leached from the latter under the same conditions. On another level, for economic reasons, HPOXAL is not an alternative approach for depleted copper slags containing <4% Cu, <0.5% Co, and <2% Ni [15].

In our previous studies, the processes of sulfation-sulfate decomposition roasting, water leaching, re-dissolving of the nickel-cobalt mixed precipitate, solvent extraction, and crystallization have been discussed in detail and each optimum condition has been determined under laboratory-scale. This work focus on the intermediate purification and separations processes of the leaching liquors. The recovery of copper and removal of impurities have been investigated. Finally, a conceptual process has been proposed based on laboratory-scale test and was performed to verify the feasibility of that under experimental conditions.

Table 1 XRF elemental analysis of the copper slag

Element	O	Ca	Mg	Al	Si	Fe	Co	Ni	Cu	S
Content (wt.%)	34.47	2.35	3.53	3.90	18.50	36.98	0.16	0.53	0.26	1.14

Materials and Methods

Materials

Copper slag used in this study was cooled by water quenching, from a copper smelting plant, China Minmetals Corporation. The copper slag was first crushed and ground into −200 mesh powder before the experiment. XRF-1800 X-ray Fluorescence spectrometry (XRF) (Shimadzu, Japan) chemical analysis result is shown in Table 1. The X-ray diffraction (XRD) patterns of the solid state were recorded by a Rigaku D/MAX 2200 Diffractometer (Rigaku, Japan).

Methods

Copper slag was roasted in a muffle furnace equiped with predetermined temperature function. The leaching of calcine and the precipitations of metals from liquors were carried out in a 500 mL magnetic agitated flask heated by an electric hot jacket. The produced slurry was filtered by vacuum filtration. Solvent extraction was conducted by contacting aqueous and organic solutions in a 60 mL pear-shaped funnel and mechanical shaking. The concentrations of metal ions in the solutions were all determined by ICP-AES (PS-6, U.S.A.). The pH values of the aqueous solutions were measured with a pH/mV meter (Model PHS-3C, China).

Results and Discussion

Roasting and Leaching

In the previous work [16], the optimal conditions for two-step roasting and water leaching have been determined by single factor experiment. Combined sulfation roasting (80 wt.% H_2SO_4, 200 °C, 1 h) and sulfate decomposition roasting (650 °C, 1 h) showed that the phase of iron undergoes a transition from fayalite to hematite, as indicated in Fig. 1. Subsequently, leaching process can effectively extract nickel, cobalt, and copper with 2 mL water per gram of solids at 90 °C for 0.5 h.

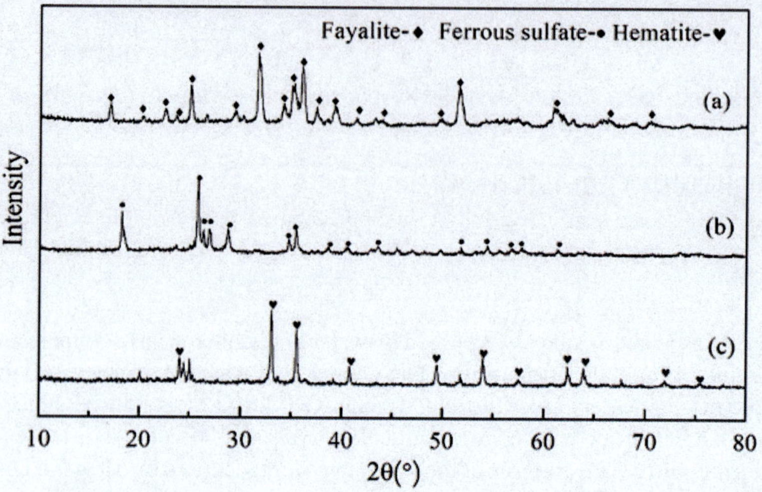

Fig. 1 XRD patterns of solids from **a** Raw slag, **b** Sulfation roast. **c** Sulfate decomposition roast

Chemical Precipitation

After water leaching, the iron powder was used to replace copper from liquors at home temperature, during which the reactions of (1) and (2) might occur.

$$Cu^{2+} + Fe = Fe^{2+} + Cu \quad (1)$$

$$Fe + 2H^{+} = Fe^{2+} + H_2 \quad (2)$$

Figure 2 shows the effect of the Fe/Cu ratio on the precipitation of copper, nickel, and cobalt with reaction time of 30 min. As can be seen, the reduction of Cu^{2+} rises from 46.12 to 99.89% as the Fe/Cu ratio rises up to 1. Further additions of iron power resulted in an insignificant increase in the precipitation of copper, but the excess iron powder will react with hydrogen ions according to Eq. (2), resulting in an increase in the pH of the solution. Simultaneously, it should be noticed that the loss of nickel and cobalt also increased slowly with an increase in Fe/Cu ratio, which was caused by the increase of solution pH, resulted in the hydrolysis of metal cations and inevitable physical adsorption. In order to increase the recovery of copper as much as possible and decrease the loss of nickel and cobalt, 1.2 Fe/Cu ratio was selected. The copper precipitation rate was 99.91%, and the loss rate of nickel and cobalt was 0.32% and 0.25%, respectively.

Figure 3 shows the effect of the reaction time on the deposition efficiencies of copper, nickel, and cobalt with a Fe/Cu ratio of 1.2. The deposition efficiency of copper increased from 92.67 to 99.96% with the increase of reaction time from 10 to 30 min and then approximately reached the optimal condition. On the other hand,

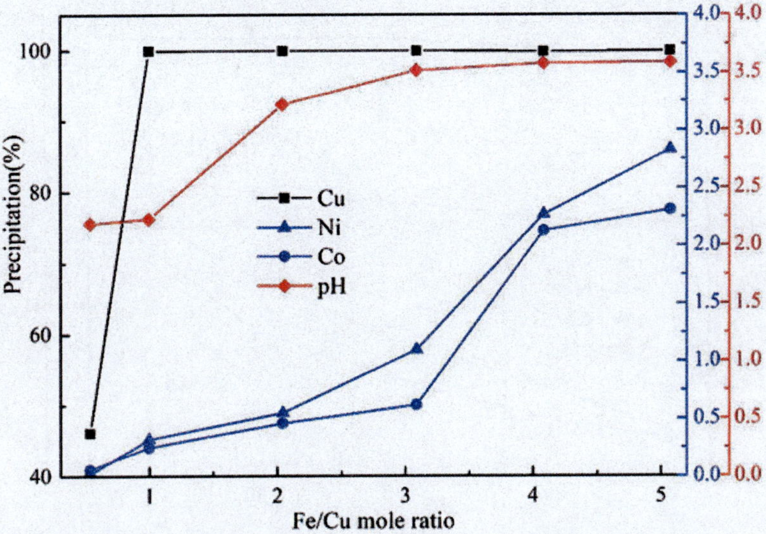

Fig. 2 Effect of Fe/Cu ratio on the precipitation efficiency of metals

the loss of nickel and cobalt increased slowly from 0.14 to 0.62% and from 0.12% to 0.58%, respectively, varying the reaction time from 10 to 50 min. Similarly, it was observed that the pH value increased slightly with the reaction time increasing, which resulted in the loss of nickel and cobalt. Therefore, the reaction time was set at 30 min in further precipitation experiments to optimize copper recovery whilst reducing the losses of nickel and cobalt.

Subsequently, the iron ions were removed by precipitation in the form of sodium jarosite, as expressed by Eq. (3). Under the most suitable conditions of 1.1 times stoichiometry Na_2SO_4 addition, reaction at 95 °C for 4 h maintaining pH of the precipitation process at 1.4–1.9 using magnesia as the neutralizer, the removal of iron was 99.99%, whilst the loss of nickel or cobalt was both below 1%.

$$6FeSO_4 + 3H_2O_2 + Na_2SO_4 + 6H_2O = Na_2Fe_6(SO_4)_4(OH)_{12} \downarrow + 3H_2SO_4 \quad (3)$$

Figure 4 illustrates different precipitation pH for metals. Over 99.9% of aluminum was removed in pH range of 4.5–5 using magnesia. Similarly, nickel and cobalt were co-precipitated over calcium and magnesium in the range of 7.5–8.0 at 60 °C for 4 h.

Solvent Extraction and Crystallization

Prior to the extraction of cobalt, the mixed nickel-cobalt precipitate was re-dissolved using HCl solution and then purified by extractant P204(D2EHPA). Subsequently, extractant N235 (R_3N, R = C_8–C_{10}, tertiary amine) was employed to selectively

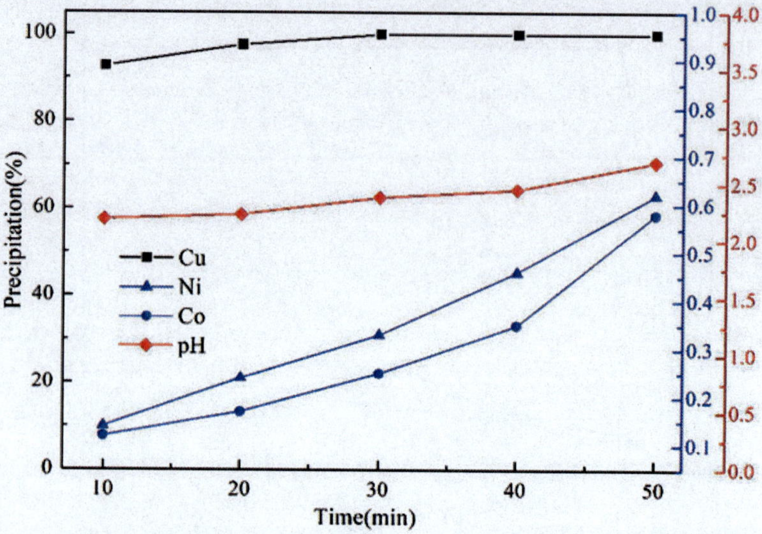

Fig. 3 Effect of reaction time on the precipitation efficiency of metals

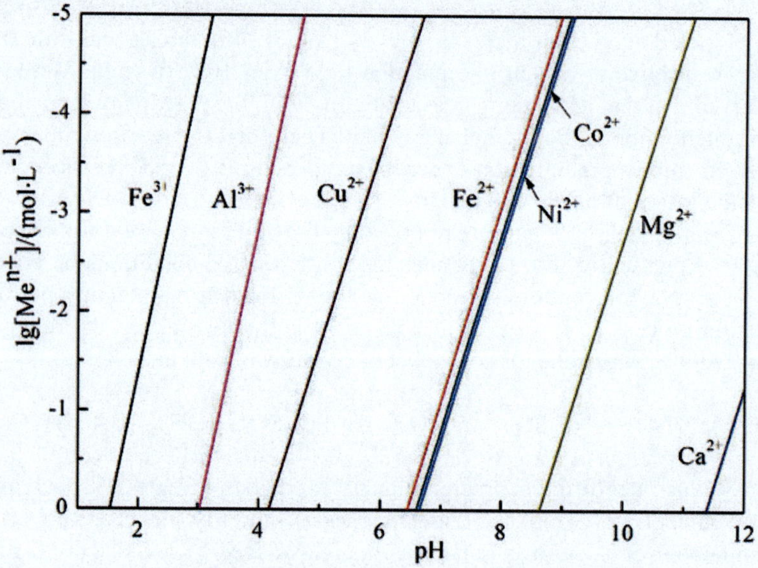

Fig. 4 Lg[Me^{n+}]-pH curves of different metal cations

Table 2 Optimal conditions and results of extraction, scrubbing, and stripping

Element ($g\cdot L^{-1}$)	Extraction (30 vol.% N235 + 15 vol.% TBP + 55 vol.% kerosene; O/A = 2; 5 min)			Scrubbing (6 mol/L HCl;O/A = 20; 5 min)	Stripping (water;O/A = 5;5 min)	
	Feed	Raffinate	Loaded organic	Scrubbed organic	Stripped organic	Strip liquor
Co	20.38	0.002	12.18	11.9	0.001	69.57
Ni	62.6	65.4	0.03	0.002	–	0.01
Mg	3.06	3.05	0.01	0.001	–	0.006

extract cobalt over nickel and magnesium. The extraction mechanism can be expressed as the Eqs. (4) and (5), respectively. The suitable conditions and results of extraction of cobalt were shown in Table 2 [17].

$$Me^{n+} + n(HX) = MeX_n + nH^+ \tag{4}$$

$$2R_3NHCl + (CoCl_4)^{2-} = (R_3NH)_2CoCl_4 + 2Cl^- \tag{5}$$

By using HCL gas, nickel chloride was subsequently crystallized over magnesium from raffinate. The pregnant liquor can be recycled to the dissolving circuit for reuse.

Proposed Process and Products

Based on the above experiment results, a comprehensive utilization and conceptual process for the recovery of nickel, cobalt, and copper and further reuse of iron source are proposed in Fig. 5. Firstly, the purity of crystallized $NiCl_2 \cdot 4H_2O$ (Product 1) meets the standard specification. The selective treatment of high purity cobalt chloride solution (Product 2) can be selled as semi-finished product or be electrolyzed to produce metallic cobalt. The use of the sponge copper (Product 3) mixed with little iron as a alloying additive is also particularly attractive as it can be used to produce copper oxide after the removal of iron. And the hematite (Product 4) generated during water leaching can be further treated by flotation, reverse flotation, or reduction-magnetic process as reported by Mowla and Ponomar [18, 19], in order to make the entire process more economical.

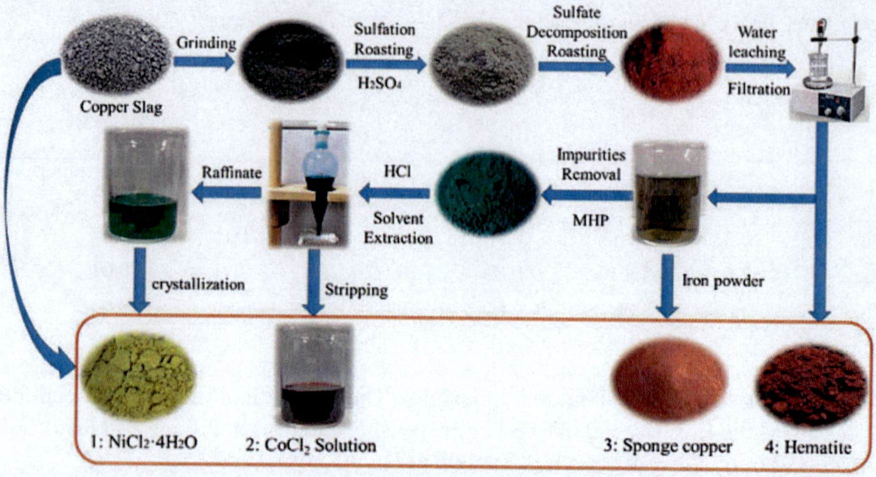

Fig. 5 Flowsheet of valuable metals extraction from copper slag

Conclusions

A pyro-hydrometallurgical process for the treatment of copper slags has been developed combining with roasting, leaching, chemical precipitation, solvent extraction, and crystallization. Over 99.9% copper was selectively recovered by adding 1.2 times stoichiometry amounts of iron powder, reaction for 30 min from the leach liquor. The selective removal of iron and aluminum as sodium jarosite and aluminum hydroxide, respectively, was conducted by chemical precipitation. Nickel-cobalt mixed hydroxide precipitate was obtained from magnesium-rich sulfate liquors by adjusting pH in the range of 7.5–8.0, and precipitation duration of 4 h at 60 °C. Under the optimized conditions in the overall process, over 70% nickel, cobalt, and copper can be recovered from copper slags at laboratory-scale and produced four marketable products: hematite, sponge copper, concentrated cobalt chloride solution, and nickel chloride which make the process economically favourable with sustainable benefits.

Acknowledgements This study was financially supported by the Natural Science Foundation of Hunan Province (2020JJ4728; 2022JJ60063) and the Fundamental Research Funds for the Central Universities of Central South University (2021zzts0618; 2021zzts0607).

Conflict of Interest Statement On behalf of all authors, the corresponding author states that there is no conflict of interest.

References

1. Tian H, Guo Z, Pan J et al (2021) Comprehensive review on metallurgical recycling and cleaning of copper slag. Resour Conserv Recycl 168:105366
2. Shi C, Meyer C, Behnood A (2008) Utilization of copper slag in cement and concrete. Resour Conserv Recycl 52:1115–1120
3. Sarfo P, Das A, Wyss G et al (2017) Recovery of metal values from copper slag and reuse of residual secondary slag. Waste Manag 70:272–281
4. Kim BS, Jo SK, Shin D et al (2013) A physico-chemical separation process for upgrading iron from waste copper slag. Int J Miner Process 124:124–127
5. Li S, Pan J, Zhu D et al (2019) A novel process to upgrade the copper slag by direct reduction-magnetic separation with the addition of Na_2CO_3 and CaO. Powder Technol 347:159–169
6. Roy S, Datta A, Rehani S (2015) Flotation of copper sulphide from copper smelter slag using multiple collectors and their mixtures. Int J Miner Process 143:43–49
7. Sarrafi A, Rahmati B, Hassani HR et al (2004) Recovery of copper from reverberatory furnace slag by flotation. Miner Eng 17:457–459
8. Muravyov MI, Fomchenko NV, Usoltsev AV et al (2012) Leaching of copper and zinc from copper converter slag flotation tailings using H_2SO_4 and biologically generated Fe2(SO4)3. Hydrometallurgy 119–120:40–46
9. Banza AN, Gock E, Kongolo K (2002) Base metals recovery from copper smelter slag by oxidising leaching and solvent extraction. Hydrometallurgy 67:63–69
10. Perederiy I, Papangelakis VG, Buarzaiga M et al (2011) Co-treatment of converter slag and pyrrhotite tailings via high pressure oxidative leaching. J Hazard Mater 194:399–406
11. Huang F, Liao Y, Zhou J et al (2015) Selective recovery of valuable metals from nickel converter slag at elevated temperature with sulfuric acid solution. Sep Purif Technol 156:572–581
12. Li Y, Papangelakis VG, Perederiy I (2009) High pressure oxidative acid leaching of nickel smelter slag: characterization of feed and residue. Hydrometallurgy 97:185–193
13. Perederiy I, Papangelakis VG (2017) Why amorphous FeO–SiO_2 slags do not acid-leach at high temperatures. J Hazard Mater 321:737–744
14. Baghalha M, Papangelakis VG, Curlook W (2007) Factors affecting the leachability of Ni/Co/Cu slags at high temperature. Hydrometallurgy 85:42–52
15. Jena S (1983) Pressure leaching of copper converter slag using dilute sulphuric acid for the extraction of cobalt, nickel and copper values. Hydrometallurgy
16. Zhang CD, Hu B, Wang HG, Wang MY, Wang XW (2020) Recovery of valuable metals from copper slag. Mining Metall Explor 37:1241–1251
17. Zhang CD, Hu B, Zhao SJ, Liao ZQ, Wang MY (2022) Separation of Co(II) and Ni(II) from hydrochloric acid leach liquor by solvent extraction and crystallization. J Sustain Metall 8:91–101
18. Mowla D, Karimi G, Ostadnezhad K (2008) Removal of hematite from silica sand ore by reverse flotation technique. Sep Purif Technol 58:419–423
19. Ponomar VP, Brik OB, Cherevko YI et al (2019) Kinetics of hematite to magnetite transformation by gaseous reduction at low concentration of carbon monoxide. Chem Eng Res Des 148:393–402

Critical Review of Chemical Metallurgy of Tungsten

Raj P. Singh Gaur

Abstract Tungsten is one of the most important refractory metals as it is used in the production of numerous end use items such as tungsten and tungsten carbide parts. Tungsten is also used in the manufacturing of steels, super alloys, and catalysts. Other applications of tungsten are in the manufacturing of tungsten-copper and tungsten-silver composites. Tungsten from its concentrates and secondary sources is processed as high-purity ammonium para tungstate (APT). Currently, there are two main APT production processes: one uses solid ion exchange and the other uses liquid ion exchange. Both processes use multiple process steps and generate large amount of process wastewater. Other serious problem of APT production is the difficulty in the availability of tungsten sources especially to the plants outside China. The ore concentrates available to some tungsten processors in the market may be low grade ores and/or the high grade ores containing large amount of critical impurities such as Mo and/or radioactive elements. This paper will discuss some of these issues and also some process-related issues such as equipment scaling, loss of LIX solvents, tungsten losses in the overall process, and large volume of process waste water, especially in the solid ion exchange method. Due to the proprietary nature of the work, literature from only presentations and publications will be presented. This paper will also discuss the recent literature, related to new conceptual APT process which could lead to increased efficiency, energy reduction, and waste minimization.

Keywords Tungsten · Chemical processing · Tungsten ore concentrates · Tungsten scraps · Sodium tungstate · Ammonium para tungstate · Digestion · LIX · Crystallization

R. P. S. Gaur (✉)
SH Nano Energy Powders, LLC, 359 Gateway Industrial Park Rd, 18810 Athens, PA, Greece
e-mail: raj.singhgaur@gmail.com

R. G. Reddy et al. (eds.), *New Directions in Mineral Processing, Extractive Metallurgy, Recycling and Waste Minimization*, The Minerals, Metals & Materials Series,
https://doi.org/10.1007/978-3-031-22765-3_26

Introduction

Tungsten is one of the critical and strategic refractory metals. It is used in the manufacturing of super alloy, hard metals (cemented carbide), and important catalysts [1]. Other applications of tungsten are in the manufacturing of W-Cu and W-Ag composites [2] with typical applications in high-, medium-, and low-voltage circuit breakers, resistance welding electrodes, electrode materials for electrical discharge machining, and heat sink materials for microelectronic packaging. In military applications, tungsten is employed in the manufacturing of high kinetic energy penetrators [1]. Tungsten chemicals also have important applications in the fields other than catalysts as reported by Christian et al. [3].

Until 1930, the largest consumption of tungsten was in the steel industry and only a small percentage was used in lamps and cemented carbides [1]. Today cemented carbides/metal cutting tools use more than 50% of the produced tungsten. Since 1910 tungsten production has also been continuously growing to over 30 thousand ton per year in recent years. In a recent article, Zeiler et al. [4] reported on the global breakdown of first-use segments for tungsten in 2016. According to this article in 2016, 65% of tungsten was used in tungsten carbide products, 17% was used in steel and superalloys, 10% was used in W-metal powders and 8% was used in chemicals and other applications. In the following section, we will briefly review the tungsten chemical process including available tungsten sources for feeding the tungsten producing plants.

Tungsten Feeds Used in Tungsten Chemical Plants

Most of the raw materials currently used in tungsten chemical production comprised high grade tungsten ore concentrates and synthetic concentrates obtained from recyclable tungsten scraps [4–11]. Table 1 presents the approximate compositions of tungsten carbide scraps and two high grade ore concentrates of tungsten namely wolframite and scheelite. As can be seen in Table 1, all four types of tungsten feeds have 50% or higher W. Scraps samples in addition to W also has Co and Ta which are separated during tungsten processing of these feeds. Cobalt and tantalum are processed separately.

Tungsten Ores

Two main economic minerals of tungsten are: Wolframite [(Fe, Mn)WO_4] and scheelite($CaWO_4$). Approximately two-thirds of the world tungsten reserves consist of scheelite deposits [1]. The wolframite group consists of three minerals: ferberite, wolframite, and hubnerite. The iron-rich tungstate ($FeWO_4$, WO_3 content 76,3%)

Table 1 Approximate compositions of main tungsten feeds (Ref. [6])

Element	Soft scrap (% Weight)	Hard scrap (% Weight)	Wolframite (% Weight	Scheelite (%Weight)
Al	0.1	–	0.2	0.4
Bi	–			0.4
Ca	–		0.8	13
Co	6.8	6.0		–
Fe	0.6	0.6	12	1.8
K	–			0.05
Mg	–			0.5
Mn	–		2.4	0.2
Mo	0.1	–		0.04
Nb	<0.2	0.4		–
Ni	0.8	0.2		–
P	–			0.14
S	–			2.0
Si	–		3.9	2.0
Ta	2.5	2.8		–
Ti	1.2	1.8		0.05
W	84.6	67.9	52	56

is ferberite, the manganese-rich tungstate ($MnWO_4$, WO_3 content = 76.6%) is hubnerite, and the iron-manganese mixed tungstate ((Fe, Mn)WO_4, WO_3 content = 76.5%) is wolframite. Wolframite contains between 20 and 80% each of $FeWO_4$ and $MnWO_4$ in their pure form. Concentration of wolframite from its ores is typically carried out by gravity and magnetic methods, as wolframite occurs mainly in vein-type deposits whose mineralization is much coarser than to most scheelite ores.

The mineral scheelite (calcium tungstate, $CaWO_4$) is normally found in quartz veins and in contact with skarn ores of complex mineralogical composition. These skarn minerals include garnets, pyroxene, and amphibole, and other minerals such as calcite, apatite, and quartz. Scheelite ores can be classified into five different categories: (1) simple scheelite ore, (2) scheelite-sulfides ore, (3) scheelite-cassiterite ore, (4) scheelite-calcite-apatite ore, and (5) scheelite-powellite ore. From an ore, scheelite is typically concentrated via a flotation process [12–14].

In a typical scheelite ore concentration process, the ore (containing 0.4–0.6% W) from the mine is brought to the plant and is fed to the grinders to grind below 200 mesh. The ground scheelite is treated in different processing steps to produce two ore concentrates: high grade ore and flot grade ore as can be seen in Fig. 1. Normally, the content of scheelite in a scheelite ore concentrate is expected to be more than 80% (tungsten content ≈ 56%). However, flot grade with WO_3 content as low as 13.9% is also reported [1, 8]. Scheelite ($CaWO_4$) in flot grade ore concentrates is associated

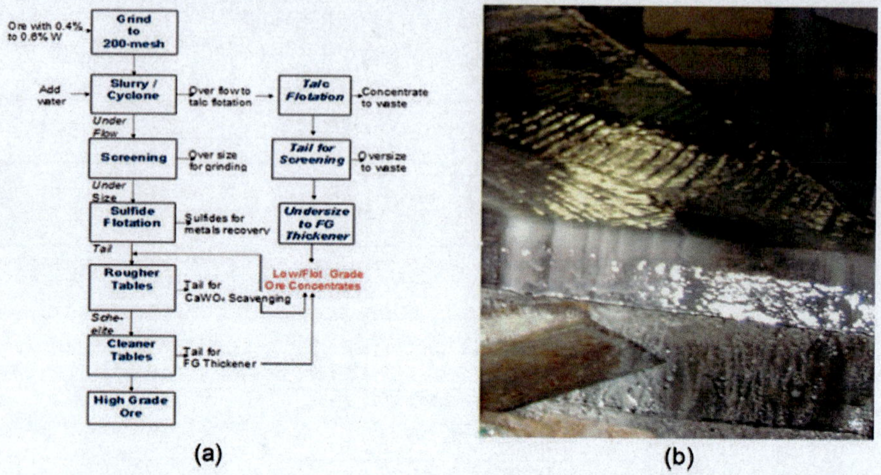

Fig. 1 **a** Scheelite ore to scheelite ore concentrates. Taken from Ref. [6]. **b** A glimpse of the industrial sulfide flotation process

with other calcium phases (such as $CaCO_3$ and apatite), iron sulfides, and silicate minerals (phlogopite/mica, etc.).

It is clear from the flow diagram presented in Fig. 1 that during scheelite ore concentration process, both ore concentrate: high grade and flot grade are produced. The approximate composition of a typical flot grade ore along with high grade ore is presented in Tables 2 and 3. Table 3 presents the comparison of high grade and flot grade ore concentrates impurity phases contents. In flot grade ore, following crystalline phases were determined by XRD: Scheelite: $CaWO_4$, Phlogopite: $KMg_3(Si_3Al)O_{10}(OH)_2$ and Talc: $Mg_3Si_4O_{10}(OH)_2$, Calcite ($CaCO_3$), Chalcopyrite: $CuFeS_2$ and Pyrrhotite: $Fe_{1-x}S$ as well as apatite ($Ca_5(PO_4)_3(OH)$). Out of these compounds, iron sulfide phases have most negative effect on their processing in a tungsten chemical plant.

Some tungsten producing plants have tighter spec on sulfur in tungsten ore concentrate and would not accept out of spec concentrates for processing. Presence of iron sulfide phases in flot grade concentrates could be further reduced by reverse flotation [15] or by producing a mid-grade concentrate instead two separate flot and high grade concentrates.

Another problem in processing flot grade concentrates is their organics content. Flot grade concentrates typically are not heat-treated to burn the flotation chemicals. Upon request from the APT plant, ore processors could heat-treat (calcining at about 1000°F) flot grade ore concentrate similar to the high grade ore concentrate and can also perform magnetic separation for additional removal of sulfide phases.

Table 2 XRF analysis of a typical high grade and flot grade scheelite ore concentrates

Elements	High grade scheelite ore concentrate (a) weight (%)	High grade scheelite ore concentrate (a) weight (%)	Flot grade scheelite ore concentrate weight (%)
Al	0.21	0.4	0.5
As	0.2	–	–
Ca	12	13	10
Cu	0.03	–	1.6
Fe	2.5	1.8	8.8
K	0.26	0.05	0.44
Mg	0.14	0.5	4.5
Mn	0.13	0.2	0.12
Mo	0.22	0.04	<0.01
P	0.16	0.14	0.74
S	–	2.0	8.6
Si	1	2.0	6.6
W	58	56	20.6

Table 3 Comparison of high grade and flot grade ore concentrates

Component	High grade ore weight (%)	Flot grade ore weight (%)
$CaWO_4$ (scheelite)	88	32
Phlogopite and talc	8	33
Calcite	1–2	14
Chalcopyrite and Pyrrhotite	–	17
Apatite	2–3	4

Tungsten Scraps

There are two types of tungsten carbide scrap available in the market [15]. Typically, these are known as hard scrap and soft scrap. XRF analysis of typical samples of soft and hard scrap is presented in Table 1. A detailed article on currently available tungsten scraps and other related topics is recently reported by Zeiler et al. [4].

As a definition, soft scrap pertains to all tungsten containing scrap that does not have a defined shape; it typically contained the rejects of the manufacturing of tungsten products such as wires, coils, powders, and turnings [7]. Soft scrap also contains some organics. However, presence of organics does not create problems in processing, as most soft scraps are calcined at high temperature to convert it to WO_3 and $CoWO_4$.

Hard scrap is made of worn tools such as cutting and drilling tool bits of tungsten carbide hard metals. The major component of hard scrap is tungsten carbide and cobalt; however, it does contain some other elements such as Ti, Ta, Nb, and V along with WC–Co. Processing of hard scrap is mostly done by zinc reclaim process [16].

Possessing of Tungsten from Its Feeds

There are three important tungsten compounds produced during the processing of tungsten feeds. These are sodium tungstate (Na_2WO_4), tungstic acid (H_2WO_4), and ammonium paratungstate (APT: $(NH_4)_{10}(H_2\,W_{12}O_{42})\cdot 4H_2O$). In the old method of tungsten processing, typically $CaWO_4$ (scheelite) was treated with HCl to convert it to tungstic acid which was then converted into APT after dissolution in NH_4OH followed by evaporation crystallization (Fig. 2).

In the old method of wolframite processing, concentrate was first converted to sodium tungstate via alkali roasting which was then converted to scheelite, followed by HCl treatment to make H_2WO_4 and APT similar to scheelite concentrate processing. Figure 3 below presents the steps involved in the old process. The chemical Eqs. (1)–(6) illustrate the chemistry of the procedures involved.

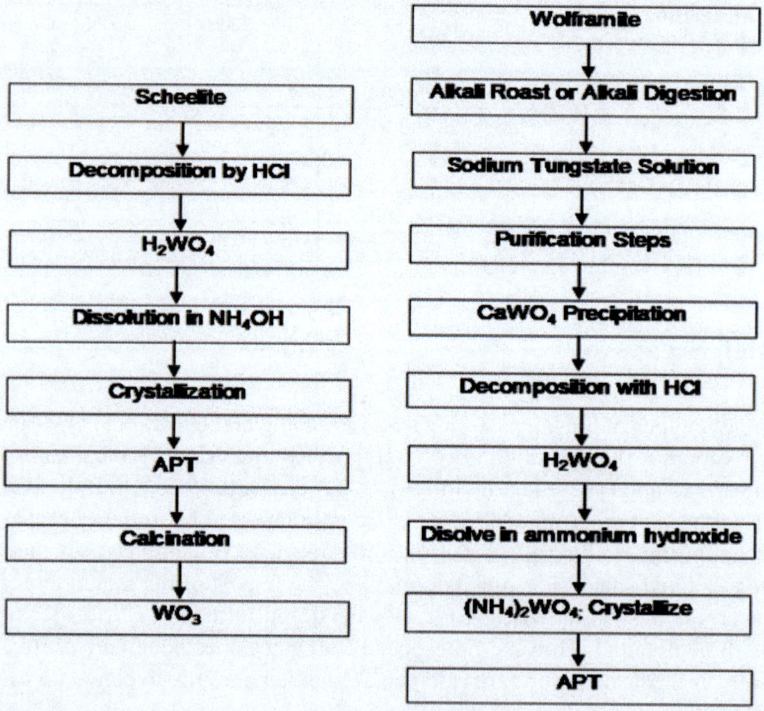

Fig. 2 Flow diagram of old method of chemical processing of tungsten from scheelite and wolframite. Ref. [8]

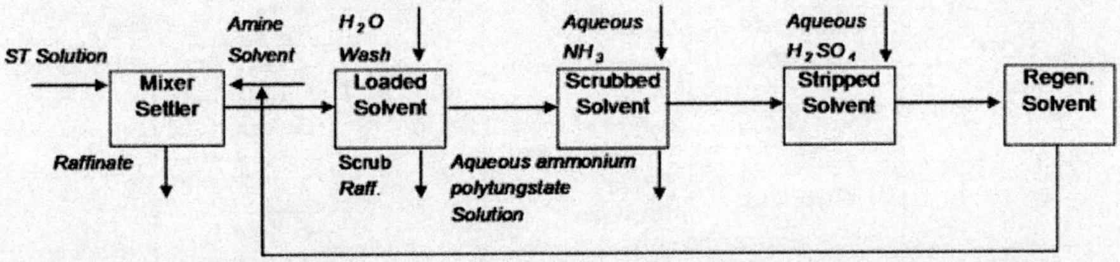

Fig. 3 Modern LIX process (solvent extraction) step of APT production. Ref. [8]

$$MnWO_4 + 2NaOH \rightarrow Mn(OH)_2 + Na_2WO_4 \quad (1)$$

$$FeWO_4 + 2NaOH \rightarrow Fe(OH)_2 + Na_2WO_4 \quad (2)$$

$$Na_2WO_4 + CaCO_3 \rightarrow CaWO_4 + Na_2CO_3 \quad (3)$$

$$CaWO_4 + 2HCl \rightarrow CaCl_2 + H_2WO_4 \quad (4)$$

$$H_2WO_4 + 2NH_4OH \rightarrow (NH_4)_2WO_4 + 2H_2O \quad (5)$$

$$12(NH_4)_2WO_4 \rightarrow (NH_4)_{10}(H_2W_{12}O_{42}) \cdot 4H_2O + 2H_2O + 14NH_3 \quad (6)$$

In the modern hydrometallurgical process of tungsten, all tungsten feeds are digested in sodium hydroxide (or sodium carbonate). Digestion is a quite significant step of this process as significant tungsten losses may occur in this step. Chemical reactions of digestion of different feeds are illustrated in the following chemical equations:

$$CaWO_4 + 2NaOH \rightarrow Ca(OH)_2 + Na_2WO_4 \quad (7)$$

$$CaWO_4 + Na_2CO_3 \rightarrow CaCO_3 + Na_2WO_4 \quad (8)$$

$$MnWO_4 + 2NaOH \rightarrow Mn(OH)_2 + Na_2WO_4 \quad (9)$$

$$FeWO_4 + 2NaOH \rightarrow Fe(OH)_2 + Na_2WO_4 \quad (10)$$

$$WO_3 + 2NaOH \rightarrow H_2O + Na_2WO_4 \quad (11)$$

$$CoWO_4 + 2NaOH \rightarrow Co(OH)_2 + Na_2WO_4 \quad (12)$$

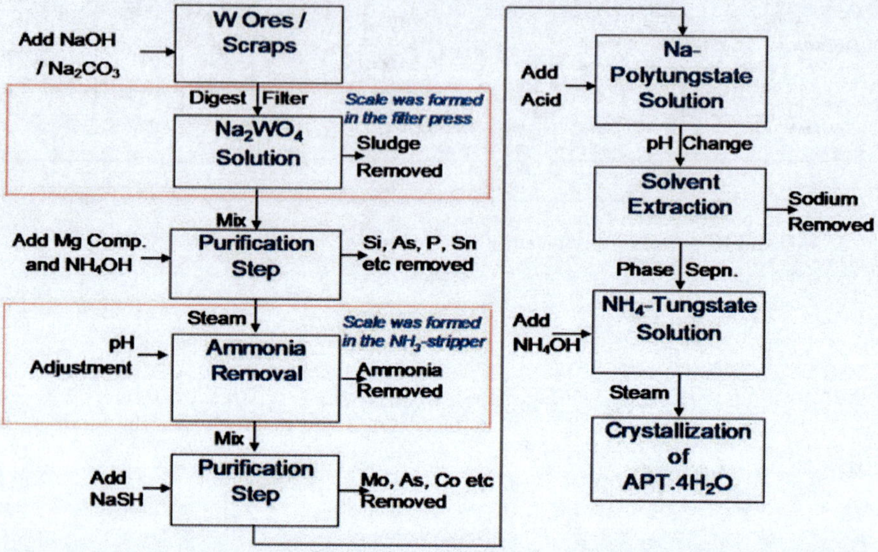

Fig. 4 Flow diagram of modern method of APT process using solvent extraction. Ref. [6]

Sodium tungstate obtained after digestion step is purified in three steps to obtain high-purity ammonium tungstate solution which is crystallized to make APT crystals (Figs. 3 and 4). In some plants (especially in China), solid ion exchange (SIX) is also employed for the separation of sodium and other impurities from sodium tungstate solution.

Processing of Soft and Hard Scrap Oxides

Both scarps are calcined to produce a mixture WO_3 and $CoWO_4$ (with small amount of tantalum oxide) before processing for APT production as per the following chemical equation: WC–Co + $3.5O_2 \rightarrow WO_3 + CoWO_4$. Typically, scraps are digested separately to keep cobalt and tantalum values (cobalt hydroxide sludge) for the recovery of cobalt and tantalum. Sodium tungstate obtained from oxidized scrap is processed with the sodium tungstate obtained from tungsten ore feed to make APT. Most of the hard scrap currently is processed through zinc reclaim process [16].

Crystallization of Ammonium Paratungstate (APT)

Composition of sodium tungstate solutions and its purification steps for the separation of As, As, P, Si, Sn, F, V, etc., and Mo (in a separate step) is reported in Ref. [7]. This review paper [7] also reported the details of solvent extraction step in which separation of sodium from purified sodium tungstate solution is obtained. The resulted tungsten solution after solvent extraction is strong ammonium tungstate solution which is crystallized to make APT. Crystallization of APT is reported by van Put [17] and van Put et al. [18] in two papers. This work is very interesting: reviewing the literature, discussing the formation of metastable phase: $APT{\cdot}6H_2O$ which dissolve to convert to $APT{\cdot}4H_2O$. They also studied the effect of seeds to avoid metastable phase formation. However, not much is reported on the mechanism, morphology, particle size, and purity of APT during crystallization. Details on seeded-crystallization and high-purity APT crystallization have also not been reported.

The function of APT crystallization in APT process is twofold: it is the final purification step in the tungsten chemical process and APT is the final product which has to have some desirable attributes such as crystal purity, morphology, and conversion to tungsten oxides with desirable characteristics. AS reported in Ref. [8], a substantial purification of tungsten occurs during APT crystallization step. Morphology of APT crystals is depicted in Fig. 5. A perusal of APT crystal morphology suggests that crystal growth of APT may have taken place according to the BCF theory of spiral growth.

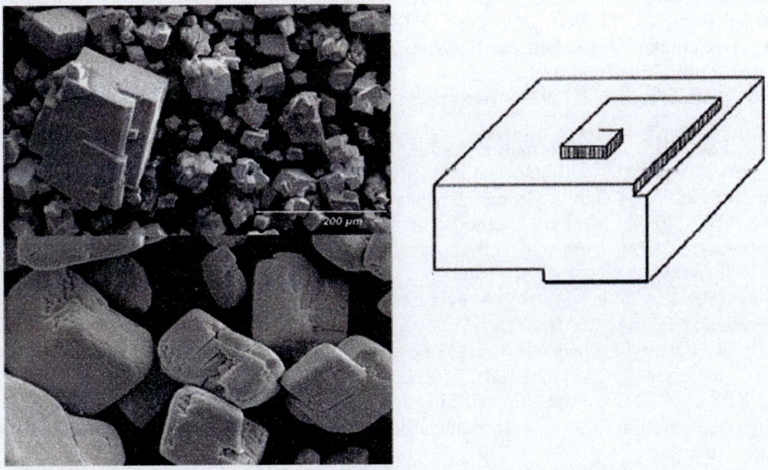

Fig. 5 Morphology of APT crystals showing spiral crystal growth

Challenges in Current APT Processing

Although most APT plants function efficiently, there are several important processing challenges. A few of them are processing of low grade tungsten ores, processing of high grade ores and scraps with high concentrations of critical impurities such as Mo, Sb, V, and radioactive elements, formation of scaling in the process equipments, solvent loss of LIX organics, crud formation in the LIX step, and overall tungsten loses in various process steps. Some of these problems such as equipment scaling and LIX related problems are common so not much can be done except additional research to find better methods of critical impurities separation as scaling and crud formation may be related to impurity elements present in sodium tungstate. Recently, two interesting papers [19, 20] are published one of which demonstrated selective digestion of scheelite using solvometallurgical process and other demonstrated selective dissolution of tungsten from scheelite as a complex. The chemistry seems to be attractive and maybe with more work could result in a better APT process.

References

1. Lassner E, Schubert W-D (1999) Tungsten, properties, chemistry, technology of the element, alloys, and chemical compounds. Kluwer Academic/Plenum Publishers, New York
2. Gaur RP, Braymiller SA, Wolfe TA, Pierce MR, Houck DL (2006) Electrochemical displacement-deposition method for making composite metal powders, US. Patent, 7,041151
3. Christian J, Singh Gaur RP, Wolfe TA, Trasorras JRL (2011) Tungsten chemicals and their applications. ITIA (TUNGSTEN), pp 1–11
4. Zeiler B, Schubert W-D, Andreas B, Wien TU (2018) Recycling of tungsten, current share, economic limitations and further potential. ITIA Tungsten Newsletter, pp 1–19
5. Yih G, SWH, Wang CT (1979) Tungsten: sources, metallurgy, properties, and applications. Plenum Press, New York
6. Gaur RPS (2015) Scale formation in tungsten hydrometallurgical process. In: Amjad Z, Demadis K (eds) Mineral scales and deposits Elsevier, pp 657–677
7. Gaur RPS (2006) Modern hydrometallurgical production for tungsten. JOM 58(9):45–49
8. Gaur RPS (2011) Modern methods of hydrometallurgical production of ammonium paratungstate (APT), Tungsten, refractory & hardmetals: powder production, San Francisco, pp 02–01 to 02–16
9. Han Z, Golev A, Edraki M (2021) A review of tungsten resources and potential extraction from mine waste. Minerals 11:701–718
10. Wolfe TA, Jewett TJ, Gaur RPS (2014) Powder synthesis. In: Sarine VK, Mari D, Llanes L (eds) Comprehensive hard materials. Elsevier, pp 185–212
11. Gaur RPS et al (2013) Processing of tungsten values of flot grade tungsten ore concentrates via high temperature roasting in sodium carbonate. Paper presented at the 18 Plansee Seminar, Reutte, Austria
12. Zajic JE, Kosaric N (1977) Flotation of scheelite from calcite with a microbial based collector. US Patent 4,046,678
13. Vedova R, Grauerholz NL (1977) Method for recovering scheelite from tungsten ores by flotation. US Patent 4,054,442
14. Agar GE (1984) Scheelite flotation process. US Patent 4,488,959
15. Shangyong L, Runqing L, Wenheng L, Wei S, Yuehua H (2020) Clean desulfurization of sulfur-rich tungsten concentrates by reverse flotation. https://www.sciencedirect.com/science/article/pii/S0959652619337461. Accessed 20 Jan 2020

16. Schubert W-D, Wein TU (2019) Recycling of tungsten, the technology-history, state of the art and peculiarities. ITIA Tungsten Newsletter, pp 1–20
17. Van Put JW (1995) Crystallization and processing of ammonium paratungstate (APT). Int J Ref Metals Hard Mater 13:61–76
18. Van Put JW, Witkamp GJ, van Rosmalen GM (1993) Formation of ammonium paratungstate tetra- and hexa-hydrate. I: stability. Hydrometallurgy 34:187–201
19. Liu QS, Tu T, Guo H, Cheng HJ, Wang XZ (2021) Complexation extraction of scheelite and transformation behaviour of tungsten-containing phase using H_2SO_4 solution with $H_2C_2O_4$ as complexing agent. Trans Nonferrous Met Soc China 31:3150–3161
20. Orefice M, Nguyen VT, Raiguel S, Jones PT, Binnemans K (2022) Solvometallurgical process for the recovery of tungsten from scheelite, Ind Eng Chem Res 61:754–764

Efficient Extraction of Bismuth in a Two-Stage Leaching Process for the Recovery of Zinc and Lead from Zinc-Rich Dust

Yan Li, Huaixuan Feng, Jingsong Wang, Xuefeng She, Kepiao Ren, and Qingguo Xue

Abstract In this study, an efficient and sustainable hydrometallurgical process was developed for the recovery of zinc, bismuth, and lead from zinc-rich dust after pyrometallurgical treatment of iron and steel metallurgical dust. The feasibility of the separation of zinc, bismuth, and lead in the leaching system was confirmed by thermodynamic analysis. In the first step of the leaching process, zinc is leached and recovered as $ZnSO_4$ in the leachate. In the second leaching step, a combination of H_2SO_4 and NaCl was used as leaching agent to study the leaching behavior of bismuth and lead under different experimental conditions (e.g., H_2SO_4 concentration, NaCl concentration, liquid-solid ratio, temperature, and reaction time). Under the optimum conditions, the leaching rates of bismuth and lead were 93.5% and 0.24%, respectively. The leachate enriched in bismuth was used to prepare bismuth products, and the leach residue enriched in lead could be further processed to recover lead.

Keywords Zinc-rich dust · Resourceful recycling · Acidic-chloride leaching · BiOCl

Introduction

Bismuth is known as a green metal due to its non-toxic properties and is widely used in metallurgy, chemical industry, new materials, medicine, electronics, nuclear industry, etc., due to its excellent properties [1–3]. In recent years, the global demand for bismuth has increased rapidly [4]. With the consumption of primary resources bismuth ore, obtaining bismuth from secondary resources has attracted extensive attention from scholars at home and abroad [5, 6]. The secondary resources of

Y. Li · H. Feng · J. Wang (✉) · X. She (✉) · K. Ren · Q. Xue
State Key Laboratory of Advanced Metallurgy, University of Science and Technology, Beijing, Beijing 100083, China
e-mail: wangjingsong@ustb.edu.cn

X. She
e-mail: shexuefeng@ustb.edu.cn

R. G. Reddy et al. (eds.), *New Directions in Mineral Processing, Extractive Metallurgy, Recycling and Waste Minimization*, The Minerals, Metals & Materials Series,
https://doi.org/10.1007/978-3-031-22765-3_27

bismuth mainly include bismuth-containing by-products from the production process of copper, lead, and zinc smelters, such as lead anode sludge, copper anode sludge, and white soot [7–9]. Another source of secondary resources of bismuth is the steel industry, where bismuth is found in metallurgical dust generated from steel smelting production processes [10–12]. Little bismuth is recovered from it due to the poor comprehensive utilization of steel metallurgical dust. After the steel metallurgical dust is treated by pyrotechnical process, bismuth and zinc will be enriched in the Zn-rich dust twice, which creates good conditions for the effective recovery of valuable elements [13–16]. The current treatment process for the comprehensive utilization of Zn-rich dust mainly focuses on the recovery of elements with high contents such as zinc and lead, and the recovery of bismuth has not attracted sufficient attention [17, 18]. In view of this, a hydrometallurgical process for the recovery of zinc, bismuth, and lead from Zn-rich dust was proposed. Firstly, sulfuric acid was used as leaching agent and zinc was leached and recovered as zinc sulfate in the leachate. Then Bi and Pb are separated by selective leaching using H_2SO_4-NaCl solution. the leachate enriched with bismuth is used to prepare BiOCl, and the leach residue enriched with Pb can be further processed to recover Pb. This process achieves efficient resource utilization of Zn-rich dust, which can effectively alleviate the shortage of bismuth resources in China.

Experimental

Materials

The raw material was zinc-rich dust (ZRD) collected from a steel metallurgical dust pyroprocessing plant of a domestic steel company. Table 1 shows the chemical composition of ZRD, which contains about 49.15% Zn, 4.99% Pb, 0.24% Bi, 0.07% Cd, 0.04% In, 3.15% Na, 8.3% K, and 0.85% Fe. X-ray diffraction analysis detected the main phase composition of ZRD as shown in Fig. 1, which mainly contains ZnO, KCl, and NaCl. The content of other elements was below the detection line of the XRD instrument and failed to detect their phase forms. The reagents H_2SO_4, NaCl, and NaOH used in the experiments were all analytically pure; distilled water was used in all experiments.

Table 1 Chemical composition of the ZRD sample

Element	Zn	Pb	Bi	In	Cd	Na	K	Fe
Weight (%)	49.15	4.99	0.24	0.04	0.07	3.15	8.30	0.85

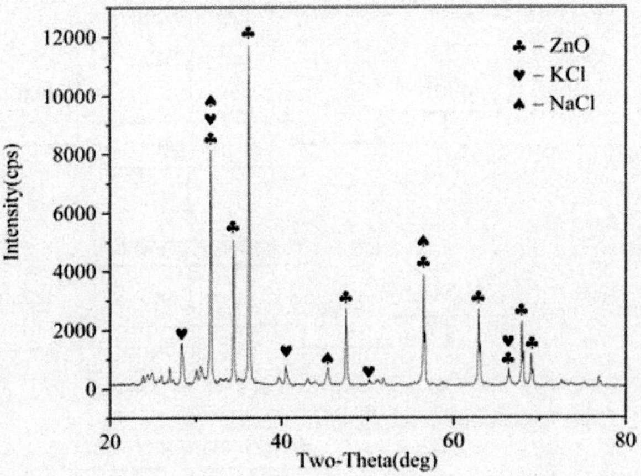

Fig. 1 XRD pattern of the ZRD sample

Experimental Procedure

The two-stage leaching hydrometallurgical process is shown in Fig. 2. KCl and NaCl in ZRD are removed in the water leaching process, and the leachate enriched with K and Na is fed to a multi-effect evaporation unit to extract KCl and NaCl products by evaporation crystallization. Zn is extracted from the water leaching residue with H_2SO_4 solution to obtain a leachate rich in zinc sulfate. Selective dissolution of Bi is in the sulfuric acid leaching residue with (H_2SO_4 + NaCl) leaching agent. Water leaching experiments and neutralized hydrolysis experiments were carried out in 1000 ml beakers. The two-stage leaching experiments were carried out in 250 mL three-necked flasks equipped with an overhead stirrer. The reaction temperature was controlled by placing the three-necked beaker in a constant temperature water bath, and the stirring speed was fixed at 200 rpm for each test.

ZRD was added to the beaker with a liquid-solid ratio of 3 (ml/g), leached in a water bath at 25 °C, and the solid-liquid separation was achieved through a vacuum filter. The water leaching filter cake was put into a three-neck flask for the sulfuric acid leaching process, and dilute sulfuric acid was used as leaching agent. After leaching was completed, the slurry was immediately vacuum filtered. The dilute leachate was sent to ICP analysis to calculate the leaching rate of elemental zinc. The leaching agent (H_2SO_4 + NaCl) was added to the three-necked flask, and a certain amount of sulfuric acid leaching residue was added to the reactor under stirring and leached in a constant temperature water bath. After the leaching was completed, the slurry was immediately vacuum filtered. The diluted leachate was sent to ICP analysis to detect Bi and Pb concentrations. The chlorinated leaching solution was added to a beaker, and sodium hydroxide solution was added under stirring, and after standing for a certain time, the solid-liquid separation was achieved by vacuum filter. the BiOCl sample was washed with de-distilled water and dried at 80 °C for 12 h, and then characterized by XRD.

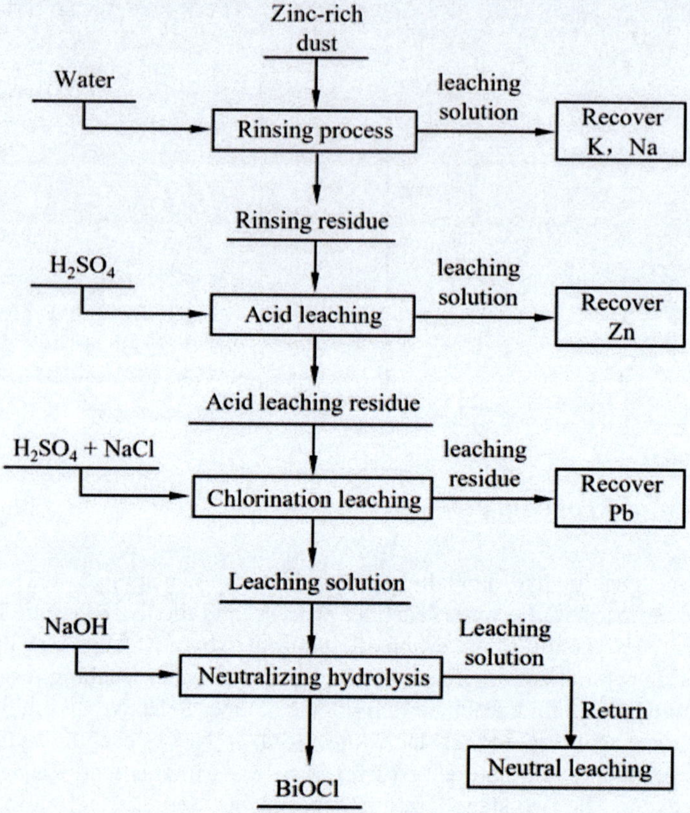

Fig. 2 Schematic diagram of the process for recovering Zn, Pb, and Bi from Zn-rich dust

Results and Discussion

Thermodynamic Analysis

The E-pH diagrams of the Zn–H_2O and Bi–H_2O systems at 298 K were calculated by HSC Chemistry 9.0, and the results are shown in Fig. 3. The dashed line indicates the stable region of water. From Fig. 3, it can be seen that Bi^{3+} in solution can be stabilized in more acidic solutions, and the dominant zone of stable presence of Zn^{2+} is much larger than that of Bi^{3+}. Therefore, the pH value of the sulfuric acid leaching process was controlled to be in the right range during the experiment in order to achieve selective leaching of zinc. PbO in the water leaching slag will react with H_2SO_4 to form $PbSO_4$, and due to the small solubility product of $PbSO_4$ (1.6×10^{-8}, 18 °C) and the large amount of SO^{4-} in the solution, the lead stays in the leaching slag as $PbSO_4$ precipitation. From Fig. 3b, it can be seen that with the increase of solution pH, the bismuth ions in the solution change in the following order: $Bi^{3+} \rightarrow BiOH^{2+}$

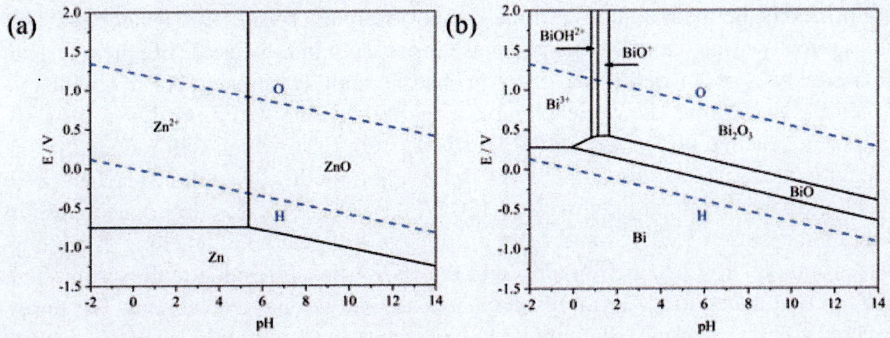

Fig. 3 E-pH of different systems at 298.15 K: **a** Zn–H_2O; **b** Bi–H_2O

Table 2 Possible reactions during the process

Element	Reaction	NO
Zn	$ZnO + H_2SO_4 = ZnSO_{4(aq)} + H_2O$	(1)
	$ZnFe_2O_4 + 4H_2SO_4 = ZnSO_{4(aq)} + Fe_2(SO_4)_{3(aq)} + 4H_2O$	(2)
Pb	$PbO + H_2SO_4 = PbSO_4 + H_2O$	(3)
	$Pb^{2+} + 2Cl^- = PbCl_2$	(4)
	$PbCl_2 + 2Cl^- \leftrightharpoons [PbCl_4]^{2-}$	(5)
Bi	$Bi_2O_3 + 6H^+ + 6Cl^- = 2BiCl_{3(aq)} + 3H_2O$	(6)
	$BiCl_3 + H_2O = BiOCl + 2HCl$	(7)

→ BiO^+ → Bi_2O_3. After the completion of the chlorination leaching process, the Bi in the leaching solution is in the form of Bi^{3+} and there is a large amount of Cl^- in the solution. With the increase of pH, $BiOH^{2+}$ and BiO^+ will undergo hydrolysis reaction and the final product is BiOCl. The reactions that may occur in the process are shown in Table 2.

Sulfuric Acid Leaching to Recover Zn

The Zn-rich dust has a high Zn content, so it makes economic sense to recover it first. On the other hand, by extracting Zn from Zn-rich dust into the leachate in the H_2SO_4 leaching process, Pb and Bi will be enriched in the leach residue, which is beneficial for subsequent process recovery. The use of H_2SO_4 as a leaching agent to recover Zn from Zn-bearing dust has been extensively investigated and is commercially available [19, 20]. H_2SO_4 has been demonstrated to be effective, cost-effective, and generally available for solubilizing valued elemental compounds [21]. Therefore, H_2SO_4 was used as leaching agent in this process. Maximizing the dissolution of Zn into the leaching solution, while ensuring that the dissolution of Bi, Pb, and Fe is minimized is the main objective of the H_2SO_4 leaching process. The leaching rate of Zn is directly related to the physical phase form of Zn in the dust. From the physical phase analysis of Zn-rich dust, it is clear that the main component is ZnO, which is beneficial to improve the Zn leaching rate. However, ZnS and $ZnFe_2O_4$ in the dust cannot be leached effectively in the H_2SO_4 system [22], which will reduce the Zn leaching rate. Also, certain amount of Cd, Fe, and In in the dust will be leached [23].

The experimental parameters were 120 g/L H_2SO_4, 300 g/L slurry density, water bath temperature of 70 °C, and reaction time of 4 h. The zinc leaching efficiency was around 80%. The zinc remaining in the H_2SO_4 leach residue mainly consisted of $ZnSO_4$ attached to the leach residue and ZnS that was not leached out. The unrecovered zinc in the slag will continue to be leached in the chloride leaching process and finally recycled back to the H_2SO_4 leaching process to ensure high recovery of zinc in this process.

Chlorination Leaching to Recover Bi

The separation of elements (Pb and Bi) from Zn-rich dust by chlorination leaching was investigated using a single-factor condition test method. The leaching behavior of Bi and Pb under different experimental conditions (e.g., H_2SO_4 concentration, NaCl concentration, liquid-to-solid ratio, temperature, and reaction time) was systematically investigated. The aim was to find the optimal conditions at laboratory scale for maximum Bi leaching efficiency and minimum amount of Pb in the aqueous phase.

Effect of NaCl Concentration

Thermodynamic analysis showed that increasing the chloride concentration promotes the dissolution of Bi and affects the formation of $PbCl_2$ [24]. Therefore, the effect of NaCl concentrations ranging from 10 to 120 g/L on the leaching efficiency of Bi and Pb was investigated. According to the results in Fig. 4, the leaching efficiency of Bi was more influenced by the NaCl concentration, while the leaching rate of Pb was less influenced by the same conditions. The leaching efficiency of Bi increased from 78.76 to 91.46% when the NaCl concentration was increased from 10 to 60 g/L. Increasing the NaCl concentration to 90 g/L did not change the leaching efficiency of Bi significantly because the chloride requirement of Bi was satisfied. However, increasing the NaCl concentration to 120 g/L, the leaching rate of Bi decreased instead, which shows that too high chloride ion concentration has a negative effect on the leaching of Bi. The same phenomenon was found in the study by Che et al. [24]. When the NaCl concentration was increased from 10 to 120 g/L, the solubility of Pb increased from 0.06 to 0.68%, i.e., the increase in the amount of chloride ions in the solution led to the dissolution of Pb in the system. This is due to the fact that $PbCl_2$ has a certain solubility in the NaCl solution and when the concentration of

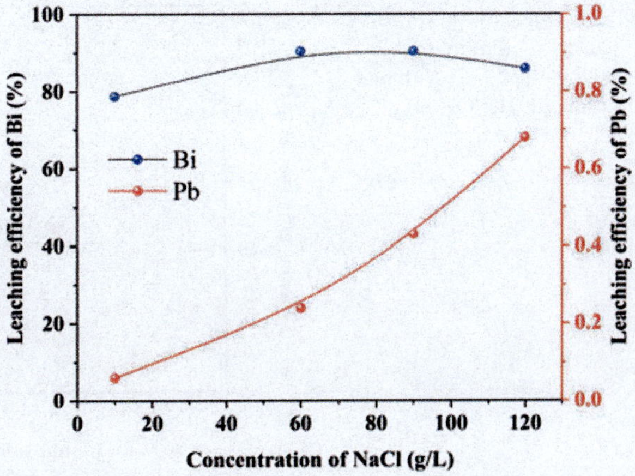

Fig. 4 Effect of NaCl concentration on the leaching efficiency of Bi and Pb (160 g/L H_2SO_4, L/S ratio of 5 mL/g, 60 °C, 4 h)

Cl^- is high, $PbCl_2$ increases its solubility by forming complex ions with Cl^- and the reaction mechanism is shown in Eq. 5 [25].

Effect of Liquid-to-Solid Ratio

The liquid-to-solid ratio is an important parameter that affects leaching kinetics, process cost, and downstream treatment processes [26]. To determine the effect of this process parameter on the leaching behavior of Bi and Pb, experiments were conducted in the range of 3–9 mL/g, and the results are shown in Fig. 5. The leaching rate of Pb increased from 0.09 to 0.84% when the liquid-to-solid ratio increased from 3 to 9 mL/g. The leaching rate of Bi increased significantly from 74.7 to 91.46% by increasing the liquid-to-solid ratio from 3 to 5 mL/g. This is due to the fact that the increase in the liquid-solid ratio improves the surface contact between the solution and the dust, thus facilitating the dissolution of Bi and Pb [27]. Increasing the liquid-solid ratio from 5 to 9 mL/g slowly increased the leaching rate of Bi to 94.05%, at which point the liquid-solid ratio was no longer the main factor affecting the leaching reaction. Increasing the liquid-solid ratio has advantages and disadvantages; it usually accelerates mass transfer and promotes the reaction, but it also increases the equipment volume and energy consumption and reduces the production efficiency. On the other hand, increasing the liquid-solid ratio will reduce the concentration of Bi and Pb in the leachate, which is not conducive to the subsequent processing of the production process. Therefore, a liquid-to-solid ratio of 5 mL/g was selected as appropriate after comprehensive consideration.

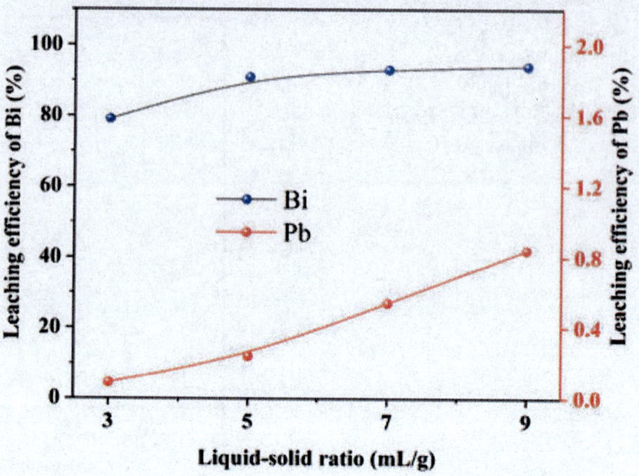

Fig. 5 Effect of liquid-to-solid ratio on the leaching efficiency of Bi and Pb (160 g/L H_2SO_4, 60 g/L NaCl, 60 °C, 4 h)

Effect of Temperature

When other operating parameters were kept constant (160 g/L H_2SO_4, 60 g/L NaCl, L/S ratio of 5 mL/g, 4 h), the effects of different reaction temperatures on the leaching of Bi and Pb are shown in Fig. 6. With the increase of temperature from 25 to 80 °C, the leaching rate of elemental Pb fluctuated between 0.23 and 0.27%. This indicates that the temperature has almost no effect on the leaching rate of elemental Pb in the experimental temperature range. As the temperature increased from 25 to 60 °C, the leaching rate of elemental Bi increased from 87.31 to 91.46%, and thereafter, the temperature continued to increase to 80 °C, and the leaching rate of Bi remained around 90.46%. Therefore, temperature plays a secondary role in the leaching process, and 60 °C was chosen as a suitable experimental leaching temperature in order to obtain a high reaction rate.

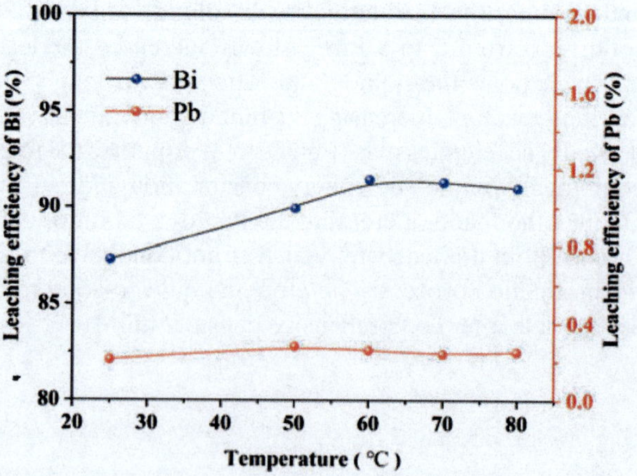

Fig. 6 Effect of temperature on the leaching efficiency of Bi and Pb (160 g/L H_2SO_4, 60 g/L NaCl, L/S ratio of 5 mL/g, 4 h)

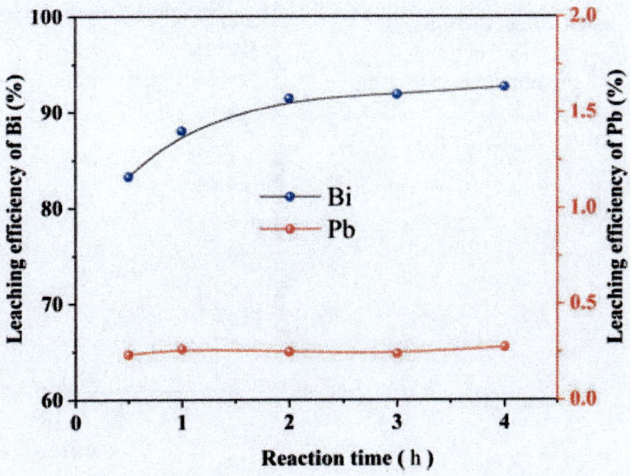

Fig. 7 Effect of reaction time on the leaching efficiency of Bi and Pb (160 g/L H_2SO_4, 60 g/L NaCl, L/S ratio of 5 mL/g, 60 °C)

Effect of Reaction Time

When other operating parameters were kept constant (160 g/L H_2SO_4, 60 g/L NaCl, L/S ratio of 5 mL/g, 60 °C), the effect of reaction time on the leaching of Bi and Pb is shown in Fig. 7. The leaching rate of Bi increased from 83.29 to 91.46% when the leaching time was extended from 0.5 to 2 h. The leaching rate of Bi did not increase significantly when the leaching time was increased from 2 to 4 h. In this process, the reaction time affects the contact time between solid and liquid, and the contact time between reactants and leaching agent is relatively long to maximize the extraction of Bi [27]. In the experimental leaching time range, the leaching rate of Pb element was less affected by the leaching time, with the reaction time extended from 0.5 to 4 h, the leaching rate of Pb element increased from 0.21 to 0.25%, which always remained at a low level. Considering that excessive increase of reaction time would reduce the production efficiency, therefore, the leaching time of 2 h was determined as the optimal reaction time.

Effect of H_2SO_4 Concentration

When other operating parameters were kept constant (60 g/L NaCl, L/S ratio of 5 mL/g, 60 °C, 2 h), the effect of H_2SO_4 concentration on the leaching of Bi and Pb was set for four groups with different experimental H_2SO_4 concentrations as shown in Fig. 8. The leaching of Bi was divided into two stages, when the H_2SO_4 concentration interval was 40–160 g/L, the leaching of Bi was mainly influenced by the H_2SO_4 concentration, and the leaching rate of Bi increased from 70.92 to 91.46% with the increase of H_2SO_4 concentration. When the H_2SO_4 concentration interval was 160–280 g/L, the H_2SO_4 concentration was no longer the main factor affecting the leaching of elemental Bi, and the leaching rate of elemental Bi increased slowly. The sensitivity of the Bi leaching efficiency suggests that Bi may be present

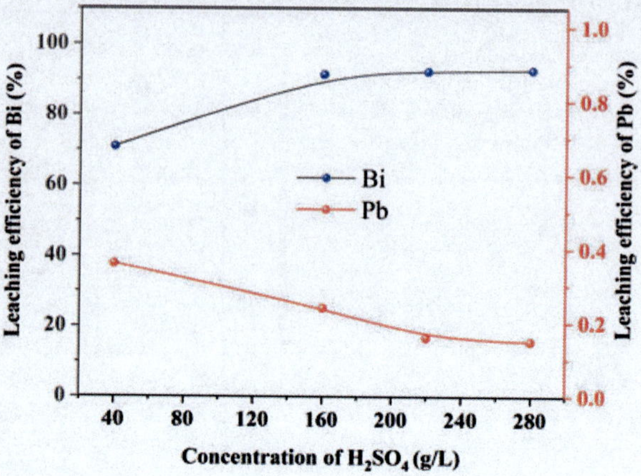

Fig. 8 Effect of H_2SO_4 concentration on the leaching efficiency of Bi and Pb (60 g/L NaCl, L/S ratio of 5 mL/g, 60 °C, 2 h)

in the dust as an oxide phase [28]. The increase in H_2SO_4 concentration from 40 to 280 g/L decreased the leaching rate of Pb from 0.36 to 0.15%. The leaching rate of Pb remained low throughout the range of experimental H_2SO_4 concentrations. Considering that too high H_2SO_4 concentration increases acid consumption, the H_2SO_4 concentration of the leaching agent was selected as 160 g/L was considered to be the most suitable.

Neutralization and Hydrolysis for Bi Recovery

The chlorination leaching process completes the initial leaching and separation of Bi, and the neutralization and hydrolysis process completes the transformation from bismuth-rich leachate to Bi product. The bismuth-rich leachate contains not only Bi but also other impurity ions such as Zn and Fe. According to the different precipitation intervals of different elements at different pH, the best precipitation of Bi is completed. Chen Ya et al. found that when the pH of the solution was less than 0.8, the precipitation efficiency of Bi was less than 5%, and as the pH increased from 0.8 to 3.0, the precipitation efficiency of Bi rapidly increased to about 99% [9]. Therefore, sodium hydroxide solution was chosen as the precipitant and the pH of the bismuth-rich leach solution was adjusted to increase, and $BiCl_3$ would undergo a hydrolysis reaction to form BiOCl, and the reaction equation is shown in Eq. (7). It was found that the precipitation efficiency of Bi reached 95.16% when the pH of the bismuth-rich leachate was increased to 2. The precipitation of $Fe(OH)_3$ increased when the pH of the leachate was continued to be adjusted to 3. In order to reduce the content of impurity ions in BiOCl, the optimal precipitation end point was selected when the pH was increased to 2. After the completion of the neutralization and hydrolysis process, the product BiOCl was obtained by solid-liquid separation,

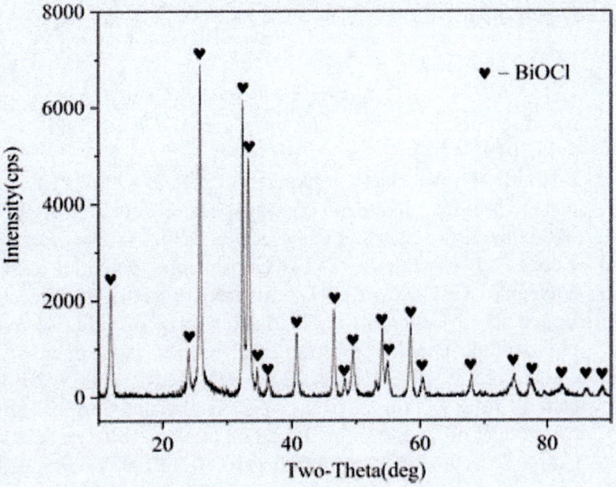

Fig. 9 XRD of BiOCl prepared by neutralization and hydrolysis process

and the XRD analysis is shown in Fig. 9. The obtained product BiOCl can be sold to non-ferrous smelters as a raw material for the production of metallic Bi.

Conclusions

A two-stage leaching hydrometallurgical process was proposed for the recovery of Zn, Bi, and Pb from Zn-rich dust. The feasibility of the separation of Zn, Bi, and Pb in the leaching system was confirmed by thermodynamic analysis. Zn was first leached and recovered as $ZnSO_4$ in the leachate, and the Pb-enriched leach residue could be further processed to recover Pb. The influence of various technical parameters on the selective leaching of Bi was studied for the chlorination leaching process. The results showed that the H_2SO_4 concentration, NaCl concentration, and liquid-solid ratio were the main factors affecting the leaching rate of Bi. Under the optimal conditions, the leaching rates of Bi and Pb were 91.5% and 0.24%, respectively. The BiOCl product could be prepared by adjusting the pH of the bismuth-rich leachate to increase to 2.

Acknowledgements This work was supported by the National Natural Science Foundation of China (2019YFC1905705) and the Guangdong Provincial Science and Technology Plan Funding Projects (GDKJ2020002).

References

1. Liao W-W, Yang L, Chen J et al (2019) Realizing Bi-doped α-Cu2Se as a promising near-room-temperature thermoelectric material. Chem Eng J 371:593–599. https://doi.org/10.1016/j.cej.2019.04.081
2. Pistofidis N, Vourlias G, Konidaris S et al (2007) The effect of bismuth on the structure of zinc hot-dip galvanized coatings. Mater Lett 61:994–997. https://doi.org/10.1016/j.matlet.2006.06.029
3. Keogan D, Griffith D (2014) Current and potential applications of bismuth-based drugs. Molecules 19:15258–15297. https://doi.org/10.3390/molecules190915258
4. Rocchetti L, Beolchini F (2015) Recovery of valuable materials from end-of-life thin-film photovoltaic panels: environmental impact assessment of different management options. J Clean Prod 89:59–64. https://doi.org/10.1016/j.jclepro.2014.11.009
5. Nan T, Yang J, Tang C et al (2022) Hydrometallurgical process for extracting bismuth from by-product of lead smelting based on methanesulfonic acid system. Trans Nonferrous Met Soc China 32:319–332. https://doi.org/10.1016/S1003-6326(22)65797-2
6. Deady E, Moon C, Moore K et al (2022) Bismuth: economic geology and value chains. Ore Geol Rev 143:104722. https://doi.org/10.1016/j.oregeorev.2022.104722
7. Wang Z, Ding F, Zhan J, Zhang C (2016) Solvent extraction mechanism and precipitation stripping of bismuth (III) in hydrochloric acid medium by tributyl phosphate. J Cent South Univ 23:3085–3091. https://doi.org/10.1007/s11771-016-3373-9
8. Shen Y, Zhang H, Cao H et al (2019) Efficient extraction of bismuth from hydrochloric acid solution by copper powder. Hydrometallurgy 189:105121. https://doi.org/10.1016/j.hydromet.2019.105121
9. Chen Y, Liao T, Li G et al (2012) Recovery of bismuth and arsenic from copper smelter flue dusts after copper and zinc extraction. Miner Eng 39:23–28. https://doi.org/10.1016/j.mineng.2012.06.008
10. Wang J, Li Y, Feng H et al (2021) Progress in treating difficult-to-handle dust and sludge and full-scale resource utilization in an iron and steel industry cluster. Chinese J Eng 43:1737–1746. https://doi.org/10.13374/j.issn2095-9389.2021.09.15.004
11. Liu G, Wu Y, Tang A et al (2020) Recovery of scattered and precious metals from copper anode slime by hydrometallurgy: a review. Hydrometallurgy 197:105460. https://doi.org/10.1016/j.hydromet.2020.105460
12. Chang F, Wu S, Zhang F et al (2015) Characterization of sintering dust, blast furnace dust and carbon steel electric arc furnace dust. In: Characterization of minerals, metals, and materials 2015. Springer International Publishing, Cham, pp 83–90
13. Wang J, Zhang Y, Cui K et al (2021) Pyrometallurgical recovery of zinc and valuable metals from electric arc furnace dust—A review. J Clean Prod 298:126788. https://doi.org/10.1016/j.jclepro.2021.126788
14. Lin X, Peng Z, Yan J et al (2017) Pyrometallurgical recycling of electric arc furnace dust. J Clean Prod 149:1079–1100. https://doi.org/10.1016/j.jclepro.2017.02.128
15. Wang J, Yang H, She X et al (2011) Study on mechanism of Zn, K, Na removal and flue gas formation in RHF direct reduction treat metallurgical dust process. J Chongqing Univ 34:83–87
16. Xiao X, Zhang S, Sher F et al (2021) A review on recycling and reutilization of blast furnace dust as a secondary resource. J Sustain Metall 7:340–357. https://doi.org/10.1007/s40831-021-00377-9
17. Li Y, Feng H, Wang J et al (2022) Current status of the technology for utilizing difficult-to-treat dust and sludge produced from the steel industry. J Clean Prod 367:132909. https://doi.org/10.1016/j.jclepro.2022.132909
18. Chengyan W, Shuang S, Baozhong M et al (2019) Status and development of antimony and bismuth metallurgy technologies in China. Nonferrous Met Metall 8:11–16. https://doi.org/10.3969/j.issn.1007-7495.2019.08.002

19. Ghasemi SMS, Azizi A (2017) Investigation of leaching kinetics of zinc from a low-grade ore in organic and inorganic acids. J Min Environ 8:579–591. https://doi.org/10.22044/jme.2016.767
20. Kukurugya F, Vindt T, Havlík T (2015) Behavior of zinc, iron and calcium from electric arc furnace (EAF) dust in hydrometallurgical processing in sulfuric acid solutions: thermodynamic and kinetic aspects. Hydrometallurgy 154:20–32. https://doi.org/10.1016/j.hydromet.2015.03.008
21. Kaya M, Hussaini S, Kursunoglu S (2020) Critical review on secondary zinc resources and their recycling technologies. Hydrometallurgy 195:105362. https://doi.org/10.1016/j.hydromet.2020.105362
22. Lampinen M, Laari A, Turunen I (2015) Kinetic model for direct leaching of zinc sulfide concentrates at high slurry and solute concentration. Hydrometallurgy 153:160–169. https://doi.org/10.1016/j.hydromet.2015.02.012
23. Jha MK, Kumar V, Singh RJ (2001) Review of hydrometallurgical recovery of zinc from industrial wastes. Resour Conserv Recycl 33:1–22. https://doi.org/10.1016/S0921-3449(00)00095-1
24. Che J, Chen Y, Ma B et al (2020) Recovery of metallic Bi from Pb Bi slag: an integrated process of chloride leaching and electrowinning. Hydrometallurgy 193:105321. https://doi.org/10.1016/j.hydromet.2020.105321
25. Adebayo AO, Olasehinde EF (2015) Leaching kinetics of lead from galena with acidified hydrogen peroxide and sodium chloride solution. Trans Institutions Min Metall Sect C Miner Process Extr Metall 124:137–142. https://doi.org/10.1179/1743285515Y.0000000001
26. Ha TK, Kwon BH, Park KS, Mohapatra D (2015) Selective leaching and recovery of bismuth as Bi_2O_3 from copper smelter converter dust. Sep Purif Technol 142:116–122. https://doi.org/10.1016/j.seppur.2015.01.004
27. Qiang L, Pinto ISS, Youcai Z (2014) Sequential stepwise recovery of selected metals from flue dusts of secondary copper smelting. J Clean Prod 84:663–670. https://doi.org/10.1016/j.jclepro.2014.03.085
28. Fanzhen K (2000) Recover of bismuth from bismuth materials contening arsenic. Hydrometall China 19:53–58

Efficient Utilization of Alloy Powder During Limonitic Laterite Sintering Process

Yikang Tu, Yuanbo Zhang, Zijian Su, and Tao Jiang

Abstract Alloy powder is a typical by-product produced from stainless steel processing, which contains a significant level of valuable metals, such as iron, chrome, and nickel. About 1 million tons of alloy powder are generated in China annually, but as mixed with impurities such as Al and Si, alloy powder is not suitable to be directly smelted into stainless steel by refining. In this study, alloy powder was efficiently utilized during limonitic laterite sintering process. The effects of adding alloy powder on green bed permeability, the yield and quality indexes of the sinters, and the microstructure of the sinters were evaluated. The results show that the concentrated distribution degree of granules and green bed permeability was improved after adding alloy powder. As the additional amount of alloy powder increased from 0 to 3%, the tumble index and yield of the sinters rose from 50.07% to 54.69% and from 62.04% to 67.35%, respectively.

Keywords Alloy powder · Limonitic laterite · Sintering

Introduction

Alloy powder is a typical solid waste produced in the turning and grinding process of stainless steel, which contains a significant level of valuable metals, such as iron, chromium, and nickel [1]. However, due to the complex relationship of various phase compositions, the comprehensive utilization of valuable metals from alloy powder has encountered several problems. In China, about 1 million tons of alloy powder are discharged annually. At present, alloy powder was mainly used as cement ingredients, building materials, etc. The secondary iron-bearing resources usually contain abundant iron, chromium, nickel, etc., which have huge potential value for comprehensive utilization [2]. Acid leaching methods with low concentration of strong acids such as aqua regia, hydrochloric acid, and sulfuric acid were proposed

Y. Tu · Y. Zhang (✉) · Z. Su · T. Jiang
School of Minerals Processing & Bioengineering, Central South University, Changsha 410083, Hunan, China
e-mail: zybcsu@126.com

R. G. Reddy et al. (eds.), *New Directions in Mineral Processing, Extractive Metallurgy, Recycling and Waste Minimization*, The Minerals, Metals & Materials Series,
https://doi.org/10.1007/978-3-031-22765-3_28

to recover valuable metals from alloy powder in previous studies. After removing impurities and valuable metals, the leaching residue was used as ironmaking raw material [3]. However, the leaching selectivity was rather poor, a large amount of iron was dissolved out as well, and it is possible that secondary pollution would be generated. Iron ore sintering and pelletizing are common processes for large-scale treatment of alloy powder. Nevertheless, alloy powder usually contains a considerable amount of chromium which will obviously increase the coke ratio of blast furnace [4]. Therefore, most enterprises are reluctant to add alloy powder during the steel production process.

In the limonitic laterite, the high content of water of crystallization and MgO are negative to the formation of high-quality liquid phase, resulting in poor strength and low yield of limonitic laterite sinter [5–7]. To improve the sinter quality, some strengthening technologies were investigated during limonitic laterite sintering process. The external pressure field helped reduce the sinter porosity, increasing the formation of SFCA (composite calcium ferrite composed of SiO_2, Fe_2O_3, FeO, CaO, and Al_2O_3), which made a tight interlocking structure between hercynite ((Mg, Fe)(Fe, Al)$_2$O$_4$, which was from the reaction of Fe_2O_3, FeO, Al_2O_3, and MgO during sintering process) and SFCA [8]. Hot waste gas recirculation during the sintering process of limonitic laterite explained that the increase of oxygen potential contributed to the formation of more SFCA. With the increase of high temperature holding time, the wettability of liquid phase bonding relative to hercynite was improved, so a closely connected microstructure with higher strength was formed [9]. Nevertheless, although there are many studies on improving the strength of sinter, the yield and quality indexes of limonitic laterite sinters still need to be improved.

Limonitic laterite sinters are mainly used for smelting 200 series or 300 series stainless steel after adding ferrochromium alloy [10]. If alloy powder is added to limonitic laterite sintering process, it is not necessary to pay attention to the increase of coke ratio, because alloy powder can reduce the addition of ferrochrome which also needs coke consumption during the smelting process.

In this study, the influence of alloy powder on limonitic laterite sinters was systematically studied via sintering pot tests. The effects of alloy powder on the size distribution and permeability of the sintering green bed, yield, quality, and microstructure of limonitic laterite sinters were investigated. This work has great guiding significance for resource utilization of alloy powder on large-scale and strengthening the limonitic laterite sintering process.

Experimental

Materials

The limonitic laterite used in this study was derived from Philippines. The alloy powder with 87.55% Fe_2O_3, contained 3.21% Cr_2O_3 and 1.30% NiO. The Fe_2O_3 and

Al_2O_3 content were 75.16% and 6.19%, respectively. The special phase composition of limonitic laterite meant that it contained high water of crystallization. In addition, the content of NiO, Cr_2O_3, FeO, CaO, SiO_2, and MgO in limonitic laterite was 0.88%, 3.45%, 2.51%, 0.06%, 2.22%, and 0.59%, respectively. The LOI (Loss on ignition) reached up to 10.72% due to high water of crystallization. High water of crystallization content would consume more energy in the sintering process. The large LOI meant that the material layer shrank seriously, and consequently more solid fuel needed to be added. The flux used throughout this study was quicklime with 72.21% CaO content. The TFe content of the return ore was 50.92%, in which the FeO content was 23.25%. Besides, the content of NiO, Cr_2O_3, CaO, MgO, SiO_2, and Al_2O_3 in the return ore was 1.03%, 4.52%, 7.69%, 2.17%, 4.35%, and 6.82%, respectively. The determination of Fe_{total} and FeO depended on ISO 2597-1:2006 EN and ISO 9035:1989 EN, respectively. The content of other elements was obtained by ICP measurement after sample dissolution.

Methods

Experimental Procedure

The pilot-scale sintering tests were carried out in a laboratory sintering pot as shown in Fig. 1. Firstly, preparing mixture: limonitic laterite, alloy powder, return ore, anthracite, and flux were weighed according to the designed parameters and mixed evenly for 2 min. All samples were rotated for 5 min again in a drum granulator with a diameter of 600 mm and a length of 300 mm after adding water, and the resulting granular sintering mixture was obtained. In the second step, exhaust sintering: the mixture was distributed to a sintering pot with a diameter of 100 mm and a height of 500 mm for ignition and exhaust sintering. Until the end of the sintering process, the exhaust gas temperature reached the maximum value. The ignition negative pressure and sintering negative pressure were 5 kPa and 7 kPa, respectively [8]. After reaching the sintering end point, the negative pressure was adjusted to -5 kPa and kept for 5 min to cool the sinter, and then, the sinter was poured out from the sintering pot and further cooled to the room temperature. At last, treatment of the sinters: the yield, tumble index, particle size distribution, and other properties of the sinters were measured after the sinters were cooled according to the previous literatures [8–11].

Characterization

The phase of the sample was analyzed by X-ray diffraction (XRD, D/Max 2500, RIGAKU, Japan) under the conditions of radiation: Cu Kα, tube current and voltage: 250 mA, 40 kV, scanning range: 10–80°/2θ, step size: 0.02°/2θ, and scanning speed: 5°/min. Element content in solution was tested by using a plasma emission spectrometer (ICP-AES, Icap7400 Radial, Thermo Fisher Scientific, USA). Topographic

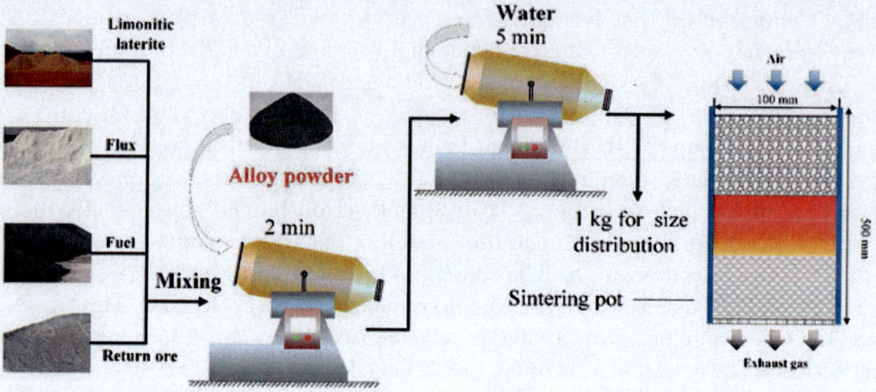

Fig. 1 Flowsheet of the limonitic laterite sintering process

analysis was examined by scanning electron microscope (SEM-EDS, JSM-6490LV, JEOL, Japan).

Assessment Indexes

The yield was calculated by Eq. (1). The tumbler index (TI) of the sinter was tested by 1/10 ISO 3271-2007 and calculated by Eq. (2). Sintering speed ($V_{\perp}$) was calculated by Eq. (3).

$$P = \frac{m_1 - 0.25}{m_0 - 0.25} \times 100\% \tag{1}$$

$$P = \frac{M_1}{M_0} \times 100\% \tag{2}$$

$$V^{\perp} = \frac{H}{t} \tag{3}$$

In Eq. (1), m_1 is the weight of the sinter with particle size greater than 5 mm (kg); m_0 is the total weight of sinter (kg); and 0.25 is the weight of hearth layer (kg). In Eq. (2), M_0 is the total weight of sinter before tumbler (kg); and M_1 is the weight of sinter with particle size greater than 5 mm after tumbler (kg). In Eq. (3), H is the height of sinter green bed (mm); and t is the sintering time (min).

Results and Discussion

Effect of the Addition of Alloy Powder on Limonitic Laterite Sintering Performance

According to the preliminary experimental exploration, the optimized experimental conditions of 1.8 basicity and 19.5% moisture of the mixture were determined. The influence of the addition of alloy powder on size distribution and permeability of the sintering green bed, yield, and quality of limonitic laterite sinters index was systematically discussed.

Effect on Size Distribution and Permeability of the Sintering Green Bed

The effect of the addition amount of alloy powder on size distribution of the sintering green bed is shown in Fig. 2. With the addition of alloy powder increasing from 0 to 9%, the mass fraction of −0.5 and 0.5–1 mm granules decreased persistently. The + 3 mm granules kept rising accordingly. Likewise, the mass fraction of 5–8 mm and +8 mm particles increased rapidly from 4.85% to 11.80% and from 0.14% to 3.20%, respectively. It shows that the addition of alloy powder could significantly increase the proportion of +3 mm particles and improve the particle size distribution of the sintering green bed. The main reason was that the alloy powder had high density and small particle size (90% less than 150 μm). It could be evenly distributed among the limonitic laterite with low density, which promoted the formation of large particles during the granulation process.

It can be observed from Fig. 3 that the permeability of the sintering green bed increased with the addition of alloy powder. This was in result of that the alloy powder reduced the mass fraction of −1 mm, and there would be more gaps between sintering green material for air passing.

Effect on Sintering Indexes

Figure 4 displays the effect of the addition of alloy powder on the limonitic laterite sintering performance. The yield and quality index were extremely terrible at 0% alloy powder dosage, and the tumble index and yield were 50.04% and 62.04%, respectively. As the dosage of alloy powder increased to 3%, the tumble index and yield rapidly rose to 54.69% and 67.35%, respectively. Overall, with the increase of alloy powder dosage, the yield and quality indexes of the sinters were getting better. However, the sintering speed decreased continuously. This was in result of the addition of alloy powder improved the permeability of the sintering green bed, which promoted the full combustion of solid fuel during the limonitic laterite sintering process. The sintering burning zone became winder, and the duration of sintering high

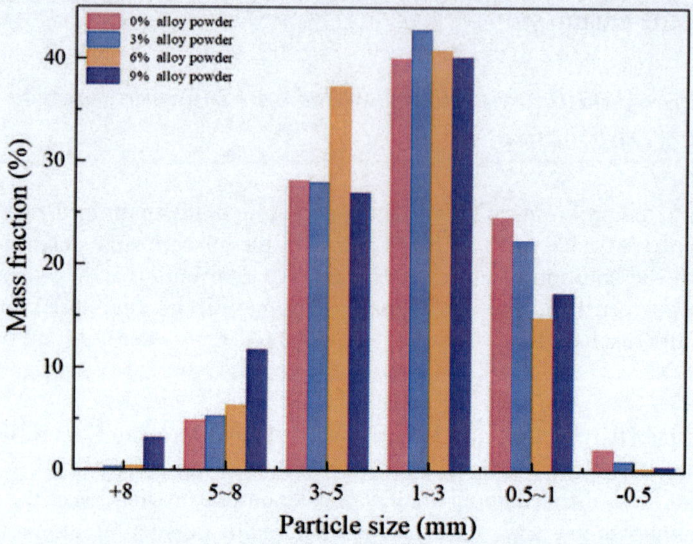

Fig. 2 Effect of the addition of alloy powder on size distribution of the sintering green bed

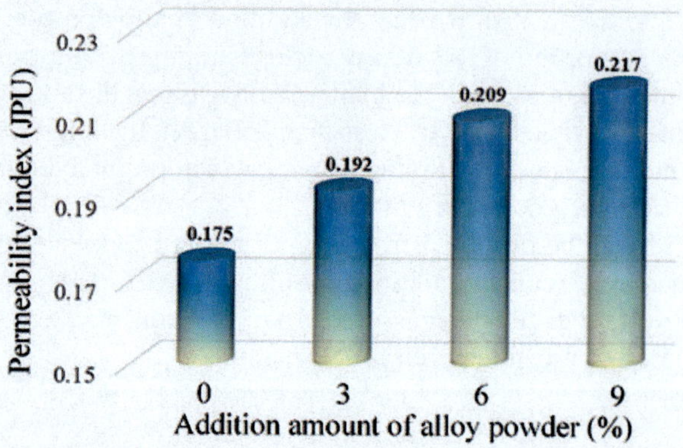

Fig. 3 Effect of the addition of alloy powder on permeability of the sintering green bed

temperature became longer. Therefore, the sintering speed decreased, and the sinter strength increases.

Analysis of the Obtained Sinters

Two samples with different alloy powder dosage were chosen for phase analysis. The XRD analysis of Fig. 5 demonstrates that the main phases of the two samples were wustite and spinel containing Mg, Al, Cr, and Ni. Moreover, the diffraction peak intensities of 36.5 and 42.1° constantly increased with the increase of alloy powder,

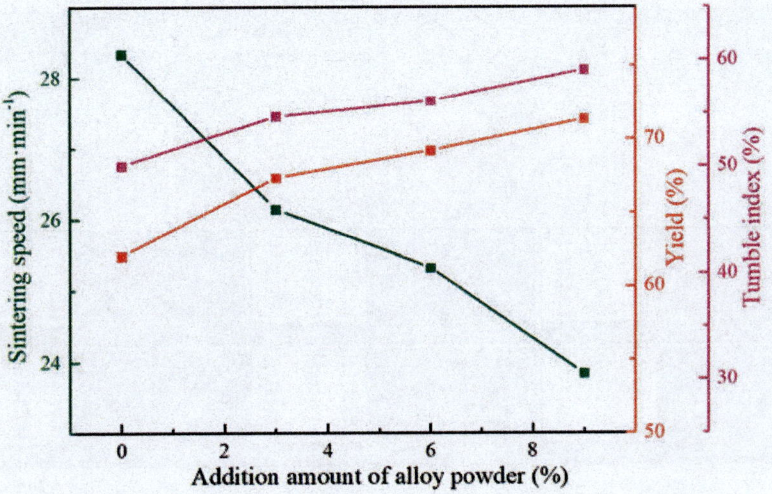

Fig. 4 Effect of the addition of alloy powder on the limonitic laterite sintering performance

and it might be owing to the extended high temperature holding time caused by the improvement of permeability by alloy powder.

Sinters with different alloy powder additions were selected for microstructure analysis. As shown in Fig. 6a, b, the sample without alloy powder was mainly composed of nearly spherical minerals, and the particle size was about 20–50 μm. The reticular liquid phase filled in the gap between granular minerals and acted as

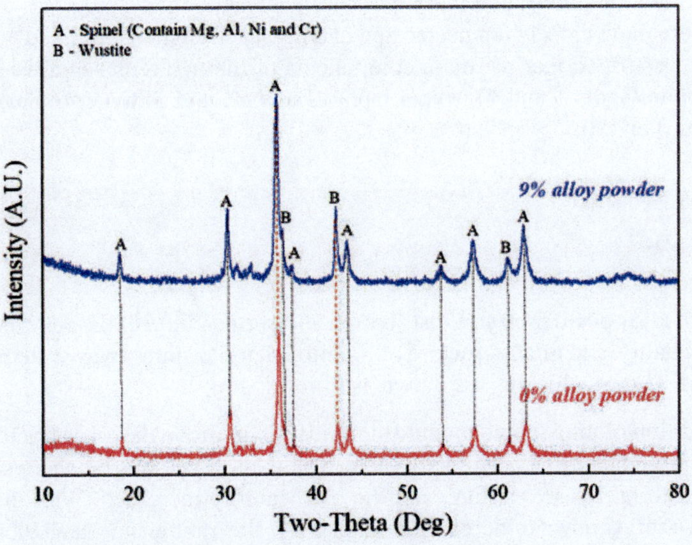

Fig. 5 XRD pattern of the representative limonitic laterite sinters

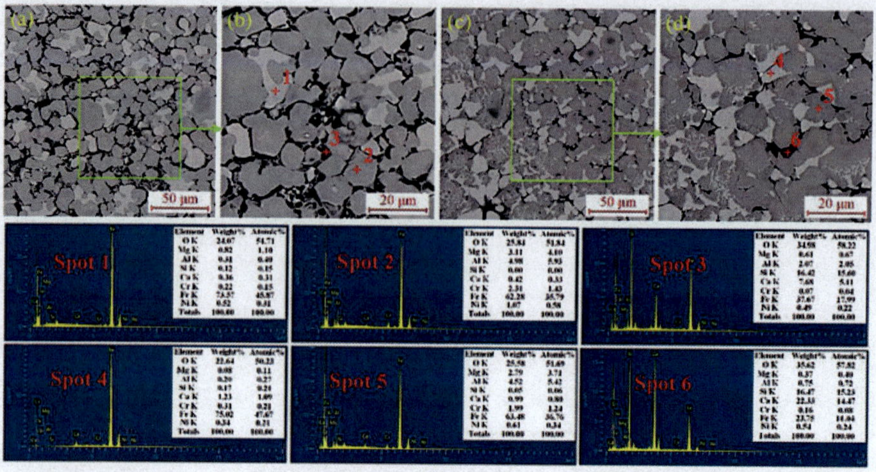

Fig. 6 SEM images of the representative limonitic laterite sinters (**a**, **b**—sinters added 0% alloy powder; **c**, **d**—sinters added 9% alloy powder)

bonding phase. Loose mineral particle arrangement was difficult to be fully wetted by reticular liquid phase, which might be the principal reason for the poor strength of the sinters. Figure 6c, d indicates that the boundary of the sinters with 9% alloy powder was not obvious, and there was less particle gap compared with no alloy powder. The main reason might be that the addition of alloy powder strengthened the granulation of the mixture and the full combustion of solid fuel during the limonitic laterite sintering process was promoted. The sintering burning zone became winder, and the duration of sintering high temperature became longer. Therefore, the sinter particles were more closely connected and had higher strength. Combined with EDS analysis, it was found that the main components of the two representative samples were boehmite (spots 1 and 4), spinel (spots 2 and 5), and SFCA acted as binding phase (spots 3 and 6).

Conclusions

The effect of alloy powder on size distribution and permeability of the sintering green bed, yield, quality, and microstructure of limonitic laterite sinters was systematically investigated. The conclusions are drawn as follows:

1. The addition of alloy powder could significantly improve the mass fraction of + 3 mm of the sintering green bed. There would be more gaps between sintering green material for air passing, and the permeability increased. With the added amount of alloy powder increased from 0 to 3%, the tumble index and yield of the sinters rosed from 50.07% to 54.69% and from 62.04% to 67.35%, respectively.

The sinter strength would be further increased by increasing the dosage of alloy powder to the sintering green bed.

2. The alloy powder improved the permeability of the sintering green bed, which promoted the full combustion of solid fuel. The sintering burning zone became winder and the duration of sintering high temperature became longer. Therefore, the sintering time was prolonged, and the sintering speed was reduced. The particles gap of sinters becomes smaller, the boundary became inconspicuous, and the sinter strength increased. This study provided a novel process of resource utilization of alloy powder on large-scale and strengthened the limonitic laterite sintering process.

Acknowledgements The authors would like to express their sincere thanks to National Natural Science Foundation of China (No. U1960114), Research Innovation Foundation for Postgraduate of Central South University (No. 2021zzts0304), and Xuchang Talent Plan (No. 2021-178).

References

1. Kim MS, Lee JC, Park HS, Jun MJ, Kim B (2018) A multistep leaching of nickel-based superalloy scrap for selective dissolution of its constituent metals in hydrochloric acid solutions. Hydrometallurgy 176:235–242. https://doi.org/10.1016/j.hydromet.2018.02.002
2. Tian Q, Gan X, Yu D, Cui F, Guo X (2021) One-step and selective extraction of nickel from nickel-based superalloy by molten zinc. T Nonferr Metal Soc 31(6):1828–1841. https://doi.org/10.1016/S1003-6326(21)65620-0
3. Han J, Wang Y, Mao X, Chang X, Zeng H, Qin W (2022) Efficient extraction of nickel from sintered alloy by stepwise leaching: thermodynamic and kinetic studies. Miner Eng 187:107776. https://doi.org/10.1016/j.mineng.2022.107776
4. Liu Y, Lv X, Xu J, Zhang S, Bai C (2016) Preparation of stainless steel master alloy by direct smelting reduction of Fe–Ni–Cr Sinter at 1600 °C. Ironmak Steelmak 43(8):600–606. https://doi.org/10.1080/03019233.2016.1209816
5. Oliveira VA, Rodrigues MLM, Leao VA (2021) Reduction roasting and bioleaching of a limonite ore. Hydrometallurgy 200:1–8. https://doi.org/10.1016/j.hydromet.2021.105554
6. Xue Y, Zhu D, Pan J, Guo Z, Yang C, Tian H, Huang Q, Pan L, Huang X (2021) An investigation into the co-sintering process of limonitic nickel laterite and low-grade chromite via multi-force fields. J Mater Res Technol 12:1816–1831. https://doi.org/10.1016/j.jmrt.2021.03.106
7. Gao J, Li W, Ma S, Du Z, Cheng F (2021) Spinel ferrite transformation for enhanced upgrading nickel grade in laterite ore of various types. Miner Eng 163:1–13. https://doi.org/10.1016/j.mineng.2021.106795
8. Tu Y, Zhang Y, Su Z, Jiang T (2022) Mineralization mechanism of limonitic laterite sinter under different fuel dosage: effect of FeO. Powder Technol 398:117064. https://doi.org/10.1016/j.powtec.2021.117064
9. Xue Y, Zhu D, Pan J, Guo Z, Yang C, Tian H, Duan X, Huang Q, Pan L, Huang X (2020) Effective utilization of limonitic nickel laterite via pressurized densification process and its relevant mechanism. Minerals 10:1–18. https://doi.org/10.3390/min10090750

10. Zhu D, Xue Y, Pan J, Yang C, Guo Z, Tian H, Wang X, Huang Q, Pan L, Huang X (2020) Co-benefits of CO_2 emission reduction and sintering performance improvement of limonitic laterite via hot exhaust-gas recirculation sintering. Powder Technol 373:727–740. https://doi.org/10.1016/j.powtec.2020.07.018
11. Zhu D, Xue Y, Pan J, Yang C, Guo Z, Tian H, Liao H, Pan L, Huang X (2020) An investigation into the distinctive sintering performance and consolidation mechanism of limonitic laterite ore. Powder Technol 367:616–631. https://doi.org/10.1016/j.powtec.2020.03.066

Experimental Study on Solubility of Metal Oxides in Novel Deep Eutectic Solvents of Choline Chloride-Organic Acid

Peng Yang, Xiaoping Liang, Guodong Cui, and Chen Yang

Abstract In this work, novel deep eutectic solvent (DES) was prepared from the choline chloride as a hydrogen-bond acceptor in combination with organic acids (i.e., phthalic acid, 2,2-dimethylol propionic acid, glutamic acid, tartaric acid) as a hydrogen-bond donor. Only at the preparation temperature of 110 °C, two novel of DESs were prepared: choline chloride:2,2-dimethylol propionic acid with molar ratios of 1:0.67 and the choline chloride:tartaric acid at a 1:0.5 molar ratio. The thermal stability of DESs was studied. DESs choline chloride:2,2-dimethylol propionic acid (1:0.67) and choline chloride:tartaric acid (1:0.5) were found stable at temperatures lower than 211 and 195 °C respectively. The solubility of metal oxides (MOs) ZnO, Fe_2O_3, Fe_3O_4, and V_2O_5 was determined in choline chloride:2,2-dimethylol propionic acid (1:0.67) and choline chloride: tartaric acid (1:0.5). The obtained MOs solubility values indicated these novel DESs with the interesting ability to dissolve the selected MOs.

Keywords Deep eutectic solvent · Metal oxide solubility · Choline chloride · Organic acid

Introduction

Metal oxides (MOs) are used as functional materials as well as starting substances for the synthesis of diverse task-specific materials; other than that they are also the

P. Yang · X. Liang (✉) · G. Cui · C. Yang
College of Materials Science and Engineering, Chongqing University, Chongqing 400030, China
e-mail: xpliang@cqu.edu.cn

P. Yang
e-mail: 20210901064@cqu.edu.cn

G. Cui
e-mail: 3288731409@qq.com

C. Yang
e-mail: 202009131259@cqu.edu.cn

R. G. Reddy et al. (eds.), *New Directions in Mineral Processing, Extractive Metallurgy, Recycling and Waste Minimization*, The Minerals, Metals & Materials Series,
https://doi.org/10.1007/978-3-031-22765-3_29

form from which most metals are extracted [1]. MOs are present in natural ores and many industrial residues and end-of-life products, and their efficient processing is of great scientific and economic significance [2, 3]. At present, pyrometallurgy or hydrometallurgy are mainly methods for MOs extraction, but issues such as environmental pollution and energy consumption have led to explore more environmental and effective methods that have attracted worldwide interest [4–6].

Solvometallurgy is an environmental and effective method for extracting MOs, and the main advantages of solvometallurgy are its higher selectivity and lower water consumption compared to other types of metallurgy [7, 8]. Conceptually, solvometallurgy is similar to hydrometallurgy in terms of a single operation and operating conditions [9]. However, solvometallurgy attempts to replace almost completely aqueous solutions with organic solvents instead of acidic or alkaline aqueous solutions [10]. Ionic liquids (ILs) are attractive for use in solvent metallurgy, and many researchers have demonstrated their excellent properties in extracting MOs [11–16]. However, the complex preparation process of ionic liquid leads to the low economy of ILs, which is the main disadvantage of its industrial scale implementation [17].

Deep eutectic solvent (DES) is an ionic liquid analogue whose great appeal comes from its low synthetic cost, low environmental toxicity, and wide solubility of Mos [18]. DES as synthesized and defined by Abbot et al. [24], is a binary or ternary eutectic mixture system composed of hydrogen bond acceptor (HBA) and hydrogen bond donor (HBD) through intermolecular hydrogen bonding, and its melting point is lower than that of any of its components [19, 20]. The DES can offer or accept electrons or protons to form hydrogen bonds, making them very efficient dissolution properties [21]. More interestingly, extensive studies prove the solubility of MOs in a variety of DESs, thus opening up "green" strategies for metal separation and recycling. Abbott et al. studied the solubility of MOs in DESs [22–28], and it was indicated that MO solubility was higher in DESs with acidic HBDs [28, 29]. It has been suggested that more acidic DESs, such as the mixture of malonic acid-ChCl, exhibit higher solubility of MOs because the protons present act as an oxygen acceptor, breaking the metal-oxide bond. At present, a mixture of ChCl and acidic HBDS such as phthalic acid, 2,2-dimethylol propionic acid, glutamic acid, and tartaric acid has not been used for the preparation of DES.

Here, ChCl is selected as the HBA, and carboxylic acids (i.e., phthalic acid, 2,2-dimethylol propionic acid, glutamic acid, tartaric acid) as the HBD tried to prepare novel DESs of ChCl-organic acid in this paper. The thermal stability of novel DESs was determined, and solubility of MOs in the novel DESs was researched. This study enriches the DES system that can be used to dissolve MOs.

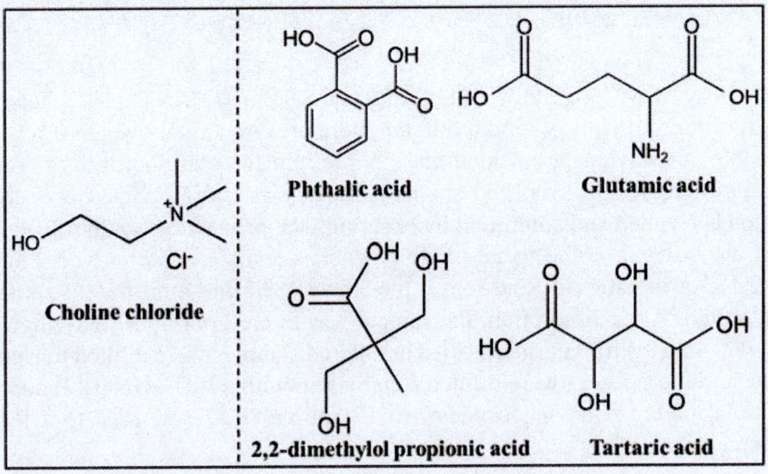

Fig. 1 Molecular structure of ChCl and organic acids used in this work

Experimental

Materials

Choline chloride (ChCl) with 99% purity was purchased from Macklin. For the four organic acids, phthalic acid and tartaric acid (TA) with 99.5% purity, and 2,2-dimethylol propionic acid (DMPA) and glutamic acid (GLU) with 98% purity were both purchased from Chron Chemicals. The MOs ZnO, Fe_2O_3, V_2O_5, and Fe_3O_4 with 99% purity were purchased from Macklin. The chemicals used in this work were used as received without further purification. The chemical structure of ChCl and organic acids in this work is presented in Fig. 1.

DESs Preparation

The DESs were prepared in this work by following the method previously described in many articles [30]. The ChCl and organic acids were weighed at a certain molar ratio and placed together in a sealed beaker, and then, the beaker was placed in a constant temperature magnetic stirrer for magnetic stirring at a certain temperature until a stable homogeneous transparent liquid was formed. Prepared DESs were stored in a sealed sample bottle for later use.

Metal Oxide Dissolution

The solubility of MOs (ZnO, Fe_2O_3, V_2O_5, Fe_3O_4) in DES was tested according to the procedure previously reported in the literature. Accurately weigh 0.1 g metal oxide and 5 ml solvent in a sealed glass bottle, put the glass bottle in a constant temperature magnetic stirrer, and stir continuously at 100 °C for 24 h or longer. Saturation is reached and confirmed by observing the presence of undissolved metal oxide in the solvent. If the metal-oxide powder is completely dissolved, continue to add 0.1 g of metal oxide and repeat the above operation until the dissolution is supersaturated. After dissolution, the sample was filtered out by PES syringe filter (pore size 0.45 μm) for later use [10]. The filtered samples were diluted to a certain multiple, and the samples were diluted step by step with 2 vo 1%HNO_3. Inductively coupled plasma emission spectrometer (ICP, Optima 8000) was used to detect the metal content in the low eutectic solvent.

Thermal Stability

Thermal decomposition temperature of the prepared DESs was measured using a thermogravimetric differential analyzer (TGA/DSC 1/1100 SF) to examine their decomposition temperature and thermal stability. Here, the samples were placed in an alumina pan under the nitrogen environment (flow rate—100 mL min^{-1}). The heating rate was 5 °C/min, and temperature range is the 40–500 °C at a thermal gradient of an inert gas (N_2) used as a protective gas.

Results and Discussion

Preparation of DESs

ChCl is selected as the HBA, and carboxylic acids (i.e., phthalic acid, 2,2-dimethylol propionic acid, glutamic acid, tartaric acid) as the HBD tried to prepare novel DESs of ChCl-organic acid in this paper. When choline chloride and organic acid can form a stable and uniform transparent liquid, it indicates that DES is successfully prepared. The detailed experimental parameters of novel DESs preparation experiment are shown in Table 1.

The experimental results showed that in experiments 1–4, ChCl and organic acids did not form stable and uniform transparent liquid under the corresponding preparation conditions. In experiments 1 and 2, white turbid solution was formed by ChCl and phthalic acid or DMPA. In experiment 3, ChCl and GLU were still white solids. ChCl and TA formed a transparent liquid with high viscosity and contains suspended white particles, in experiment 4.

Table 1 Detailed experimental parameters of DES preparation experiment

Number	ChCl: organic acid	Molar ratios	Preparation temperature/°C
1	ChCl: phthalic acid	1:1	80
2	ChCl: DMPA	1:2	80
3	ChCl: GLU	1:1	80
4	ChCl: TA	1:1	80
5	ChCl: phthalic acid	1:1	95
6	ChCl: DMPA	1:2	95
7	ChCl: GLU	1:1	95
8	ChCl: TA	1:1	95
9	ChCl: DMPA	1:2	110
10	ChCl: DMPA	1:1	110
11	ChCl: DMPA	1:0.67	110
12	ChCl: TA	1:1	110
13	ChCl: TA	1:0.5	110

In experiments 1–4, neither ChCl nor phthalic acid, DMPA, GLU, and TA formed a stable homogeneous transparent liquid, presumably because the preparation temperature was too low, so the preparation temperature was increased and experiments 5–8 were performed. According to the experimental results of 5–8, it can be seen that when the preparation temperature is increased, ChCl and phthalic acid or GLU formed transparent liquid with high viscosity and containing white suspended particles, ChCl and DMPA or TA form a stable transparent liquid with high viscosity. Because it is thought that ChCl and DMPA or TA more likely to form DES.

Temperature and molar ratio are important factors in the preparation of DES. Based on the results of experiments 5–8, the molar ratios of ChCl and DMPA or TA were reset in this study, while the preparation temperature was raised to 110 °C and experiments 9–12 were performed.

The pictures of liquid formed with ChCl as the hydrogen bond acceptor and DMPA or TA as the hydrogen bond donor are shown in Fig. 2. It can be seen from Fig. 2 that the clarification of the solution was improved with the increase of ChCl in ChCl-DMPA and ChCl-TA systems. As shown in Fig. 2(9)–(11), when the molar ratio of ChCl to DMPA is 1:0.67, stable and uniform transparent liquid is formed. From Fig. 2(12), (13), it can be concluded that, when the molar ratio of ChCl to TA is 1:0.5, stable and uniform transparent liquid is formed.

It can be observed from the preparation results of DESs that a stable and uniform transparent liquid can be formed only when the preparation temperature is 110 °C and the molar ratio of ChCl:DMPA is 1:0.67 or the molar ratio of ChCl:TA is 1:0.5. In other words, DESs can be formed when the preparation temperature is 110 °C, the molar ratio of ChCl:DMPA is 1:0.67 or the molar ratio of ChCl:TA is 1:0.5. Therefore,

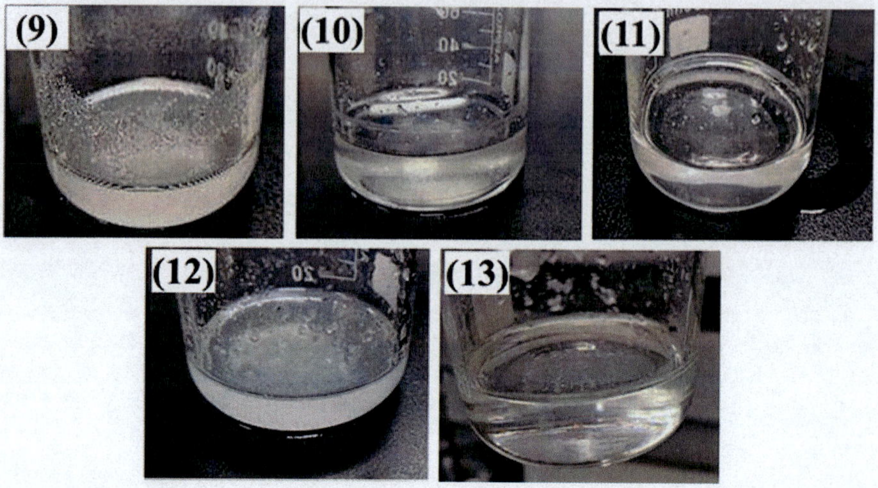

Fig. 2 Photographs of liquid: (9) ChCl: DMPA (1:2), (10) ChCl: DMPA (1:1), (11) ChCl: DMPA (1:0.67), (12) ChCl: TA (1:1), (13) ChCl: TA (1:0.5)

the thermal stability of these two novel DESs was explored and the dissolution of MOs was carried out with the two DESs.

Thermal Stability Analysis of DES

The stability of solvents directly affects their reliability and performance in large-scale applications. DES has relatively good thermal and chemical stability, and its thermal stability of DES can be assessed by decomposition temperature [31]. The thermogravimetric (TG) analysis results of DESs ChCl: DMPA (1:0.67) and ChCl: TA (1:0.5) are presented in Fig. 3.

In Fig. 3, it can be observed that these DESs ChCl: DMPA (1:0.67) and ChCl: TA (1:0.5) exhibit excellent thermal stability, and the decomposition temperatures were 211 and 195 °C, respectively. It is also noticeable that the thermal decomposition curve for all the prepared DESs exhibited two stages. First, a desorption/drying region was observed, where a large mass loss was followed by a plateau. This behavior is associated with the evaporation of a volatile component, namely, water [10]. Second, a typical single stage decomposition curve was observed. When the temperature was higher than 211 °C, ChCl: DMPA (1:0.67) began to decompose, and when the temperature was higher than 195 °C, ChCl: TA (1:0.5) began to decompose, which is due to the decomposition of organic acids. When the temperature is above 305 °C, the mass of these DESts tends to be stable. The similar trend of the thermal stability was previously reported by several researchers for the ChCl-based DESs [32, 33].

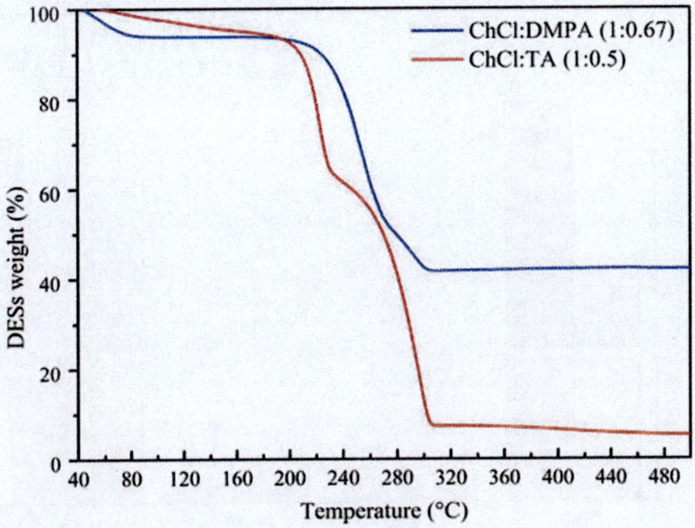

Fig. 3 Dynamic TGA analysis of the investigated DESs

Dissolution of Metal Oxides

The solubility (g/L) of MOs ZnO, Fe_2O_3, Fe_3O_4, and V_2O_5 in DESs ChCl: DMPA (1:0.67) and ChCl: TA (1:0.5) was determined. The obtained solubility values are depicted in Fig. 4.

From Fig. 4, it can be observed that the solubilities of ZnO, Fe_2O_3, Fe_3O_4, and V_2O_5 in ChCl: DMPA (1:0.67) were 164.781 g/L, 2.347 g/L, 1.217 g/L, and 8.794 g/L, respectively. The solubilities of MOs in ChCl: TA (1:0.5) from largest to smallest was ZnO > V_2O_5 > Fe_3O_4 > Fe_2O_3. Among them, the solubility of ZnO in the two DESs is the highest. MOs are generally stable compounds because they form ionic bonds between oxygen negative ions and metal cations, the strength of which is measured by lattice energy. The larger the lattice energy, the stronger the ionic bond and the more stable the crystal [34]. Both Fe_3O_4 and Fe_2O_3 have high lattice energy, which makes them very stable, which can explain the low solubility of Fe_3O_4 and Fe_2O_3 in the two DESs. The solubilities of V_2O_5 in ChCl: DMPA (1:0.67) and ChCl: TA (1:0.5) are 8.794 g/L and 19.78 g/L, respectively. From what has been discussed above indicate that these two novel ChCl-organic acid DESs are excellent solvents for the solubility of MOs.

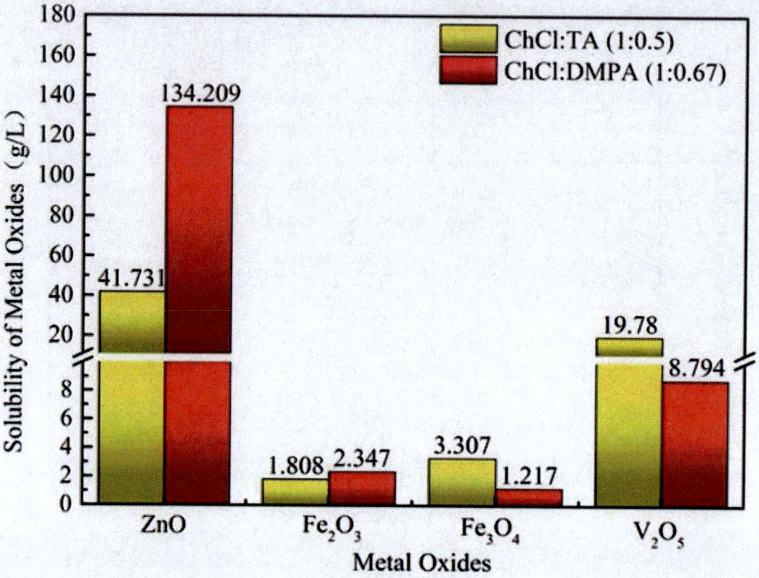

Fig. 4 Solubility of the metal oxides in the prepared DESs

Conclusions

Combinations of ChCl with different organic acids (i.e., phthalic acid, 2,2-dimethylol propionic acid, glutamic acid, tartaric acid) were tested for novel ChCl-organic acid DES formation, resulting in two novel ChCl-organic acid DESs: (1) ChCl: DMPA (1:0.67) and (2) ChCl: TA (1:0.5). The thermal stability of the newly reported DESs was determined. DESs ChCl: DMPA (1:0.67) and ChCl: TA (1:0.5) were found stable at temperatures lower than 211 and 195 °C, respectively. The solubility of MOs ZnO, Fe_2O_3, Fe_3O_4, and V_2O_5 in DESs ChCl: DMPA (1:0.67) and ChCl: TA (1:0.5) was measured. The solubilities of ZnO, Fe_2O_3, Fe_3O_4, and V_2O_5 in ChCl: DMPA (1:0.67) were 164.781 g/L, 2.347 g/L, 1.217 g/L, and 8.794 g/L, respectively. The solubilities of MOs in ChCl: TA (1:0.5) from largest to smallest was ZnO > V_2O_5 > Fe_3O_4 > Fe_2O_3. The results indicate that these two novel ChCl-organic acid DESs are excellent solvents for the solubility of MOs.

Author Contributions Peng Yang contributed in data curation and wrote the paper; Xiaoping Liang contributed in review and editing; Guodong Cui contributed in investigation; Chen Yang contributed in resources.

Funding This research did not receive any specific grant from funding agencies in the public, commercial, or not-for-profit sectors.

Conflicts of Interest The authors declare that they have no conflict of interest.

References

1. Richter J, Ruck M (2020) Synthesis and dissolution of metal oxides in ionic liquids and deep eutectic solvents. Molecules 25:78
2. Guo CT, Jian LI, Yi XH (2010) Application of ionic liquids in hydrometallurgy of nonferrous metals. T Nonferr Metal Soc 20:513–520
3. Binnemans K, Jones PT, Manjón Fernández Á et al (2020) Hydrometallurgical processes for the recovery of metals from steel industry by-products: a critical review. J Sustain Metall 6:505–540
4. Nithya R, Sivasankari C, Thirunavukkarasu A (2021) Electronic waste generation, regulation and metal recovery: a review. Environ Chem Lett 19:1347–1368
5. Hao JJ, Wang YS, Wu YF, Guo F (2020) Metal recovery from waste printed circuit boards: a review for current status and perspectives. Resour Conserv Rec 157:104787
6. Ilankoon IMSK, Ghorbani Y, Chong MN, Herath G, Moyo T, Petersen J (2018) E-waste in the international context–a review of trade flows, regulations, hazards, waste management strategies and technologies for value recovery. Waste Manage 82:258–275
7. Binnemans K, Jones PT (2017) Solvometallurgy: an emerging branch of extractive metallurgy. J Sustain Metall 3:570–600
8. Capello C, Fischer U, Hungerbühler K (2007) What is a green solvent? a comprehensive framework for the environmental assessment of solvents. Green Chem 9:927–934
9. Abbott AP, Frisch G, Gurman SJ et al (2011) Ionometallurgy: designer redox properties for metal processing. Chem Commun 47:10031–10033
10. Rodriguez NR, Machiels L, Binnemans K (2019) p-toluenesulfonic acid-based deep-eutectic solvents for solubilizing metal oxides. ACS Sustain Chem Eng 7:3940–3948
11. Hartley JM (2013) Ionometallurgy: the processing of metals using ionic liquids. University of Leicester
12. Abbott AP, Frisch G, Hartley J, Ryder KS (2011) Processing of metals and metal oxides using ionic liquids. Green Chem 13:471–481
13. Nockemann P, Thijs B, Parac-Vogt TN, Van HK, Van ML et al (2008) Carboxyl-functionalized task-specific ionic liquids for solubilizing metal oxides. Inorg Chem 47:9987–9999
14. Nockemann P, Thijs B, Pittois S, Thoen J, Glorieux C, Van HK, Van ML, Kirchner B, Binnemans K (2006) Task-specific ionic liquid for solubilizing metal oxides. J Phys Chem B 110:20978–20992
15. Davris P, Balomenos E, Panias D, Paspaliaris I (2016) Selective leaching of rare earth elements from bauxite residue (Red Mud), using a functionalized hydrophobic ionic liquid. Hydrometallurgy 164:125–135
16. Dupont D, Raiguel S, Binnemans K (2015) Sulfonic acid functionalized ionic liquids for dissolution of metal oxides and solvent extraction of metal ions. Chem Commun 51:9006–9009
17. Zhang Q, Vigier KDO, Royer S, Jérôme F (2012) Deep eutectic solvents: syntheses, properties and applications. Chem Soc Rev 41:7108
18. Schaeffer N, Passos H, Billard I, Papaiconomou N, Coutinho J (2018) A. Recovery of metals from waste electrical and electronic equipment (WEEE) using unconventional solvents based on ionic liquids. Crit Rev Environ Sci Technol 48:859–922
19. Abbott AP, Boothby D, Capper G, Davies DL, Rasheed RK (2004) Deep eutectic solvents formed between choline chloride and carboxylic acids: versatile alternatives to ionic liquids. J Am Chem Soc 126:9142–9147
20. Wu K, Su T, Hao D, Liao W, Zhao Y, Ren W, Deng C, Lü H (2018) Choline chloride-based deep eutectic solvents for efficient cycloaddition of CO_2 with propylene oxide. Chem Commun 54:9579–9582
21. Singh MB, Kumar VS, Chaudhary M, Singh P (2021) A mini review on synthesis, properties and applications of deep eutectic solvents. J Indian Chem Soc 98:100210
22. Söldner A, Zach J, König B (2019) Deep eutectic solvents as extraction media for metal salts and oxides exemplarily shown for phosphates from incinerated sewage sludge ash. Green Chem 21:321–328

23. Abbott AP, Capper G, Davies DL, Rasheed RK, Shikotra P (2005) Selective extraction of metals from mixed oxide matrixes using choline-based ionic liquids. Inorg Chem 44:6497–6499
24. Abbott AP, Capper G, Davies DL, Rasheed RK, Tambyrajah V (2003) Novel solvent properties of choline chloride/urea mixtures. Chem Commun 9:70–71
25. Abbott AP, Capper G, Davies DL, Shikotra P (2006) Processing metal oxides using ionic liquids. Miner Process Extr Metall 115:15–18
26. Francisco M, van den Bruinhorst A, Kroon MC (2013) Low-transition-temperature mixtures (LTTMs): a new generation of designer solvents. Chem Int Ed 52:3074–3085
27. Omar KA, Sadeghi R (2022) New chloroacetic acid-based deep eutectic solvents for solubilizing metal oxides. J Mol Liq 347:118393
28. Abbott AP, Capper G, Davies DL et al (2006) Solubility of metal oxides in deep eutectic solvents based on choline chloride. J Chem Eng Data 51:1280–1282
29. Pateli IM, Thompson D, Alabdullah SM, Abbott AP, Jenkin GRT, Hartley JM (2020) The effect of pH and hydrogen bond donor on the dissolution of metal oxides in deep eutectic solvents. Green Chem 22:5476
30. Kareem MA, Mjalli FS, Hashim MA, Alnashef IM (2010) Phosphonium-based ionic liquids analogues and their physical properties. J Chem Eng Data 55:4632–4637
31. Yuan ZW, Liu H, Yong WF, She QH, Esteban J (2022) Status and advances of deep eutectic solvents for metal separation and recovery. Green Chem 24:1895
32. Chemat F, Anjum H, Shariff AM, Kumar P, Murugesan T (2016) Thermal and physical properties of (choline chloride+ urea+ larginine) deep eutectic solvents. J Mol Liq 218:301–308
33. Delgado-Mellado N, Larriba M, Navarro P, Rigual V, Ayuso M, García J, Rodríguez F (2018) Thermal stability of choline chloride deep eutectic solvents by TGA/FTIR-ATR analysis. J Mol Liq 260:37–43
34. Abbott AP, Alabdullah SSM, Al-Murshedi AYM, Ryder KS (2018) Brønsted acidity in deep eutectic solvents and ionic liquids. Faraday Discuss 206:365–377

In-Situ Microscopy Observations of Oxide Phases Formation During High-Temperature Oxidation of End-of-Life Ni/Cu/Ni-Coated-NdFeB Permanent Magnets

D. C. Nababan, R. Mukhlis, Y. Durandet, M. I. Pownceby, L. Prentice, and M. A. Rhamdhani

Abstract An understanding of the reaction phases formed during the high-temperature oxidation of rare earth permanent magnets is vital for developing appropriate strategies for their pyrometallurgy-based recycling processes aimed at the concentration and potential recovery of the rare earth elements. The current study uses in-situ high-temperature optical microscopy combined with ex-situ scanning electron microscopy to analyze oxide growth on the surface of an end-of-life Ni/Cu/Ni-coated NdFeB permanent magnet oxidized at 900 and 1100 °C for up to 4 h under ambient air conditions. Distinct oxide morphologies were observed at the two oxidation temperatures over time ranging from blister- and spike-like structures. At the longest oxidation times, visible cracking of the surface of the magnet and protrusion of Fe_2O_3 phase were observed. The presence of a Ni/Cu/Ni coating on the magnets was found to significantly affect the oxidation growth; a nickel ferrite phase was formed at 1100 °C after 4 h while at 900 °C, there was still discrete NiO and Fe_2O_3.

Keywords Rare earth · NdFeB magnet · High-temperature oxidation · Hot-stage microscopy

D. C. Nababan (✉) · R. Mukhlis · Y. Durandet · M. A. Rhamdhani
Fluid and Process Dynamics (FPD) Group, Department Mechanical and Product Design Engineering, Swinburne University of Technology, Melbourne, VIC 3122, Australia
e-mail: dnababan@swin.edu.au

M. A. Rhamdhani
e-mail: arhamdhani@swin.edu.au

M. I. Pownceby
CSIRO Mineral Resources, Clayton, Melbourne, VIC 3169, Australia

L. Prentice
CSIRO Manufacturing, Clayton, Melbourne, VIC 3169, Australia

R. G. Reddy et al. (eds.), *New Directions in Mineral Processing, Extractive Metallurgy, Recycling and Waste Minimization*, The Minerals, Metals & Materials Series,
https://doi.org/10.1007/978-3-031-22765-3_30

Introduction

Neodymium-iron-boron (NdFeB) rare earth permanent magnets (REPM) have a wide range of applications including in electric vehicles [1], wind power turbines [2], and computer hard disk drives [3]. Typical NdFeB magnets contain up to 32 wt% of rare earth elements (REEs) which is dominated by the light REE, neodymium (Nd) [4]), but with minor praseodymium (Pr) and dysprosium (Dy) also present. Recycling of end-of-life (EOL) NdFeB permanent magnets will be vital to support the expected growth in global demand for Nd and Dy, due to the fact that these metals had significant supply shortfalls in the recent past. In addition, direct mining of REEs from ores generates many issues with regard to the environment as their minerals commonly contain trace to minor radioactive elements such as uranium and thorium [5].

The recycling of NdFeB magnets can be carried out through pyrometallurgy, hydrometallurgy, and their combined routes [6–8]. Despite the challenges associated with its high-temperature requirement, the pyrometallurgy route offers some advantages especially when renewable energy sources such as concentrated solar energy are applied to supply the heat. In the pyrometallurgy approach, the oxidation behavior of the EOL NdFeB magnet at high temperatures will govern how the materials will be processed in the subsequent steps, as well as its overall recycling process. The aim of this oxidation step is to transform the REEs form into REE-oxides where further processing would more easily take place.

In the context of recycling, there have been several studies regarding the high-temperature oxidation of NdFeB magnets available in the literature. Firdaus et al. [9–12] conducted oxidation studies in the range of 700–1200 °C. using new uncoated magnets as the starting samples, observed the formation of Fe_2O_3 on the outer oxide layer while the Nd was concentrated mainly in the form of $NdFeO_3$ underneath, particularly during the oxidation at the range of 1000–1200 °C. Using the same temperature range, a similar pattern was also observed by Nababan et al. [5, 13] which used Ni/Cu/Ni-coated EOL NdFeB magnets. This suggested that the Fe was diffused outward while the Ni and Cu diffused inward and reacted with the formed Fe_2O_3. The formation of Fe_2O_3 on the surface was a significant finding which has an important implication on the general mechanism of concentrating both Fe and REEs.

More detailed mechanistic studies of the oxide growth on the surface of NdFeB magnets are still limited. This current study is focused on directly observing the growth of oxide(s) on the surface of Ni/Cu/Ni-coated EOL NdFeB magnets in the early stage of oxidation at 900 and 1100 °C. In-situ high-temperature optical microscopy combined with *ex-situ* scanning electron microscopy was used to analyze the oxide growth on the surface for an oxidation period of four hours under ambient air conditions.

Table 1 Chemical composition of typical REPM samples (wt%)

Item	Element									
	Nd	Pr	Dy	Fe	B	Nb	Co	Al	Ni	Cu
EOL NdFeB magnet used in this work	17.0	5.4	4.51	68.43	1.0	0.16	0.93	1.0	0.85	0.72

Methodology

Sample Preparation

The EOL NdFeB magnet samples used in this study were obtained from Note Printing Australia Ltd., Australia, and were previously used in banknote machine printing. Inductively Coupled Plasma Optical Emission Spectrometry (ICP-OES) was used to determine the chemical composition of the magnets, and the results are shown in Table 1.

The samples were firstly cleaned in an ultrasonic cleaner in acetone medium to ensure any contaminants were removed from the surface. This was followed by thermally demagnetization of the samples at 300 °C for 30 min in an oven. This step was not found to oxidize the samples or change their microstructure.

Sample Characterization

The microstructure observation and surface analysis of the original magnet and the oxidized EOL NdFeB magnet samples were carried out using Scanning Electron Microscope (SEM) equipped with Energy Dispersive X-ray (EDX) spectroscopy (for generating elemental distribution maps and providing semi-quantitative chemical composition analyses). Figure 1 shows the cross-sectional microstructure of the original NdFeB magnet used in this study. It shows the Ni/Cu/Ni coating layer on the surface of the magnet with a thickness of approximately 20 μm as well as the two main matrix phases in the magnet, i.e. the Φ phase ($Nd_2Fe_{14}B$) as the main matrix phase and an Nd-rich phase typically at grain boundaries.

Experimental Techniques

The oxidation experiment was conducted using Linkam Scientific hot-stage coupled with an optical microscope (Olympus). The experiments were conducted at 900 and 1100 °C with the longest duration heating experiment being four hours. The setup can be seen in Fig. 2. The sample was placed under a lens in the hot stage surrounded by heating elements. Two flanges were kept open to allow the ambient air to circulate.

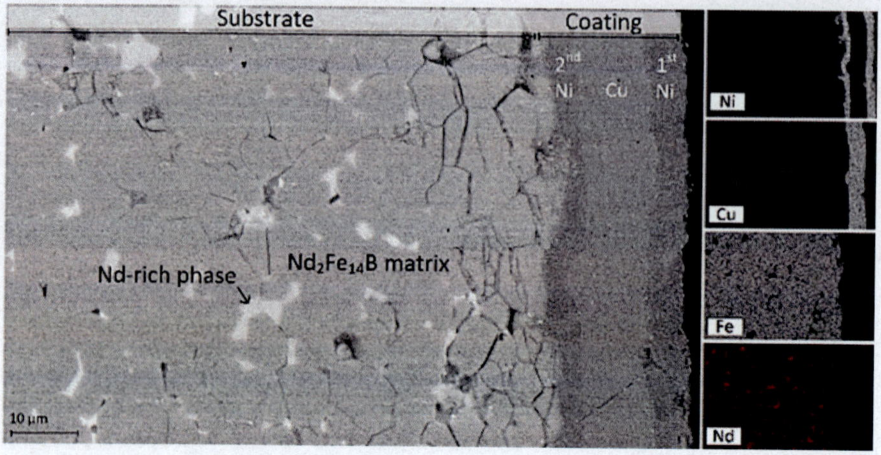

Fig. 1 SEM back-scattered image showing the typical microstructure of the EOL NdFeB magnet used in this study. The image shows the Ni/Cu/Ni coating layer, the intergranular Nd-rich phase, and the main Φ matrix phase ($Nd_2Fe_{14}B$). Also included are corresponding EDX elemental maps showing the distribution of the elements Fe, Cu, Ni, and Nd

A lid with a water-cooling system was installed to protect the direct exposure of the heat on the lens. The heating rate on each experiment was set to 150 °C/min.

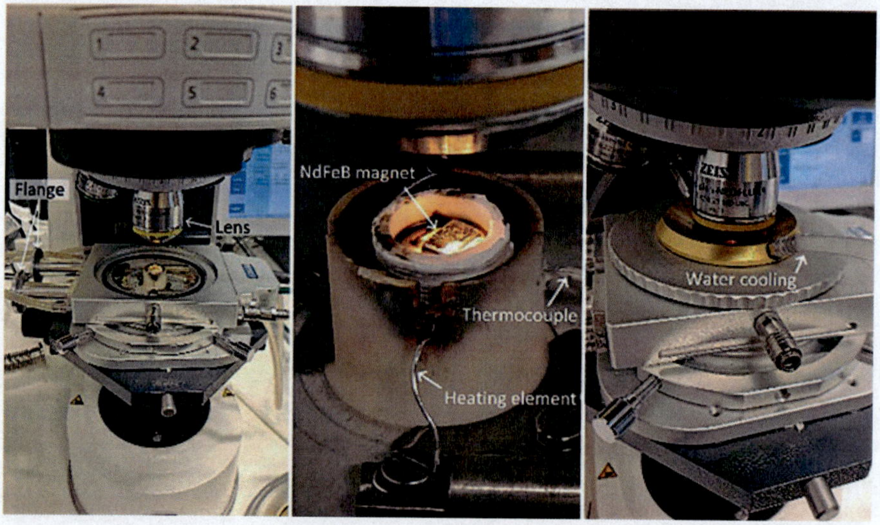

Fig. 2 Hot-stage microscopy apparatus used in this study

Results and Discussion

Previous works by the authors [5, 13] studied the high-temperature oxidation of the EOL NdFeB magnet through microstructural observations of cross-sections of oxidized-samples. This current work aims to carry out a complementary, but more detailed observation of magnet's oxidation behavior by focusing on the oxide growth on the surface in the early stages. Two oxidation temperatures, 900 and 1100 °C, were selected in the current study.

Oxide Growth Observation During the High-Temperature Oxidation of NdFeB Permanent Magnets at 900 °C

Figure 3 shows the optical images obtained from oxidation of a NdFeB magnet at 900 °C. The heating time to reach the target temperature was approximately six minutes. It is important to note that the untreated surface of the magnet had some surface imperfections (valleys) formed during its manufacturing process. Once the temperature reached 900 °C, the surface was already fully covered by NiO which formed due to the oxidation of the 1st Ni layer on the coating surface according to Reactions (1) and (2) below:

$$Ni(s) + {}^{1}/{}_{2}O_2(g) \rightarrow NiO(s)\ \Delta G^{\circ} = -135\,kJmol^{-1}\left(900^{\circ}C\right) \tag{1}$$

$$Ni(s) + {}^{1}/{}_{2}O_2(g) \rightarrow NiO(s)\ \Delta G^{\circ} = -118\,kJmol^{-1}\left(1100^{\circ}C\right) \tag{2}$$

As the oxidation proceeded, after about 13 min, another oxide (predicted to be Fe_2O_3) was observed to begin growing on the surface. The former nickel oxide continued to grow bigger while some Fe_2O_3 was formed at other locations. A crack on the NiO surface was observed after 50 min while at the same time, a distending surface was observed. The crack continued to form and grow larger, particularly on the distending surface. As clearly indicated in Fig. 3 at 170 and 240 min, the NiO surface was cracked and significantly distended caused by the growth of Fe_2O_3 underneath.

The ex-situ SEM examination of the surface of the oxidized NdFeB magnet after 240 min of heating at 900 °C is shown in Fig. 4a. Also shown are results from with the EDX mapping from a smaller region in Fig. 4d. Results show the surface was covered by a NiO layer with some cracks present. Figure 4c shows spiky Fe_2O_3 growing out from underneath the NiO, thereby generating further cracks on the surface. The growth of spiky Fe_2O_3 phase is shown in Fig. 4e where the Fe_2O_3 continued to grow outward and to the side. Figure 4b, d shows the Fe_2O_3 that has grown along, and penetrated through, the surface cracks.

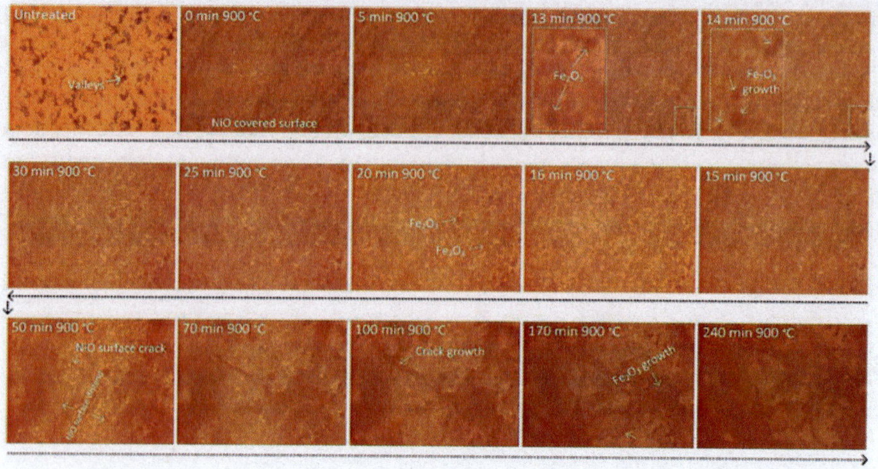

Fig. 3 Hot-stage microscope images of EOL NdFeB magnet's surface oxidized at 900 °C for 4 h in air. The heating rate is 150 °C/min

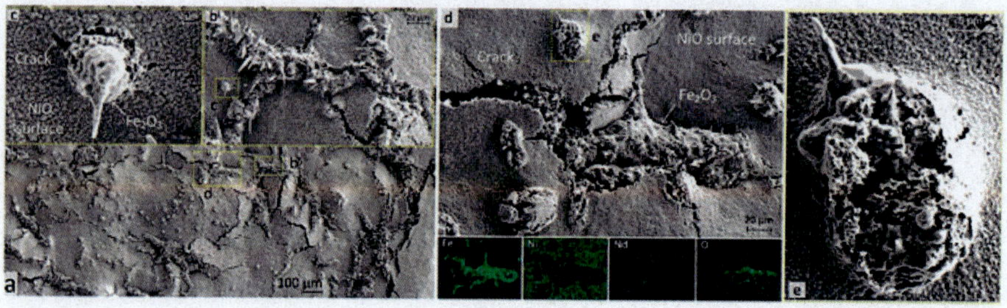

Fig. 4 SEM secondary electron images showing the surface of EOL NdFeB magnet sample oxidized at 900 °C for 4 h in air. Also included are EDX element distribution map results showing the distribution of O, Fe, Ni, and Nd across a selected imaged area (shown in Fig. 4d)

Oxide Growth Observation During the High-Temperature Oxidation of NdFeB Permanent Magnets at 1100 °C

Figure 5 shows the transformation over time of the surface morphology of the NdFeB magnet during the oxidation heating process up to 1100 °C under air atmosphere. The hot-stage microscopy images were taken in one heating sequence with one sample. The SEM images, however, were taken separately following a discontinued heating process, i.e. one representative image was taken from each sample after heating to a particular temperature.

Similar to the observations at 900 °C, the untreated surface of the magnet had some surface imperfections (valleys) formed during its manufacturing process. At the early oxidation stages, up to 500 °C, reflections with distinct colors were captured which

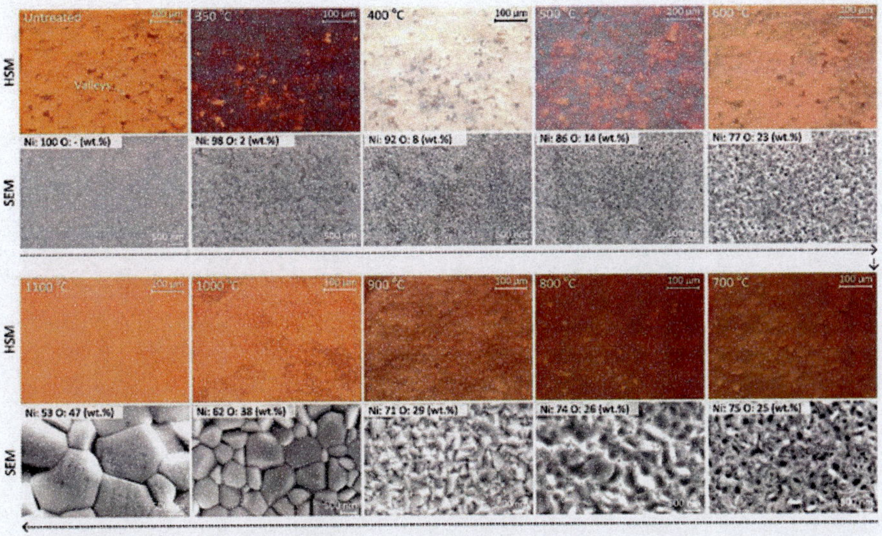

Fig. 5 Hot-stage microscopy coupled with SEM images of EOL NdFeB magnet's surface heated-oxidized at 350–1100 °C in air. The heating rate was 150 °C/min

might indicate changes in the surface's morphology and/or chemical composition. **The SEM images indicated significant grain growth (fine-grained) on the surface, and the EDX analysis confirmed an increase of oxygen concentration. It is plausible** to suggest that the NiO was already formed during the heating stage. At temperatures **higher than 600 °C, the surface reflections showed a relatively homogeneous color** which might indicate that the whole surface was fully covered by the NiO. As the heating and oxidation continued to proceed, the oxygen concentration was increased up to 47 wt.% at 1100 °C where NiO grain growth was pronounced.

Figure 6 shows the microscopy images during the oxidation of NdFeB magnets at 1100 °C for 4 h, a continuation of the experiment presented in Fig. 5. It can be seen that another oxide (indicated by arrows) grew immediately after 1 min oxidation. The previous study [5] observed that the NiO layer on the surface was dissolved and reacted with Fe_2O_3 as the oxidation proceeded which grew outward to form the spinel phase $NiFe_2O_4$. **For simplification, this study assumed that the formed oxide** was $NiFe_2O_4$ whereas the EDX result presented in Table 2 shows the approximate composition of the oxide as, 36 wt.% O, 52 wt.% Fe, and 11 wt.% Ni. The reaction might follow the Reaction (3):

$$NiO(s) + Fe_2O_3(s) \rightarrow NiFe_2O_4(s) \quad (3)$$

The nucleation of $NiFe_2O_4$ occurred in several different locations after a certain period of time (e.g. 5 min), and the $NiFe_2O_4$ continued to grow and merge with the

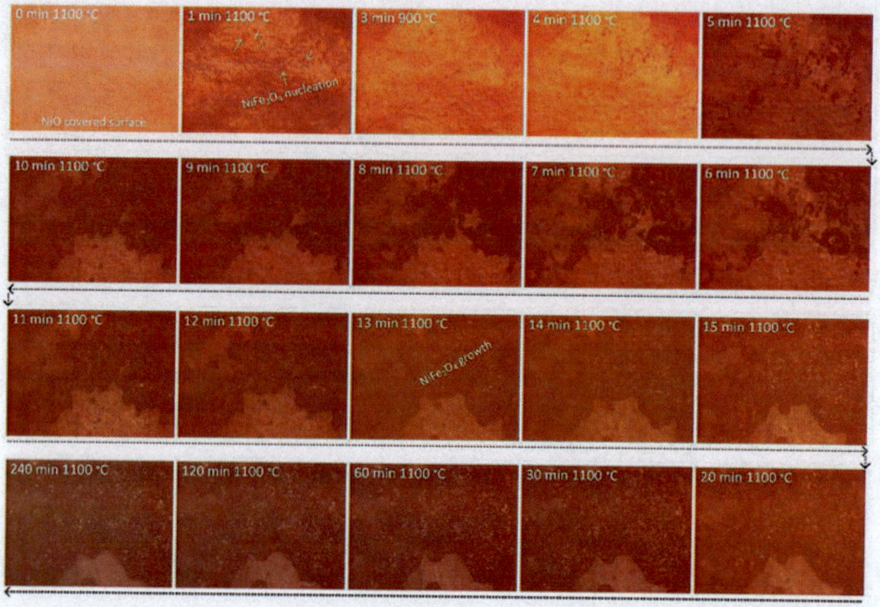

Fig. 6 Hot-stage microscope images of EOL NdFeB magnet's surface oxidized at 1100 °C for 4 h

Table 2 EDX semi-quantitative composition analysis including the deduced phase from points 1–4 in Fig. 7

Element (wt%)	Point			
	1	2	3	4
O	28.88	36.09	30.92	30.06
Fe	12.62	52.72	10.69	21.74
Ni	58.50	11.19	58.39	48.20
Deduced main phase	NiO	$NiFe_2O_4$	NiO	–

Table 3 EDX semi-quantitative composition analysis including the deduced phase from points 1–10 in Fig. 9

Element (wt%)	Point			
	1	2	3	4
O	27.42	32.16	27.70	30.25
Fe	5.23	66.80	27.70	4.96
Ni	67.35	1.04	5.12	64.79
Deduced main phase	NiO	$NiFe_2O_4$	Fe_2O_3	NiO

other surrounding $NiFe_2O_4$. The $NiFe_2O_4$ layer almost fully covered the surface of the magnet after 60 min of reaction time.

Figure 7 shows SEM images of the surface of a NdFeB magnet oxidized at 1100 °C for 1 min, using SE and back-scattered (BSE) modes and with semi-quantitative EDX point analysis. The locations of the EDX point analyses are also given in Fig. 7, and the results are presented in Table 2. Figure 7b–e illustrates the early stage of $NiFe_2O_4$ formation. In Fig. 7d, Point 1 shows that the surface is dominated by Ni, 58.5 wt% with the Fe being 11.19 wt.% whereas Point 3 has a similar composition. For simplicity, this phase was called NiO (with Fe in solution). It is interesting to see another phase, with black color, formed on the NiO phase, marked by Point 4 with a higher amount of Fe, 21.74 wt.%. This might indicate the outward diffusion of Fe to react with NiO. As the Fe continued to diffuse outwardly and react with NiO, the $NiFe_2O_4$ formed bigger sized crystals and distended in shape (particle labelled as 2 in Fig. 7d). EDX analysis confirmed an increasing amount of Fe up to 52.72 wt.% was associated with this phase. Furthermore, Fig. 7e shows that some blisters coalesced with their surroundings. Figure 8 shows the surface of an oxidized NdFeB magnet at 1100 °C for 5 min where it shows that the $NiFe_2O_4$ blisters were formed at different spots and with the former blisters continuing to grow over time as shown in Fig. 8b.

The microscopy images provided in Fig. 6 show that after being oxidized for 4 h, the formed oxides, which are dominated by $NiFe_2O_4$, had partially covered the surface. Figures 9a, b show the SE/BSE SEM observation of the sample's surface. Figure 9a shows a layer of spiky oxide covering some area on the surface of the magnet with Fig. 9c showing a bigger magnification of a specific area in Fig. 9a. Based on the EDX result, the outer oxide layer was dominated by Fe, marked by Point 2, with a composition of Fe up to 66.80 wt% whereas the surface area covered

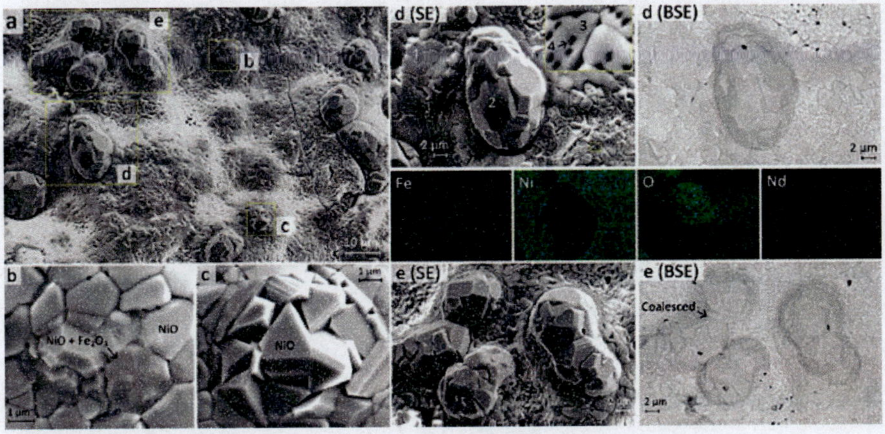

Fig. 7 SEM secondary electron and back-scattered electron images showing the surface of EOL NdFeB magnet sample oxidized at 1100 °C for 1 min. Also included are EDX element distribution maps for O, Fe, Ni, and Nd across the selected imaged area

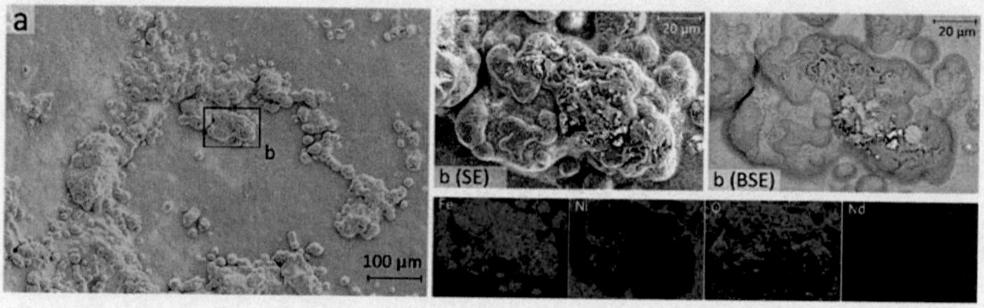

Fig. 8 SEM secondary electron and back-scattered electron images showing the surface of EOL NdFeB magnet sample oxidized at 1100 °C for 5 min. Also included are EDX element distribution maps for O, Fe, Ni, and Nd across the selected imaged area

by NiO was still present, marked by number 1, with a composition of Ni up to 67.35 wt%. The other area provided in Fig. 9b shows a similar observation to the oxidation at 900 °C. Instead of reacting with the NiO, the Fe_2O_3 cleaved the NiO to grow with a spiky shape. This might occur due to the spalling of NiO which allowed the Fe_2O_3 to grow outward directly instead of reacting with the NiO.

The previous works [5, 13] showed that at higher temperatures (900–1200 °C), the Cu and Ni (from the 2nd Ni layer), as the components of the coating, diffused inward through the Nd-rich matrix phase in the substrate where the 1st Ni layer on the surface was oxidized immediately to NiO. With longer oxidation time at 900 °C, the thin dense layer of NiO still remained on the surface. The observation

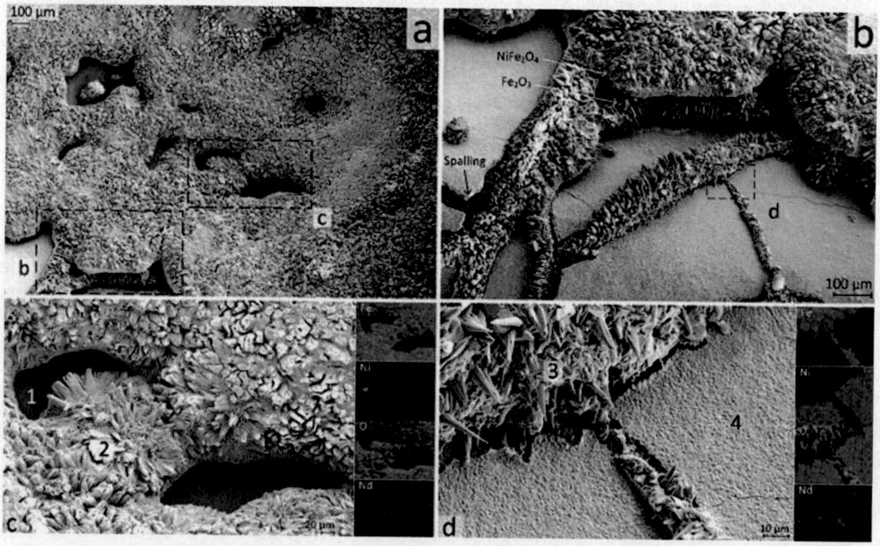

Fig. 9 SEM secondary electron images showing the surface of EOL NdFeB magnet sample oxidized at 1100 °C for 4 h. Also included are EDX element distribution maps for O, Fe, Ni, and Nd across the selected imaged area

in the current study is in line with these, where at 900 °C, the outer surface was, in general, homogenous with NiO but with some spiky Fe_2O_3 grown outward breaking the NiO layer. Furthermore, with longer oxidation time at 1100 °C, the NiO layer was diminished as it reacted with the Fe_2O_3 to form the spinel phase $NiFe_2O_4$ as its reaction is more thermodynamically favorable at higher temperatures.

Conclusion

The growth of oxides on the surface of the EOL NdFeB permanent magnet during the high-temperature oxidation at 900 and 1100 °C was investigated using in-situ hot-stage microscopy. Ex-situ scanning electron microscopy was utilized to supplement the analysis and to characterize the oxides. It was observed that at a lower temperature of 900 °C, the surface was covered by a NiO layer with some growth of a spiky Fe_2O_3 from inside which penetrated the NiO layer. This observation indicated that at 900 °C, the reaction between NiO and Fe_2O_3 was minimal. After four hours of oxidation, the NiO layer remained while the Fe_2O_3 continued to grow outward. At a higher temperature of 1100 °C, the Fe_2O_3 also grew from inside but at this temperature it immediately reacted with the NiO layer resulting in the formation of a $NiFe_2O_4$ layer. A longer period of oxidation at this temperature led to the $NiFe_2O_4$ entirely covering the surface. Both NiO at 900 °C and $NiFe_2O_4$ at 1100 °C, once covered the surface, appeared to significantly reduce further oxidation.

Acknowledgements This work was conducted under the Swinburne-Automotive Engineering Graduate Program (AEGP) scholarship, funded by Australian Government through the Department of Industry, Science, Energy and Resources (previously known as the Department of Industry, Innovation, and Science (DIIS)). The work was also co-funded by the Commonwealth Scientific and Industrial Research Organisation (CSIRO), Australia.

Conflict of Interest The authors declare that they have no conflict of interest.

References

1. Chau K, Chan CC, Liu C (2008) Overview of permanent-magnet brushless drives for electric and hybrid electric vehicles. IEEE Trans Industr Electron 55(6):2246–2257
2. Lacal-Arántegui R (2015) Materials use in electricity generators in wind turbines—state-of-the-art and future specifications. J Clean Prod 87:275–283
3. München DD, Veit HM (2017) Neodymium as the main feature of permanent magnets from hard disk drives (HDDs). Waste Manag 61:372–376
4. Yang Y, Walton A, Sheridan R, Güth K, Gauß R, Gutfleisch O, Buchert M, Steenari B-M, Van Gerven T, Jones PT, Binnemans K (2017) REE recovery from end-of-life NdFeB permanent magnet scrap: a critical review. J Sustain Metall 3(1):122–149

5. Nababan DC, Mukhlis R, Durandet Y, Pownceby MI, Prentice L, Rhamdhani MA (2021) Mechanism and microstructure evolution of high temperature oxidation of end-of-life NdFeB rare earth permanent magnets. Corros Sci 182:109290
6. Zhang Y, Gu F, Su Z, Liu S, Anderson C, Jiang T (2020) Hydrometallurgical recovery of rare earth elements from NdFeB permanent magnet scrap: a review. Metals 10(6):841
7. Elwert T, Goldmann D, Roemer F, Schwarz S (2017) Recycling of NdFeB magnets from electric drive motors of (hybrid) electric vehicles. J Sustain Metall 3(1):108–121
8. Bailey G (2016) Life cycle assessment of the pyrometallurgical and hydrometallurgical recycling routes used in rare earth recycling: a case study of NdFeB magnets, Recycling 2016 Berlin, World Congress and Expo on Recycling, Date: 2016/07/25–2016/07/27. Berlin, Germany
9. Firdaus M, Rhamdhani MA, Rankin WJ, Pownceby M, Webster NAS, D'Angelo AM, McGregor K (2018) High temperature oxidation of rare earth permanent magnets. Part 1—microstructure evolution and general mechanism. Corros Sci 133:374–385
10. Firdaus M, Rhamdhani MA, Durandet Y, Rankin WJ, McGregor K (2018) High temperature oxidation of rare earth permanent magnets. Part 2–Kinetics. Corros Sci 133:318–326
11. Firdaus M, Rhamdhani MA (2018) High temperature recovery of rare earth ortho-ferrites from permanent magnets. In: Davis BR, Moats MS, Wang S, Gregurek D, Kapusta J, Battle TP, Schlesinger ME, Alvear Flores GR, Jak E, Goodall G, Free ML, Asselin E, Chagnes A, Dreisinger D, Jeffrey M, Lee J, Miller G, Petersen J, Ciminelli VST, Xu Q, Molnar R, Adams J, Liu W, Verbaan N, Goode J, London IM, Azimi G, Forstner A, Kappes R, Bhambhani T (eds) Extraction 2018. Springer International Publishing, Cham, pp 805–813
12. Firdaus M, Akbar Rhamdhani M, Durandet Y, John Rankin W, McGregor K, Webster NAS (2017) Microstructure observation of oxidation of Nd-magnet at high temperatures. In: Kim H, Alam S, Neelameggham NR, Oosterhof H, Ouchi T, Guan X (eds) Rare metal technology 2017. Springer International Publishing, Cham, pp 65–74
13. Nababan DC, Mukhlis R, Durandet Y, Pownceby MI, Prentice L, Rhamdhani MA (2021) Kinetics of high temperature oxidation of end-of-life Ni/Cu/Ni coated NdFeB rare earth permanent magnets. Corros Sci 189:109560

Introducing Membrane Percrystallisation Technology for Hydrometallurgical Applications

Siti Nurehan Abd Jalil, Julius Motuzas, and James Vaughan

Abstract Crystallisation is an important industrial unit operation for raw product formation, purification, and waste removal. This process typically requires multiple steps (crystallisation, solid liquid separation, and product drying) to attain the final product. Here, we introduce a novel process: membrane percrystallisation which achieves complete separation of solvent from solute in a single step. Liquid is vaporised at the inorganic membrane permeate surface, whilst dry solid product is continuously ejected from the permeate surface, under vacuum. This novel technology has been demonstrated at the laboratory scale in metal recovery and waste brine treatment applications. Membrane percrystallisation intensifies the process offering a simple, compact, and closed system. Within this study, key percrystallisation operating parameters in salt crystallisation that employ supported continuous saccharides-based carbon membranes will be discussed with examples of potential applications in hydrometallurgical processes.

Keywords Carbon membrane · Saccharide · Percrystallisation

Introduction

Many industries including pharmaceutical [1], mining [2], bulk commodity chemical industry [3], food ingredient production [4], and desalination [5] use crystallisation in their processes to produce final products. The crystallisation process involves

S. N. A. Jalil · J. Motuzas · J. Vaughan (✉)
Materials and Membrane Laboratory, School of Chemical Engineering, The University of Queensland, FIM2Lab – Functional Interfacial, Brisbane, QLD 4072, Australia
e-mail: james.vaughan@uq.edu.au

S. N. A. Jalil
Centre for Chemical Engineering Studies, CawanganPulau Pinang, Universiti Teknologi MARA, Permatang Pauh, 13500 Penang, Malaysia

J. Vaughan
Hydrometallurgy Research Group, School of Chemical Engineering, The University of Queensland, Brisbane, QLD 4072, Australia

R. G. Reddy et al. (eds.), *New Directions in Mineral Processing, Extractive Metallurgy, Recycling and Waste Minimization*, The Minerals, Metals & Materials Series,
https://doi.org/10.1007/978-3-031-22765-3_31

concentrating a solution to its saturated state which causes the crystallisation of solid particles. Conventional crystallisation processes such as solar evaporative crystallisation using ponds are utilised in mining wastewater and salt mining purification applications [2, 6]. Solar evaporation is a relatively technologically simple process with low energy requirements. However, the evaporation process is time consuming and requires a large area. The process is also dependent on having dry and sunny weather. Conventional evaporative crystallisation has limits in the type of crystal particle characteristic that can be obtained. This is due to limitations in controlling supersaturation conditions due to imperfect mixing. This process also requires high energy usage relative to solar evaporation [7].

In this context, a new concept and approach in crystallisation process using membrane technology is being investigated to overcome the limitations of conventional crystallisation technologies. The membrane crystallisation process usually allows better control of crystal growth compared with current commercial methods [7–9]. Industrial membrane crystallisation inherently produces crystals on the feed side of the membrane. Subsequently, there is a need for two additional processing steps to filter and dry the wet crystal particles after the crystallisation process [10]. These extra processes increase system complexity, increases the plant size, and generates low crystallisation rates [11–13]. Therefore, a novel technology, termed membrane percrystallisation, has been proposed to overcome the current drawbacks of membrane crystallisation processes. Figure 1 exhibits the schematic process flow diagram of membrane percrystallisation and membrane crystallisation, respectively.

Polymeric membranes are commonly used in membrane assisted crystallisation. They are also popular among researchers, especially in the field of water treatment, due to their high selectivity, permeability, and low production costs [15]. For instance, Drioli and co-workers used polypropylene membranes to recover sodium chloride, magnesium sulphate hydrate, and lithium chloride [16]. However, polymeric membranes do exhibit disadvantages that reduce their attractiveness such as

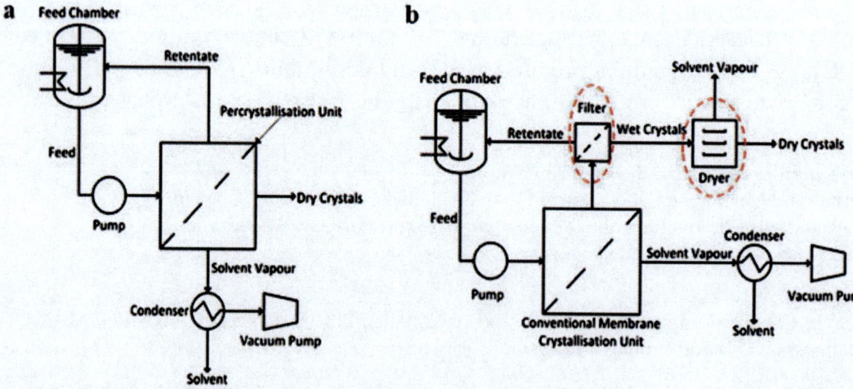

Fig. 1 Schematic diagram of **a** membrane percrystallisation and **b** membrane crystallisation [14]

limited long term stability, fouling, and poor chemical and thermal stability [17]. Inorganic membranes, specifically carbon membranes, have advantages including low fouling rates (and are relatively easy to clean), long service life, and most importantly temperature, pressure, and chemical high stability [18]. However, carbon membranes formed from polymeric sources must achieve a superior performance to polymeric membranes to compensate for their higher cost. Therefore, application of a low-cost precursor is recommended to overcome this challenge. Using saccharides to create carbon membranes has many advantages such as low environmental impact, low cost, and the material is abundant. So far, researchers have started to explore the production of carbon membranes using saccharide precursor materials. Most of this research has been in the development of gas separation technology [18].

This work will review the concept of using membrane percrystallisation in hydrometallurgical applications and will include the promising results obtained thus far. The inorganic carbon-based membranes investigated in this study utilise sucrose as precursor. In addition, this work also presents the uses of monosaccharides such as glucose and fructose to observe their potential as carbon-sources for inorganic membranes for percrystallisation. There are several of characterisation techniques are performed including SEM for membrane morphology analysis and TGA as carbon materials characterisation. These characterisation works were correlated to the performance of carbon membranes during percrystallisation application and discovered the influence of salt produced. The final salt product has also been characterised using XRD to determine the types of crystalline phases formed. Crystal size distribution, crystal shape, and product purity are often the properties targeted by crystallisation processes.

Perspective on Membrane Percrystallisation

In 1917, Kober discovered the percrystallisation phenomena. He stated that when a dialysable solute within the membrane container reaches saturation, it crystallises on the outside of the membrane. Initially, experiments were carried out with sodium chloride solution, ammonium, sulfate and starch digestion. As the crystals are practically dry and free from mother liquor, he concluded that percrystallisation should be beneficial, in many chemical operations including those that use conventional crystallisation as well [19]. Furthermore, experiments conducted by Tauber and Kleiner have also shown that sodium chloride crystals (from enzyme-salt mixture) form on the outside (permeate side) of a collodion bag. This process occurs by percrystallisation [20]. Overall, membrane percrystallisation can be defined as a process whereby a membrane separates solvents from solutes in a single step, contrary to conventional membrane crystallisation which requires three processing steps. No further investigation is reported after this discovery until it was proposed by Motuzas and co-workers in early 2018 [21]. However, in this case the researchers investigated inorganic, rather than organic membranes.

Motuzas et al. [21] reported the first demonstration of continuous percrystallisation phenomena using food grade sugar as a precursor material for the carbon membrane. The results indicate that these carbonised sugar membranes can create a supersaturated condition which leads to the percrystallisation of compounds on the permeate side of the membrane, under vacuum. Interestingly, this novel technology can be used in chemical, mining, or hydrometallurgical operations to recover substances. In addition, this process has the added benefit of recovering water/solvent which complies with new environmental regulations of zero liquid discharge. As a result, dry NaCl crystals are continuously ejected off the membrane surface at a rate of 48,000 kg/year/m^2 of membrane surface. The authors proposed that the percrystallisation process involves surface thin-film evaporation. This occurs due to the formation of a wet thin-film on the permeate side of the membrane as illustrated in Fig. 2. During percrystallisation, the wet thin-film is essential for the fast crystallisation and water evaporation. Thin films are known to improve heat transfer, thus promoting evaporation of water close to the solid–liquid-vapor contact line [22].

Further investigation of this phenomena was carried out by Madsen et al. [23]. Membrane morphological features and operating parameters were studied for the percrystallisation of NaCl across sucrose derived carbon membranes calcined at different carbonization temperatures. It was discovered that the highest carbonisation temperature (750 °C) produced both the highest membrane pore volume and wet contact angle. The carbon membrane with the highest flux delivered the smallest NaCl crystallite sizes, the smallest particle sizes, and the narrowest particle size distribution. Therefore, some control over the particle size, morphology, and crystallite size can be obtained by tailoring the membrane structure. With high flux and water evaporation, NaCl crystals were ejected from membrane surface, thus reducing their particle aggregation and reducing crystal growth restriction. Madsen and coworkers also explored the membrane percrystallisation process for nickel sulphate

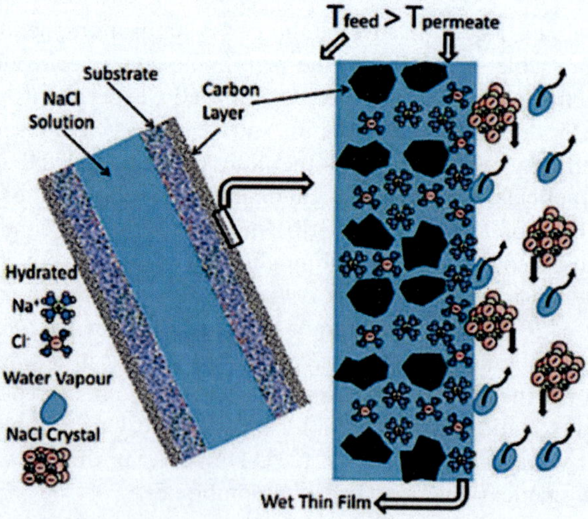

Fig. 2 Proposed schematic of continuous percrystallisation via wet thin-film [21]

Fig. 3 SEM images of shape of crystal produced from different sucrose solution concentrations during membrane preparation [14]

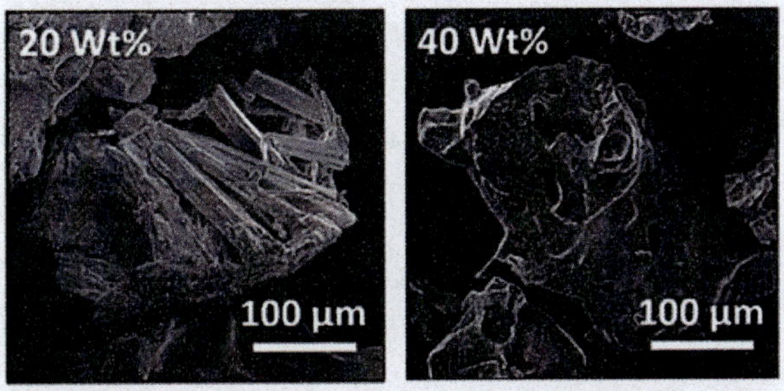

production using the same sucrose-based carbon membranes [14]. The transport of solution through the membrane affected the hydration state of nickel sulphate, as well as its crystal type and shape as demonstrated in Fig. 3. The salt produced is pure nickel sulphate heptahydrate with elongated and laminar crystal particles. Moreover, pure water recovery is attained and low activation energy also generated. These studies highlight the promise of the membrane percrystallisation process in the field of hydrometallurgy.

While the previous work only focused on a single type of salt solution, Moltedo and co-workers [24] looked at more complex experimental studies. They coupled percrystallisation with electrodialysis for solutions containing multiple salts and brine effluent from a pulping mill. They also developed a computational fluid dynamic (CFD) model to study the effect of temperature and vacuum pressure during the process. The authors found that the feed water temperature is critical in this process and vacuum pressure must be sufficiently lower than the water vapor pressure to allow the evaporation of the permeate solution at a rate that is sufficient to form dry crystals. However, coupling percrystallisation with electrodialysis can result in high energy consumption in industrial applications. Similarly, Meng et al. [25] discovered that the feed temperature and operating pressure had a significant effect on the morphology of NaCl salt produced. The salt flux reported was higher than the capacity of existing solar-driven evaporator crystallisation technologies.

Inorganic carbon membranes offer many advantages such as high stability, low fouling, and the ability to withstand harsh thermal, chemical, and high-pressure environments. These characteristics are beneficial in hydrometallurgical processes. Research thus far has highlighted the potential of membrane percrystallisation to be useful in hydrometallurgical applications. Investigation into the use of different sugar-types, including glucose and fructose, may give further insight into membrane percrystallisation's relevance to the sector. Furthermore, the cost of sugar is cheap for membrane making, and salt can be generated much faster than the conventional crystallisation process.

Experimental, Results and Discussion

Material Preparation and Characterisation

Saccharides of fructose (F) supplied by Fisher Chemicals and glucose (G) provided by Chem-Supply were dissolved in deionized water to produce 10 and 20 wt.% solution, respectively. Subsequently, the solutions were transferred into a petri dish and placed into a desiccator, then left for 2 h under vacuum. A Mettler Toledo TGA–DSC apparatus was employed to determine structural and mass changes of material where suitable profile of the thermal treatment applicable in the formation of carbon membranes. The analysis was conducted with a heating rate of 5 °C/min up to 1000 °C using continuous flow of nitrogen gas (65 mL/min). The thermal analysis (Fig. 4) revealed the existence of several mass change steps in the carbon formation from fructose and glucose. [26, 27]. In the first stage, a partial reduction in weight of 10 wt.% is noticed up to 110 °C. The change was attributed to the removal of adsorbed moisture. The second mass loss occurring between 140 °C and 220 °C was associated to the primary degradation of saccharide [28]. The third major mass loss (220—700 °C) is related to the thermal decomposition of functional groups and opening of cyclic rings within the structures. These TGA results corresponded to the DSC measurements which were attributed to changes in the saccharide structures [29]. The sample mass stabilises above 600 °C and changes very little up to 900 °C suggesting that functional groups were completely eliminated [30].This result indicates that in order to get stable carbon materials, the samples must be treated above 600 °C. Thus, the carbon membranes were thermally treated at 700 °C for 4 h under a continuous flow of nitrogen with a heating/cooling rate of 5 °C/min.

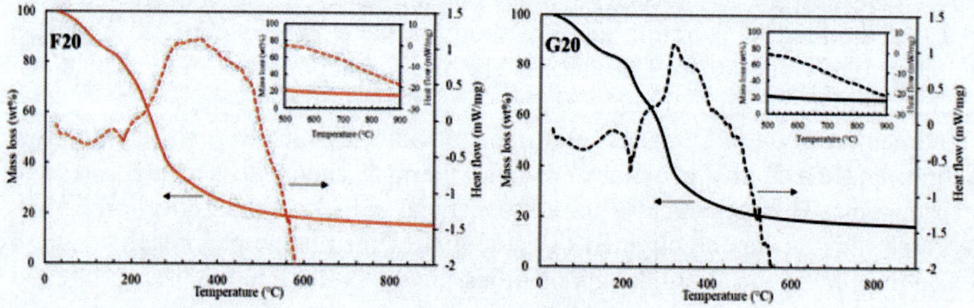

Fig. 4 TGA and DSC analysis of 20 wt.% of fructose and glucose where solid line is TGA and broken line is DSC, 5 °C/min heating/cooling rate, N_2 gas environment

Membrane Preparation and Characterisation

The membranes were prepared using solution impregnation method as reported elsewhere [31]. Porous α-alumina tubes (supplied by Ceramic Oxide Fabricators, Melbourne, Australia) with ID/OD = 6/10 mm were used as supports to prepare infiltrated membranes. The support was immersed completely in the saccharide solution and placed in the desiccator with vacuum (<0.3 mbar) for 5 min. Subsequently, all impregnated substrates were dried in an oven overnight at 60 °C and followed with calcination at 700 °C (with a heating/cooling rate of 5 °C/min) for 4 h in nitrogen atmosphere. The membranes were name accordingly such as F20 corresponds to the membrane (M) produced from 20 wt.% of fructose solution.

The cross section and surface morphology (Fig. 5) of the membranes were analysed by a JEOL JSM-7001F scanning electron microscope (SEM) with a hot (Schottky) electron gun at an accelerating voltage of 2.5 kV at a working distance of 10 mm. Both membranes illustrated the impregnation of saccharides solution within the alumina grains of support where membrane from 10 wt.% of glucose solution was able to penetrate deeper than 20 wt.% of glucose solution. This can be related to the viscosity of the saccharide's solution. The surface morphology for G20 looks smooth and no void as compared to G10.

Fig. 5 SEM images of **a** cross section and **b** surface morphology of carbon membranes where (1) G10 and (2) G20

Membrane Performance

For percrystallisation performance testing, the carbon membrane was placed in the vacuum chamber. 10 wt.% of NaCl solution was heated on a hot plate with a chemical transfer pump used to circulate the solution throughout the system. The solution was supplied through the inner shell of the membrane while the outside of the membrane was exposed to vacuum (0.15±0.03 mbar). A condenser was immersed into liquid nitrogen to collect the permeated water vapor through the membrane. The water flux (L/(m^2 hr)) was determined from the weight of collected water in the cold trap. The schematic diagram of percrystallisation testing setup is presented in Fig. 6. The salt in the vacuum chamber was collected for structural analysis. The mass of salt permeate is measured by dissolving it and recording its conductivity. The conductivity of the resulting solutions was measured using a conductivity meter (Model TPS labCHEM), and the amount of collected salt was calculated from the standard calibration curve. X-ray diffraction (XRD) patterns were performed on the salt crystals via A Rigaku SmartLab X-Ray Diffractometer equipped with 9 kW rotating (Cu) anode X-Ray generator operating at 45 kV and 200 mA and with a HyPix 3000 detector. The tested range of 2θ was from 10° to 90° with a step size of 0.01° 2θ.

The solution flux using membranes prepared with fructose and glucose at 10 and 20 wt.% of sugar concentration is shown in Fig. 7a, while XRD results of salt collected from F20 and G20 are illustrated in Fig. 7b. All testings are conducted with the same process temperature (50 °C) and feed concentration (10 wt.% NaCl solution).

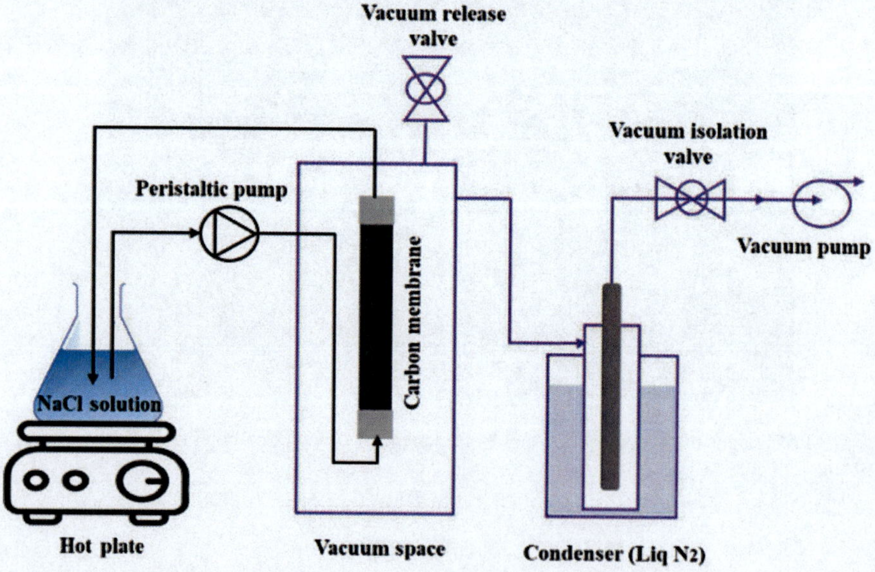

Fig. 6 Membrane testing in continuous percrystallisation using NaCl solution

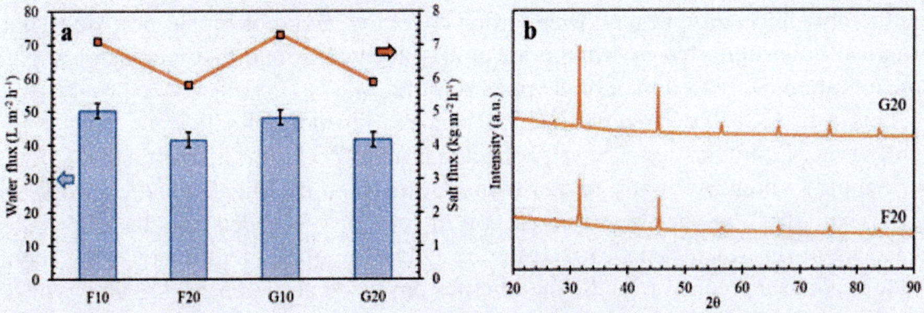

Fig. 7 **a** Solution flux of carbon membranes from fructose and glucose using different sugar concentration and **b** XRD patterns of salt collected from F20 and G20, respectively

Glucose and fructose both belong to the same group of saccharides (monosaccharides) and have the same molecular weight (180.16 g/mol) of anomer monohydrate structures [32]. As a result, each category of fructose and glucose shows similar results. For instance, the water flux for F20 and G20 are 41 and 42 L m^{-2} h^{-1}, respectively. This similarity is repeated for salt flux. In terms of sugar concentration for the membrane preparation, the highest flux recorded was with the lowest sugar concentration. Total flux (water and salt) for the G10 membrane is 55 L m^{-2} h^{-1}, higher than the 47 L m^{-2} h^{-1} recorded for the G20 membrane. Overall, the reduction in solution (combined water and salt) fluxes from G10 to G20 is 15%. Strong peaks in the XRD graphs demonstrate the highly crystalline NaCl that is produced. There are two major peaks at 2θ 31° and 45°, and several minor peaks (56, 66, 75 and 84°) which attributed to the cubic (Fm-3 m) structure with space group number 225 (pdf file 01-077-2064).

Correlation Discussion

The impregnated fructose and glucose films form a hydrophobic surface, as evidenced by TGA/DSC results shown in Fig. 4. Motuzas et al. [21] showed that a hydrophobic layer is required to operate membrane percrystallisation. That is, a hydrophobic surface facilitates the evaporation of water, leaving dry salt crystals, which are ejected from the membranes permeate surface during operation. SEM micrographs show that the surface morphology of G10 and G20 membranes differs (Fig. 5). The G10 membrane shows a patchy surface with areas of only partial film coverage. Meanwhile, the G20 membrane surface shows glucose completely covering the alumina support. Similarly, the 20 wt.% glucose solution shows a more homogenous impregnation of the alumina than the 10 wt.% solution. These observations correlate well with the results of the membrane testing which show that G10 and F10 membranes exhibit higher fluxes than the G20 and F20 membranes. It is proposed that this is due to the G10 and F10 membranes having a lower mass transfer resistance through the membrane because of the lower carbon coverage, and thus higher porosity. This proposition is supported by Mukherjee et al. [33], who found that increasing sugar concentration decreased membrane pore volume.

This process achieves zero liquid discharge. In addition, all water recovered contains no salt impurities. The salt collected is up to 7.3 kg/m^2 in 1 h (64 000 kg/m^2 per annum) which is slightly higher than that reported by Madsen and co-workers [23]. Their study of sucrose-based carbon membranes recorded a highest salt flux of 6.9 kg/m^2/h. Meanwhile, Motuzas et al. [21] developed membranes using food grade sugar as a precursor. These membranes produced around 5.5 kg/m^2/h of NaCl salt. The results from this study compare favourably to the 41 000 kg/m^2/annum salt flux reported from conventional evaporation crystallisation [34]. Furthermore, solar pond evaporation crystallisation takes on the order of 6 months or longer, relies on appropriate hot/dry weather, and produces wet salt crystals that required further processing to obtain final salt product.

Conclusions

Saccharide-based carbon membranes are prepared using low-cost precursors and can function in a membrane percrystallisation process under elevated temperatures and pressures. This study has shown that crystallised salt crystals are successfully ejected from the membranes permeate surface, allowing for largely continuous operation. This result opens the door for testing this technology in hydrometallurgical applications involving a wide range of other metal feedstocks. It has been shown that fructose and glucose, used as precursor materials, produce membranes with promising product yields when operated in a percrystallisation mode. The salt yield is up to 64,000 kg/year/m^2 which higher than reported for conventional evaporation crystallisation. These results show that membrane percrystallisation, operating as a compact, closed system, without the need for subsequent processing steps, shows potential for use in hydrometallurgy applications. Further process required can be investigated in future to accommodate the industry demand.

Acknowledgements The authors acknowledge the financial support for this research from the Australian Research Council (ARC) Discovery Project Grant (DP190102502) Inorganic Membrane Percrystallisation in Hydrometallurgy. Siti Nurehan Abd Jalil acknowledges The University of Queensland for providing an UQ International Scholarship and gratefully thanks for the study leave opportunity given by Universiti Teknologi MARA to further her study at The University of Queensland.

References

1. Jones EC, Bimbo LM (2020) Crystallisation behaviour of pharmaceutical compounds confined within mesoporous silicon. Pharmaceutics 12:214
2. Nathoo J, Randall DG (2016) Thermodynamic modelling of a membrane distillation crystallisation process for the treatment of mining wastewater. Water Sci Technol 73:557–563
3. Yazdanpanah N, Nagy ZK (2020) The handbook of continuous crystallization. Royal Society of Chemistry
4. Deora N, Misra N, Deswal A, Mishra H, Cullen P, Tiwari B (2013) Ultrasound for improved crystallisation in food processing. Food Eng Rev 5:36–44
5. Ahmad MM (2012) Assessment of freezing desalination technologies. Swansea University (United Kingdom)
6. Loganathan P, Naidu G, Vigneswaran S (2017) Mining valuable minerals from seawater: a critical review. Environ Sci: Water Res Technol 3:37–53
7. Drioli E, Di Profio G, Curcio E (2012) Progress in membrane crystallization. Curr Opin Chem Eng 1:178–182
8. Choi Y, Naidu G, Nghiem LD, Lee S, Vigneswaran S (2019) Membrane distillation crystallization for brine mining and zero liquid discharge: opportunities, challenges, and recent progress. Environ Sci: Water Res Technol 5:1202–1221
9. Hickenbottom KL, Cath TY (2014) Sustainable operation of membrane distillation for enhancement of mineral recovery from hypersaline solutions. J Membr Sci 454:426–435
10. Rundquist EM, Pink CJ, Livingston AG (2012) Organic solvent nanofiltration: a potential alternative to distillation for solvent recovery from crystallisation mother liquors. Green Chem 14:2197–2205
11. Chabanon E, Mangin D, Charcosset C (2016) Membranes and crystallization processes: state of the art and prospects. J Membr Sci 509:57–67
12. Ter Horst JH, Schmidt C, Ulrich J (2015) Fundamentals of industrial crystallization. In: Handbook of crystal growth. Elsevier, pp 1317–1349
13. Ji X, Curcio E, Al Obaidani S, Di Profio G, Fontananova E, Drioli E (2010) Membrane distillation-crystallization of seawater reverse osmosis brines. Sep Purif Technol 71:76–82
14. Madsen RS, Motuzas J, Julbe A, Vaughan J, da Costa JCD (2019) Novel membrane percrystallisation process for nickel sulphate production. Hydrometallurgy 185:210–217
15. Ladewig B, Al-Shaeli MNZ (2017) Fundamentals of membrane bioreactors. Springer Trans Civ Environ Eng
16. Mariah L, Buckley CA, Brouckaert CJ, Curcio E, Drioli E, Jaganyi D, Ramjugernath D (2006) Membrane distillation of concentrated brines—role of water activities in the evaluation of driving force. J Membr Sci 280:937–947
17. Pabby AK, Rizvi SS, Requena AMS (2015) Handbook of membrane separations: chemical, pharmaceutical, food, and biotechnological applications. CRC press
18. Bhave R (2012) Inorganic membranes synthesis, characteristics and applications: synthesis, characteristics, and applications. Springer Science & Business Media
19. Chen X, Chong JJE, Fah ZWC, Hong L (2017) Glucose-derived carbon molecular sieve membrane: an inspiration from cooking. Carbon 111:334–337
20. Kober PA (1917) Pervaporation, perstillation and percrystallization. J Am Chem Soc 39:944–948
21. Tauber H, Kleiner IS (1932) Needle-shaped crystals of sodium chloride obtained by percrystallization. J Am Chem Soc 54:2392–2393
22. Motuzas J, Yacou C, Madsen RS, Fu W, Wang DK, Julbe A, Vaughan J, da Costa JCD (2018) Novel inorganic membrane for the percrystallization of mineral, food and pharmaceutical compounds. J Membr Sci 550:407–415
23. Stephan P, Brandt C (2003) Advanced capillary structures for high performance heat pipes. In: International conference on nanochannels, microchannels, and minichannels, pp 69–75

24. Madsen RS, Motuzas J, Vaughan J, Julbe A, da Costa JCD (2018) Fine control of NaCl crystal size and particle size in percrystallisation by tuning the morphology of carbonised sucrose membranes. J Membr Sci 567:157–165
25. Moltedo JJ, Schwarz A, Gonzalez-Vogel A (2022) Evaluation of percrystallization coupled with electrodialysis for zero liquid discharge in the pulping industry. J Environ Manage 303:114104
26. Meng Y, Li K, Yu L, Zhang Y, Liu H, Wang J (2021) Preliminary study on a novel vacuum membrane percrystallization process, Chinese. J Environ Eng 15:2314–2321
27. Ismail AF, Rana D, Matsuura T, Foley HC (2011) Carbon-based membranes for separation processes. Springer Science & Business Media
28. Centeno TA, Fuertes AB (2000) Carbon molecular sieve gas separation membranes based on poly (vinylidene chloride-co-vinyl chloride). Carbon 38:1067–1073
29. Latham K, Jambu G, Joseph S, Donne S (2014) Nitrogen doping of hydrochars produced hydrothermal treatment of sucrose in H_2O, H_2SO_4, and NaOH. ACS Sustain Chem Eng 2:755–764
30. Hurtta M, Pitkänen I, Knuutinen J (2004) Melting behaviour of D-sucrose. D-Glucose D-Fruct, Carbohydr Res 339:2267–2273
31. Song Y, Wang DK, Birkett G, Martens W, Duke MC, Smart S, da Costa JCD (2016) Mixed matrix carbon molecular sieve and alumina (CMS-Al_2O3) membranes. Sci Rep 6:30703
32. Jalil A, Nurehan S (2017) Investigation of vacuum-assisted preparation methods of inorganic membranes
33. Ibrahim M, Alaam M, El-Haes H, Jalbout AF, Leon AD (2006) Analysis of the structure and vibrational spectra of glucose and fructose. Ecletica quimica 31:15–21
34. Mukherjee I, Mylonakis A, Guo Y, Samuel SP, Li S, Wei RY, Kojtari A, Wei Y (2009) Effect of nonsurfactant template content on the particle size and surface area of monodisperse mesoporous silica nanospheres. Microporous Mesoporous Mater 122:168–174
35. Lwanyaga JD (2013) Optimization of solar energy to foster the evaporative crystallization process at Lake Katwe. Uganda

Leaching of Arsenopyrite Contained in Tailings Using the TU-OX System

Erick Jesús Muñoz Hernández, Aislinn Michelle Teja Ruiz, Martin Reyes Pérez, Gabriel Cisneros Flores, Miguel Pérez Labra, Francisco Raúl Barrientos Hernández, and Julio Cesar Juárez Tapia

Abstract Iron (Fe), being such an abundant element, is used in various industries such as the steel industry. Similarly, arsenic (As) can be treated to obtain an industrially useful by-product. For this reason, the leaching of arsenopyrite (FeAsS) from mining tailings is silvered, using Thiourea (CH_4N_2S)-Oxalate ($C_2O_4{}^{2-}$) system. The presence of FeAsS was confirmed by XRD and SEM–EDS, while the ICP analysis indicated a content of 5.47% Fe and 0.93% As. Likewise, the parameters analyzed were [CH_4N_2S] and [$C_2O_4{}^{2-}$]. As a result, the highest recovery of Fe was 21.29% and As of 24.04%, in concentrations of Thiourea of 0.03 mol and Oxalate of 0.1 mol, at 60 min.

Keywords Mineral tailings · Leaching · Oxalate · Thiourea · Iron · Arsenic

Introduction

Tailings are a by-product of mineral processing; most of these inert solids contain non-ferrous species and abundant metal sulfides [1]. Continuous exposure to weathering factors and improper handling generates acid mine drainage, which is caused

E. J. M. Hernández · A. M. T. Ruiz · M. R. Pérez · G. C. Flores · M. P. Labra · F. R. B. Hernández · J. C. J. Tapia (✉)
Universidad Autónoma del Estado de Hidalgo, Área Académica de Ciencias de La Tierra Y Materiales, Cd. Del Conocimiento Km 4.5, C.P. 42180, Mineral de La Reforma Hidalgo, México
e-mail: jcjuarez@uaeh.edu.mx

E. J. M. Hernández
e-mail: mu449806@uaeh.edu.mx

A. M. T. Ruiz
e-mail: aislinn_teja@uaeh.edu.mx

M. R. Pérez
e-mail: mreyes@uaeh.edu.mx

G. C. Flores
e-mail: ci336682@uaeh.edu.mx

R. G. Reddy et al. (eds.), *New Directions in Mineral Processing, Extractive Metallurgy, Recycling and Waste Minimization*, The Minerals, Metals & Materials Series,
https://doi.org/10.1007/978-3-031-22765-3_32

by the oxidation of metal sulfides, such as arsenopyrite [2–4]. Arsenopyrite is mainly constituted by arsenic (46%), iron (34.3%), and sulfur (19.7%) [5, 6]. However, this species is also related to low-grade gold content, due to the inclusions it can have in its crystal structure [7, 8]. On the other hand, iron is the fourth most abundant element in the earth; it is found in several geological deposits and in an important range of minerals [9, 10]. It is used in the steel industry for the manufacture of various steels and is considered the pillar of modern society [11–13]. While arsenic is associated as one of the major contaminants due to the adverse health effects it can cause [14, 15]. At the same time, it can be processed to obtain a saleable by-product for the semiconductor industry, wood preservatives, among others [16, 17]. Deng et al. [18], studied the influence of pyrite on arsenopyrite leaching, using sulfuric acid, by leaching experiments and electrochemical analysis, in order to provide information for the treatment of refractory gold ores. As a result of leaching, the addition of pyrite facilitated the extraction of arsenic, while electrochemical studies indicated that adding pyrite increased the conductivity of the electrodes and electrolytes, verifying the catalytic effect of galvanic interaction on arsenopyrite leaching [18]. In a study by Rogozhnikov et al. [8], the kinetics and oxidation mechanism of arsenopyrite were evaluated, as well as the influence of pyrite on the reaction in order to determine the mechanism of interaction and the efficiency of ferric ions in the dissolution of arsenopyrite [8]. The objective of this paper is to evaluate the dissolution of iron and arsenic present in arsenopyrite, contained in mineral tailings, using Thiourea (TU) (CH_4N_2S) and Oxalate (Ox) ($C_2O_4^{2-}$) ion as leaching substances.

Methodology

The sample used for this work was donated by the mining company "El Espiritu", which is located in the Zimapán mining district and is characterized by the exploitation of Pb, Zn, and Ag minerals [19]. In order to identify the mineralogical species present, the sample powders were analyzed by the X-ray diffraction (XRD) technique, using an EQUINOX 2000 X-ray diffractometer, with Co − $K\alpha 1$ radiation (1.789010 Å) operating at 30 mA, 20 kV, and voltage of 220 V.

Subsequently, the powders of the sample were absorbed in epoxy resin, to obtain a tablet, which was ground and polished to a mirror finish to be analyzed by the technique of scanning electron microscopy with energy dispersive analysis (SEM–EDS), using a JEOL brand equipment model JSM-6610LV, in order to confirm the presence of iron and arsenic species.

To know the concentration of Fe and As, 1 g of the powders were used, which were subjected to an acid attack composed of 20 ml of HCL and HNO_3 in proportion 3:1 for 60 min, this digestion was carried out in triplicate. The solution obtained was analyzed by the technique of inductively coupled plasma emission spectrometry (ICP), using a Perkin Elmer model 8300 equipment.

The equipment used for the experimental leaching tests was a Thermo Scientific HP88857190 heating grill, on which a 0.5 L reactor was placed, in order to keep the particles in suspension; an IKA model EW 20 agitation motor was used with a PTFE propeller, and a Thermo Scientific Orion Ultra Triode Combination Electrodes potentiometer was also used along with its pH/ATC electrode to monitor pH and temperature throughout the dissolution process. The monitoring of the reaction was performed by the ICP technique.

Results and Discussion

The results obtained by the different characterization techniques applied and the dissolution curves corresponding to the leaching process are presented.

X-Ray Diffraction Analysis

Using the Match 3 program, the mineralogical species present in the mineral sample were identified. Figure 1 shows the analysis of the diffractogram, where iron and arsenic sulfides were recognized as pyrite [(FeS_2) (96-901-0012)] and arsenopyrite [(FeAsS) (96-210-4742)], as well as such as the presence of gangue species such as calcium carbonate [Calcite ($CaCO_3$) (96-101-0963)], gypsum [($CaSO_4 \cdot 2H_2O$) (96-901-3172)], iron oxide [siderite ($FeCO_3$) (96-901-5963)], and feldspar [Microcline ($Al_{1.03}K_{0.986}Na_{0.014}Si_{2.97}$ O_8)(96-900-5304)].

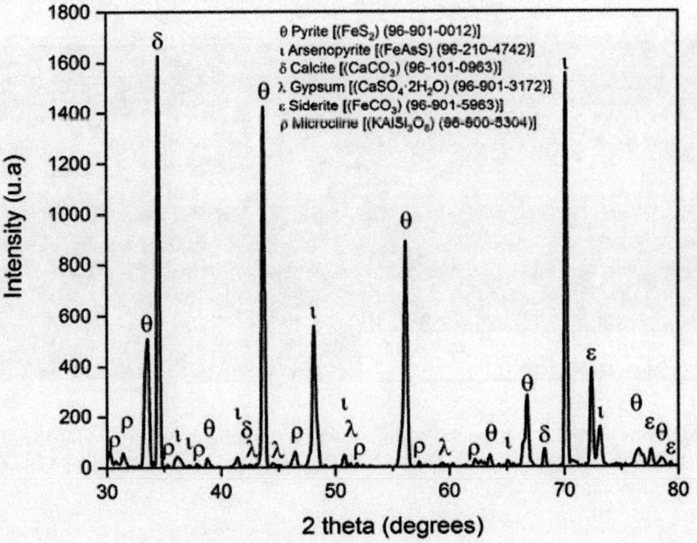

Fig. 1 X-ray diffractogram of the tailings from the "El Espíritu" mine

Analysis by Scanning Electron Microscopy with Dispersive Energy Analysis

To further the chemical–mineralogical characterization, the sample was analyzed by scanning electron microscopy (SEM) with analysis of energy dispersive X-rays (EDS), performing several point analyses. In Fig. 2a, the presence of calcite ($CaCO_3$) is observed, while, in other points, feldspar ($KAlSi_3O_8$) (Fig. 2b) and arsenopyrite (Fig. 2c, d) respectively, using the latter as an extraction source for the leaching process.

A metallic associate was also found that includes four mineral species related to the content of iron and lead, as shown in the micrograph of Fig. 3, which correspond to pyrite (FeS_2), wustite (FeO), galena (PbS), and cerussite ($PbCO_3$); the last 3 are considered minority species, which is why they were not identified by XRD. Figure 4 presents the X-ray EDS analyses of the 4 species containing iron and lead.

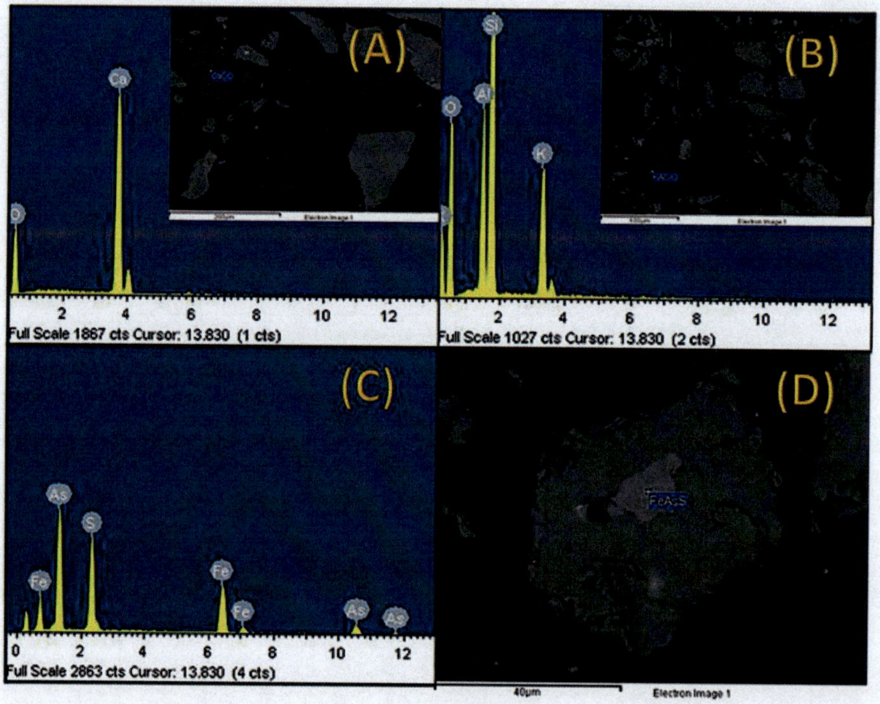

Fig. 2 Micrographs obtained with backscattered electrons and their corresponding X-ray analysis. **a** Calcite ($CaCO_3$). **b** Feldspar ($KAlSi_3O_8$). **c** X-ray EDS of the arsenopyrite (FeAsS) species. **d** FeAsS micrograph obtained at 40 μm

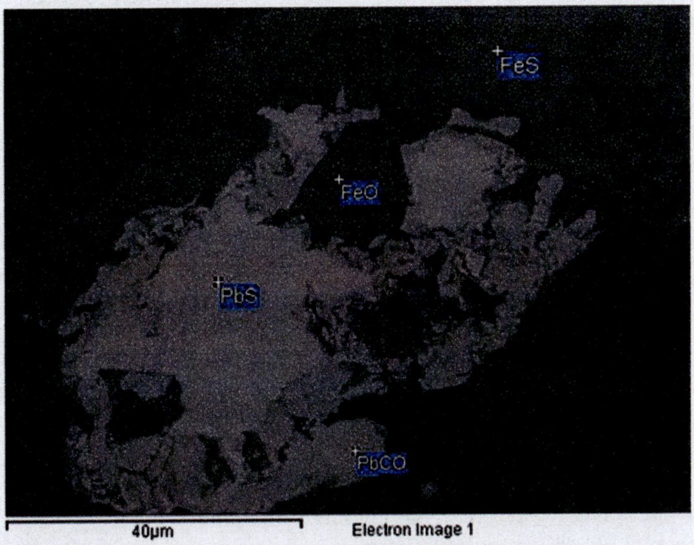

Fig. 3 Micrograph obtained by backscattered electrons of iron and lead species

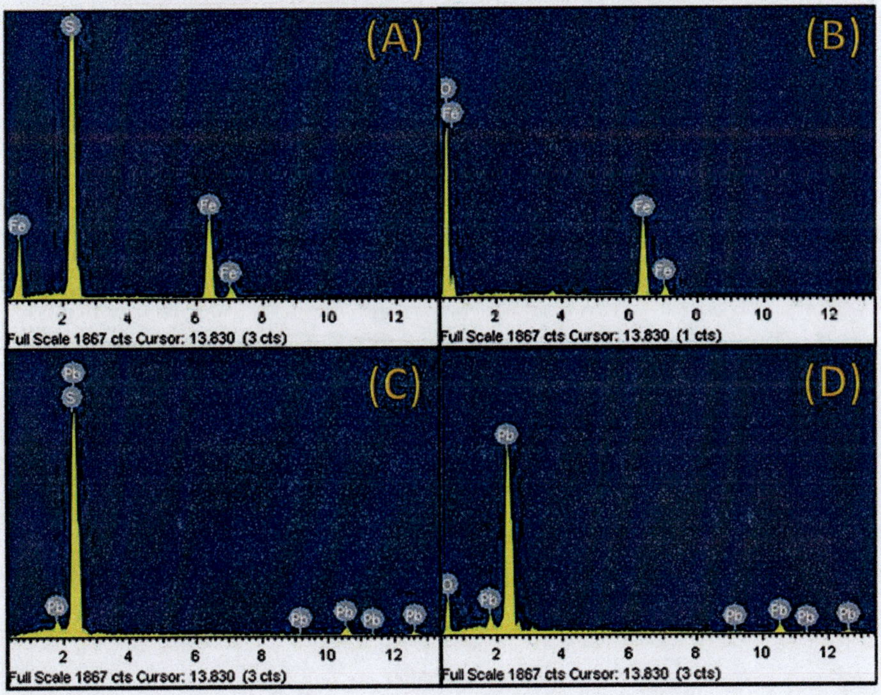

Fig. 4 Analysis of dispersive energies of the mineral species related to the content of iron and lead. **a** Pyrite (FeS_2). **b** Wustite (FeO). **c** Galena (PbS). **d** Cerussite ($PbCO_3$)

Chemical Analysis by Inductively Coupled Plasma Emission Spectrometry

The results of the chemical analysis reported 5.57% iron and 0.93% arsenic in the sample; these results were obtained by the ICP technique.

Thiourea Concentration

To evaluate the effect of Thiourea concentration, the following experimental parameters were established: [TU] = 0.03, 0.01, 0.007, and 0.001 mol L^{-1}, [OX] = 0.1 mol L^{-1}, T = 303.15 (K), pH = 2, solution volume = 0.5 L, sample content = 10 g, r_o = 37 μm, ω = 600 min^{-1} y t = 60 min. The monitoring of the reaction reported that both elements show a similar behavior (Figs. 5 and 6), having a period of progressive conversion that tends to stability. The maximum recovery for iron was 21.29% at 55 min; however, when using the lowest proposed concentration of 0.001 mol L^{-1}, only 12.07% was reached at 20 min of reaction. When using concentrations of 0.01 and 0.007 mol L^{-1}, the recoveries obtained were similar.

In the case of As, a recovery of 24.04% is obtained at 35 min, using a concentration of 0.03 mol·L^{-1}. While using concentrations of 0.01 and 0.007 mol L^{-1} after 30 min, dissolution of 22.69 and 20.67%, respectively, is obtained. Similarly, when the concentration of Thiourea decreased to 0.001 mol L^{-1}, a recovery of 13.81% was reached at the same reaction time, increasing to 15.60% at 50 min.

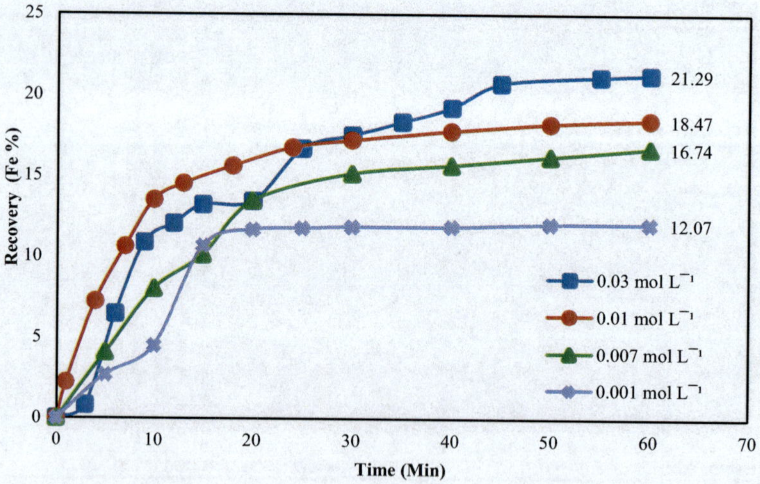

Fig. 5 Fe recovery percentage in the solution ([TU])

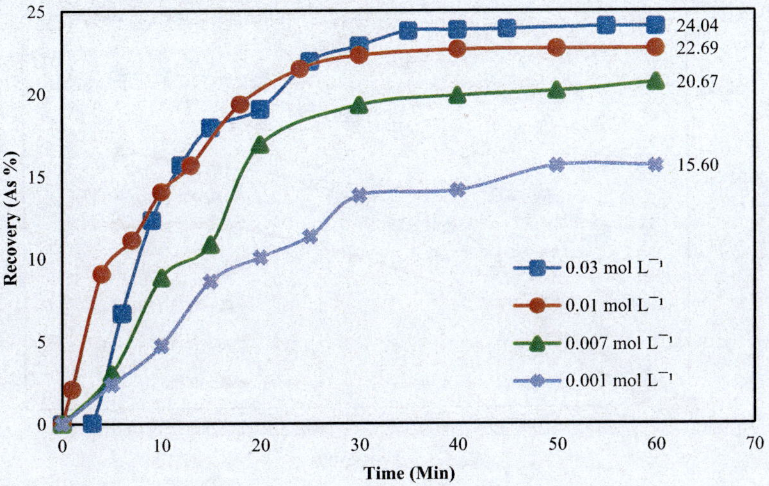

Fig. 6 Recovery percentage of As in the solution ([TU])

Oxalate Concentration

To evaluate the effect of sodium Oxalate concentration on iron dissolution, the following experimental conditions were used: [TU] = 0.03 mol L^{-1}, [OX] = 0.1, 0.0012, 0.0007 and 0 mol L^{-1}, T = 303.15 (K), pH = 2, solution volume = 0.5 L, sample content = 10 g, r_o = 37 μm, ω = 600 min^{-1} y t = 60 min. The graphs in Figs. 7 and 8 show the results obtained from the recovery of Fe and As after the leaching process. When the Oxalate concentration decreases, the leaching curves for both elements present a similar behavior, during the first minutes there is a brief induction period, and later they go through a conversion process, ending in a stability stage.

On the other hand, the recoveries of Fe and As increase by increasing the concentration of Oxalate up to 0.1 mol L^{-1}, however, it was also observed that in the absence of OX and using a [TU] of 0.03 mol L^{-1}, it can dissolve both elements forming stable complexes [20], reaching recoveries of 11.56 and 9.25% respectively under certain analysis conditions.

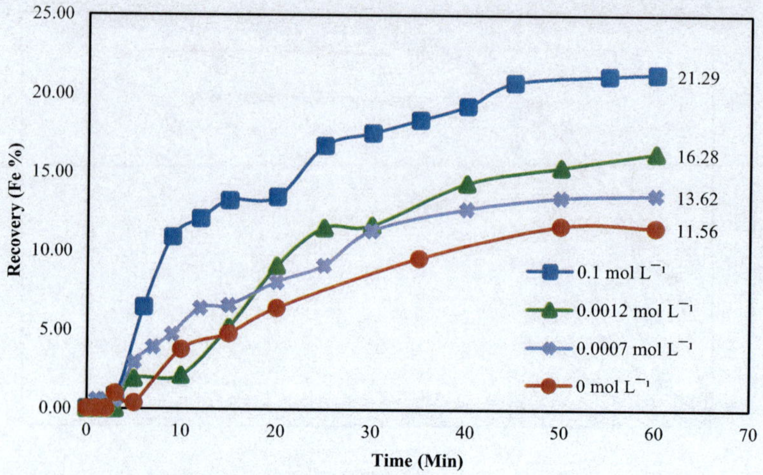

Fig. 7 Fe recovery percentage in the solution ([OX])

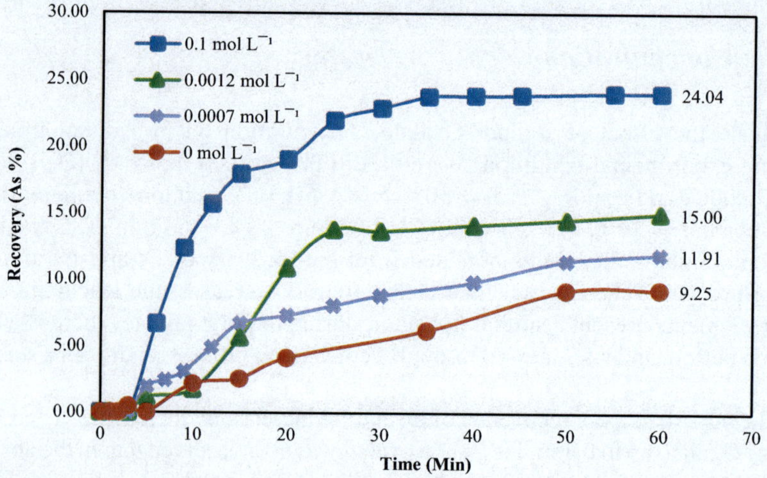

Fig. 8 Recovery percentage of As in the solution ([OX])

Conclusions

Through the characterization techniques used, the presence of arsenopyrite (FeAsS) was confirmed, mineralogical species that contains Fe and As. Additionally, pyrite (FeS_2), siderite ($FeCO_3$), and wustite (FeO), which contain Fe, were found.

According to the results obtained by increasing the concentration of Thiourea and Oxalate, a greater recovery of Fe and As is achieved; it can be seen that the reaction of both elements behaves stable after 30 min, so it is not necessary to continue up to the proposed 60 min of reaction. In addition, it was confirmed that Thiourea acts as a leaching substance for Fe and As.

References

1. Hancock GR, Coulthard TJ (2022) Tailings dams: assessing the long-term erosional stability of valley fill designs. Sci Total Environ 849:157692. https://doi.org/10.1016/j.scitotenv.2022.157692
2. Lorenzo-Tallafigo J, Iglesias-González N, Romero-García A, Mazuelos A, Ramírez del Amo P, Romero R, Carranza F (2022) The reprocessing of hydrometallurgical sulphidic tailings by bioleaching: the extraction of metals and the use of biogenic liquors. Minerals Eng 176:107343. https://doi.org/10.1016/j.mineng.2021.107343
3. Hu X, Yang H, Tan K, Hou S, Cai J, Yuan X, Lan Q, Cao J, Yan S (2022) Treatment and recovery of iron from acid mine drainage: a pilot-scale study. J Environ Chem Eng 10(1):106974. https://doi.org/10.1016/j.jece.2021.106974
4. Berberich J, Li T, Sahle-Demessie E (2019) Biosensors for monitoring water pollutants: a case study with arsenic in groundwater. Sep Sci Technol 11:285–328
5. Halda SK, Tišljar J (2014) Chapter 1—rocks and minerals: introduction to mineralogy and petrology. s.l. Elsevier, pp 1–37 https://doi.org/10.1016/B978-0-12-408133-8.00001-8
6. Habashi F (2001) Arsenic, antimony, and bismuth production: encyclopedia of materials: science and technology, 2nd edn. Pergamon, pp 332–336. https://doi.org/10.1016/B0-08-043152-6/00069-3
7. Merkulova M, Mathon O, Glatzel P, Rovezzi M, Batanova V, Marion P, Boiron M-C, Manceau A (2019) Revealing the chemical form of "Invisible" gold in natural arsenian pyrite and arsenopyrite with high energy-resolution X-ray absorption spectroscopy. ACS Publcations. https://doi.org/10.1021/acsearthspacechem.9b00099
8. Rogozhnikov D, Karimov K, Shoppert A, Dizer O, Naboichenko S (2021) Kinetics and mechanism of arsenopyrite leaching in nitric acid solutions in the presence of pyrite and Fe(III) ions. Hydrometallurgy 199:105525. https://doi.org/10.1016/j.hydromet.2020.105525
9. Manuel JR, Clout JMF (2022) Mineralogical, chemical, and physical metallurgical characteristics of iron ore. In: Lu L (ed) Iron ore, 2nd edn. Woodhead Publishing, pp 59–108
10. Vaughan J, Sulfides D, Encyclopedia of geology, 2nd edn. Academic, pp 395–412
11. Liu Y, Li H, Huang S, An H, Santagata R, Ulgiati S (2020) Environmental and economic-related impact assessment of iron and steel production: a call for shared responsibility in global trade. J Clean Prod 269:122239. https://doi.org/10.1016/j.jclepro.2020.122239
12. Babich A, Senk D (2013) Coal use in iron and steel metallurgy. In: Osborne D (ed) The coal handbook: towards cleaner production. vol 2. Woodhead Publishing, pp 267–311
13. Yoro KO, Daramola MO (2020) CO_2 emission sources, greenhouse gases, and the global warming effect. Adv Carbon Capture Chap 1:3–28
14. Enfermedades, Agencia para Sustancias Tóxicas y el Registro de (2010) Agencia para Sustancias Tóxicas y el Registro de Enfermedades. Agencia para Sustancias Tóxicas y el Registro

de Enfermedades. [En línea] Agencia para Sustancias Tóxicas y el Registro de Enfermedades, [Citado el: 10 de 08 de 2022.] https://www.atsdr.cdc.gov/es/csem/arsenic/efectos_fisiologicos.html

15. Hussain CM, Keçili R (2020) Chapter 1—Environmental pollution and environmental analysis. Elsevier, Modern environmental analysis techniques for pollutants. pp 1–36
16. Organization, World Health (2018) World Health Organization. World Health Organization. [En línea] World Health Organization (WHO), 2021. All Rights Reserved. [Citado el: 10 de 08 de 2022.] https://www.who.int/news-room/fact-sheets/detail/arsenic
17. Bruce A, Fowler C-H, Selene J, Chou Robert L, Costa JM, Chen C-J (2022) Chapter 3–Arsenic. handbook on the toxicology of metals. s.l., Volume II: Specific Metals. Academic
18. Deng S, Gu G, He G, Li L (2018) Catalytic effect of pyrite on the leaching of arsenopyrite in sulfuric acid and acid culture medium. Electrochimica Acta 263:8–16. https://doi.org/10.1016/j.electacta.2018.01.043
19. Mexicano, Servicio Geológico (2021) Panorama Minero del Estado de Hidalgo. Mexicano, Servicio Geológico. [En línea]. [Citado el: 2022 de 08 de 11.] https://www.gob.mx/sgm/es/articulos/consulta-los-panoramas-mineros-estatales.
20. Calla-Choque D, Nava-Alonso F, Fuentes-Aceituno JC (2016) Acid decomposition and thiourea leaching of silver from hazardous jarosite residues: effect of some cations on the stability of the thiourea system. J Hazard Mater 317:440–448. https://doi.org/10.1016/j.jhazmat.2016.05.085

Optimization of Citric Acid Leaching Conditions for Zinc-Containing Electric Furnace Dust Based on Orthogonal Experiment

Zhihui Guo, Chengbo Wu, and Xuefeng Bai

Abstract Based on the harmless and resourceful treatment of zinc-containing electric furnace dust, the orthogonal design of 4 factors and 3 levels was used to explore the effect of citric acid leaching experimental conditions on the leaching rate of Zn, Fe, Pb, and other components. When the acid concentration is 0.8 mol/l, the leaching temperature is 55 °C, the liquid–solid ratio is 8, and the leaching time is 60 min, the leaching rates of Zn and Fe are 82.2 and 13.71%, and the removal rates of Pb, S, and P are 93.08, 86.4, and 73.9%. According to experimental and theoretical research, zinc can be extracted from zinc-containing electric furnace dust, and a leaching residue with Fe content of 48.2% can be produced, which can be returned to the iron and steel process for recycling. The test results can provide reference for the comprehensive development and utilization of zinc-containing electric furnace dust.

Keywords Electric furnace dust · Citric acid · Zinc recovery rate · Leaching

Introduction

The increasing global demand for zinc resources and the depletion of global zinc resources necessitate the development of secondary resources, and electric furnace (EAF) dust is a high-value secondary resource. Electric furnace dust is one of the wastes produced by ironworks. This waste is a by-product produced mainly in electric arc furnaces. Currently, as much as 20–40 kg of EAF dust is generated per ton of steel [1]. Steel produced by this method produces a huge waste by-product. In zinc-containing EAF dust treatment technology, while pyrometallurgical processes are dominant in EAF dust recovery processes, there is a growing contemporary interest in achieving sustainable and greener recovery methods to compensate for pyrometallurgy Toxic heavy metals in the process and environmental issues caused by chlorine and sulfur oxide emissions during electrolysis in traditional hydrometallurgy, selective leaching of metals is seen as one of the most lucrative solutions to problems

Z. Guo · C. Wu (✉) · X. Bai
College of Materials Science and Engineering, Chongqing University, Chongqing 400044, China
e-mail: wuchengbo@cqu.edu.cn

R. G. Reddy et al. (eds.), *New Directions in Mineral Processing, Extractive Metallurgy, Recycling and Waste Minimization*, The Minerals, Metals & Materials Series,
https://doi.org/10.1007/978-3-031-22765-3_33

associated with EAF dust recovery, with the advantage of dissolving Metallic zinc and lead can be recovered as products separately from the solution, while iron-rich slag can be used as a secondary raw material for the electric arc furnace process [2].

Zinc is mainly present in the form of zinc oxide and zinc ferrite. Wet treatment of zinc-containing EAF dust is superior to pyrometallurgical processes; in addition, the hydrometallurgical treatment process has good scalability and offers the possibility of a selective metal recovery process route, inorganic acids such as sulfuric, nitric, and hydrochloric. It has been extensively studied previously for the selective dissolution of metals in EAF dust [3]. Among the three inorganic acids, sulfuric acid is the most studied and promising option [4, 5] Although positive results have been observed, most inorganic acids are expensive. For example, nitric acid and hydrochloric acid require extensive water cleaning of nitrates and chlorides, respectively, and to prevent these substances from entering the furnace. On the other hand, organic acids such as citric acid are highly selective for zinc while containing elements that are fully compatible with electric arc furnaces [6]. Mineral acid leaching carries associated problems and limitations, such as poor selectivity, low zinc recovery, and leaching residue issues. To overcome these problems, a method that is both selective and produces a recyclable iron-rich residue was investigated [7]. In this paper, citric acid is selected for leaching in the leaching stage. Based on the harmless and resourceful treatment of zinc-containing electric furnace dust, the orthogonal design of 4 factors and 3 levels is adopted to explore the effect of citric acid leaching experimental conditions on Zn, Fe, Pb, and other components. The effect of leaching rate,

Experimental

Experimental Materials and Preprocessing

After mixing the dried electric furnace dust, all elements are detected. After the electric furnace dust was analyzed by X-ray fluorescence spectroscopy (XRF), its chemical composition was obtained as shown in Table 1.

It can be seen from Table 1 that Zn, Fe, Ca, Mn, and Pb constitute a large proportion of the zinc-containing electric furnace dust, of which Zn element accounts for 25.56%, and Fe element accounts for 20.04%. After washing, Zn and Fe elements are obtained. The enrichment accounts for 31.01 and 24.64% respectively, and 94% of Cl, 85.8% of K, and 73.6% of S are removed by water washing, which greatly reduces the production cost of the subsequent electrodeposition recovery process of zinc.

Table 1 Chemical analysis of major metal elements in the studied EAF dust samples

Element	The raw EAF dust sample (wt.%)	Washed EAF dust sample (wt.%)
Zn	25.56	31.01
Fe	20.04	24.64
Ca	12.25	11.38
Si	2.35	3.25
Cl	3.76	0.25
K	2.61	0.41
Mn	2.04	2.41
Mg	1.38	1.37
S	0.89	0.26
Pb	1.01	1.22
Al	0.65	0.89
Cr	0.39	0.47
P	0.09	0.12
Cu	0.08	0.1

Phase Composition of Electric Furnace Dust

X-ray diffraction (XRD) was used to detect it to determine its phase composition. The detection results are shown in Fig. 1.

As can be seen from Fig. 1, the preliminary characterization shows that the zinc element in the electric furnace dust mainly exists in the form of zinc ferrite ($ZnFe_2O_4$) and zinc oxide (ZnO), while the iron element is in the form of zinc ferrite ($ZnFe_2O_4$) and tetroxide. It exists in the form of three iron (Fe_3O_4), and $ZnFe_2O_4$

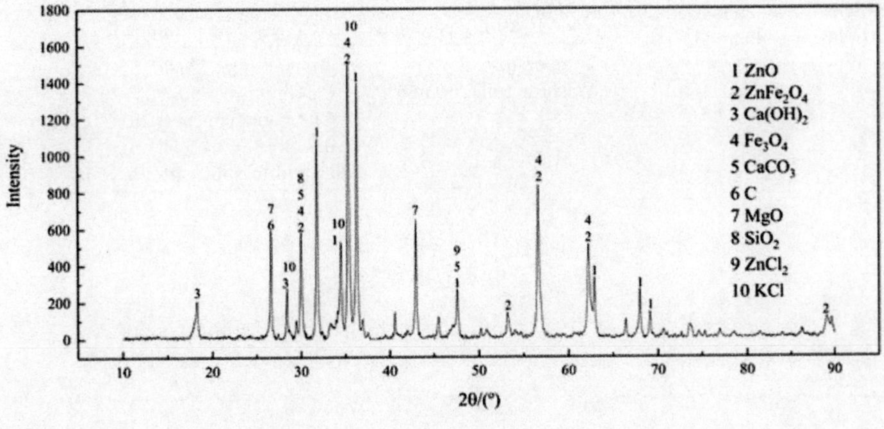

Fig. 1 XRD analysis results of electric furnace dust

is formed by combining ZnO and Fe_2O_3 under the high temperature of electric furnace steelmaking. Pb exists as PbO, Mn exists as MnO, and Cr exists as Cr_2O_3. Other species include alkali metals; potassium, magnesium, sodium, and calcium. Aluminum, silicon, and carbon are also present in large quantities.

Test Methods and Parameters

Before pretreatment, the zinc-containing electric furnace dust was homogenized, dried in an oven at a temperature of 200 °C for 24 h, and a 1000 ml beaker was used to perform water washing and leaching experiments in a water bath, which was used to maintain. The leaching temperature, during the water wash and leaching, was maintained at a constant stirring speed of 750 rpm using a laboratory stirrer. After the leaching experiment, the dried leaching residue was analyzed by XRD.

Put 50 g of zinc-containing electric furnace dust pretreated with water into a beaker, add a certain amount of citric acid to the zinc-containing electric furnace dust to prepare a slurry, and then perform leaching at a set temperature. After the leaching is completed, filter and fully wash the filter residue. Dry back for analysis. The leaching rate of zinc was calculated by the slag meter method (Table 2).

The research is carried out in two stages. The first stage is the pretreatment of zinc-containing electric furnace dust, namely the water washing stage, and the second stage is the experiment of leaching zinc with citric acid as the leaching medium.

It can be seen from Table 1 that KCl and other substances can be removed by water washing pretreatment, which can greatly reduce the content of alkali metals in the leaching residue and chloride ions in the leaching solution, and at the same time enrich Zn, Fe, etc. Elements of value, meeting the requirements of environmental protection and craftsmanship.

Table 2 Parameters used in EAF dust washing and leaching experiments

Water washing parameters	Acid leaching parameters
Temperature: 25 °C Washing time: 60 min	Temperature: 25, 40, 55°C Pickling time: 60, 90, 120 min Citric acid concentration: 0.4, 0.6, 0.8 mol/l Liquid-to-solid ratio: 6, 8, 10

Table 3 Factor level

Factor level	A Citric acid concentration	B Leaching temperature	C Liquid-to-solid ratio	D Leaching time
1	0.4	25	6	60
2	0.6	40	8	90
3	0.8	55	10	120

Results and Discussion

Test Design

Fourth, the zinc-containing electric furnace dust after washing is used as the leaching raw material. The effects of different concentrations of citric acid leaching, leaching temperature, liquid–solid ratio, and leaching time on the zinc leaching rate of zinc-containing electric furnace dust were investigated. The orthogonal design with 4 factors and 3 levels was adopted for the dust of electric furnace containing zinc, and L9 (34) was selected to design the experiment. In the experiment, 750 rmp was used as the stirring speed. Its 4 factors and 3 levels are shown in Table 3.

According to the factor levels in Table 3, 9 experiments were arranged. The experimental conditions and results are shown in Table 4.

Orthogonal Test Results

It can be seen from Table 4 that the zinc leaching rate obtained in Test 9 is the highest at 82.2%, and the weight loss of electric furnace dust is also the largest. The zinc leaching rate obtained in Test 1 is the lowest, which is 60.1%, and the weight loss of electric furnace dust is also the smallest. The weight loss rate of electric furnace dust is proportional to the zinc leaching rate. The zinc leaching rates obtained in tests 2, 4, and 7 were all between 70 and 80%. The zinc leaching rates obtained in tests 3, 5, 6, 8, and 9 were relatively high, ranging from 80 to 90%.

Range Analysis

Since the test adopts the orthogonal test method of four factors and three levels L9 (3^4), each level of each factor participates in three tests, so in the range analysis, the E value is the sum of the three levels of zinc leaching rate, and each level The difference between the maximum value and the minimum value of the E value is the

Table 4 Orthogonal experimental conditions and results

Serial number	Factor				Quality before leaching/g	Quality after leaching/g	Weight loss rate/%	Zinc recovery/%
	Citric acid concentration A	Leaching temperature B	L/S C	Leaching time D				
1	A1	B1	C1	D1	50	31.641	36.7	60.1
2	A1	B2	C2	D2	50	27.222	45.6	73.3
3	A1	B3	C3	D3	50	23.338	53.3	81.3
4	A2	B1	C2	D3	50	25.692	48.6	77.2
5	A2	B2	C3	D1	50	23.009	53.9	81.0
6	A2	B3	C1	D2	50	24.557	50.8	80.3
7	A3	B1	C3	D2	50	23.316	53.4	79.8
8	A3	B2	C1	D3	50	23.168	53.7	81.6
9	A3	B3	C2	D1	50	22.089	55.8	82.2

range value (r). The range analysis results are shown in Table 5, and the relationship between the primary and secondary factors is shown in Fig. 2.

It can be seen from Table 5 and Fig. 2 that for the leaching rate of acid leaching zinc, the range of factors A, B, C, and D is 9.63, 8.9, 6.7, and 5.6%, respectively, and **the range of factor A is the largest., is a significant influencing factor, indicating that** the leaching rate of zinc is closely related to the concentration of citric acid when leaching electric furnace dust by citric acid method. The extreme difference value of **factor B is the second most significant factor, indicating that the leaching rate of nickel** is closely related to the leaching temperature. The extreme difference between factor C and factor D is small, and the liquid–solid ratio and leaching time are the general **influencing factors. The significance of the factors affecting the zinc leaching rate is** A (citric acid concentration) > B (leaching temperature) > C (liquid–solid ratio) > D **(leaching time). The concentration of citric acid was a significant factor affecting the leaching rate of zinc, the leaching temperature was the second most significant factor** affecting the leaching rate of zinc, and other factors had little effect on the leaching

Table 5 Range analysis

E /%	Influencing factors			
	Citric acid concentration A	Leaching temperature B	L/S C	Leaching time D
E1	214.71	217.11	222	223.29
E2	238.5	235.89	232.71	233.4
E3	243.6	243.81	242.1	240.09
E1 average	71.57	72.37	74	74.43
E2 average	79.5	78.63	77.57	77.8
E3 average	81.2	81.27	80.7	80.03
Range r/%	9.63	8.9	6.7	5.6

Fig. 2 Relationship between primary and secondary factors

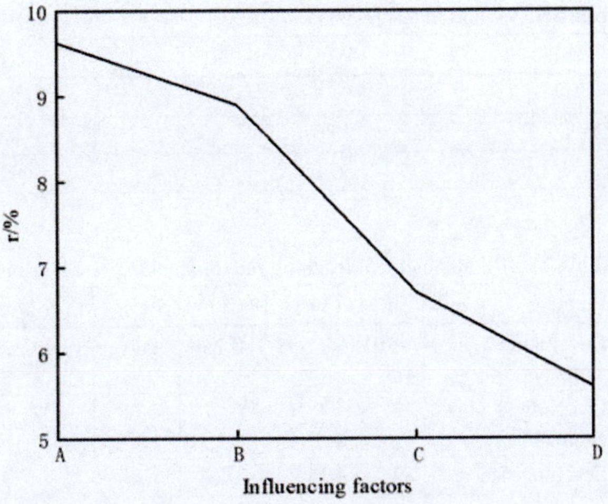

rate of zinc. This result is consistent with the trend of the influence of various factors in Fig. 2. Since the concentration of citric acid and the leaching temperature have a significant effect on the leaching rate of nickel, in order to ensure the leaching rate of zinc, 3 levels and 3 levels should be selected for the two, namely A3 and B3 are the optimal conditions. C-factor and D-factor can choose C3 and D3. That is A3B3C3D3. That is, Test 9 is the optimal group. Therefore, the optimal conditions for leaching zinc-containing electric furnace dust with citric acid are: liquid–solid ratio of 10, leaching temperature of 55 °C, sulfuric acid concentration of 0.8 mol/L, and leaching time of 2 h.

The Verification Test

Comprehensively considered factors such as range analysis results, economic costs and operational indicators, and determined the optimal leaching scheme for zinc in zinc-containing electric furnace dust as follows: citric acid concentration 0.8 mol/L, liquid–solid ratio 8, leaching time 1 h, leaching temperature 55 °C. Three verification tests were carried out under this condition, and the results are shown in Table 7

It can be seen from Table 6 that the effect of zinc leaching rate greater than 82% can be obtained in the three verification tests, and the data is stable, indicating that the test results are repeatable and reliable under the optimal combination of orthogonal test conditions.

However, compared with Test 9, increasing the liquid–solid ratio and leaching time, the leaching rate of zinc only slightly increased, mainly because zinc oxide has basically completely entered the solution, and citric acid is difficult to leach zinc in zinc ferrite, so consider When it comes to production cost, 2 levels of liquid–solid ratio and 1 level of leaching time are selected, namely A3B3C2D1. That is, Test 9

Table 6 Validation test results

Test No	Weight loss rate	Zinc recovery rate
1	55.8	82.2
2	56.0	82.49
3	56.1	82.6

Table 7 XRF and element leaching rate of leaching residues in the optimal group test

Element	Zn	Fe	Cl	S	P	Mn	Pb	Mg	K
Raw material element/%	31.01	11.38	0.25	0.26	0.12	2.41	1.22	1.37	0.41
Leaching residue element/%	12.46	48.13	0	0.08	0.07	4.43	0.19	1.52	0.15
Leaching rate/%	82.2	13.7	100	86.4	73.9	18.8	93.1	51	83.4

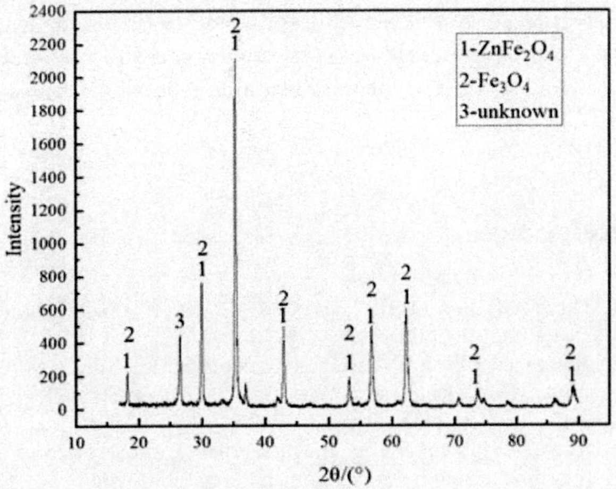

Fig. 3 XRD of leaching residues in the optimal group test

is the optimal group. Therefore, the optimal conditions for leaching zinc-containing electric furnace dust with citric acid are: liquid–solid ratio 8, leaching temperature 55 °C, sulfuric acid concentration 0.8 mol/L, and leaching time 1 h (Fig. 3).

The XRD examination of the zinc-containing electric furnace dust after leaching under optimized conditions, namely Test 9, showed that there were mainly two phases, namely zinc ferrite ($ZnFe_2O_4$) and ferric tetroxide (Fe_3O_4) in the form. After XRF analysis, the Fe content in the slag is 48.2%, which can be returned to the iron and steel process for recycling. Whereas all the zinc in the form of ZnO was transferred into the solution, the iron identified in the solution was derived from zinc ferrite and ferrous oxide slightly decomposed by citric acid.

Conclusion

(1) Use citric acid leaching to pretreat zinc-containing electric furnace dust. Different factors in the leaching process have different influences on the zinc leaching rate. The concentration of citric acid has the greatest influence on the zinc leaching rate, which is the most significant factor. The second influence is the second most significant factor, while the leaching time and the liquid–solid ratio have little influence on the zinc leaching rate, which is an insignificant factor.

(2) Through variance analysis and range analysis of orthogonal test results, it is determined that the optimal acid leaching test conditions are when the concentration of citric acid is 0.8 mol/l, the leaching temperature is 55 °C, the liquid–solid ratio is 8, and the leaching time is 60 min, and Zn, Fe The leaching rate was 82.2% and 13.71%, and the indicators were ideal.

(3) The study revealed the influence of different factors on the zinc leaching rate during the citric acid leaching process of zinc-containing electric furnace dust and optimized the acid leaching test conditions.

References

1. Dutra AJB, Paiva PRP, Tavares LM (2006) Alkaline leaching of zinc from electric arc furnace steel dust. Miner Eng 19(5):478–485
2. Hamuyuni J, Halli P, Tesfaye F, Leikola M, Lundström M (2018) A sustainable methodology for recycling electric arc furnace dust. In: Energy technology 2018 carbon dioxide management and other technologies, pp 233–240
3. .Kaya M, Hussaini S, Kursunoglu S (2020) Critical review on secondary zinc resources and their recycling technologies. Hydrometallurgy 195:105362
4. Havlik T, Turzakova M, Stopic S, Friedrich B (2005) Atmospheric leaching of EAFdust with diluted sulphuric acid. Hydrometallurgy 77(1):41–50
5. Oustadakis P, Tsakiridis PE, Katsiapi A, Agatzini-Leonardou S (2010) Hydrometallurgical process for zinc recovery from electric arc furnace dust (EAFD): Part I: characterization and leaching by diluted sulphuric acid. J Hazard Mater 179(1):1–7
6. Halli P, Hamuyuni J, Revitzer H, Lundström M (2017) Selection of leaching media for metal dissolution from electric arc furnace dust. J Clean Prod 164:265–276
7. Halli P, Hamuyuni J, Leikola M et al (2018) Developing a sustainable solution for recycling electric arc furnace dust via organic acid leaching. Miner Eng 124:1–9

Optimization of the Ratio of Air and Fuel in Ignition Chamber of Sintering Machine

Yapeng Zhang, Wen Pan, Shaoguo Chen, Huaiying Ma, Xiaochen Zhang, Jie Liang, and Sida Ren

Abstract The concept of heating strength of sintering surface is introduced, which is used as the evaluation standard of sintering ignition effect. At the same time, ignition system in Shougang Jingtang sintering machine (550 m^2) was studied and analyzed. The results showed that, with the increase of air–fuel ratio (A/F), the high temperature (above 1000 °C) holding time was extended on the sintering bed surface at the condition of fixed gas flow in ignition chamber. The maximum temperature also rose, which means that the heating strength of sintering bed increased and the ignition effect improved. By increasing air–fuel ratio (A/F), the coke oven gas consumption decreased by 247 Nm^3/h, and the sinter quality index remained stable in Shougang Jingtang Company.

Keywords Ignition system · Air–fuel ratio · Heating intensity · Energy saving · Sinter quality

Introduction

Energy consumption of sintering process accounts for about 12% of the total energy consumption in iron and steel producers, and the ignition accounts for about 10% of the sintering energy consumption [1–3]. Reducing the energy consumption of ignition has a positive significance for energy saving and emission reduction of iron and steel producers.

The purpose of sintering ignition is to supply sufficient heat to the surface layer of the sintering bed, so that the solid fuel on the surface layer can be ignited and burned. The surface blend can be dried, preheated, and sintered under the action of high-temperature flue gas in the igniter [4]. The sintering ignition is a process in

Y. Zhang (✉) · W. Pan · S. Chen · H. Ma · X. Zhang · J. Liang · S. Ren
Research Institute of Iron & Steel, Shougang Group Co., LTD Research Institute of Technology, Beijing 100043, People's Republic of China
e-mail: zyp1989zyp@163.comc

Beijing Key Laboratory of Green Recyclable Process for Iron & Steel Production Technology, Beijing 100043, People's Republic of China

R. G. Reddy et al. (eds.), *New Directions in Mineral Processing, Extractive Metallurgy, Recycling and Waste Minimization*, The Minerals, Metals & Materials Series,
https://doi.org/10.1007/978-3-031-22765-3_34

which the igniter heats the surface layer of the sintering mixture through combustion of the exhaust gas which radiates heat, and the combustion of carbon in the blend jointly heats the surface layer of the sintering bed [5]. The process directly affects the thermal state of sintering, and ultimately affects the sinter quality and energy consumption [6]. So the reasonable control of ignition process and the temperature field of sintering bed is particularly important [7].

The sintering ignition system mainly includes the control and selection of parameters such as gas flow, air flow, air–fuel ratio (A/F), etc. Most of the systems are adjusted and optimized based on the internal temperature or heating strength of the furnace. In order to further study the heat distribution within the sintering bed surface, this paper introduces the heating intensity index to optimize the system of two 550 m^2 sintering machines of Shougang Jingtang.

Research Methods and Theory

Tracking Measurement of Ignition Furnace Temperature

An online furnace temperature tracker was used in the experiment, and K-type thermocouple was used. The temperature measuring range is between 25 °C and 1300 °C.

During the experiment, the black box was placed above the sintering bed surface from the entrance of the ignition furnace to ensure that the measuring point of the thermocouple was in contact with the surface of the bed. The equipment was placed in the ignition furnace during the operation of the sintering machine until removal. The temperature change of the bed surface was recorded in real time with a recording interval of 12 s, and the data was stored in the black box.

Heating Strength of Sintering Bed Surface

The measured sintering bed surface temperature curve in the ignition furnace is shown in Fig. 1. In order to quantitatively measure the ignition effect, the concept of the heating intensity of the bed surface is introduced.

First, we determine the temperature baseline of the high-temperature and mid-high temperature areas of the sintering bed surface. The areas higher than the temperature baseline are the high-temperature zone and the mid-high temperature zone of the sintering bed surface. In the high temperature zone, the solid fuel in the sintering blending combusts completely. In the mid-high temperature zone, the solid fuel in the sintering blend reaches the ignition point and starts to combust. The larger these two areas, the better the ignition effect. The areas of these two zones are defined as the high-temperature absolute heating strength (AS_1) and the mid-high temperature

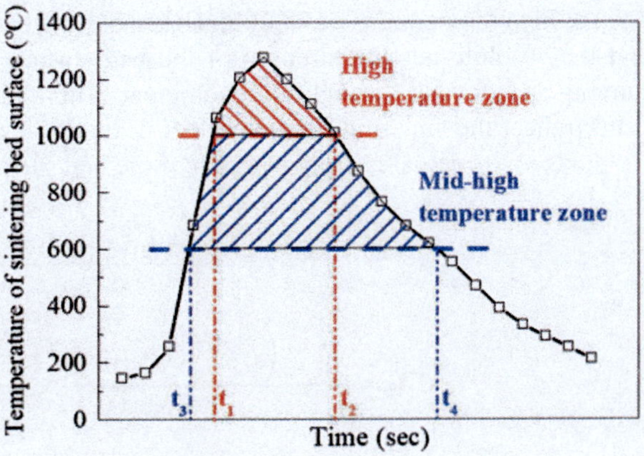

Fig. 1 Temperature curve of the sintering bed in ignition furnace

absolute heating strength (AS_2) of the sintering bed surface, respectively. Equation 1 and Eq. 2 are the calculation formulas:

$$AS_1 = \int_{t1}^{t2} T(t) \times dt \tag{1}$$

$$AS_2 = \int_{t3}^{t4} T(t) \times dt - \int_{t1}^{t2} T(t) \times dt \tag{2}$$

where

AS_1	Absolute heating strength of sintering bed surface at high temperature, °C·s;
AS_2	Absolute heating strength of sintering bed surface at mid-high temperature, °C·s;
t_1	Intersection of the baseline of the high-temperature area and the rising section of the sintering bed surface temperature curve, s;
t_2	Intersection of the baseline of the high-temperature area and the falling section of the sintering bed surface temperature curve, s;
t_3	Intersection of the baseline of the mid-high temperature area and the rising section of the sintering bed surface temperature curve, s;
t_4	Intersection of the baseline of the mid-high temperature area and the falling section of the sintering bed surface temperature curve, s;
t	Time to enter the ignition furnace, s;
$(t_2 - t_1)$	Duration of high temperature, s;
$(t_4 - t_3)$	Duration of mid-high temperature, s.

Because of the influence of running speed of the sintering trolley on the ignition time, the absolute heating intensity of the bed surface cannot fully reflect the ignition effect of the sintering ignition furnace. Therefore, the actual heating strength (CS_1,

CS_2) of high temperature and mid-high temperature of the bed surface is introduced, and the absolute heating strength of the bed surface is modified with the trolley running speed V as a weighting coefficient. This factor is introduced in order to better reflect the sintering ignition effect. Equation 3 and Eq. 4 are the calculation formulas of the actual heating strength of the bed surface respectively.

$$CS_1 = \frac{V \times \int_{t1}^{t2} T(t) \times dt}{60} \tag{3}$$

$$CS_2 = \frac{V \times \left(\int_{t3}^{t4} T(t) \times dt - \int_{t1}^{t2} T(t) \times dt\right)}{60} \tag{4}$$

where

CS_1 Actual heating strength of sintering bed surface at high temperature, °C·m;
CS_2 Actual heating strength of sintering bed surface at mid-high temperature, °C·m;
V Running speed of sintering trolley, m/s.

Results and Discussion

The air–fuel ratio (A/F) of the ignition furnace of sintering machine of Shougang Jingtang was adjusted step by step between 4.5 and 5.5. The bed surface temperature was tested after the adjustment was stable. Tables 1 and 2 are the ignition and sintering parameters under different A/F ratios, and the bed surface heating strength parameters are calculated according to the bed surface temperature curve.

Effect of Air–Fuel Ratio on the Maximum Temperature of Sintering Bed Surface

When the gas flow rate does not change much (air–fuel ratio 5.15, 5.30, and 5.45), the maximum temperature of the bed surface tends to rise with the increase of the

Table 1 Parameters of ignition and sintering

A/F	Ignition furnace	Air flow	Gas flow	Speed of sintering machine
	°C	Nm^3/h	Nm^3/h	m/min
5.45	1234	13,060	2405	2.30
5.30	1205	13,566	2549	2.28
5.15	1222	13,036	2530	2.31
4.30	1207	12,913	3022	2.28

Table 2 Parameters of heating intensity of sintering bed surface

A/F	AS_1	AS_2	CS_1	CS_2	t_2–t_1	t_4–t_3	Maximum temperature of sintering surface
	°C·s	°C·s	°C·m	°C·m	s	s	°C
5.45	12,903	41,654	495	1597	72	136	1295
5.30	11,250	40,020	429	1524	67	135	1265
5.15	8868	39,123	341	1506	63	135	1237
4.30	11,787	40,172	447	1524	69	137	1287

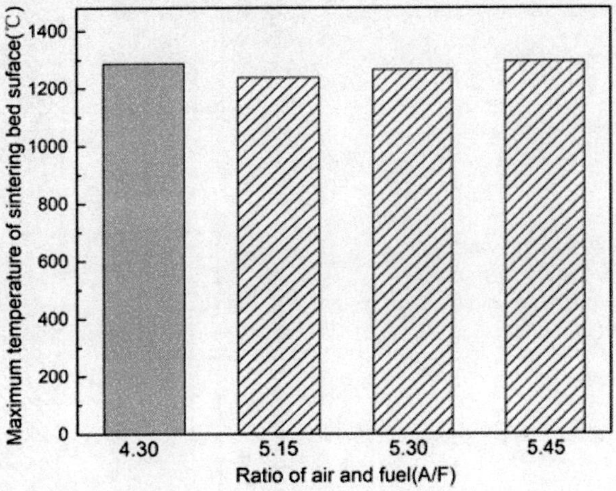

Fig. 2 Relationship between the maximum temperature of sintering surface and the ratio of air and fuel

air–fuel ratio. Especially when the air–fuel ratio is 5.45, the gas flow rate is the lowest, and the maximum temperature of the bed surface is obtained, as shown in Fig. 5.

Under the condition of low air fuel ratio, the maximum temperature of the bed surface could only be ensured by great increase of the gas flow. Taking the air–fuel ratio of 4.30 as an example, to ensure that the maximum temperature of the bed surface is equivalent to that of 5.30, it is necessary to increase the gas flow from 2549 Nm^3/h to 3022 Nm^3/h in operation (Fig. 2).

Effect of Air–Fuel Ratio on High Temperature Duration

As shown in Fig. 3, when the gas flow rate does not change much, the high temperature duration increases linearly with the increase of the air–fuel ratio, which indicates that increasing the air–fuel ratio improves the temperature uniformity in the ignition furnace, thus expanding the high temperature area. Similarly, under the condition of low air–fuel ratio, only by greatly increasing the gas flow can the bed surface obtain a high temperature retention time comparable to that of high-altitude fuel ratio.

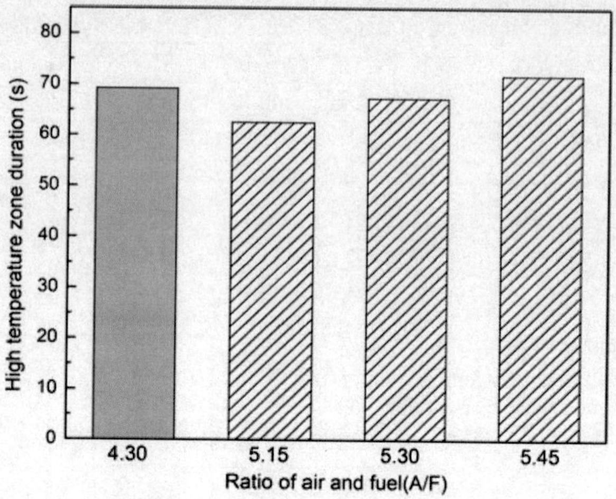

Fig. 3 Relationship between the duration of high temperature and the ratio of air and fuel

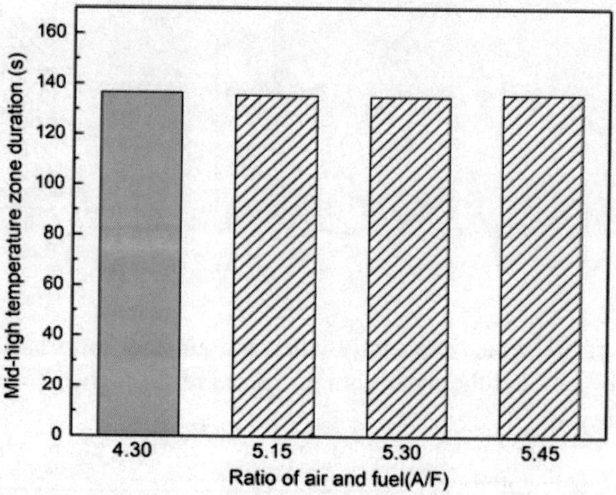

Fig. 4 Relationship between the duration of mid-high temperature and the ratio of air and fuel

It can be seen from Fig. 4 that neither changing the air–fuel ratio nor the gas flow rate has any effect on it, which indicates that the size of the mid-high temperature area of the sintering bed surface in the ignition furnace is fixed within a certain range, and adjusting the ignition system has little effect on this area.

Effect of Air–Fuel Ratio on Heating Strength of Charge Surface

The heating intensity of the sintering bed surface in the ignition furnace integrates the temperature of the bed surface, the duration of the high temperature zone, and other

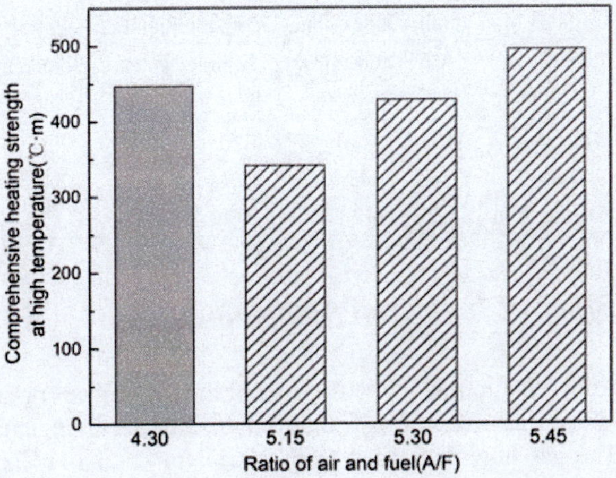

Fig. 5 Relationship between actual heating intensity and the ratio of air and fuel at high temperature

factors, which reflect on the ignition effect. Slight increase of the oxygen content of ignition flue gas can accelerate the combustion process of solid fuel under the igniter [8].

In general, the influence of air–fuel ratio on heating intensity is consistent in Figs. 5 and 6. That is, when the gas flow rate does not change much, the actual heating strength of the charge surface increases with the increase of air–fuel ratio, and the ignition effect improves. In comparison, the air–fuel ratio has a significant impact on the high-temperature actual heating strength, and a relatively small impact on the mid-high temperature actual heating strength.

Under the condition of low air–fuel ratio, the high-temperature heating strength of the bed surface can be obtained by greatly increasing the gas flow rate. That is, by increasing the air–fuel ratio and appropriately reducing the gas flow, the ignition effect can be improved and the consumption of ignition gas can be reduced.

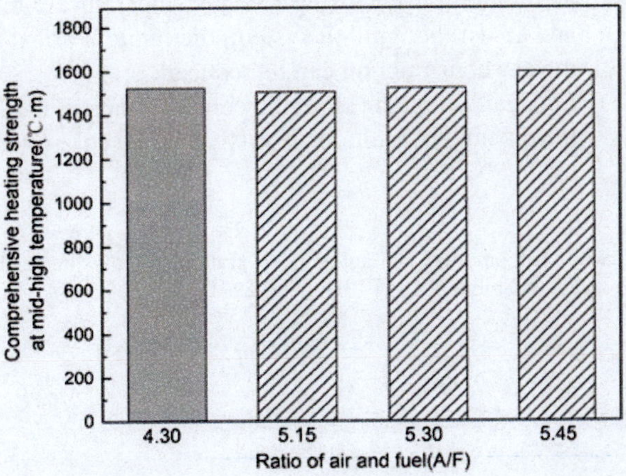

Fig. 6 Relationship between actual heating intensity and the ratio of air and fuel at mid-high temperature

Table 3 Main quality index changes of sintered ore

Item	A/F	Gas flow	Tumbler index	Screening index	Average particle size
		Nm^3/h	%	%	mm
Before	5.1	1661	82.6	2.25	21.27
After	6.37	1414	82.71	2.2	21.52
Comparison	1.27	−247	0.11	−0.05	0.25

Effect of Industrial Application

According to the applied results of the heating strength of sintering bed surface, the ignition air–fuel ratio of sintering machine of Shougang Jingtang was raised by 1.27. The gas flow decreased from 1661 Nm^3/h to 1414 Nm^3/h. The changes of sinter quality before and after implementation are shown in Table 3.

The ignition schedule was optimized based on the heating strength of the bed surface and the fuel characteristics. The sinter drum strength, screening index, and average particle size remained stable, improving the ignition effect and reducing the gas consumption.

Conclusions

(1) When the gas flow rate is constant, with the increase of the air–fuel ratio, the duration of the high-temperature area above 1000 °C on the sintering bed surface and the maximum temperature increase, the high-temperature actual heating strength of the bed surface also increases, and the ignition effect improves.

(2) Taking the high-temperature actual heating intensity of the bed surface as a supplementary parameter to evaluate the ignition effect of the ignition furnace, and combining it with the average temperature of the ignition furnace to optimize and adjust the ignition system, the ignition effect can be further improved and the gas consumption can be reduced.

(3) Shougang Jingtang sintering company increased the air–fuel ratio by increasing the opening of ignition air valve, and the coke oven gas consumption decreased by 247 Nm^3/h.

Acknowledgements The authors are grateful for the financial support of National Key R&D Program of China (2017YFB0304300&2017YFB0304302).

References

1. Hong Z, Jing L (2022) Enlightenment based on green and low carbon development practice of global steel enterprises. Shandong Metall 44(01):9–14
2. Liu C, Ye H, Wei J et al (2021) Analysis of green low carbon technology in iron making process of iron and steel industry. Sintering Pelletizing 46(06):1–7
3. Li X, Xiong C, Jiang X et al (2021) Accelerating the green and low-carbon development of iron and steel by improving self generation. China Metall 31(07):1–5
4. Ding J, Li C (1987) Research and practice of low temperature and low negative pressure ignition system. Sintering Pelletizing 3(20):20–21
5. Ren Z, Wang F (1990) Study on sintering ignition process system. Sintering Pelletizing 24(2):24–25
6. Fudong L (2009) Research on intelligent control method of sintering ignition and combustion based on optimization setting of ignition intensity. Central South University
7. Wanhong L (2013) Study on sintering ignition temperature control and bed layer temperature distribution. Chongqing University, pp 5–6
8. Perikov AH (1992) Optimization of ignition and heat preservation system of sinter layer. Sintering Pelletizing 5(38):24–25

Preparation of Slag Wool Fibers Using Casting Residue Slag Based on Gas Quenching Technology

Wen-feng Gu, Jiang Diao, Jin-An Wang, Wen-Feng Tan, Hong-Yi Li, and Bing Xie

Abstract In order to solve the problems of high cost, high risk, and serious environmental pollution in the current treatment of molten casting residue slag and some of the casting residue slag still cannot be effectively utilized, the molten casting residue slag is treated by gas quenching technology to prepare slag wool fibers in this paper. The mechanism of fiberization of molten casting residue slag in the process of gas quenching is analyzed. The results show that the acidity coefficient of the prepared casting residue slag fibers is 1.25. The total content of SiO_2, Al_2O_3, CaO, and MgO was 93.25%. The slag wool fibers with fibers diameter less than 3 μm is 54.57%. This proportion decreases as the fibers diameter increases. 84.07% of slag wool fibers diameter is less than 7 μm. The results show that the molten casting residue slag can be prepared into the slag wool fibers that meet the application requirements by gas quenching technology.

Keywords Casting residue slag · Gas quenching technology · Slag wool fibers · Fiberization · Acidity coefficient

Introduction

Slag wool fibers are widely used in various industrial and civil buildings due to their light weight, low thermal conductivity, good sound absorption performance, insect resistance, chemical corrosion resistance, and low price [1–3]. The sensible heat of

W. Gu · J. Diao (✉) · J.-A. Wang · W.-F. Tan · H.-Y. Li · B. Xie
College of Materials Science and Engineering, Chongqing University, Chongqing 400044, People's Republic of China
e-mail: diaojiang@163.com

W. Gu · J. Diao · J.-A. Wang · H.-Y. Li · B. Xie
Chongqing Key Laboratory of Vanadium-Titanium Metallurgy and Advanced Materials, Chongqing University, Chongqing 400044, People's Republic of China

J. Diao
College of Materials Science and Engineering, Chongqing University, Shapingba District, No.174 Shazheng Street, Chongqing 400044, China

R. G. Reddy et al. (eds.), *New Directions in Mineral Processing, Extractive Metallurgy, Recycling and Waste Minimization*, The Minerals, Metals & Materials Series,
https://doi.org/10.1007/978-3-031-22765-3_35

molten slag directly injected into fibers can be recycled, which is consistent with the global policy on energy saving and emission reduction [4]. It is an effective way to realize the high added value utilization of slag. The preparation methods of slag wool fibers mainly include injection method, centrifugal method, and centrifugal injection method. It is prepared from molten slag such as blast furnace slag, copper slag, and aluminum slag [5, 6]. Li et al. [7] conducted a pilot study on the modification of blast furnace slag with quartz as a modification agent and the production of slag wool fibers with high acidity coefficient using a high-speed centrifugal method. Four samples of slag wool fibers were successfully prepared. The results showed a 50–70% reduction in total energy consumption and production costs compared to the traditional cupola furnace method. Zhao et al. [8] successfully prepared slag wool fibers by high-speed air injection method using molten blast furnace slag modified by coal ash as raw material. The results show that compared with the traditional production mode, the energy consumption and pollutant emission are reduced by 70 and 90%, respectively.

The casting residue slag is a mixture of molten steel and slag left in the ladle after continuous casting or ingot casting [9, 10]. At present, the casting residue slag in China is often treated by direct hot pouring method, which forms slag blocks. After repeated screening and magnetic separation of the casting residue slag, some of the slag blocks are still discarded and cannot be effectively utilized [11, 12]. At present, the treatment of casting residue slag has the disadvantages of high cost, high risk, and serious environmental pollution. It is neither economical nor environmental friendly.

The casting residue slag has the characteristics of high Al_2O_3 content, which can replace SiO_2 to form the matrix framework in the preparation of slag wool fibers. At the same time, it will increase the viscosity of the fibers and make the mineral wool fibers thinner, which has a positive effect on increasing the chemical stability of slag wool fibers [13, 14]. In addition, the casting residue slag contains a small amount of TiO_2 and MnO. The existence of TiO_2 facilitates the formation of longer fibers and also improves the chemical stability of the fibers, the surface tension, and viscosity of the melt. MnO can reduce the melting temperature of casting residue slag, increase its surface tension, and increase the chemical stability of slag wool fibers [15–18].

The gas quenching technology is safe and efficient for the treatment of molten casting residue slag, simple in process, and low in investment. In addition, the injected gas can be recycled, which is environmentally friendly. It can realize the sensible heat recovery through the gas medium [19]. Therefore, based on the gas quenching technology, the molten casting residue slag was treated to prepare slag wool fibers, and a new way of utilizing the casting residue slag is developed to promote the high added value utilization of its resources.

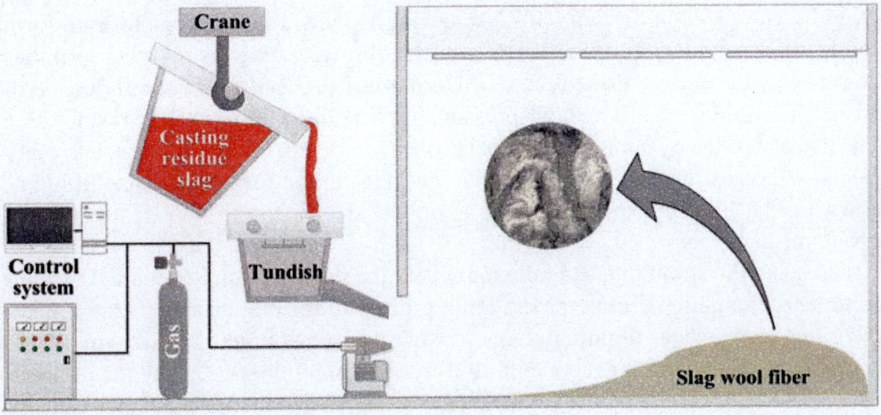

Fig. 1 Schematic of the gas quenching system

Principle of Gas Quenching Technology

In the gas quenching technology, the free falling molten casting residue slag stream is broken by blowing high-speed gas flow and stretched into fibers. Due to the effect of high-pressure gas flow and its viscosity, the slag flow generates melt film and splits into thin streams. The turbulence of the gas flow is intensified, and the thin streams form melt filaments. Under the action of surface tension, the molten slag filaments are stretched, grown, and thinned. Finally, the molten slag filaments are separated from the surface of the melt film by the root contraction. Under the action of gas flow cooling, the molten slag filaments are rapidly solidified and stretched to form slag wool fibers (Fig. 1).

Fibers Forming Mechanism of Casting Residue Slag

In the process of treating molten casting residue slag by gas quenching technology, the molten casting residue slag contacts with the high-pressure gas flow and rapidly forms a melt film on the gas–melt contact surface. On the surface of molten slag melt film, due to Rayleigh–Taylor instability, the surface wavelength has different growth rates. Combined with the pressure fluctuation of low frequency and limited amplitude, the wave below a certain cutoff wavelength will be fully attenuated by its own force. The shear stress at the higher wavelength is greater than the resistance. The wavelength of the melt film is multiplied and has different growth rates, thus causing the instability of the melt film on the gas–melt contact surface. Due to the instability of the melt film and uneven force on the surface, the protrusion is formed. The protrusion is subjected to strong tangential stress along the gas flow direction. Under the action of the gas on the surface of the melt film, the protrusion is rapidly elongated in the semi-solid state and is divided into fibrous form to form melt filaments, as shown in Fig. 2a. Finally, the fibrous casting residue slag solidifies and moves with the gas flow direction. During this process, under the influence of impact and cooling caused by high-pressure gas flow, the molten casting residue slag may also be broken by continuous impact, collision, shear, mutual friction, and other effects of the gas flow, so as to form slag particles in the form of drop splitting, as shown in Fig. 2b. This splitting mode is not conducive to the fibers formation of casting residue slag.

Increasing the strain rate can help to prevent the melt filament from breaking and promote its elongation during the dragging process of the molten slag melt filament. According to the shear thinning characteristic of viscous fluid, the viscosity of the slag melt filament decreases, preventing it from forming fibers. During the drawing process of the melt filament, the gas flow takes away most of the heat, and the viscosity increases sharply, which leads to the enhancement of the deformation resistance of the casting residue slag and the reduction of the fibers forming rate. The existence of the shear thinning characteristic delays the viscosity increasing trend to a certain extent. Therefore, setting a reasonable injection parameter, regulating the gas flow rate, and stabilizing a reasonable viscosity range of molten slag will help to conduct fibers formation and improve the fibers formation efficiency. In the process of gas

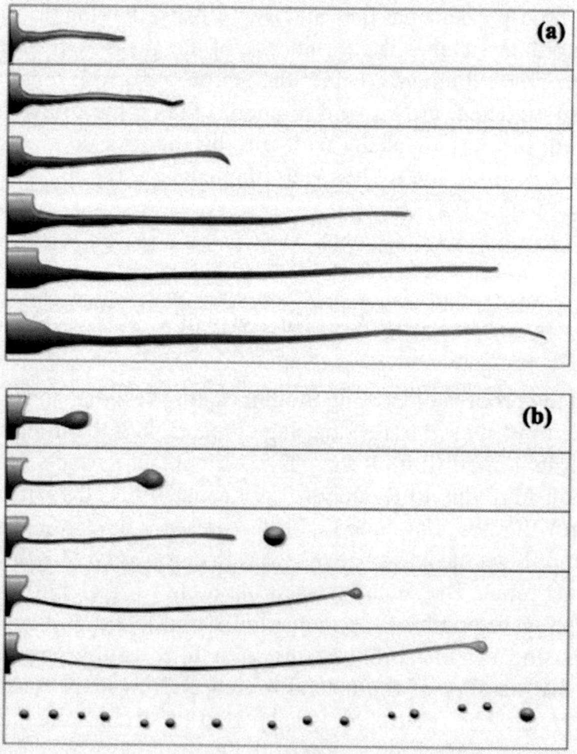

Fig. 2 The morphology of the gas quenching of casting residue slag in molten state: **a** Fibrous splitting; **b** Droplet splitting

quenching, the diameter of the nozzle, the flow rate and pressure of the gas flow, the temperature, viscosity, and surface tension of the casting residue slag in the process of high-speed gas will have a great impact on the fibers forming effect.

The Casting Residue Slag Wool Fibers

The chemical composition of casting residue slag wool fibers is characterized by the content of SiO_2, Al_2O_3, CaO, and MgO greater than 90 wt%. Acidity coefficient is an important parameter to determine whether the casting residue slag can meet the requirements of slag wool fibers raw materials. It refers to the mass ratio of acid oxide and basic oxide contained in the casting residue slag [20], as follows:

$$\text{acidity coefficient} = \frac{m_{Al_2O_3} + m_{SiO_2}}{m_{CaO} + m_{MgO}} \tag{1}$$

When the acidity coefficient is too high, the fibers may be longer, the chemical stability is improved, the service temperature is increased, but casting residue slag is difficult to melt and the fibers are thicker. The chemical composition of slag wool fibers of casting residue slag is shown in Table 1, and the acidity coefficient is 1.25. The total content of SiO_2, Al_2O_3, CaO, and MgO was 93.25%. Al_2O_3 and SiO_2 in slag wool fibers can not only affect their chemical stability, but also help to produce long fibers. However, when the acidity coefficient gradually increases, the viscosity of the slag also increases, resulting in a tendency for the fibers diameter to increase. CaO and MgO belong to basic oxides, which have adverse effects on the formation of solid fibers skeleton, thereby reducing the chemical stability of fibers. If the content of CaO and MgO is reduced, it is beneficial to strengthen the firmness of the fibers skeleton and improve the chemical stability. The raw materials are easier to melt and obtain fine fibers.

The slag wool fibers produced by the gas quenching process of casting residue slag are light yellow and filamentous, as shown in Fig. 3a. In addition, some fine slag balls are mixed in the slag wool fibers. SEM is used to observe the fibers fracture section, and it is found that the fibers of slag wool fibers are solid, as shown in Fig. 3b. It can be seen that the fibers of casting residue slag are smooth and cylindrical. The distribution of their fibers diameters is shown in Fig. 4. The slag wool fibers with diameter less than 3 μm reach 54.57%. With the increase of fibers diameter, its proportion gradually decreased. The GB/T 5480–2008 [21] requires fibers diameter of less than 7 μm, and 84.07% of slag wool fibers prepared in the experiment meets

Table 1 Chemical compositions of casting residue slag wool fibers (wt%)

CaO	TFe	SiO_2	MgO	Al_2O_3	MnO	TiO_2	Other
34.40	4.00	10.95	6.99	40.91	0.86	0.79	1.09

this requirement. Therefore, under the condition of reasonable gas quenching process parameters, slag wool fibers meeting the national standard can be produced by casting residue slag.

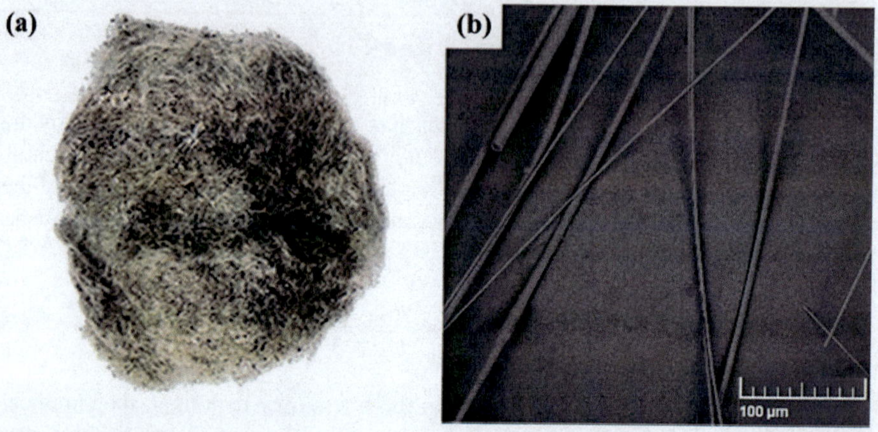

Fig. 3 Topography of casting residue slag wool fibers: **a** macroscopic topography; **b** SEM image

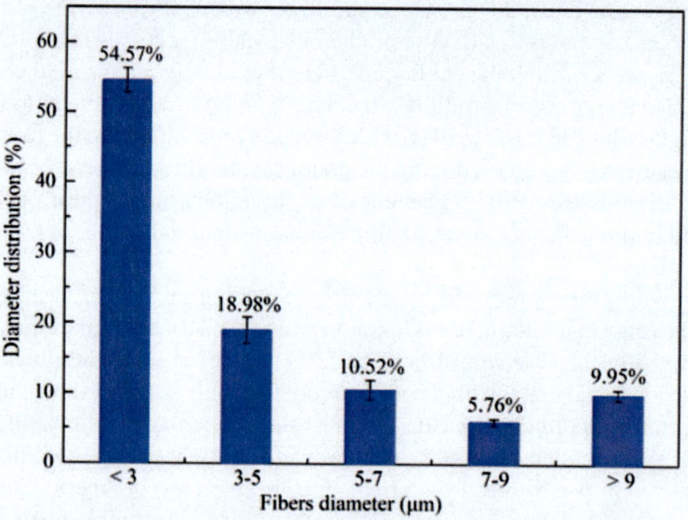

Fig. 4 Diameter distribution of casting residue slag wool fibers

Conclusion

Based on the summary of the mechanism of casting residue slag fibers forming, the molten casting residue slag is processed to prepare slag wool fibers based on the gas quenching technology in this paper. It is clear that the treatment of molten casting residue slag by the gas quenching technology can produce slag wool fibers that meet criteria for their use. However, there are still deficiencies in the preparation of slag wool fibers products from casting residue slag, such as high slag ball content and relatively uneven fibers diameter. The research in the future should also be combined with the mechanism of slag fibers forming and a large number of experiments to achieve the perfection and optimization of technology parameters. Furthermore, the preparation and performance of slag wool fibers products should be further studied. In the industrial production of slag wool fibers, the study of its fibers forming mechanism has a profound impact on the effective saving of energy and the improvement of the quality and productivity of slag wool fibers products.

Acknowledgements This work was supported by the National Natural Science Foundation of China (No.51974047), the Natural Science Foundation of Chongqing, China (cstc2022ycjh-bgzxm0003, cstc2020jcyj-msxmX0043), and the Large Instrument Foundation of Chongqing University (No.202203150207).

References

1. Otakulov BA, Karimova MIQ, Abdullayev IA (2021) Use of mineral wool and its products in the construction of buildings and structures. Sci Progress 2(6):1880–1882
2. Zhukov AD, Smirnova TV, Zelenshchikov DB, Khimich AO (2014) Thermal treatment of the mineral wool mat. Adv Mater Res 838:196–200
3. Krasnovskih MP, Maksimovich NG, Vaisman YI, Ketov AA (2014) Thermal stability of mineral-wool heat-insulating materials. Russ J Appl Chem 87(10):1430–1434
4. Zhang LJ, Yue L, Li ZH, Du PP (2016) Effect of jeting parameters on slag wool's quality. J Mater Metall (China) 15(1):20–32
5. Chen Z, Wang H, Wang M, Liu L, Wang X (2022) Investigation of cooling processes of molten slags to develop multilevel control method for cleaner production in mineral wool. J Clean Prod 339:130548
6. Öhberg I (1987) Technological development of the mineral wool industry in Europe. Ann Occup Hyg 31(4B):529–545
7. Li J, Zhang L, Zhao G, Cang D (2018) Pilot trial of direct modification of molten blast furnace slag and production of high acidity coefficient slag wool fibers. In: Characterization of minerals, metals, and materials 2018. Springer, Cham, pp 113–120
8. Zhao DW, Zhang ZT, Tang XL, Liu LL, Wang XD (2014) Preparation of slag wool by integrated waste-heat recovery and resource recycling of molten blast furnace slags: from fundamental to industrial application. Energies 7(5):3121–3135
9. He JP, Liu XT, Wang CF, Qian L, Jin K (2014) Hot recycling of continuous casting slag from ladles. Res Iron Steel (China) 42(2):55–58
10. Salman M, Cizer Ö, Pontikes Y, Vandewalle L, Blanpain B, Van Balen K (2014) Effect of curing temperatures on the alkali activation of crystalline continuous casting stainless steel slag. Constr Build Mater 71:308–316

11. Salman M, Cizer Ö, Pontikes Y, Snellings R, Blanpain B, Vandewalle L, Van Balen K (2012). Recycling of continuous casting stainless steel slag into construction material. In: Proceedings of 4th international conference on engineering for waste and biomass valorization, pp 935–940
12. Guo DR (2014) Produce on the hot recycling of casting residue from ladles. Steelmaking (China) 30(6):18–21
13. Li ZH, Zhang YJ, Zhang YZ, Du PP, Ren QQ (2018) Effect of content of Al_2O_3 and MgO on crystallization of blast furnace slag during fiber formation. J Central South Univ 25(10):2373–2379
14. Eastes W, Potter RM, Hadley JG (2000) Estimating rock and slag wool fiber dissolution rate from composition. Inhalation Toxicol 12(12):1127–1139
15. Yan ZM, Lv XW, He WC, Xu J (2017) Effect of TiO_2 on the liquid zone and apparent viscosity of SiO_2-CaO-8wt%MgO-14wt% Al_2O_3 system. ISIJ Int 57:31–36
16. Il S, Wang WL, Matsuura H, Tsukihashi F, Min DJ (2012) Influence of TiO_2 on the viscous behavior of calcium silicate melts containing 17 mass% Al_2O_3 and 10 mass % MgO. ISIJ Int 52:158–160
17. Pang ZG, Xing XD, Zheng JL, Du YL, Ren S, Lv M (2021) The effect of TiO_2 on the thermal stability and structure of high acidity slag for mineral wool production. J Non-Cryst Solids 571:121071
18. Wang WL, Dai SF, Zhang TS, Zhang HL, Li ZM, Xie YJ (2021) The effect of CaO on the crystallization properties and viscosity of synthetic silicomanganese waste slag for mineral wool production. J Clean Prod 288:125603
19. Gu WF, Diao J, Liu L, Ge WS, Li HY, Xie B (2021) Investigation of properties of air-quenched steel slag as sandblasting abrasive. JOM 73(10):2995–2999
20. Zhao GZ, Zhang LL, Cang DQ (2019) Fundamental and industrial investigation on preparation of high acidity coefficient steel slag derived slag wool. J Ceram Soc Jpn 127(3):180–185
21. GB/T 5480–2008, Test methods for mineral wool and its products (China)

Rare Earth Reduction: A Technological Overview of State-of-the-Art Technology and Novel Developments

Robert G. Rush and Patrick R. Taylor

Abstract Rare earths are critical to the future of green technology, especially for wind turbines and electric vehicle motors. However, reducing these rare earths into the metals is challenging metallurgically and economically. An overview will be given on the state-of-the-art processes for reducing rare earths from the rare earth oxides to the rare earth metal. The benefits and the challenges of the state-of-the art processes will be presented. An overview of novel processes being developed in the field will also be given. These novel processes range from innovations in metallothermic reduction and molten fluoride electrolysis to ionic liquid electrolysis of rare earth oxides and attempts of reducing rare earth oxides using solid oxide membrane electrolysis. This lays the foundation for direction of future trends on commercial rare earth reduction technology.

Keywords Rare earths · Metallothermic reduction · Electrolytic reduction · Rare earth metal production

Introduction

Rare earth elements (REEs) are comprised of the lanthanides, scandium, and yttrium, which have similar characteristics. Among the REEs, neodymium (Nd) is a key critical material in the manufacturing of neodymium-iron-boron (NdFeB) magnets. These magnets have the highest magnetic strength among other magnets and enable high energy efficiency in energy technologies. These magnets, typically consist of 31% of Nd and 5.5% of another REE, Dysprosium, are used in wind turbines and motors for electric vehicles (EVs) with a total magnet weight of up to 600 kg/MW and 2 kg/vehicle respectively [1]. To describe this critical need in real life, about

R. G. Rush · P. R. Taylor (✉)
Colorado School of Mines, 1500 Illinois Street, Golden, CO 80401, USA
e-mail: prtaylor@mines.edu

R. G. Rush
e-mail: rgrush@mines.edu

R. G. Reddy et al. (eds.), *New Directions in Mineral Processing, Extractive Metallurgy, Recycling and Waste Minimization*, The Minerals, Metals & Materials Series,
https://doi.org/10.1007/978-3-031-22765-3_36

0.75% of the United States' 270 million cars are electric [2]. Therefore, a significant amount of REEs is necessary for the US to migrate to EVs.

The world's rare earth supply chain security is in danger of becoming an economical monopsony (a single buyer). China's mine production in the last decade had decreased from a 97% share in 2012 to a 57% share in 2020. The rare earth mine supply has become increasingly diversified through new production in the USA and Myanmar [3, 4]. Despite this, Chinese authorities are pursuing centralization of the domestic and international REEs supply chain by consolidating the REEs sector in China and promoting international standardization of REEs products [5]. Moreover, China is using their current wealth to invest in their downstream supply chain, which is almost 70% of worldwide rare earth consumption as of 2020 [3].

The important segment of the downstream supply chain involves the reduction of the individual rare earth oxides into the rare earth metals. This is not only very energy intensive but also difficult, since rare earth oxides are extremely stable to the point that they cannot be reduced in normal reducing conditions [6]. This section of the supply chain also has potential environmental hazards due to the production of toxic perfluorocarbon gases (PFC), which is equivalent to 7,390–12,200 CO_2-equivalents [7]. China possesses about 90% of the rare earth reduction industry, which promotes a worldwide production 9 million tons of PFC gases per year of rare earth electrolysis in a medium-emission scenario [8]. It is a strategic importance in western countries to secure domestic downstream rare earth production, especially rare earth reduction. This should be pursued to meet the increasing demand of green technology, to avoid an economic monopsony, and to promote a greener rare earth production.

Brief Background

REEs are not "rare" in that they are not rare in the Earth's crust, but "rare" in that economically viable mineable concentrations of REEs are less common. Although there're many ores that bear REEs, 95% of the REEs ores in the world are bastnasite, monazite, and xenotime [6, 9]. There are REEs ores that can also be found as mixtures in rock formations such as basalts, granites, gneisses, shales, and silicate rocks. REEs are widely distributed in low concentrations. This, along with difficulties of obtaining the ore with relative ease and the feasibility of extracting the metal from the ore, contributes to categorizing these metals as "rare" [6].

Mining REEs can be relatively complex process. A schematic of processing REEs is seen in Fig. 1. REEs are either extracted from hard rock deposits or placer deposits. In hard rock deposits, the dominant source of REEs are carbonates followed by pegmatites. Monazite is contained in sand-based placer deposits. Open-pit mines extract hard rock deposits through drilling and blasting the ore. The ore is then loaded and hauled to the mill. Mines that exploit placer deposits extract REEs as a byproduct. These deposits are extracted through dredging. The extracted ore undergoes multiple beneficiations to increase the REEs concentration in the ore. This involves separation of the ore and the gangue in the concentration through gravity separation, magnetic

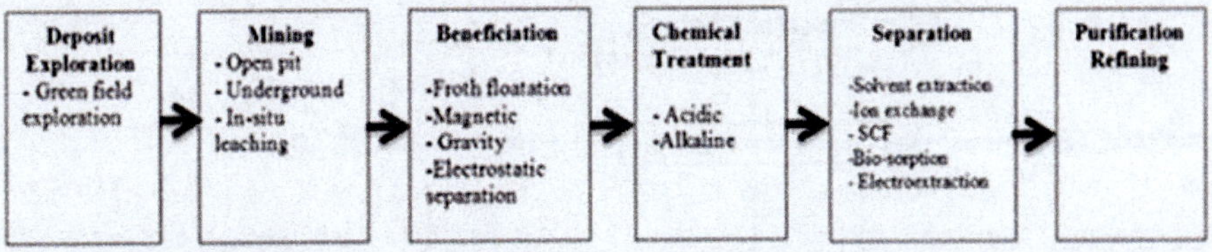

Fig. 1 Schematic of processing REE [6]

separation, and froth flotation. After the ore undergoes beneficiation, chemical treatments are applied to the beneficiated ore to upgrade the rare earth oxide (REO) concentrate of the ore and to remove unwanted impurities from the ore. A sulfuric acid treatment can be used to selectively solubilize thorium, a radioactive material, from the concentrate. Alkali treatments can also be used to recover phosphates as a byproduct. A hydrochloric acid treatment upgrades a bastnasite concentrate to 70% REOs and removes calcium, thorium, and strontium carbonates. An upgraded REO concentrate can also be calcined to further upgrade the concentrate to 85–90% [6].

The mixed chemically processed REO is separated into its individual oxides through multiple processes. Cerium, praseodymium, and terbium are removed through selective oxidation [6]. Samarium, europium, and ytterbium can also be separated by reducing the oxides to a divalent state. The rest of the mixed earths undergo solvent extraction. Solvent extraction involves dividing groups of REOs or individual REOs using two immiscible liquid phases. One of the liquid phases is a non-aqueous organic phase while the other liquid phase is an aqueous solution. When the liquid phases come in contact, the desired rare earths are distributed between the two liquid phases [6]. This process can be seen in Fig. 2 which was used by Indian rare earths. When purities greater than 99.9% are desired, Ion exchange is used [6].

Separated are earth oxides are the typical product of multiple rare earth processing operations. These products are the reactants for metal conversion in most rare earth reduction operations. The oxides are extremely stable, and consequently their reduction reaction becomes very difficult to perform [6]. This difficulty is even intensified by the REOs and the REEs physical properties such as their melting points. Currently, metallothermic reduction and molten fluoride electrolysis are the state-of-the-art processes used to reduce these oxides.

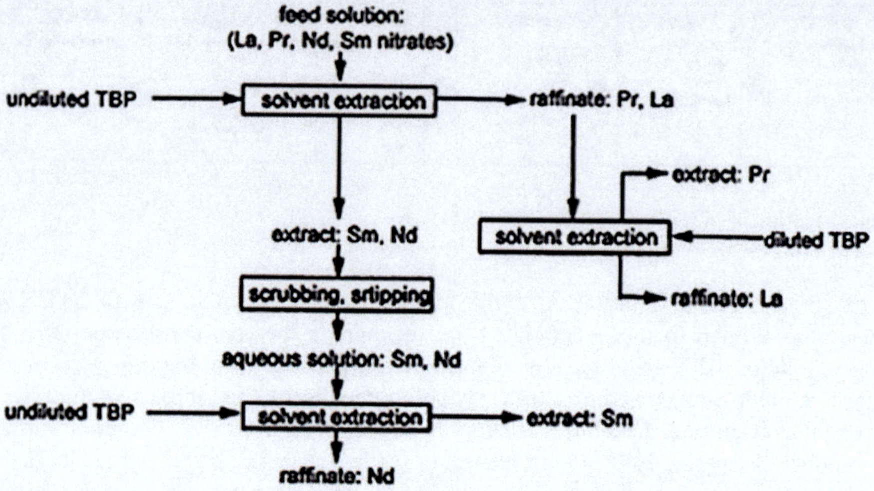

Fig. 2 Schematic of solvent extraction to separate individual REOs [6]

Metallothermic Reduction

Metallothermic reduction, which involves the reduction of a metal compound by another metal, is one of the older technologies used for reducing rare earths and the widely used reduction process for metal in general [10]. The fundamental thermodynamics of metallothermic reduction reactions are expressed in the form shown below [6]:

$$M(I)X_n + iM(II) = M(I) + iM(II)X_{n/i} \tag{1}$$

where M(I) is metal produced, X is the anion on the desired metal to be reduced, which is a halide or an oxide, and i and n are the stoichiometric coefficients. Though a reaction's feasibility is preliminarily determined by the free energies of formation, the products and reactants' boiling point, melting point, vapor pressure, density, viscosity, chemical reactivity, and alloying behavior are important to the success of the reaction in a metallurgical and economic standpoint. RE metallothermic reduction struggles to achieve these economic characteristics. The reactions usually involve dispensing the metal as a powder. The reaction usually occurs at low pressures to make to help avert additional refining. Metallothermic reduction may allow production of REEs of higher purity than electrolysis because the metal product from electrolysis can be contaminated from electrodes and trapped electrolytes [11, 12].

Reduction Process

There are various ways at which fluoride-based metallothermic reduction can be executed. A $CaCl_2$-based process will be described in detail. The flowsheet of the process is summarized in Fig. 3 [13]. The feed involves a two-phase melt at 850 °C and a charge of rare earth fluoride. The first melt is of calcium chloride while the other melt is a higher density metal alloy containing rare earths and iron. The initial charge of $CaCl_2$ is a variable in this process that the salt product would be about 1.5–2 times the theoretical weight of the rare earth metal produce. This low ratio is preferred for the reaction to facilitate a rapid reduction reaction and separation of a clean product. The heavy iron-rare earth charge can also be adjusted to add desired alloys to form rare earth products, such as magnets. The two-phase melt and the charge are mixed with calcium metal and are fed into the reduction chamber. Once the reduction chamber is charged, the reaction proceeds and the charge is mechanically stirred for no more than 45 min to complete the reaction. Then the stir rate is slowed or ceased to allow the heavier metal phase, containing the newly produced rare earth, to separate and gather below the salt layer of the reduction chamber. The reaction produces a rare earth-iron ingot and a mixture of calcium fluoride and calcium chloride, which the calcium chloride is dried and recycled into the reduction reaction and the calcium fluoride is used to prepare rare earth nitrates, which is the initial feed into the reaction, into rare earth fluorides.

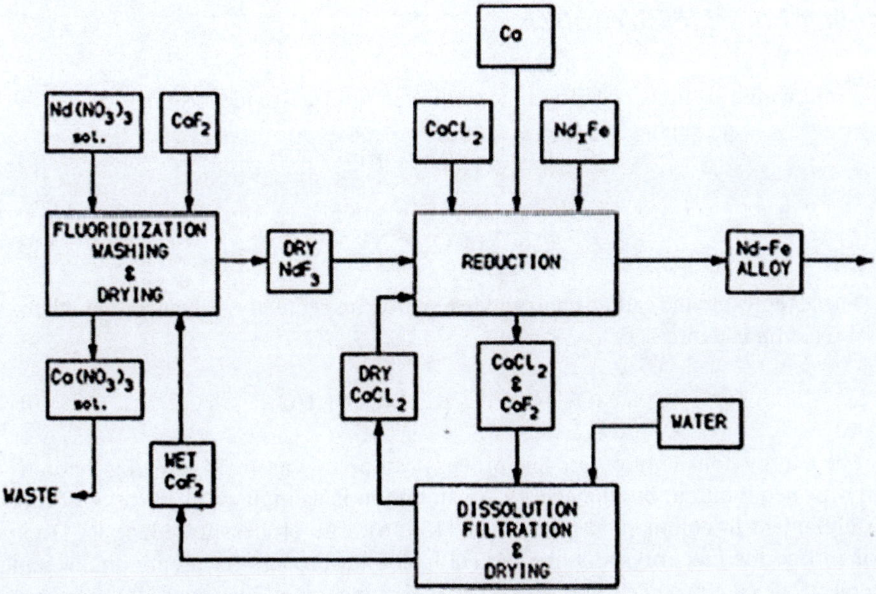

Fig. 3 General flowsheet of the chloride-fluoride reaction [13]

Advantages and Disadvantages of Metallothermic Reduction

The advantages of metallothermic reduction are that the process produces REEs with high purity and has lower energy costs [11]. Despite these advantages of metallothermic reduction, the process usually requires vacuum distillation [6, 14]. Metallothermic reduction processes are batch processes, hence the low productivity [6, 14]. Nevertheless, due to the process' ability to have products with high purity, metallothermic reduction has an important niche in the global market [11, 14].

Electrolytic Reduction

Metallothermic processes, along with other chemical reduction processes, fundamentally rely on the availability of its reducing agents to reduce the rare earth compounds to the metals and are mostly batch processes. However, molten salt electrolytic reduction is not reliant on chemical stability considerations and is continuous processes [5, 8]. This process of reducing metals from their salts involves depositing a mixture of salts into a molten bath. Rare earth metals are commercially produced at large quantities throughout the world through this process.

Molten Fluoride Electrolysis

The following reactions occur within the molten fluoride-oxide system [15]:

$$RE^{3+} + 3e^{-} = RE \tag{2}$$

$$3O^{2-} + 2C = CO + CO_2 + 6e^{-} \tag{3}$$

These represent the cathode and anode reaction respectively. When combined, the total reaction is expressed:

$$RE_2O_3 + 2C = 2RE + CO + CO_2 \tag{4}$$

The energy consumption for this process can be low as 1.334 kWh/kg theoretically for neodymium; but the energy consumption is as high as 11 kWh/kg of rare earth element in commercial production [15, 16]. The process to reduce REO via molten fluoride electrolysis is like the Hall–Heroult process for aluminum, except the electrodes are in a vertical position. In one arrangement, the electrodes are about 2 cm in diameter. A tungsten cathode and a carbon anode are used to conduct the electrolysis. A graphite ring anode surrounds the cathode and reacts with the ions to produce CO and CO_2. The anode–cathode distance (ACD) is determined by the cell set up and is non-adjustable. The electrolyte is 85 wt% RE fluoride and 15 wt% lithium fluoride on average [16, 17]. A crucible is included at the bottom of the cell to capture the metal and is periodically tapped out of the cell during operation. The system can be run at temperatures as high as 1050 °C to deposit neodymium. Iron cathodes are used for higher melting REEs, which deposit a low melting heavy RE-Fe alloy [6]. Table 1 and Fig. 4 show the model and the processing features of a 3 kA version of a molten fluoride electrolytic cell. The typical cell voltage and operational current of these systems are 8–12 V and 2.5–3 kA respectively, with a current efficiency between 70 and 75% [15, 17]. In the industrial scale, most molten fluoride cells utilize 5–25 kA currents to obtain higher current efficiency and reduce energy consumption [15].

In one arrangement, the chemicals may react to air, the whole process commonly enclosed in a gas-tight pressure chamber made of steel as seen in Fig. 5. Glove and view ports are included in the chamber for the operator. The cell is also attached to pumps, which removes air and the product gases out of the system. A refrigeration unit is included in the chamber to control the gas temperature increase to less than 50 °C [6]. A controlled flow of cool air or helium through a copper coil is located beneath the cell, which is used to maintain a desire temperature for tapping and the frozen electrolyte skull in the collection zone of the system [6].

The standard procedure begins by vacuum drying the solvent phase electrolyte and depositing this solvent into the cell chamber. Once the chamber is depressurized and backfilled with noble gases, the electrodes are lowered to solvent to melt the

Table 1 Characteristics of an average 3 kA electrolytic cell, adapted from [18]

	3 kA cell (model)
Cell voltage	8.2 V
Cell current	2,200 A
Anode current distance	7.5 cm
Active anode area	1,600 cm^2
Anode current density	1.38 A/cm^2
Reverse decomposition voltage	1.32 V
Ohmic resistance anode	0.061 mΩ
Ohmic voltage drop anode	0.134 V
Ohmic resistance cathode	0.07 mΩ
Ohmic voltage drop cathode	0.155 V
Ohmic resistance electrolyte	2.71 mΩ
Ohmic voltage drop electrolyte	5.35 V
Anodic concentration corrected charge transfer overvoltage	1.08 V
Anodic product diffusion overvoltage	0.13 V
Cathodic overvoltage	0.1 V
Total cell resistance	3.7 mΩ

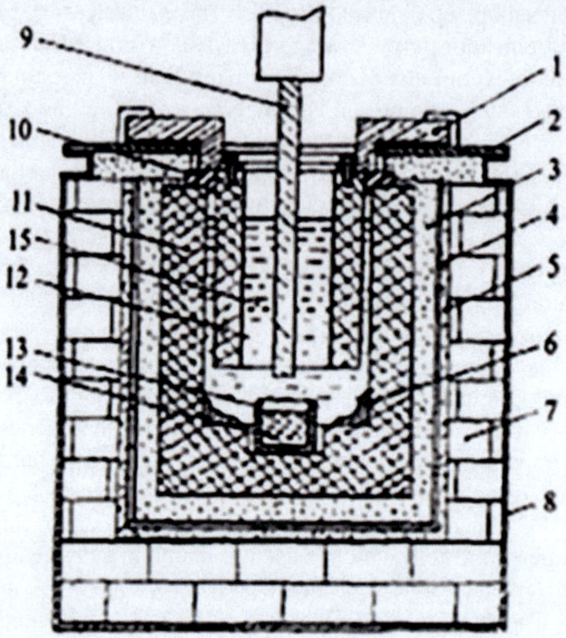

Fig. 4 Sketch of a 3 kA rare earth electrolysis cell, labelled as the following: 1—anode conductive plate, 2—cover, 3—thermal insulation layer, 4—iron layer, 5—asbestos fiber layer, 6—electrolyte crust, 7—insulating bricks, 8—shell, 9—W cathode, 10—corundum gasket, 11—graphite crucible, 12—graphite anode, 13—Mo crucible, 14—liquid Nd metal, 15—liquid electrolyte [17]

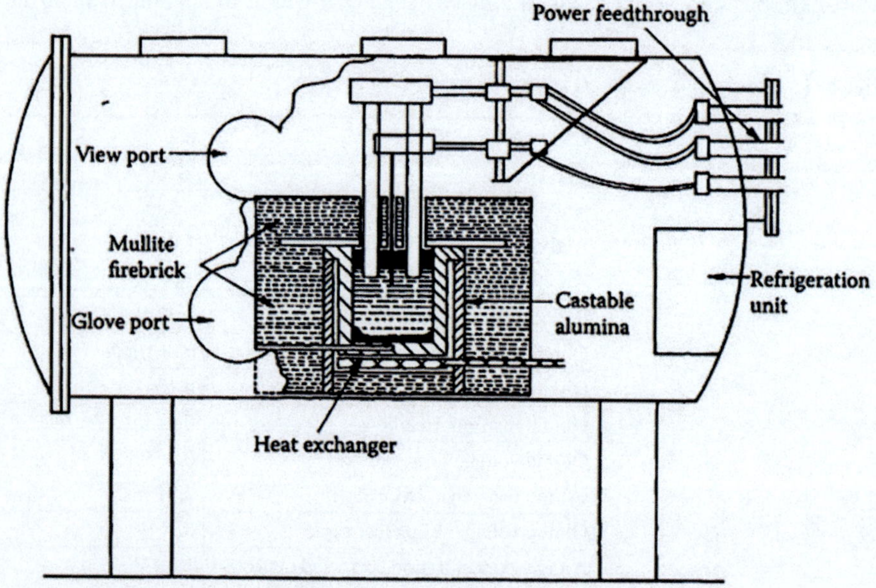

Fig. 5 Protective atmosphere chamber used for RE electrolysis [6]

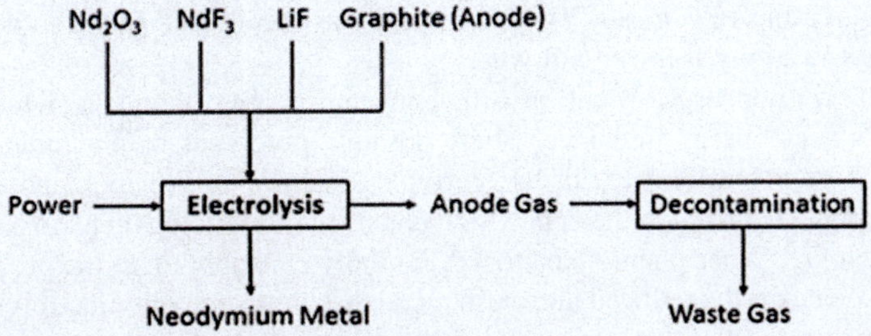

Fig. 6 Process diagram of Nd_2O_3 molten fluoride electrolysis [6]

solvent and heat it to temperatures sufficient for electrolysis for the desired metal. At this time the "skull" or "frozen bath" is formed electrolytes on the inner surface of the vessel. This is used to prevent the reduced rare earth metal to contact with the graphite crucible [6]. This part of the process is critical in that the geometry of the electrode and the cell are designed to specifically maintain this electrolytic skull. Once the cell is ready for electrolysis, the oxides are fed into the cell and the cell switches from AC to DC power to undergo electrolysis. Some rare earths are fed to produce a continuous reduction lasting for days, with occasional anode changes, tapings, and recharges. Neodymium has difficulty in continuous feeding because of its lower solubility. Hence, the oxide is intermittently fed into the cell, to control the system at the cost of making the system semi-continuous. These cells are hand fed manually and manually tapped with a titanium spoon [17]. The tapped metal in some designs can also flow through a pipe into an inert casting chamber to be produced as an ingot [6]. A simplified diagram is displayed in Fig. 6.

Challenges

The rare earth metal produced from the processes is in liquid form. Many issues persist in this kind of extraction. The first issue concerning this type of extraction is that REEs metals themselves dissolve in the metals of their halides, which consumes the metal product. This not only decreases the cell productivity by decreasing the yield, but the dissolved metal itself also imparts electronic conductivity, which corrupts the fluoride to act as a proper electrolyte on the system. The metal solubility in molten salts also increases with temperature. Therefore, care is required to maintain the reduction cell at a consistent temperature.

Like the Hall–Heroult process, when the molten fluoride electrolytic process low on oxides in the system, the fluorides began reacting to the graphite anode to create instead PFCs, which ranges from 7,390 to 12,200 CO_2 equivalents [17]. The high voltage, high current densities, and lack of automation are factors perpetuating even more PFC production. In some systems, ineffective off-gas suction hoods are employed to capture the processes off gasses, but there's no available technological solution to treating PFC gases [7]. However, it was shown that PFC can be emitted at low concentrations at lower cell voltages [7].

The cell experiences a low current efficiency and a high cell voltage, which makes the process very energy intensive. Many possible processes or phenomena in the process that may contribute to this low current efficiency. One phenomenon involves a back reaction occurring in the cell, which results in re-reducing metal to oxide back into the melt [7]. Other phenomena that potentially contributing to the process' low current efficiency is the reduced rare earth metal dissolving into the electrolyte, which would decrease the recoverable metal yield and cause electric conduction through the electrolyte. The system could also be also reducing lithium and calcium in the system [19]. Molten fluoride's low oxide solubility of about 2–3% is an issue with the system, as well as the slow oxide dissolution within the cell [6]. Nevertheless, since molten fluoride electrolysis is a continuous process.

Rare Earth Reduction—Novel Processes

As discussed in the previous sections concerning metallothermic reduction and electrolytic reduction, there are many issues present in the processes. Concerning metallothermic reduction, most processes are batch processes that have low productivity, require distillation afterwards, and the reductant materials are expensive. Molten fluoride electrolysis has many problems such as PFC generation, low current efficiency, low solubility, and high energy intensity. Therefore, a variety of new technologies and innovations are being examined to improve the economic and environmental efficiencies of the current systems, or to even create new processes to reduce rare earth oxides.

Molten Sodium Reduction

There's been attempts relating to reducing rare earth metals using calcium and sodium mixtures in metallothermic reduction. This involves using an amount of sodium with calcium in reducing process or even reducing the rare earth oxide completely with sodium. One of the most known successful processes using calcium-sodium reduction is a process devised by Sharma [6, 20, 21]. The overall reaction in the system is expressed:

$$Nd_2O_3 + 3CaCl_2 + 6Na \rightarrow 2Nd + 3CaO + 6NaCl \quad (5)$$

The reaction takes place at 710–725 °C. The process takes place in a helium atmosphere dry box apparatus, with a furnace. A schematic diagram of the reaction vessel is shown in Fig. 7. A zinc-rare earth or an iron-rare earth eutectic alloy pool to recover the rare earth from the salt phase was prepared by inserting the eutectic amounts of RE chunks and either iron or zinc in a tantalum reaction vessel. Then the furnace is heated to 800 °C to melt the eutectic alloy. This eutectic alloy is used to collect the reacted RE alloy so that the RE would be liquid at lower temperatures. For example, the melting points of the eutectic compositions of Nd-Zn and Nd-Fe are 630 °C and 640 °C respectively. After the eutectic alloy is molten, the temperature is lowered to about 700–725 °C, and additional Zn or Fe are added into the molten pool along with the $CaCl_2$ and NaCl salts. The salt composition of this is 70 wt% $CaCl_2$ and 30 wt% NaCl, though this salt can also be 95% $CaCl_2$ and 5% KCl melt (in this case pure sodium is added to the melt instead). When the desired operating temperature is achieved, then a preweighted amount of Nd_2O_3 is added into the melt. Then the stirrer is introduced, which is used to mix the oxide into the melt to assist in the reduction processed. Then the reaction melt was fast-stirred for an assigned time interval then slow-stirred for additional amount of time, both of which are necessary for good yields. During the process, additional salts and Nd_2O_3 can be added into the melt. Once the alloying time is passed, the stirrer motor is shut off and removed from the melt. Then the reaction vessel is pulled from the furnace well and quenched. Radiographs are taken of the reaction vessel, to indicate a clear, sharp interface between the salt and metal; after this the salts are washed out the vessel with warm-running water, and a RE alloy is obtained to be analyzed for chemical composition. The process obtained 95% yields through salt/neodymium ratios as low as 3.5. Although the stirrer position has little effect on the yield, the stirring pattern and increasing the temperature increase the yield.

Despite the benefits of using a cheaper reductant metal in the process, there are a few difficulties related to this process, and the first difficulty reflects having the proper amount of sodium in the reaction. Sodium metal taken is more than the stoichiometric amount necessary to compensate for the loss of sodium metal by vaporization. However, if there is too much access of sodium metal, a Na-Ca melt of higher sodium concentration will form, which will interrupt the NdOCl reaction in the salt. Therefore, care must be taken to have a higher Ca concentration in the Ca-Na melt to scavenge the rare earth oxychlorides of the system. Second of all, the system is a batch system, which makes the process difficult to use commercially.

The Nanoscale Powder's process involves the reducing neodymium chlorides with sodium metal to get Nd metal and salt [22, 23]. This is best expressed in Eq. 6, which is conducted between 150 and 700 °C:

$$NdCl_3(l) + 3Na\ (l) = Nd(s) + 3NaCl(s) \quad (6)$$

Other metals are reduced with the neodymium to create a rare earth magnet alloy:

$$2NdCl_3(l) + 14FeCl_2(l) + BCl_2(l) + 37Na(l) = Nd_2Fe_{14}B(s) + 37NaCl(s) \quad (7)$$

A schematic of the reduction vessel is displayed at Fig. 8. Like Sharma's reduction vessel, it contains a mixer to assist in the reaction. Also, a feed tube (140 in Fig. 9) is included in the reaction, which not only allows for controlled feeding during the reaction, but also allows for the feed to be deposited as a liquid into the reduction

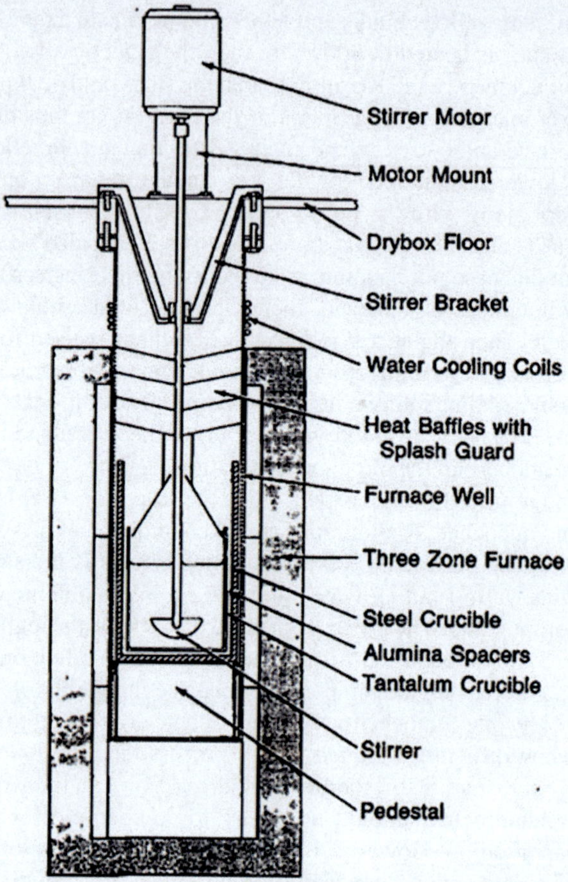

Fig. 7 Reaction apparatus for Sharma's process [21]

vessel at (130 in Fig. 8). The curved inside is the vortex the mixer would create to precipitate reaction. Lastly, (170 in Fig. 8) is a port with a freeze plugging valve, which can empty the contents of the reactor without tipping the apparatus. The reaction vessel is also within an airtight globe box (180 in Fig. 9), which permits the operator to manipulate in the vessel while managing the reaction environment; the box is filled with argon gas.

First the sodium is loaded into the furnace in a 5:1 stoichiometric excess ratio to the reactants, which will be added later. Then the system is heated to 200 °C, at which the sodium would transform into a liquid, since sodium's melting point is about 97.2 °C. Once the sodium is melted, then the vessel is heated to the desired reaction temperature mentioned earlier, and the mixer is activated to create a vortex within the molten sodium bath at speeds between 60 and 2000 rpm. Then the metal chlorides are mixed into a stoichiometric ratio for rare earth magnets and are fed into the system as liquids at small intervals. Once the metal chlorides are added to make the system a 5:1 salt-metal ratio. Once the reaction is complete, the system

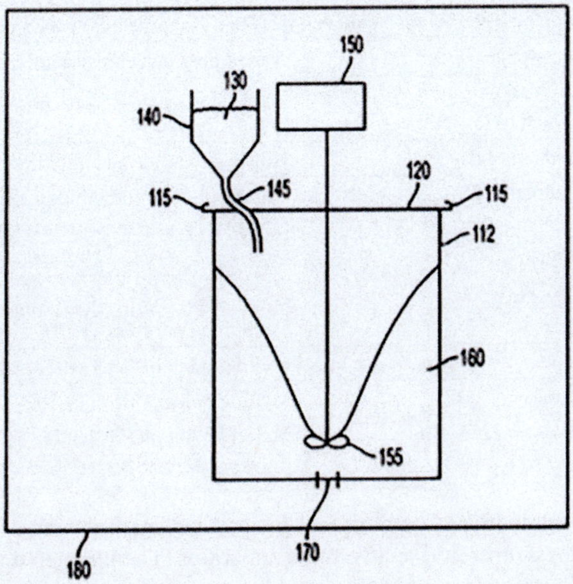

Fig. 8 Reaction apparatus for Nanoscale Powder's process [22]

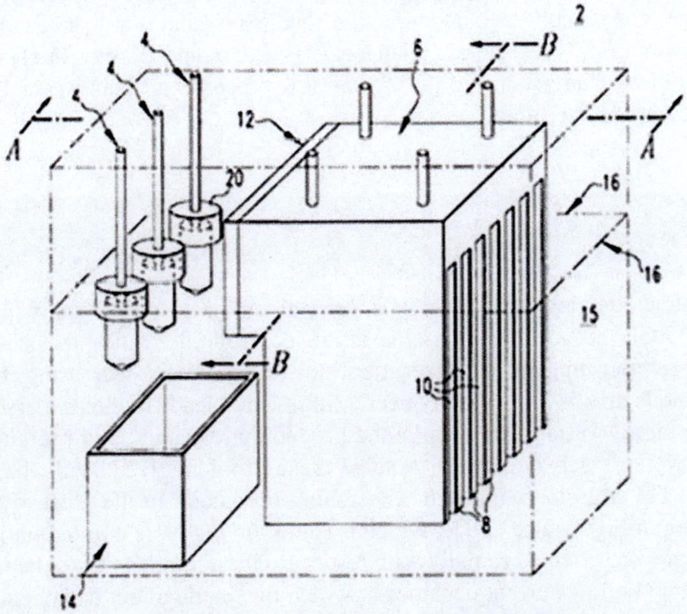

Fig. 9 Keller's electrolysis cell [26]

Table 2 Comparison between Sharma's process [21] and Nanoscale Powder's process [22, 23]

Sharma's Na system	Nanoscale Powder's system
$Nd_2O_3 + 3CaCl_2 + 6Na =$ $2Nd + 3CaO + 6NaCl$ (~750 °C)	$2NdCl_3(aq) + 14FeCl_2(l) + BCl_3(l) + 37Na(l)$ $= Nd_2Fe_{14}B + 37NaCl$ (~310–700 °C)
Nd_2O_3 administered as solid	$NdCl_3$ administered as solute with $FeCl_2(l)$
Crucible must be extracted from system (batch system)	Tap at bottom of crucible (more efficient batch system)
Melting point of Na (97.2 °C)	
Melting point of NaCl (801 °C)	
	Melting point of $FeCl_3$ (306 °C)
	Melting point of $FeCl_2$ (677 °C)
Melting point of $CaCl_2$ (772 °C)	Melting point of BCl_3 (−107 °C)
Melting point of Nd_2O_3 (2233 °C)	Melting point of $NdCl_3$ (758 °C)

is deactivated to allow the system to cool for product recovery. The bottom port is opened, and the system is drained from the apparatus. The unreacted metal is either syphoned off, evaporated, or reacted with chloride gas. The salt is removed through washing products, evaporation, a rotary furnace, or ultrafiltration. Then the metal powder is dried in a vacuum oven to produce the metallic powder. Although there're results of the process done with other metals, there's no quantifiable data concerning rare earth reduction. Nevertheless, Nanoscale Powders and Energy Fuels signed a Memorandum of Understanding [24]. Sharma's process and Nanoscale Powder's process is compared at Table 2.

A Low PFC Emissions Cell

The current anode design, which is a vertical cylinder, poses many problems concerning ACD because this variable is not controllable. Since the side of the cylinder is deteriorating from the reaction, not only ACD is decreasing, but even the active anode area is dramatically decreasing, which leads to the process temperature being highly variable, decreases the production capacity, and promotes PFC emissions. Keller's cell attempted to solve these problems [25], with the goal of preventing PFC emissions, through redesigning the anode in the electrolytic cell, which is seen in Fig. 9. The anode, which is (6) in the Fig. 9, is divided into several plates, which were oriented radially with respect to the cathodes (4), to decrease the current density bellow current densities at which the anode effect can occur. Openings were also present between the plates to allow a free circulation of electrolytes there between. The cell would undergo voltages around 2–5 V with low amperage to avoid the anode effect to oxide contents as low as 0.01–0.1 wt%. A permeable barrier is inserted between the electrodes to prevent CO and CO_2 back-reacting with the product rare earth metal and claims to be permeable to rare earth ions. Problems exist in this design. According to Keniry, critiqued that the anode design is complex, and the wear of the anode plates are highly variable and wasteful [26]. Also, there is a lack of demonstration that the barrier material between the electrodes is indeed effective for rare earth permeation in the patent's examples. Lastly, the inclusion of such barrier material would result the cell being more energy intensive from a larger ACD equivalent [26].

Dysinger's Automatic Feeding Cell

Another issue related to molten fluoride electrolysis is that the feeding process is done manually. This brings many issues regarding the ability to control the process and to avoid the anode effect of the system, which will create PFC gasses. Therefore, Dysinger designed and tested a feeder for his process, which was described in the earlier in this work [19]. The powder feeder was designed and constructed from stainless steel, which was continuously vibrated to prevent the clogging of oxide powder into the electrolytic system. Not only the current efficiency was improved as seen in Fig. 10 and the current density of the system improved from 0.6 A/cm^2 to even 1 A/cm^2, but it also assisted in the increasing the sustainability of the anodic density and helped prevent the build-up of undissolved Nd_2O_3 powder in the system by dispensing the oxide evenly throughout the cell [19]. However, the Dysinger didn't address the ability to scale the feeder to commercial sizes, which would be an issue if the process is within a vacuum chamber.

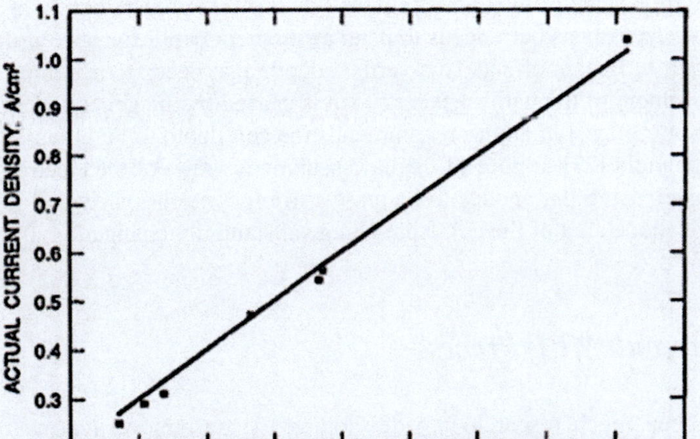

Fig. 10 Actual and theoretical current density of Dysinger's cell during continuous feeding [19]

An Open Atmospheric Cell

Carbon anodes are used to reduce rare earths oxides. However, some of the unreacted carbon from the anode deteriorates into the cell, which leads to undesirable about of carbon contamination of levels greater than 400 ppm (the maximum allowable amount of carbon for rare earth magnets) and causes a reduction in the current efficiency [27]. Tamamura pursued to solve this problem by introducing oxygen into the pressure chamber to counteract the rogue carbon in the molten salt [27], since the carbon is most likely to be present at the top of the electrolyte pool. Although this puts rare earth metal at the risk of back-reacting from an oxygen atmosphere, this can be negated if the rare earth oxide has a smaller specific gravity than the rare earth metal. If the oxide and the electrolyte have a smaller specific gravity than the rare earth metal, the metal would gather at the bottom of the electrolytic cell and the other chemicals would gather above the rare earth metal. Therefore, the metal can be tapped at the bottom of the cell, which avoids the metal from being in contact with the upper cell atmosphere. This is shown in Tamamura's experimental results, which showed that at oxygen partial pressures of 20–40 vol%, not only the carbon content was as low as 40 ppm, but even the oxygen content was as low as 70 ppm [27], which in turn the process produced a higher purity rare earth. Two trials were performed in open air (78 vol% nitrogen, 21 vol% oxygen), and the data correlated to the oxygen partial pressure trials. From the experiments, Tamamura concluded that atmospheres with oxygen concentrations form 15–30% assist in converting the rogue carbon in the cell, which floats at the top of the electrolytic bath, into CO_2 and CO gas. If the oxygen content is less than 15 vol% the powder carbon in the system increases, and if the oxygen content is greater than 40 vol%, the electrolysis becomes vigorous, and problems arise in the system. This process comes at a caveat, at which the electrolysis bath needs to be at a sufficient depth for rare earth metal's specific gravity to be an effect. This vertical depth also needs to accommodate the vertical movement of the bath surface, which is caused by the generated anode gas, also comes into play. Tamamura recommends the cell depth to be at least 10 cm to produce high yield [27]. Figure 11 displays schematic views of the electrolytic cell, which involves using flat anodes to increase current control quality and to prevent the effective anode area of the cell from being substantially changed.

A Hall–Heroult MFE Process

There has been much research into developing the current vertical state-of-the-art molten fluoride electrolysis process into a horizontal electrolytic process like the Hall–Heroult process in aluminum electrolysis. Vogel describes research being conducted in Beijing, China, ever since 2005 to develop a liquid bottom cathode cell (which composes of the already reduced molten RE metal) with horizontally oriented electrodes [18, 28]. One of the prototypes of this process is a 3 kA liquid bottom

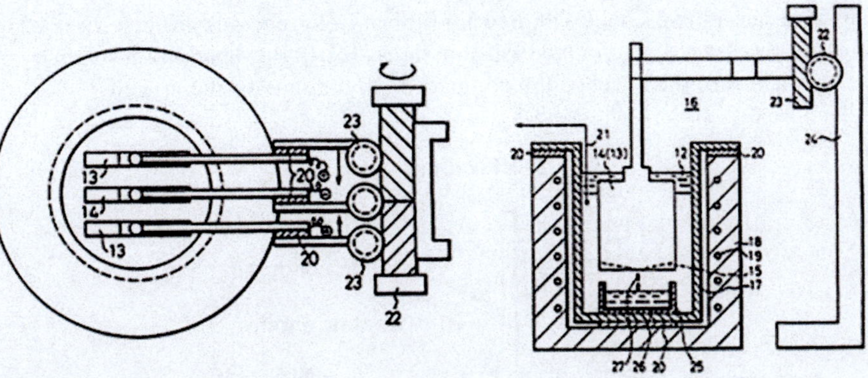

Fig. 11 Overhead schematic view of Tamamura's cell (left), presenting the anodes (13) and the cathode (14) and cross section of Tamamura's cell (right) [27]

cathode cell design, which is shown in Fig. 12. The prototype includes moveable anode blocks, at which the anode can change the electrode distance of the cell, to improve the cell's efficiency. Also, the cell is in a quasi-stationary flow condition, which is crucial for the electrolytes, the process control, and stability of the system. The anode blocks and the quasi-stationary flow condition allow the electrode distance and the free surface area to be decreased significantly, which leads to lower the cell voltages of 8–15 V in the state-of-the-art systems to as low as 5–6 V [18, 28]. This is because, according to Hass, the ion transport through the thin diffusion layer through the electrodes is the rate determining step [16]. These blocks would be able to be reinstalled with even larger electrodes without changing the sedimentation pattern of the reduced metal [16]. According to Vogel, there are many Chinese-based publications for the heat balance, flow field, and electrical field for a conceptual 10 kA liquid cathode cell; however, the design is not tested or ready to be implemented into industrial production. One of the issues of testing this process is its complicated test set up, which requires to obtain physical measurements from a running high temperature corrosive cell. Because of this, Hass stated the near impossibility of such a test setup [16]. Numerical simulation by CFD is therefore preferred to estimate the behavior [16, 28]. Hass developed a CFD simulation of both this novel process and the current state of the art processes, to simulate the flow velocities of the processes and the flow direction and nature of the oxidized carbon [16]. These simulations, as geometric models, are shown in Fig. 13. The flow diagrams of the same geometric models are shown in Fig. 14. The simulation presents undisturbed interface between the metal and electrolyte, which indicates that the system will promote high current efficiency and very little product reoxidation [16]. Also, the upward directed flow of the system counteracts the sedimentation of undissolved oxide particles, which will prevent oxide contamination in the produced RE metal. However, this process demands higher level of automation. According to Hass, the process needs a point feeder like aluminum electrolysis, or an online oxygen concentrator measurement.

This is because the product withdrawal of this novel process is difficult these additions are necessary to overcome PFC emissions [16]. Table 3 presents a summary of the parameters of the novel cell in comparison to the state-of-the-art cell.

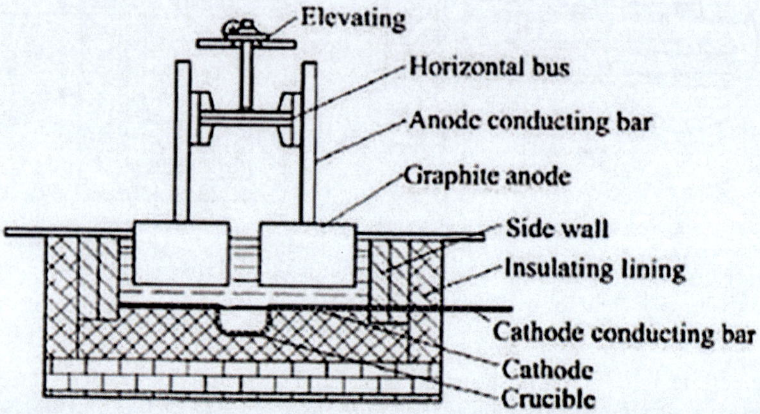

Fig. 12 Experimental 3 kA liquid bottom cathode cell design with moveable anodes from National Engineering Research Center for rare earth materials in Beijing [18]

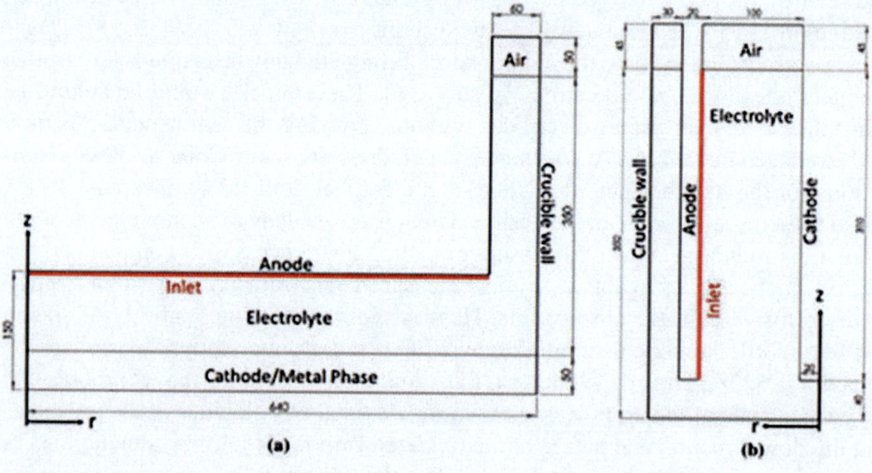

Fig. 13 Geometric model and phase distribution of **a** novel cell and **b** state of the art with lengths in millimeters [16]

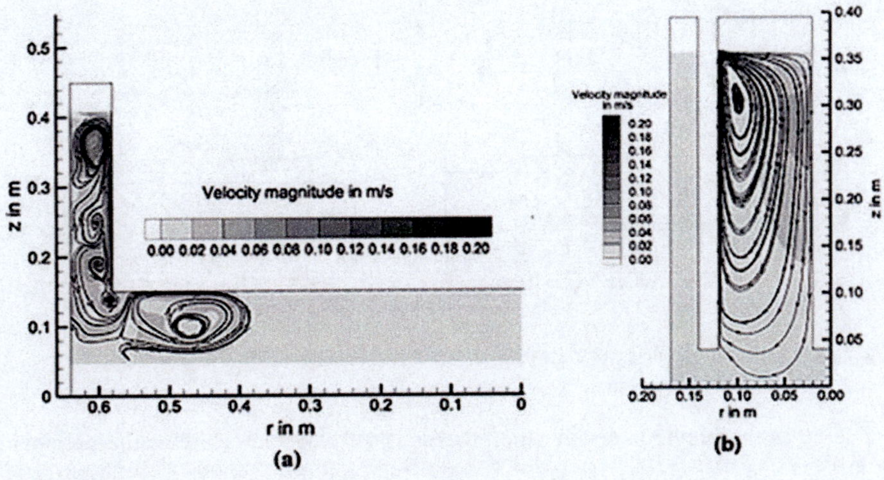

Fig. 14 Fluid velocity distribution of **a** the novel cell and **b** the state of the art [16]

Table 3 Summary of comparisons of modern and conventional cell design; adapted from [16]

Parameter	Novel cell	State-of-the-art
Product reoxidation	Low probability	High probability
Electrode distance reduction	Possible	Impossible
Dissolution rate oxide	High	High
Dwelling time gas bubbles	High	Low
Convectional mass transport	Between crucible and anode	Between anode and cathode
Free surface area	Low	High
Dead volume	Not existing	Existing
Product withdrawal	Pump system	Ladle

Lynas Processes

In 2013, Lynas Rare Earths patented a molten fluoride electrolytic rare earth reduction process. This process [26] is seen in Fig. 15. The electrodes to conduct electrolysis are oriented horizontally. The electrolysis is conducted at an angle between 5 and 20°, and the channel (22) and the cathode (24) are combined into one unit, which would promote the rare earth metal flow towards the sump, where it would be tapped. The anodes (40), which are block-shaped, are modified to accommodate a process for vertically translating the anode position, so that the process would continue to have the same ACD. The anode also does not need to be presloped since the electrolysis would make the anodes at the desired slope throughout the process. The ACD is about

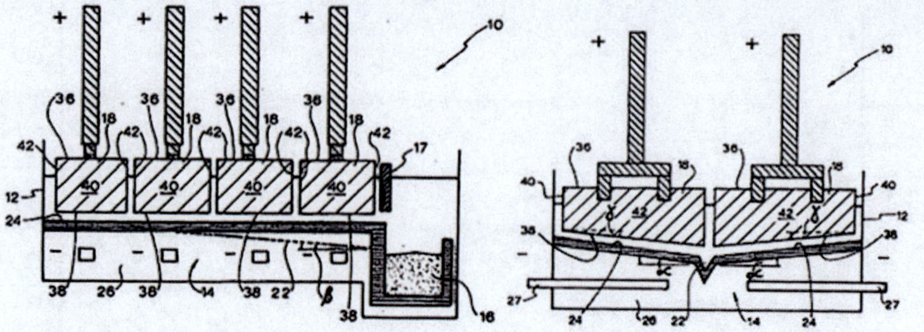

Fig. 15 Schematic of angled electrode cell (left) and cross-sectional view of cell (right) [26]

the same as the previous design; the process is virtually the same as the previous process.

The operation of this cell is like other electrolytic cells. First, the molten electrolyte is loaded in the cell, at which alternating current is supplied to raise the operating temperature to the desired temperature. The operating temperature in this process would be 5–50 °C greater than the freezing point of the metal. Then the feed material is charged into the cell, which the feed material dissolves. After this, direct current between 5 and 100 kA is supplied to the anodes, which commences the electrolysis of the feed material. When the rare earth is reduced on the cathode, the metal falls on the inclined channels at the bottom of the cell (22) and flow to the sump (16) at the end of the series of electrodes. Throughout the process, the feed material is regularly charged into the electrolytic cell at areas of high electric flow and the metal is tapped at the sump (16).

SOM Electrolysis

Over the last decade, solid oxide membrane (SOM) electrolysis gained interest due to improvements to the process [29–32]. This process involves the direct oxidation of metal oxide which is expressed in the following equation:

$$RE_2O_3 = 2RE + 3/2\,O_2 \tag{8}$$

The metal product is precipitated as either a gas, solid, or liquid, depending on the temperature of the system. Oxygen is a byproduct in the system. The critical part of the process is the anode which consists of the inert anode itself and the SOM. The SOM selectively conducts the oxygen ions, which avoids possible dissociation of molten salts at high voltages. This produces high current density and Faradaic current efficiency. The SOM also prevents the oxygen gas back-reacting into the system, which would promote high yields of the product metal.

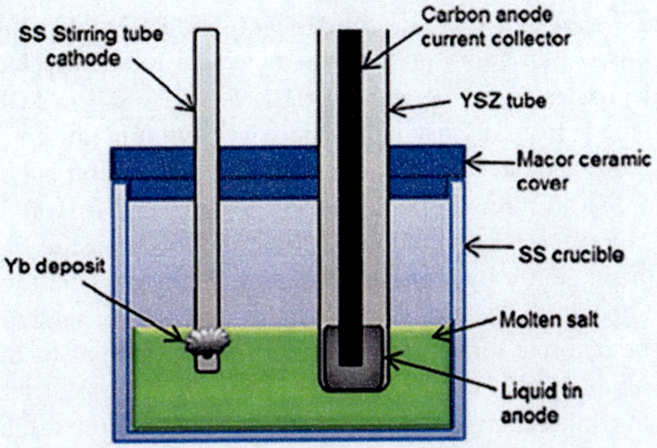

Fig. 16 Schematic diagram of a SOM process used to reduce Yb [33]

Most patents for SOM electrolysis describe the process to produce magnesium metal and only refer to the process' ability to reduce rare earths. However, Jiang developed an experimental SOM process on reducing the rare earth metal Yb [33]. A schematic of the process is presented in Fig. 16. The process involved using molten fluorides with a mixture of Yb_2O_3 and a powder of 21 wt% LiF and 79 wt% YbF_3, all of which were mixed and heated to 850 °C to create a powder flux. A liquid tin anode and a carbon anode collector was used for the process. A stainless-steel cathode was also used, which was also used to stir the solution. Since Yb has a stable +2 ion, YbF_3 was converted to YbF_2 by adding Yb metal and heating the fluoride for 4 h. This fluoride is inserted into the cell. The electrolytic reduction process is conducted at 800 °C. The process showed that the YSZ membrane was relatively stable throughout run and electrode-reduced an amount of Yb on the cathode.

Ionic Liquids/Bourbos Process

Ionic liquids electrolytes are gaining interest over the last decade in Europe as a green alternate process to classical electrolytic processes [34, 35]. This is because ionic liquids have exceptional electrolytic characteristics [38]. And their salts are generally liquid at ambient conditions less than 100 °C [36]. Ionic liquids also have high chemical and thermal stability and negligible vapor pressure [36–38]. The University of Alabama, Tuscaloosa found success in developing a process of reducing lead and zinc using ionic liquids with a 86.7% and 95.6% current efficiencies respectively [39].

Though this technology is still at its infancy, Bourbos proposed a potential electrolytic rare earth process using ionic liquids [40]. Figure 17 displays Bourbos' cell, by which he conducted the process to reduce metals such as neodymium with the ionic liquid n-butyl-n-methylpyrrolidinium bistriflimide (BMPTFSI). The cell consists of a Cu wire cathode and a Pt wire anode, which are compartmentalized in the cell. The anode is compartmentalized with a porous plug membrane, which allows a solution of 0.1 M of $NdCl_3$ in dimethylsulfoxide (DMSO) solution to be contained in the anolyte. An additional Pt wire acts as a reference electrode in the system, and the electrodes are immersed in a 0.05 M $Nd(TFSI)_3$ in a BMPTFSI solution. $Nd(TFSI)_3$ is made from dissolving hydrated neodymium nitrate with the bistriflimide salt at in a pure acetone solution, after which the solution is heated to 60 °C under vacuum for 2 h to remove the acetone, and is heated to 100 °C for 24 h to remove water. A thermostat is included in the system to measure the temperature and the cell is stirred gently by a magnetic stirrer. A current of 30 μA is used in the process at a temperature of 70 °C. $NdCl_3$ in the anolyte is separated into Nd^{3+} ions and Cl^- ions. The chlorine ions are converted into chlorine gas by electromigrating to the anode. The anolyte is compartmentalized from the rest of the reaction because rare earth salts (other than rare earth bistriflimides) have low solubility in BMPTFSI electrolyte and using bistriflimides as the sacrificial salt in the anodic reaction is very costly. At the electrolyte, $Nd(TFSI)_3$ is dissolved in the solvent through the following reaction:

$$[(x-3)][TFSI^-] + Nd(TFSI)_3 \rightarrow [Nd(TFSI)_x^-] \tag{9}$$

When the BMP^+ ions electromigrated at potentials lower than −2.2 V, the BMP^+ ions electromigrated to the cathode, carrying the $Nd(TFSI)_x{}^-$ ions in the electrolyte When the ions arrive at the electrode interface, $TSFI^-$ disassociates from Nd^{3+}. Then the Nd^{3+} ions transfer through the electrolyte-formed boundary layer, to the cathode through electromigration. Once the Nd^{3+} reach the cathode, the neodymium ion is reduced to the neodymium metal. The metal is recovered by washing the cathode with DMSO and acetone, to dissolve the electrolyte layer.

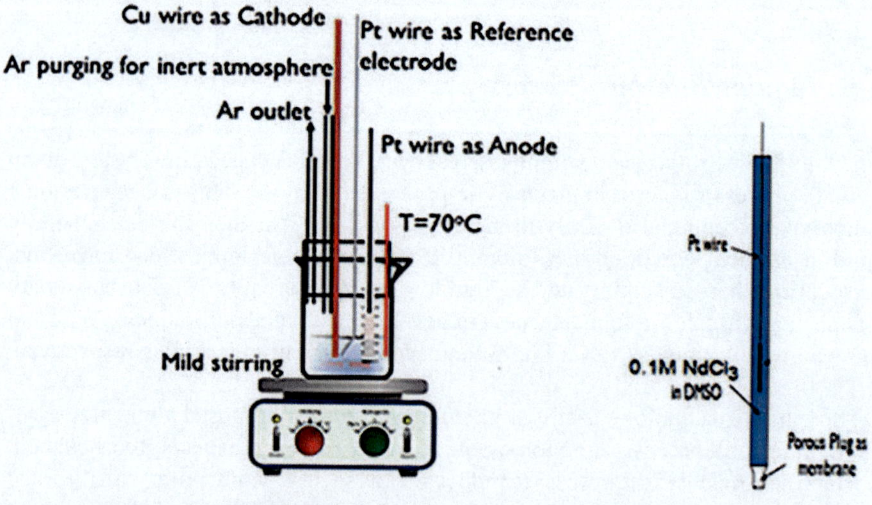

Fig. 17 Schematic Bourbos' ionic liquid experimental electrolytic cell (left) and a detailed schematic of the cell's anode (right) [40]

Although the cell is conducted at a temperature less than 100 °C, some issues are present in the process which could hinder its commercial production. The first issue is the ionic liquid itself which is being used as the electrolyte, BMPTFSI. Bistriflimides are very costly and difficult to synthesize [40], which makes the process uneconomical when this electrolyte is being washed from the cathode for recovery. Therefore, a process is needed at which the residue electrolyte is recycled to be reused in the process. Also, due to the nature of the cathodic reaction described earlier, the recovery of the rare earth metal from the cathode could be difficult and might require the system to be a batch process than a continuous process if the process is scaled to commercial production levels.

Microwave Hydrogen Plasma Reduction

Research is being conducted in the last decade to reduce REOs using hydrogen plasma. Sceats conducted an experimental investigation to reduce neodymium oxide with microwave hydrogen plasma with little success [41]. Sceats also concluded that there is an advantage of using atomic hydrogen over molecular hydrogen in his investigation [41].

Conclusion

There are various approaches to reduce rare earths which have been, or are being researched, which ranges from improving molten fluoride electrolysis, to finding new reactions in metallothermic reduction, to even exploring new innovations in electrolytic reduction. Nevertheless, much work and research is still necessary to fully understand not only the processes themselves, but the factors relating to scaling the processes up to commercial levels, the capital investment required, and the potential issues relating to these new systems. But this research and development is needed if the world wants a green future. This area must be by supported by the pillars of willing private investors and a supportive government if the world wants to stretch out to grasp their dream of an environmental and sustainable green society.

References

1. Bauer D, Diamond D, Li J, Sandalow D, Telleen P, Wanner B (eds) (2010) U.S. Department of Energy Critical Materials Strategy. https://www.energy.gov/sites/prod/files/piprod/documents/cms_dec_17_full_web.pdf. Accessed 4 Sept 2022
2. Pawar G, Ewing RC (2022) Recent advances in the global rare-earth supply chain. MRS Bull 47(3):244–249. https://doi.org/10.1557/s43577-022-00305-6
3. Roskill (2021) Rare earths market outlook report. Roskill, London
4. Schlinkert D, van den Boogaart KG (2015) The development of the market for rare earth elements: insights from economic theory. Resour Policy 56:272–280. https://doi.org/10.1016/j.resourpol.2015.10.010
5. Mancheri NA, Sprecher B, Bailey G, Ge J, Tukker (2019) A effect of Chinese policies on rare earth supply chain resilience. Resour, Conserv Recycl 142:101–112. https://doi.org/10.1016/j.resconrec.2018.11.017
6. Gupta CK, Krishnamurthy N (2016) Extractive metallurgy of rare earths. CRC Press, Boca Raton, Florida
7. Vogel H, Flerus B, Stoffner F, Friedrich B (2017) Reducing greenhouse gas emission from the neodymium oxide electrolysis. Part I: analysis of the anodic gas formation. J Sustain Metall 3(1):99–107. https://doi.org/10.1007/s40831-016-0086-0
8. Vogel H, Friedrich B (2018) An estimation of PFC emission by rare earth electrolysis. In: Martin O (ed) Light metals 2018. The Minerals, Metals, and Materials Society 2018. Springer International Publishing, pp 1507–1517. https://doi.org/10.1007/978-3-319-72284-9_197
9. Mat Suli L, Ibrahim W, Abdul Aziz B, Deraman MR, Ismail N (2017) A review of rare earth mineral processing technology. Chem Eng Res Bull 19:20–35. https://doi.org/10.3329/cerb.v19i0.33773
10. Habashi F (2018) Metallothermic reactions—past, present and future. Res Rep Metals 2(1). SciTechnol
11. Bujnovskij AS, Sofronov VL, Sachkov V, Anufrieva AV (2015) Basic stages of magnet production by fluoride technology. Adv Mater Res 1085:209–213. https://doi.org/10.4028/www.scientific.net/AMR.1085.209
12. Ivanov VA, Dedyukhin AS, Polovov IB, Volkovich VA, Rebrin OI (2018) Fabrication of rare-earth metals by metallothermic reduction: thermodynamic modeling and practical realization. AIP Conf Proc 2015(1):020033. https://doi.org/10.1063/1.5055106
13. Sharma RA (1994) Metallothermic reduction of rare earth fluorides. US Patent 5,314,526, 24 May 1994
14. Takeda O, Uda T, Okabe TH (2014) Chapter 2.9—Rare earth, titanium group metals, and reactive metals production. In: Seetharaman S (ed) Treatise on process metallurgy. Elsevier, Boston, pp 995–1069
15. Zhu H (2014) Rare earth metal production by molten salt electrolysis. In: Kreysa G, Ota K, Savinell RF (eds) Encyclopedia of applied electrochemistry. Springer, New York, pp 1765–1772
16. Haas T, Hilgendorf S, Vogel H, Friedrich B, Pfeifer H (2017) A comparison between two cell designs for electrochemical neodymium reduction using numerical simulation. Metall Mater Trans 48(4):2187–2194. https://doi.org/10.1007/s11663-017-0982-0
17. Vogel H, Friedrich B (2018) An estimation of PFC emission by rare earth electrolysis. In: Martin O (ed) Light metals. Springer International Publishing, pp 1507–1517. https://doi.org/10.1007/978-3-319-72284-9_197
18. Vogel H, Friedrich B (2015) Development and research trends of the neodymium electrolysis—a literature review. Paper presented at 8th European metallurgical conference (EMC), Düsseldorf, Germany, 14–17 June 2015
19. Dysinger DK, Murphy JE (1994) Electrowinning neodymium from a molten oxide-fluoride electrolyte. United States Department of the Interior, pp 1–8
20. Sharma RA, Seefurth RN (1988) A molten salt process for producing neodymium-iron alloys by reduction of Nd_2O_3 with sodium. In: Bautista RG, Wong (eds) Rare earths: extraction, preparation and applications: proceedings of a symposium on rare earths, extraction, preparation and applications. TMS, Warrendale, Pennsylvania, pp 345–368

21. Sharma RA (1986) Metallothermic reduction of rare earth oxides. US Patent 4,578,242, 25 Mar 1986
22. Henderson D, Matheson A, Van Lieshout R, Finnerty D, Koenitzer JW (2022) Methods for producing metal powders. US Patent US2022/0008993, 13 Jan 2022
23. Matheson A (2017) Manufacturing and applications of metal powders and alloys. US Patent 9,586,262 B2, 7 Mar 2017
24. (2022) Energy fuels announces strategic venture with nanoscale powders to develop innovative rare earth metal-making technology. Energy Fuels Inc. https://www.energyfuels.com/2021-12-15-Energy-Fuels-Announces-Strategic-Venture-with-Nanoscale-Powders-to-Develop-Innovative-Rare-Earth-Metal-Making-Technology. Accessed 27 Aug 2022
25. Keller R, Larimer KT (1998) Electrolytic production of neodymium without perfluorinated carbon compounds on the offgases. US Patent 5,810,993, 22 Sept 1998
26. Keniry J, Kjar AR (2013) Drained cathode electrolysis cell for production of rare earth metals. International Publication Number WO 2013/170310, 21 Nov 2013
27. Tamamura H (1987) Process for preparation of neodymium or neodymium-iron alloy. US Patent 5,091,065, 25 Feb 1987
28. Vogel H (2018) Process optimization of neodymium electrolysis. PhD thesis, Aachen University
29. Guan X, Pal UB, Jiang Y, Su S (2016) Clean metals production by solid oxide membrane electrolysis process. J Sustain Metall 2(2):152–166. https://doi.org/10.1007/s40831-016-0044-x
30. Pal UB, Gratz E, Guan X, Powell IV AC, Zink PA, Tucker RS, Pati S, Strauss J, Tajima A (2013) Methods and apparatuses for efficient metals production, separation, and recycling by salt and argon-mediated distillation with oxide electrolysis, and sensor device related thereto. US Patent US 2013/0152734 A1, 20 June 2013
31. Powell IV AC, Dussault D, Earlam M, Tajima A, Raymes C (2021) Method and apparatus for efficient metal distillation and related production process. International Publication Number WO 2021/022098 A1, 4 Feb 2021
32. Powell IV AC, Pal UB, Barriga S, Derezinski SJ, Earlam M (2014) Clean, efficient metal electrolysis via SOM anodes. US Patent US 2016/0376719 A1, 29 Dec 2014
33. Jiang Y, Zink PA, Pal UB (2012) (Invited) Solid oxide membrane process for the reduction of uranium oxide surrogate in spent nuclear fuel. ECS Trans 41(33):171–189. https://doi.org/10.1149/1.3702424
34. Balomenos E, Deady E, Yang J, Panias D, Friedrich B, Binnemans K, Seisenbaeva G, Dittrich C, Kalvig P, Paspaliaris I (2017) The EURARE project: development of a sustainable exploitation scheme for Europe's rare earth ore deposits. Johns Matthey Technol Rev 61(2):142–153. https://doi.org/10.1595/205651317x695172
35. Davris P, Balomenos E, Taxiarchou M, Panias D, Paspaliaris I (2017) Current and alternative routes in the production of rare earth elements. BHM Berg- Huettenmaenn Monatsh 162(7):245–251. https://doi.org/10.1007/s00501-017-0610-y
36. Bourbos E, Giannopoulou I, Karantonis A, Paspaliaris I, Panias D (2016) Chapter 13—Electrodeposition of rare earth metals from ionic liquids. In: Borges De Lima I, Filho WL (eds) Rare earths industry 2017. Elsevier, Boston, pp 199–207
37. Bourbos E, Giannopolou I, Karantonis A, Panias D, Paspaliaris I (2014) Electrodeposition of rare earth metals in ionic liquids. Paper presented at 1st European rare earth resources conference (EMC), Milos Island, Greece, 4–7 Sept 2014
38. Poole CF (2007) Ionic liquids. In: Wilson ID (ed) Encyclopedia of separation science 2007. Academic Press, Oxford, UK, pp 1–8
39. Reddy R (2012) Reduction of metal oxide to metal using ionic liquids, report. University of Alabama, Tuscaloosa. ed. University of North Texas Libraries, US
40. Bourbos E, Giannopoulou I, Karantonis A, Paspaliaris I, Panias D (2018) Reduction of light rare earths and a proposed process for Nd electrorecovery based on ionic liquids. J Sustain Metall 4(3):395–406. https://doi.org/10.1007/s40831-018-0186-0
41. Sceats JH (2018) On the reduction of metal oxides in non-equilibrium hydrogen plasmas. PhD thesis, Metallurgical and Materials Engineering, Colorado School of Mines, Golden, CO

Recovery of Bismuth in Blast Furnace Dust by Carbothermal Volatilization Reduction

Huaixuan Feng, Yan Li, Jingsong Wang, and Xuefeng She

Abstract Blast furnace dust (BFD) is a kind of solid waste generated in the iron-making process. It contains approximately 0.1% of bismuth (Bi). To reduce environmental pollution and storage problems associated with BFD, direct reduction process to recover Bi from BFD was proposed in this study. Thermodynamic calculations and thermogravimetric analysis were first used to determine the initial reaction temperature between Bi_2O_3 and C. High-purity Bi_2O_3 and graphite were evenly mixed that the carbon to oxygen ratio is 0.4, 0.6, 0.8, respectively and heated at 800–1200 °C under 50 °C gradient for 2 h. Experimental results showed that almost all bismuth volatilized at 1200 °C and Bi and BiO could be obtained in the condensation zone by carbothermal reduction. Therefore, this study provides a new and potential useful technology for recovery of Bi from BFD, as well as a certain practical approach to comprehensively utilizing such metallurgical solid waste.

Keywords Bismuth · Carbothermal reduction · Blast furnace dust (BFD) · Resource utilization

Introduction

Bismuth [1] (Bi) is a rare metal, located in group VA of the sixth period of the periodic table, and is widely used for electronics industry and medical field. With the rapid development of science and technology, Bi continues to play a critical role in semiconductors, superconductors, flame retardants, chemical reagents, electronic ceramics, and medical fields. Bismuth, and its compounds are used in the manufacture of low melting point alloys [2], the most commonly used are bismuth and lead, tin, antimony, indium, and other metals to form a series of low melting point alloys,

H. Feng · Y. Li · J. Wang (✉) · X. She (✉)
State Key Laboratory of Advanced Metallurgy, University of Science and Technology Beijing, Beijing 100083, China
e-mail: wangjingsong@ustb.edu.cn

X. She
e-mail: shexuefeng@ustb.edu.cn

R. G. Reddy et al. (eds.), *New Directions in Mineral Processing, Extractive Metallurgy, Recycling and Waste Minimization*, The Minerals, Metals & Materials Series,
https://doi.org/10.1007/978-3-031-22765-3_37

which have special and important applications in fire protection and electrical safety devices, such as for manufacture of automatic sprinklers, safety plugs for boilers, electrical fuses, and automatic device signaling equipment. Adding a trace amount of bismuth to steel can improve the processing properties of steel, and adding a trace amount of bismuth to malleable cast iron can make malleable cast iron have properties similar to stainless steel [3]. In addition, high-purity bismuth (99.999% Bi) is used as a heat carrier or coolant in nuclear industrial reactors and can be used as a material to protect atomic fission devices. In recent decades, the global amount of bismuth is around 4,000 tons and is increasing year by year [1]. Hence, the demand for Bi will grow rapidly along with the development of science and technology. Bismuth is considered a critical metal [4], but it has a low occurrence in the earth's crust [5]. The content of bismuth in the earth's crust is only $2 \times 10^{-5}\%$, and it is difficult to form individual deposit under natural conditions. Most of bismuth minerals are associated with tungsten, molybdenum, lead, tin, copper, and other metallic minerals. Bi exists in the state of elemental substance and compound in nature. The main minerals include natural bismuth (Bi), bismuth (Bi_2O_3), bismuthite ($nBi_2O_3 \cdot mCO_2 \cdot H_2O$), copper bismuthite ($3Cu_2S \cdot Bi_2S_3$), galena ($2PbS \cdot Bi_2S$), and so on. Because of the low content of Bi, these minerals cannot be directly used for Bi production.

The increasing demand for Bi has prompted researchers to search for Bi-containing resources and explore more methods of extracting it. Instead, most Bi is recovered from industrial by-products of the copper, lead and zinc smelting system and steel industry. Among them, blast furnace flue dust (BFD) is an important secondary resource containing several valuable metals like zinc, lead, iron, bismuth, and indium, and its treatment and recovery have been broadly investigated during the past decades. Therefore, bismuth has a wide range of uses, and it is very necessary to recycle the secondary solid waste resources of bismuth in the face of large demand. A very important source of by-product bismuth is the blast furnace system in the iron and steel industry. An important source of the by-product bismuth occurs in the iron ore concentrate in the strong reducing atmosphere and high temperature of blast furnace smelting, volatilizes with zinc, and is enriched in the gas ash is the iron and steel industry. The trace amount of bismuth occurs in iron concentrates volatilizes with zinc and it is concentrated in blast furnace ash under the strong reducing atmosphere and high temperature of blast furnace smelting. The content of Bi in blast furnace dust can reach approximately 1% indicating potential for recycling Bi. Unfortunately, the comprehensive utilization and harmless disposal of blast furnace ash in China is relatively poor, and the amount of bismuth recovered from blast furnace ash does not exceed 1,000 tons per year. This BFD provided by a domestic iron and steel plant, which contains valuable elements such as zinc and trace rare metals. The chemical compositions of the sample, given in Table 1, were determined by X-ray fluoroscopy (XRF) combined with chemical methods. For accurate measurement, inductively coupled plasma atomic emission spectrometry (ICP-AES) was used to determine the content of bismuth in the dust as 0.8%. Currently, much effort has been devoted to recovering valuable metals from metallurgical dust and sludge. Pyrometallurgical process, hydrometallurgical process, and their combination are usually applied for

Table 1 Chemical compositions of blast furnace ash

Fe_2O_3	SiO_2	CaO	MgO	Al_2O_3	Cl	PbO	ZnO	Bi	C
54.3381	2.0319	29.7715	5.3219	0.2716	0.7892	0.0332	2.6277	0.8	0.9

treating the waste. Among them, the direct reduction process was considered as an efficient and environmentally friendly method to separate and recover Fe, Zn, K, and Na from the metallurgical wastes [6]. More than 95% of Zn and 80% of Pb was removed from zinc-bearing dust by carbothermic reduction at 1503 K with a C/O ratio of 1.0. The reduction behavior of BF dust particles during in-flight process was studied using a high-temperature drop-tube furnace [7]. The results showed that the effects of temperature and gas composition on the reduction degree were remarkable. The coal-based reduction (carbothermic reduction) was considered as an efficient method to separate and recover heavy metals from the hazardous waste [8], in which zinc and lead oxides are reduced to volatilize by reductant or volatilized by volatile agent. Unfortunately, these studies focused on the reduction of Fe compounds and the migration of Pb and Zn components, but the removal behavior of Bi was rarely investigated.

It was found that the reduction and oxidation of bismuth and its oxide is feasible at temperatures of approximately 900 °C with the activation energies of 229.4 kJ/mol and 173.6 kJ/mol for the reduction and oxidation reactions [9], respectively, at chemical conversion of 0.7. Unlike Zn and Pb, the volatilization or removal of Bi elements was highly relevant to the reduction behavior of the BFD, because the volatile Bi compound existed the intermediate BiO, not just the final reduction product Bi. And reduction product liquid bismuth was extremely volatile ordinary liquid bismuth phase.

Therefore, in this study, the volatilization mechanisms and the migration behavior of bismuth in the BFS were clarified by characterization of the phase transformation during carbothermic reduction, that is, the element migration process in the carbothermal volatilization reduction process. The impacts of processing parameters, especially the heating temperature, on the volatilization ratio of Bi were investigated.

Experimental

Materials and Apparatus

Since the content of bismuth in metallurgical dust is very low, we first conducted a high-temperature reduction experiment with analytical pure Bi_2O_3 oxidizing reagent and analytical pure graphite reagent and then studied the carbothermic reduction process of bismuth in metallurgical dust. The bismuth oxide was fully mixed with the graphite with in different mass ratios and extruded into the tablet under the

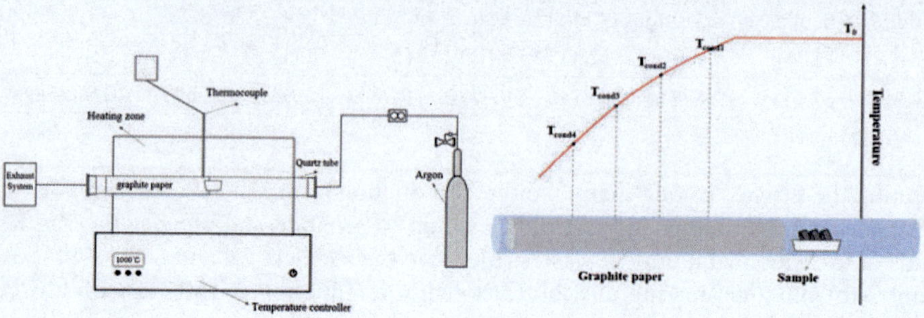

Fig. 1 **a** Schematic diagram of horizontal tube furnace. **b** The temperature distribution of quartz tube

pressure of 15–20 MPa which the mass is approximately 1 g. The diameter of the tablet is 10 mm and the mass of the table is 1 g.

The main equipment used in the experiment was horizontal tube furnace as shown in Fig. 1a. The furnace's main components are a quartz tube (Φ40 × 1200 mm), a temperature controller, argon shielding gas supply device, and a volatile collection equipment. The middle of the furnace is the heating zone, and the left part is the condensing zone, and the right part is connected with the exhaust pipe. In order to collect the products in the process of reducing volatilization, graphite paper was placed in the condensing area of the quartz tube during the experiment, and fiber filter paper was placed at the end of the furnace tube to collect soot. The temperature distribution of the whole furnace was accurately measured and calibrated by the S-type thermocouple before the high-temperature reduction experiment. A schematic of the reactant heating and product condensation from the heating process to the cooling process is shown in Fig. 1b, where T_0 represents the heating temperature and T_{cond1}, T_{cond2}, T_{cond3}, and T_{cond4} represent the condensation temperatures measured by the four thermocouples condensing zones, respectively.

Experimental Procedure

As shown in Fig. 1a, a graphite paper was placed inside the quartz tube to collect the condensation products. The horizontal tube furnace was preheated to a certain temperature and then purged with argon (The gas flow is 100 ml/min) for at least 30 min prior to placing an alumina boat with a table into the hot zone of the quartz tube. The experimental temperature was 800–1200 °C with a temperature gradient of 100 °C. The briquettes with carbon–oxygen ratios of 0.4, 0.6, and 0.8 were heated at the designated temperature for 120 min to ensure sufficient reaction occurred. Then the sample was cooled to the ambient temperature under argon flow.

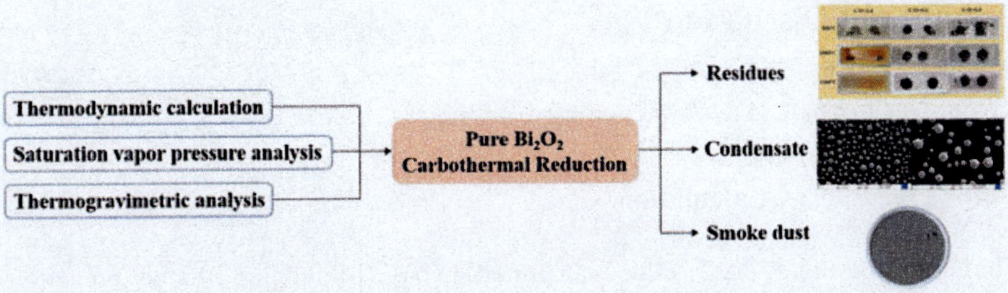

Fig. 2 The flow chart of the main ideas and research methods of this study

Figure 2 shows a flow chart of the main ideas and research methods of this study. Before the carbothermic reduction experiment of bismuth oxide and graphite, the roast temperature of the carbothermic reduction reaction was obtained through thermodynamic calculation, saturation vapor pressure analysis, and thermogravimetric analysis, and the residue, condensate, and smoke dust were collected separately during the carbothermic reduction experiment. The residues, condensate, and smoke dust were collected in the quartz boat and on the surface of the graphite paper, respectively, and were identified by X-ray diffraction (XRD, Smartlab(9kw), Rigaku, Japan). To observe the microscopic morphology of the condensate, scanning electron microscopy (SEM, Supar55, Zeiss, Germany) images were taken, and the condensate's chemical composition was determined with energy dispersive spectroscopy (EDS, Supar55, Zeiss, Germany). The carbothermal reduction reaction of Bi_2O_3 was analyzed by thermogravimetry and differential scanning calorimetry (TG-DSC, TGA55, USA).

Characterization Techniques

The effects of different reaction temperature, C/O ratio, holding time on the volatilization ratios of table were calculated by the following Eq. (1):

$$\eta = (M_0 - M_1)/M_0 \times 100\% \quad (1)$$

where M_0 is the mass of tables before reduction, M_1 is the mass of the roasted tables.

Results and Discussion

Thermodynamic Analysis

Gibbs Free Energy Calculation

The Gibbs free energy (ΔG) value is an important basis for judging whether a reaction can proceed spontaneously. In order to predict the thermodynamic temperature at which Bi_2O_3 reacts with carbon, the ΔG of reactions (2)–(6) at different temperatures is calculated by the following Eq. (2):

$$\Delta G = \Delta_r G^{\Theta} + RT\,In\,Q \tag{2}$$

where $\Delta_r G^{\Theta}$ is the standard Gibbs free energy, T is the thermodynamic temperature, R is the molar gas constant, Q is the pressure quotient.

The possible reduction reactions between Bi_2O_3 and C can be described as follows:

$$Bi_2O_3(s) + 3C(s) = 2Bi(l) + 3CO(g) \tag{3}$$

$$Bi_2O_3(s) + C(s) = 2BiO(g) + CO(g) \tag{4}$$

The possible disproportionation reaction of BiO can be described as follows:

$$BiO(g) = 1/3Bi_2O_3(s) + 1/3Bi \tag{5}$$

Figure 3a shows the relationships between ΔG and temperature change of reduction reactions (3) and (4) under the standard atmospheric pressure. When ΔG is negative, the reaction can proceed spontaneously. By contrary, when ΔG is positive, the reaction proceed spontaneously. The dashed line in Fig. 3a represents the temperature at 1139 °C which ΔG changes for reaction (4) is equal to zero. And the reduction reaction of Bi_2O_3 generating BiO occurred above the temperatures of 1139 °C. As shown in Fig. 3a, the reduction reaction of Bi_2O_3 generating Bi was more likely to occur than the reduction reaction of Bi_2O_3 generating BiO. The main gaseous product of the carbothermal reduction of Bi_2O_3 was expected to be BiO. In the high-temperature reduction process, Fig. 3b shows the relationships between ΔG and temperature change of disproportionation reaction (5) under the standard atmospheric pressure. Bi could also be produced by disproportionation of BiO at the experimental temperature 800–1200 °C, and the lower the temperature, the more negative the ΔG, and the easier the reaction occurred. Therefore, the BiO soot generated by bismuth oxide reduction at high temperature was prone to disproportionation when the temperature decreased during volatilization.

From the thermodynamic calculation results, it can be concluded that at the experimental temperature, Bi and BiO are expected to be the products of carbothermal

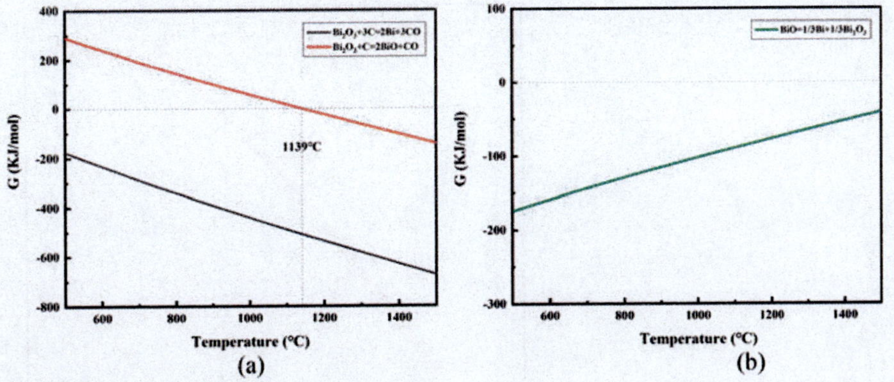

Fig. 3 Relations between ΔG and T for reactions

reduction of Bi_2O_3. Bi and Bi_2O_3 can also be produced by reaction (5). The main purpose of this research is to convert non-volatile Bi_2O_3 into gaseous volatile form of bismuth under high-temperature reduction atmosphere, so as to study the treatment of dust solid waste by carbothermal volatilization reduction process. Bismuth oxide converted from solid phase to vapor-phase BiO through reaction (4) at high temperature Therefore, the effective execution of reaction (4) is a key factor of the experiment.

Saturation Vapor Pressure Analysis

In order to analyze the degree of evaporation of reduced bismuth metal at high temperature, the relationship between temperature and saturated vapor pressure was calculated by Eq. (6).

$$\lg p* = AT^1 + B\lg T + CT + D \tag{6}$$

where p* represents the vapor pressure of the pure substance, T represents temperature, A, B, C, and D represent constants of different elements.

Figure 4 shows the saturated vapor pressure of Bi at temperatures ranging from 1850 to 2300 K [10]. The results show that as the temperature increases, the saturated vapor pressures of Bi increase. At the high temperatures, the saturated vapor pressure of bismuth is very low. At 1850 K, the saturated vapor pressure of Bi is close to zero, so the calculated results of saturated vapor pressure show that bismuth is difficult to volatilize at experimental temperature. This suggests that normal state bismuth vapors cannot reach the condensation zone.

However, by consulting the thermodynamics manual [10], we learned that in addition to the normal state bismuth, there are also volatile Bi(g) and Bi_2 phase the

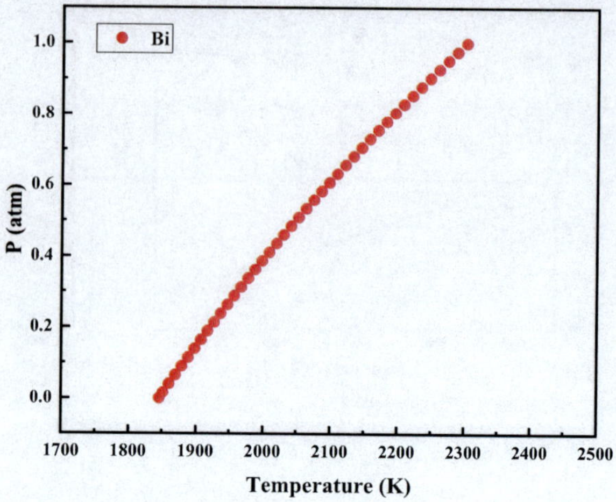

Fig. 4 Relations between ΔG and T for reactions

high temperature. Therefore, it is not excluded that the reduced liquid phase bismuth will be converted into the bismuth of the volatile phase.

Differential Thermal Analysis of Reduction Reaction

The conditions for the reduction reaction of Bi_2O_3 and graphite can be better understood by thermogravimetric analysis. Bismuth oxide and graphite analytical reagents were mixed at different mass ratios as samples for thermogravimetric analysis. The 5 mg of the mixed samples were placed in an alumina crucible and heated to 1250 °C at a heating rate of 10 °C/min. The entire measurement process was performed in an argon atmosphere.

The results of the TG curve are shown in Fig. 5. The TG curves show two stages of weight loss during the reduction process, and the ultimate weight loss is close to 100%. The first stage of weight loss occurred at temperatures 715.3–980.3 °C. According to the three weight loss curves with carbon–oxygen ratios of 0.4, 0.6, and 0.8, the larger the carbon–oxygen ratio is, the larger the slope of the weight loss curve is, the faster the weight loss rate of the sample is, and the bigger the weight loss is. And the weight loss rates of the three samples were similar, in which the carbon–oxygen ratio of 0.4 was the highest, and the carbon–oxygen ratio of 0.8 was the lowest. Thus, the reaction of Bi_2O_3 and graphite was performed at 715.3 °C initially. Combined with the previous reaction Gibbs free energy analysis, the reaction (3) occurs at this moment, and bismuth oxide is reduced to form bismuth metal element. As the sample temperature continued to rise, weight loss rate increased rapidly.

The second stage of weight loss occurred at temperatures above 980.3 °C. According to the three weight loss curves with carbon–oxygen ratios of 0.4, 0.6, and 0.8, the larger the carbon–oxygen ratio is, the larger the slope of the weight loss

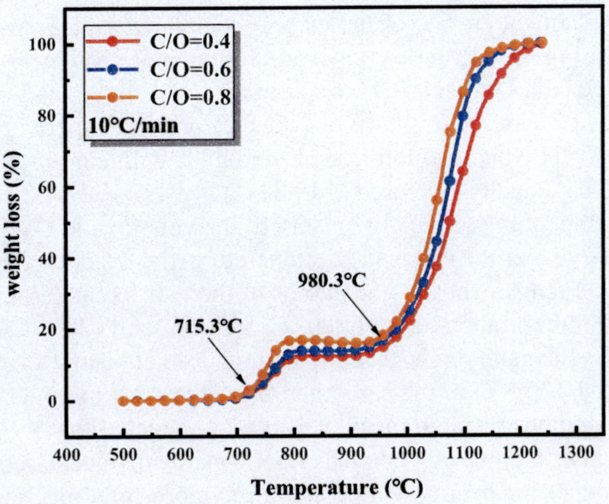

Fig. 5 TG curves of carbothermic reduction of Bi_2O_3 and graphitic at C/O ratios of 0.4, 0.6, and 0.8

curve is, the faster the weight loss rate of the sample is. The slash at the second stage is steeper than that at the first stage. It means that as the temperature increased, the reaction became more intense, and the slopes of the three weight loss curves became larger. Finally, the weight loss rates of the three samples were close to 100% with the temperature rising to 1250 °C, and almost all of them reacted and evaporated. Combined with the previous reaction Gibbs free energy analysis, because the reaction temperature of reaction (4) was higher, the carbothermic reduction reaction was still dominated by reaction (3). Due to the high volatilization rate at this stage, other reactions occur in addition to reaction (3) and generate bismuth-containing volatile phase which resulted in nearly 100% volatilization of the samples.

Carbothermic Reduction and Volatilization Behavior of Bi_2O_3

Thermogravimetric analysis results showed that the reaction of Bi_2O_3 and graphite at 715.3 °C reached its maximum rate at about 1200 °C which is slightly larger than the temperature at which the Gibbs free energy of reaction (2) is less than zero at 1139 °C. Therefore, it was necessary to study the carbothermic reduction process of bismuth oxide when the experimental temperature was 800–1200 °C.

Effect of Heating Temperature

The influence of heating temperature on the volatilization and migration process of bismuth element was first studied in the carbothermic reduction process. The tables with carbon–oxygen ratios of 0.4, 0.6, and 0.8 were heated at temperatures of 800, 900, 1000, 1100, and 1200 °C for 120 min. The relationship between temperature and volatilization rate of the carbon-containing tables was determined and was shown in Fig. 6. At the low calcination temperature of 800 °C, the volatilization rate of the tables with carbon–oxygen ratios of 0.4, 0.6, and 0.8 was 11.66%, 14.34%, and 14.07%, respectively. When the calcination temperature increased from 800 to 1000 °C, the volatilization rate of the tables with carbon–oxygen ratios of 0.4, 0.6, and 0.8 gradually increased to 19.31%, 24.94%, and 24.2%, respectively. As the heating temperature gradually increased, the volatilization rate of the tables increased, and the higher the calcination temperature, the faster the volatilization rate increased.

Figure 7 showed the photos of three different carbon quantities of tables roasted at three temperatures of 800 °C, 1000 °C, and 1200 °C, respectively. It could be seen that bismuth oxide generated liquid droplet bismuth metal at the temperature of 800 and 1000 °C in the carbothermic reduction experiment. The absence of significant luminous bismuth metal at 1200 °C indicated that the resulting bismuth metal had either evaporated completely or a volatile bismuth containing phase still existed The higher the heating temperature, the more fully reaction (3) gone on. According to the Gibbs free energy calculation in section "Gibbs Free Energy Calculation", reaction (3) occurred at 800 °C that was consistent with the results of thermogravimetric analysis in the first stage at the temperature of 715.3–980.3 °C in section "Differential Thermal Analysis of Reduction Reaction". Therefore, when the temperature was around 800–1000 °C, the main reaction that dramatically took place was the carbothermic reduction reaction (3). When the calcination temperature increased from 1000 to 1100 °C, the volatilization rate of the tables with carbon–oxygen ratios of 0.4, 0.6, and 0.8 increased rapidly, which was 53.4%, 53.63%, and 50.76%, respectively. With the increase of temperature, the volatilization rate of bismuth oxide carbothermic reduction became higher, and it was easier to form volatile phase. When the temperature was above 1000 °C, the volatilization rate increased obviously. As the calcination temperature continued to rise to 1200 °C, almost all of the tablets were

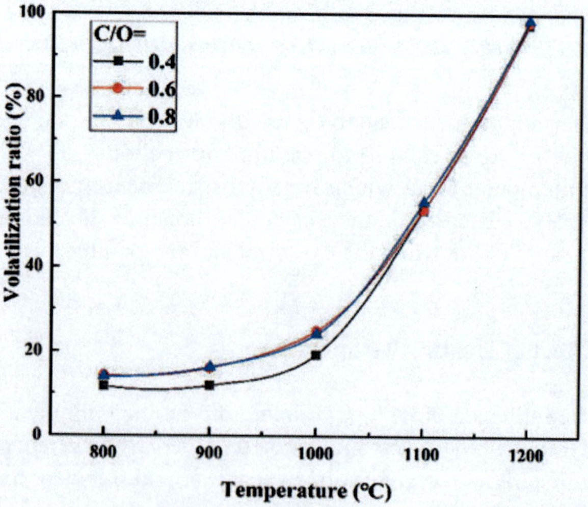

Fig. 6 The relationship between calcination temperature and volatilization rate

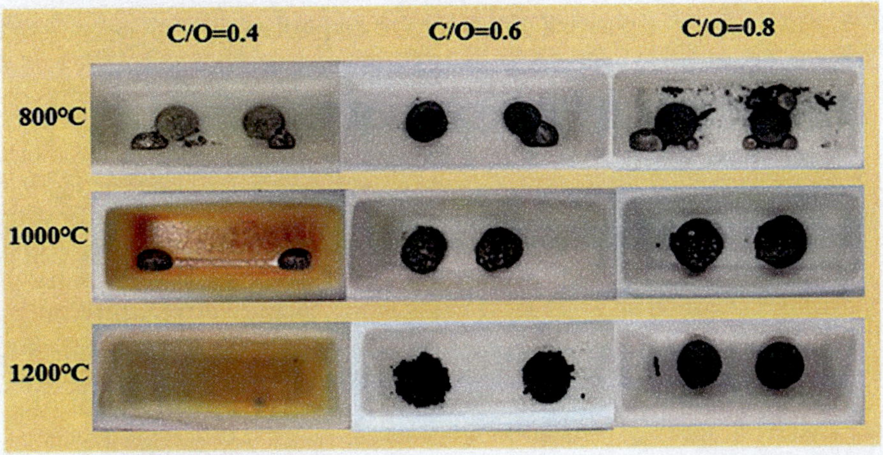

Fig. 7 The photos of three different carbon quantities of tables roasted at three temperatures of 800, 1000, and 1200 °C

volatilized. In addition, with the roasting temperature rising to 1200 °C, reaction (4) also reached the thermodynamic reaction condition. So not only was reaction (3) happening, but reaction (4) was happening. The reduction product of reaction (4) was BiO that was a volatile and unstable bismuth oxide in the low-price state [10].

The results of the carbothermic reduction experiments with heating temperatures ranging from 1000 to 1200 °C were consistent with the results of the second stage thermogravimetric experiments in section "Gibbs Free Energy Calculation". Combined with previous thermogravimetric experiments and related thermodynamic analysis, the carbothermic reduction reaction was still dominated by reaction (3) and the formation of bismuth volatile phase also emerged which resulted in nearly 100% volatilization of the samples at the end. In Fig. 7, the liquid bismuth metal gradually disappeared with the temperature elevated. It indicated that the liquid bismuth metal produced in reaction (3) formed volatile phase in reductive atmosphere.

Figure 8 showed the XRD diffraction peak of the condensate formed on graphite paper by condensation of the volatile products generated under high-temperature carbothermal conditions. By XRD analysis of the graphite paper with the experimental heating temperature of 1200 °C, bismuth metallic simple was found. Tiny liquid beads of bismuth metal were also observed by cutting the quartz tube in the condensation section after the experiment at 1200 °C. Therefore, the main volatile product was bismuth and the reaction (3) was the main reduction reaction. The reaction (4) produced less low-valence bismuth oxides without obvious diffraction peak. Then the liquid bismuth generated in reaction (3) formed volatile bismuth at high temperature, which could be volatilized in reductive atmosphere, and high the temperature was conducive to the transformation of liquid bismuth into volatile bismuth. Therefore, it can be inferred that the volatile bismuth is mainly the volatile Bi(g) elemental and Bi_2 phase. So the higher temperature was favorable for bismuth element volatilization and enrichment in the soot during the carbothermic reduction process.

Figure 9a showed the scanning electron microscope (SEM) image of condensate in the lower temperature condensation region at 500 times magnification, which the condensate was collected in the experiment at 1200 °C. Spherical particles of various sizes scattered on the graphite conductive adhesive were observed from the Fig. 9a. The spherical particles with smooth surfaces were observed in the SEM image with the higher magnification of 1000 times. Figure 10 showed the electron microscope scanning image and energy dispersive spectrum analysis of condensate in the lower temperature condensation, which the condensate was collected in the experiment at 1200 °C. Elemental analysis of these spherical particles showed that it was mainly composed of Bi. The spherical particles were probably tiny droplet of bismuth metal or some kind of the condensed phase of bismuth. It could be seen from Fig. 11 that the main element of spherical particles with smooth surfaces was only bismuth without other elements.

Figure 12a, b showed the SEM image of condensate in the higher temperature condensation region at 100 times magnification, which the condensate was collected in the experiment at 1200 °C. Irregular spherical particles with rough surfaces of various sizes scattered on the graphite conductive adhesive were observed from the Fig. 12a, b. The irregular spherical particles in Fig. 12a were littler than that in Fig. 12b. It could be inferred that the irregular particles grew from small to large. Figure 13 showed the SEM image and elemental mapping of condensate in the

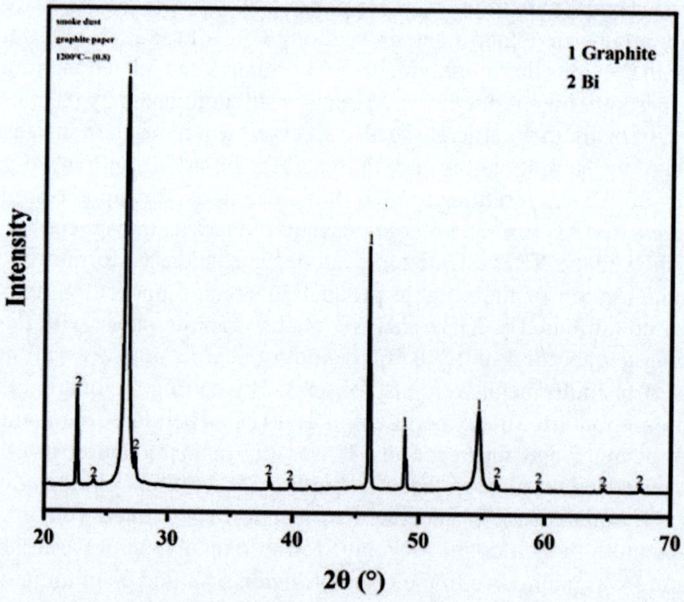

Fig. 8 XRD diffraction peak of condensate on graphite paper

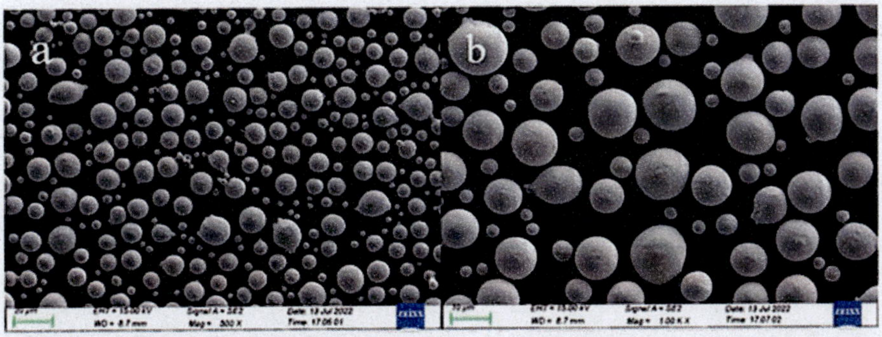

Fig. 9 The scanning electron microscope (SEM) images of condensate in the lower temperature region (1200 °C): **a** 500 times magnification and **b** 1000 times magnification

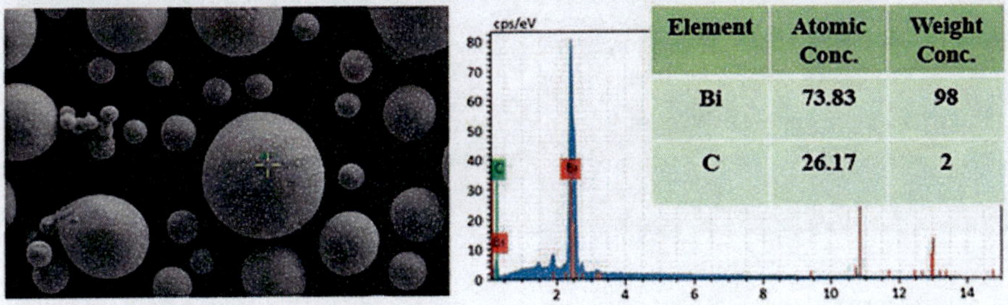

Fig. 10 The scanning electron microscopy image and EDS spectra of condensate in the lower temperature region (temperature = 1210 °C, particle size = 12 mm, reduction time = 25 min)

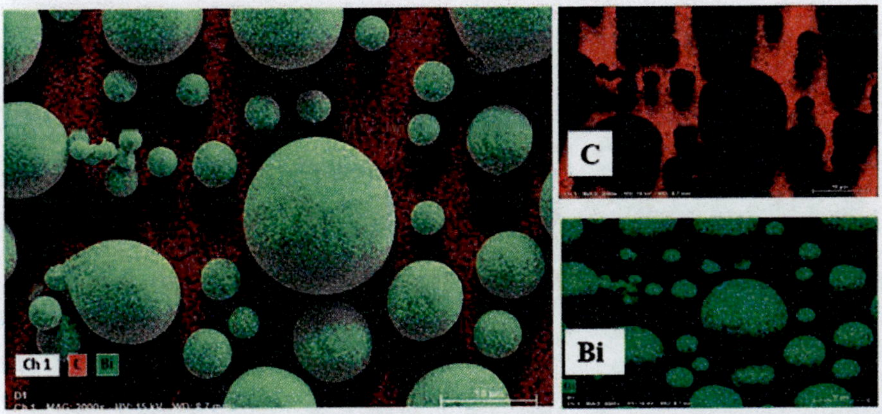

Fig. 11 The SEM image and elemental mapping of condensate in the lower temperature

higher temperature. As can be seen from the Fig. 13, the irregular spherical particles contained not only bismuth but also oxygen, indicating the existence of oxygen-containing bismuth compounds. According to the possible reduction and disproportionation reactions, the oxides of bismuth are bismuth oxide (Bi_2O_3) and bismuth low-price volatile oxide (BiO). The oxide of bismuth should be Bi_2O_3 because of the nature that BiO disproportionation reaction was easy to occur at low temperature. Therefore, the final volatile products at 1200 °C was bismuth elemental and bismuth oxide (Bi_2O_3), and the disproportionation reaction (5) occurs in BiO during the volatilization process.

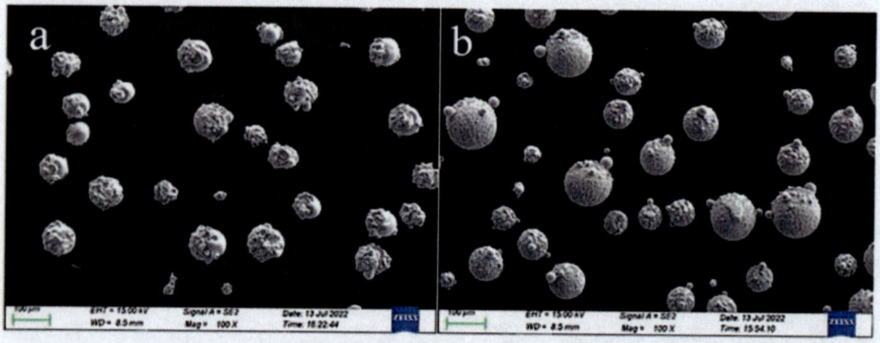

Fig. 12 The SEM images of condensate in the lower temperature region (1200 °C 100 times magnification): **a** littler particles and **b** bigger particles

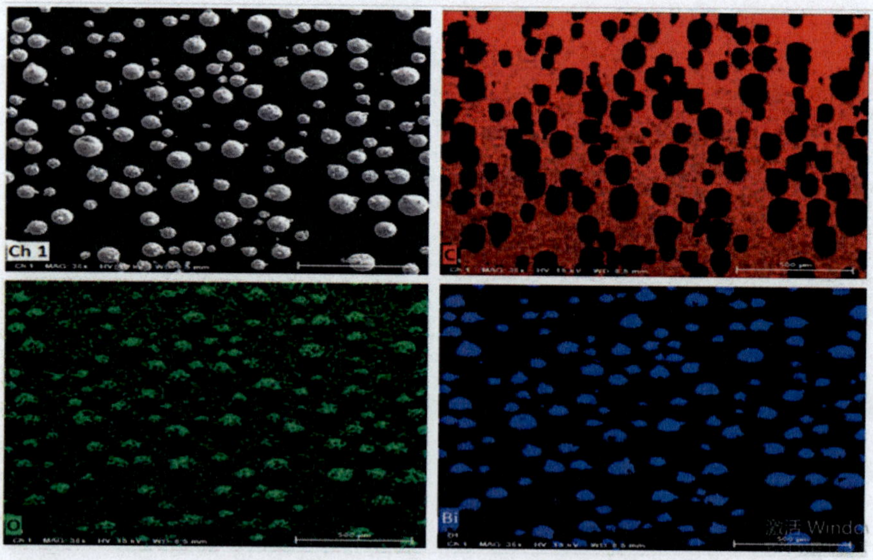

Fig. 13 The SEM image and elemental mapping of condensate in the higher temperature

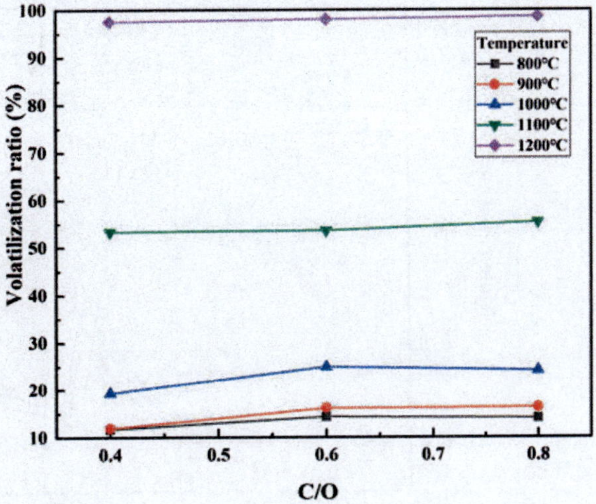

Fig. 14 Relationship between calcination temperature and carbon–oxygen ratio

Effect of Carbon Content

The effect of carbon content of the tables was investigated by varying C/O ratio of 0.4, 0.6, and 0.8. As shown in Fig. 14, with the continuous increase of C/O ratio, the influence of the volatilization rate of the tablets was not obvious, especially at 1100 and 1200 °C. In the low temperature section less than 1100 °C, the volatilization rate of the table with C/O ratio of 0.6 and 0.8 was slightly higher than that of the table with C/O ratio of 0.4, which indicated that the higher the carbon content, the higher the carbothermic reduction reaction rate of the tablet. However, as the experimental temperature gradually increased, the reduction reaction was more sufficient, and the amount of volatile phase formed at the same time was greater, so the effect of carbon content to the volatilization rate of the tablet was smaller.

Thermodynamic Simulation Calculation

The relevant thermodynamic data in thermodynamics manual and Factsage software were consulted, and the calculated results were shown in Fig. 15. The main product of bismuth oxide carbothermal reduction was the volatile Bi(g) elemental and Bi_2 phase.

Carbothermal Reduction of Table

The volatile product was green powder dust under 1200 °C in Fig. 16. After XRD phase identification, the main composition was Bi and Bi_2O_3. This also verified the occurrence of reactions (3)–(5). In the direct reduction process of rotary kiln,

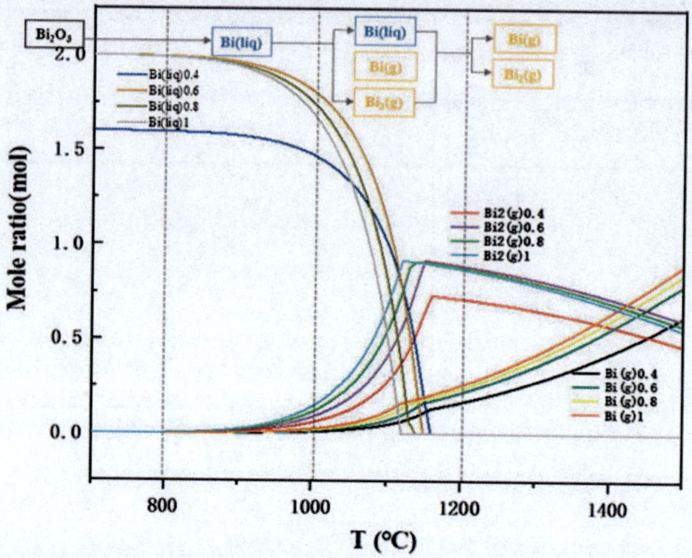

Fig. 15 Thermodynamic simulation of bismuth oxide carbothermal reduction

the enrichment rate of bismuth element in blast furnace dust sludge in secondary volatilization soot is about 90%. The bismuth elements in secondary soot are mainly bismuth and bismuth oxide in reducing atmosphere and can be oxidized to bismuth oxide in the process of secondary oxidation [1, 11].

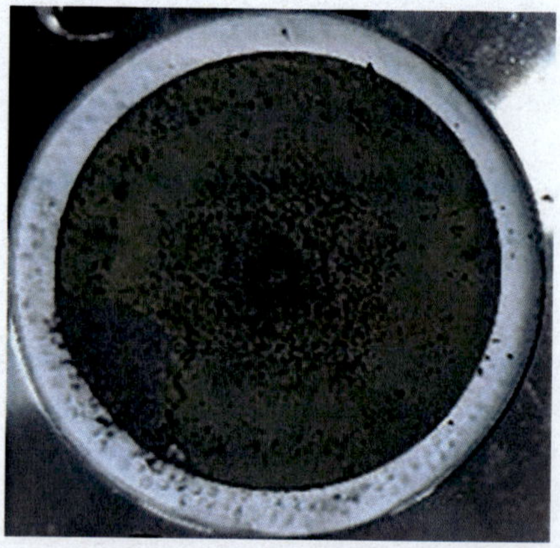

Fig. 16 The volatile product at the temperature of 1200 °C

Conclusions

In this study, we proposed a new vacuum carbothermal reduction method to recover Bi from blast furnace dust and experimentally demonstrated the method's feasibility. The migration route of bismuth element in dust during carbothermal reduction was studied in detail. The main contents and conclusions of this study were as follows.

1. The results of thermodynamic calculations under show that Bi_2O_3 reacts with carbon form Bi and BiO, respectively. The actual reaction temperature of Bi_2O_3 and graphite was subsequently obtained by thermogravimetric analysis.
2. Through experiments with pure Bi_2O_3 and graphite, Bi and bismuth containing oxide was obtained in the condensation zone. The calculation results of the saturated vapor pressure show that the presence of Bi in the condensation zone was due to either its vaporization under the high temperature, or the disproportionation reaction of BiO. The final volatile products at 1200 °C was bismuth elemental and bismuth oxide (Bi_2O_3), and the disproportionation reaction (5) occurs in BiO during the volatilization process.
3. With both our theoretical calculations and experimental results, blast furnace ash as the raw material also obtained enriched bismuth and bismuth containing oxide in the secondary soot with direct reduction process in rotary kiln. This demonstrates the feasibility of recovering Bi from blast furnace ash by direct reduction process.

References

1. Wang L-G (1986) Bismuth metallurgy. Metallurgical Industry Press, Beijing
2. Glasgow GP (1991) The safety of low melting point bismuth/lead alloys: a review. Med Dosim 16:13–18
3. Hu S-D, Wang C, Wang X-Y, Li L-X, Liu S (2022) Process optimization of planetary rolling of bismuth-containing austenitic stainless steel. J Mater Eng Perform 31:2012–2022
4. Panousi S, Harper EM, Nuss P, Eckelman MJ, Hakimian A, Graedel TE (2016) Criticality of seven specialty metals. J Ind Ecol 20:837–853
5. Hespanhol MC, Patrício PR, da Silva LHM, Vargas SJR, Rezende TCS, Campos RA (2019) A sustainable methodology to extract bismuth from secondary sources. J Braz Chem Soc 30:2376–2385
6. She X-F, Wang J-S, Xue Q-G, Ding Y-G, Zhang S-S, Dong J-J, Zeng H (2011) Basic properties of steel plant dust and technological properties of direct reduction. Int J Miner Metall Mater 18:277–284
7. Xu J, Wang N, Chen M, Xin J, Li X, Li H, Wang Y (2018) Reduction behavior of blast furnace dust particles during in-flight processes. Ind Eng Chem Res 57:111–121
8. Peng C, Zhang F, Huifang LI, Guo ZJII (2009) Removal behavior of Zn, Pb, K and Na from cold bonded briquettes of metallurgical dust in simulated RHF. 49:1874–1881

9. Sarafraz MM, Jafarian M, Arjomandi M, Nathan GJ (2019) Experimental investigation of the reduction of liquid bismuth oxide with graphite. Fuel Process Technol 188:110–117
10. Ye D-L (2002) Practical inorganic thermodynamics data book. Metallurgical Industry Press, Beijing
11. Zhu Y-P (2009) Recovery of indium, zinc and bismuth from blast furnace gas ash. Non-Ferr Metals (Smelt Part) 14–16

Recovery of Valuable Metals from Li-Ion Battery Waste Through Carbon and Hydrogen Reduction: Thermodynamic Analysis and Experimental Verification

Bintang A. Nuraeni, Katri Avarmaa, Leon H. Prentice, W. John Rankin, and M. Akbar Rhamdhani

Abstract The digitalisation of the world and electrification of transportation have been driving the increased usage of rechargeable batteries in the past decade. Recycling and recovering valuable metals from Li-ion batteries waste is critical for securing the resources to meet the future demand and production of batteries. In this study, recycling of a battery waste, a black mass containing multiple metals from the cathode and carbon anode, was evaluated using a pyrometallurgical route. Reduction of battery cathode material using carbon from the anode combined with hydrogen as reductant was systematically investigated at 400–1000 °C. The study included thermodynamic assessment using the FactSage™ thermochemical package combined with selected experimental results. The overall reduction mechanism was found to be uniquely characterized by the reduction temperature. The data and information obtained can be employed to develop and optimize the recycling process of Li-ion batteries.

Keywords Recycling · Material recovery · Secondary resource · Hydrogen reduction · Li-ion battery · Thermodynamic assessment

---

B. A. Nuraeni (✉) · K. Avarmaa · W. J. Rankin · M. A. Rhamdhani
Fluid and Process Dynamics (FPD) Group, Department of Mechanical and Product Design Engineering, Swinburne University of Technology, Melbourne, VIC 3122, Australia
e-mail: bnuraeni@swin.edu.au

M. A. Rhamdhani
e-mail: arhamdhani@swin.edu.au

L. H. Prentice
CSIRO Manufacturing, Clayton, VIC 3169, Australia

R. G. Reddy et al. (eds.), *New Directions in Mineral Processing, Extractive Metallurgy, Recycling and Waste Minimization*, The Minerals, Metals & Materials Series,
https://doi.org/10.1007/978-3-031-22765-3_38

Introduction

Li-ion batteries have widely used in the applications of energy storage systems, such as portable devices, electric vehicles, and in the grid-energy storage [1–4]. Li-ion batteries contain a variety of valuable metals (Li, Co, Mn, Ni) from the cathode, carbon from the anode, and organic compounds. Over the next two decades, the development of electric vehicles and storages will spur battery demand causing a significant need to extract minerals for battery materials. While the pressure to scale up all stages of the supply chain is high, extraction and processing are particularly critical due to long lead times [5].

Recycling valuable metals from all possible resources is critical for securing availability of metals in the future. For example, Li-ion batteries contain 5–7 wt% Li which is a much higher concentration than the lithium content in natural resources [6]. Thus, spent batteries must be seen as valuable resources. Recycling of spent Li-ion batteries is also inevitable to deal with the accumulation of spent batteries since the waste is harmful to the environment. In the field of spent Li-ion battery recycling, the pyrometallurgical route is one of the alternatives due to its simplicity and high productivity.

In this study, a high-temperature reduction process using carbon and hydrogen gas was investigated, aiming to recover valuable metals from Li-ion battery waste. A comprehensive evaluation of the thermodynamic feasibility was conducted and assessed by isothermal reduction experiments. Along with H_2, carbon was used as the supporting reducing agent as it is contained in the anode of the battery mixed with the valuable metals in the waste.

Materials and Methodology

Black Mass Characterization

The black mass that contained cathode, anode and other components of waste Li-ion batteries was supplied by Envirostream Australia Pty Ltd. The elemental composition of the black mass is presented in Table 1, and the component classification is shown in Table 2. Lithium Cobalt Oxide (LCO) and Lithium Nickel Manganese Cobalt Oxide (NMC) are the predominant Li-ion battery cathode materials contained in the black mass. The morphology of black mass powder is shown in Fig. 1a and a cross-section image of NMC-cathode powder is shown in Fig. 1b.

Table 1 Elemental analysis of as-received black mass

Element	Wt%
C	37–40
Co	8–9
Mn	1–5
Ni	14–16
Al	3–4
Cu	2–4
Fe	<5
Li	2–4
P	2–4
F	<2
O	Bal

Table 2 Component classification of as-received black mass

	Material	Wt%
Anode	Graphite	~35–50
Cathode	Lithium Nickel Manganese Cobalt Oxide (NMC)	~5–30
	Lithium Cobalt Oxide (LCO)	~10–30
	Lithium Manganese Oxide (LMO)	~0–5
	Lithium Ferro Phosphate (LFP)	~0–5
Other	Electrolyte ($LiPF_6$)	~0–5
	Binder (PVDF)	
	Substrate (Al, Cu foil)	

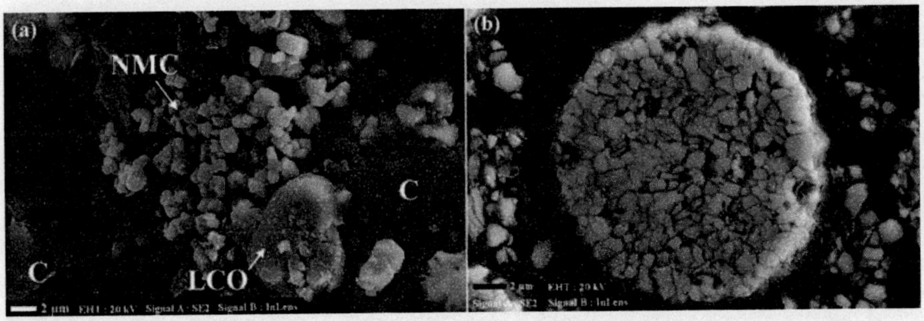

Fig. 1 **a** SEM-SE image of black mass powder, **b** cross-section of NMC-cathode grains

Thermodynamic Assessment

Thermodynamic assessment was carried out using FactSage© 7.2 (Thermfact/CRCT, Montreal, Canada), with the Equilib module. The Equilib module calculates the equilibrium state of a system under given constraints using information from the chosen database(s) by means of Gibbs energy minimisation. Two built-in databases, FactPS (database for pure substances) and FToxid (database for oxides), were applied for the thermodynamic calculations [7–9]. Since some of the thermodynamic data were not available in the databases, a customized database was generated using available thermodynamic data found in the literatures [10–23].

Reduction Experiment and Characterization

Prior to the experiments, the black mass was weighed and mixed in a ball-mill (146 mm diameter × 203 mm long) using 100 alumina balls of 8 mm diameter. The powders and balls were placed into the milling jar and milled for 3 h at 100 rpm. The mixed powders were then briquetted using a cold press with a load of 10 tonne for 5 min at room temperature. The final pellets had a dimension of approximately 13 mm in diameter and 12 mm in thickness.

The experiments were performed in a water-cooled vertical tube resistance furnace equipped with Mo_2Si heating elements. The ends of the tube were sealed with water-cooled jackets. A pellet sample was placed inside an 18 mm diameter cylindrical alumina crucible and was placed in the hot zone of the furnace and heated isothermally at 600, 800, and 1000 °C for 12 h. The furnace was purged with Ar (0.5 L/min) for 5 min before switching to H_2 (0.3 L/min) to eliminate oxygen in the system prior to placement of sample.

To identify the existing phases, X-ray Powder Diffraction (XRD) characterization was carried out using a Bruker XRD (Bruker AXS GmbH). The analyses were carried out using Cu-$K\alpha$ radiation and a 2θ between 5 and 140° at a step size of 0.02° (~2 h per scan) operated at 40 kV and 40 mA. Phase identification was performed using DIFFRAC.EVA V4 software equipped with the International Centre of Diffraction Data (ICDD) PDF2 + database.

Scanning Electron Microscope (SEM) (Zeiss SUPRA-40), in conjunction with Energy Dispersive X-ray Spectroscopy (EDX) (INCA Suite 5.05), were used to conduct analyses on microscopic images, elemental mapping, and semi-quantitative chemical compositions. The EDX results (ratio of different elements) were used to deduce the possible phases during reaction (cross checked with XRD results). Secondary Electrons (SE) and Angle Selective Backscatter (AsB) signals were used for the morphology imaging and EDX analysis, respectively.

Results and Discussion

Thermodynamic Assessment of Black Mass Reduction by C and H_2

Thermodynamic assessment was performed to investigate the temperature dependence of the equilibrium phases with H_2 and C under 1 atm total pressure at 400–1000 °C. The input of the reaction was as follows:

$$LiCoO_2 + LiNi_{0.333}Mn_{0.333}Co_{0.333}O_2 + LiMn_2O_4 + LiFePO_4 + 3C + H_2$$

The molar ratio applied for all cathode materials was assumed to be equal to H_2, aside for C content which was in larger quantity based on the common battery composition.

Figure 2a shows that formation of metallic Co and Ni was thermodynamically possible in full temperature range studied, although some CoO and NiO were present as minor phases (Fig. 2c). Instead, Mn and Fe were only possible to reduce as MnO and FeO with small formations of spinel phases: $MnFe_2O_4$ and Fe_3O_4 (Fig. 2c). Li could not be recovered as pure metal, but it formed various compounds of Li_3PO_4, Li_2CO_3, and $LiFeO_2$ (Fig. 2b). Li_3PO_4 was stable in full temperature range while a slight decrease of Li_2CO_3 occurred before $LiFeO_2$ started to form above 900 °C.

The gas by-products were predicted to include H_2, H_2O, CO, CO_2, and CH_4 (Fig. 3). Higher temperature favoured the formation of CO instead of CO_2, and H_2 instead of H_2O. The increase of CO was correlated to the full consumption of C which happened at around 750 °C (Fig. 2a), while H_2 remained stable at higher temperatures. This might indicate that C is more reactive than H_2. Low reactivity between C and H_2 was also shown by the minor formation of CH_4 below 800 °C. Overall, CO and H_2 were more utilized on the reduction process at lower temperatures (<750 °C) while at higher temperatures, an increase in CO/CO_2 and H_2/H_2O was observed which would cause to more reducing environment. However, no significant change in the reduced phases was observed with temperature increase in this range.

Experimental Reduction of Black Mass Reduction by C and H_2

Figure 4a shows the phase analyses from XRD characterization of the reduced pellets at 600, 800, and 1000 °C. At all temperatures, a significant amount of C was still present and $LiCoO_2$ existed as a non-stoichiometric $Li_{0.185}Co_{0.815}O$. Metallic Ni was recovered at all temperatures (peak no. 2); and the higher magnification of this peak is shown on the right-hand side. At 800 and 1000 °C, this peak was broadened and shifted to the left, resulting a detection of two metal phases: Co and Ni. Meanwhile, Fe was recovered as Fe_3P at all temperatures.

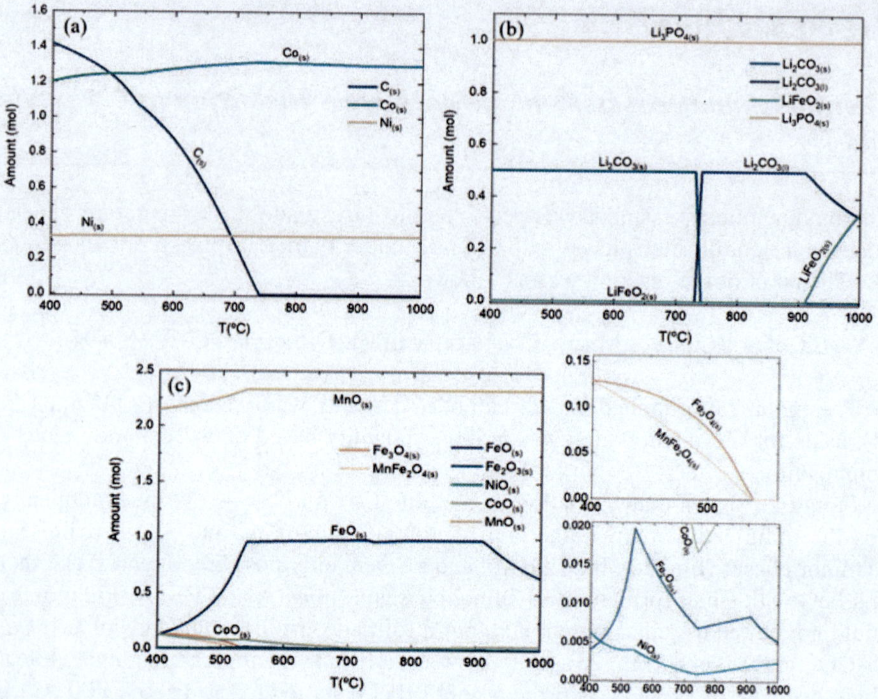

Fig. 2 Predicted equilibrium molar amounts: **a** C, Co, Ni, **b** Li compounds: Li_3PO_4, Li_2CO_3, $LiFeO_2$, **c** monoxide and spinel; left: MnO, FeO, CoO, right: magnified images of low molar amounts of $MnFe_2O_4$, Fe_3O_4, Fe_2O_3, NiO

Fig. 3 Predicted equilibrium molar amount of the gas compounds: H_2, H_2O, CO, CO_2, CH_4

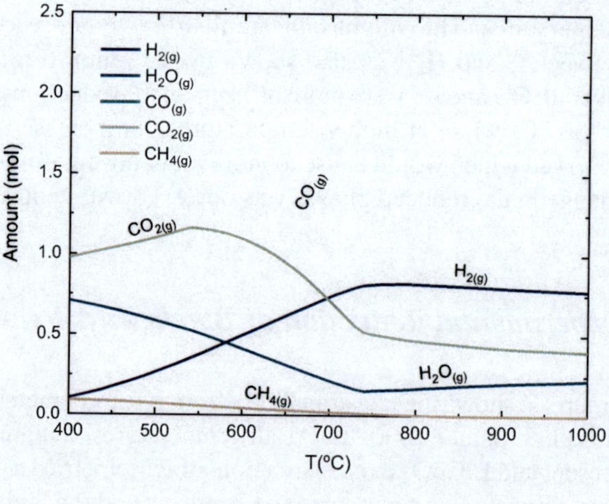

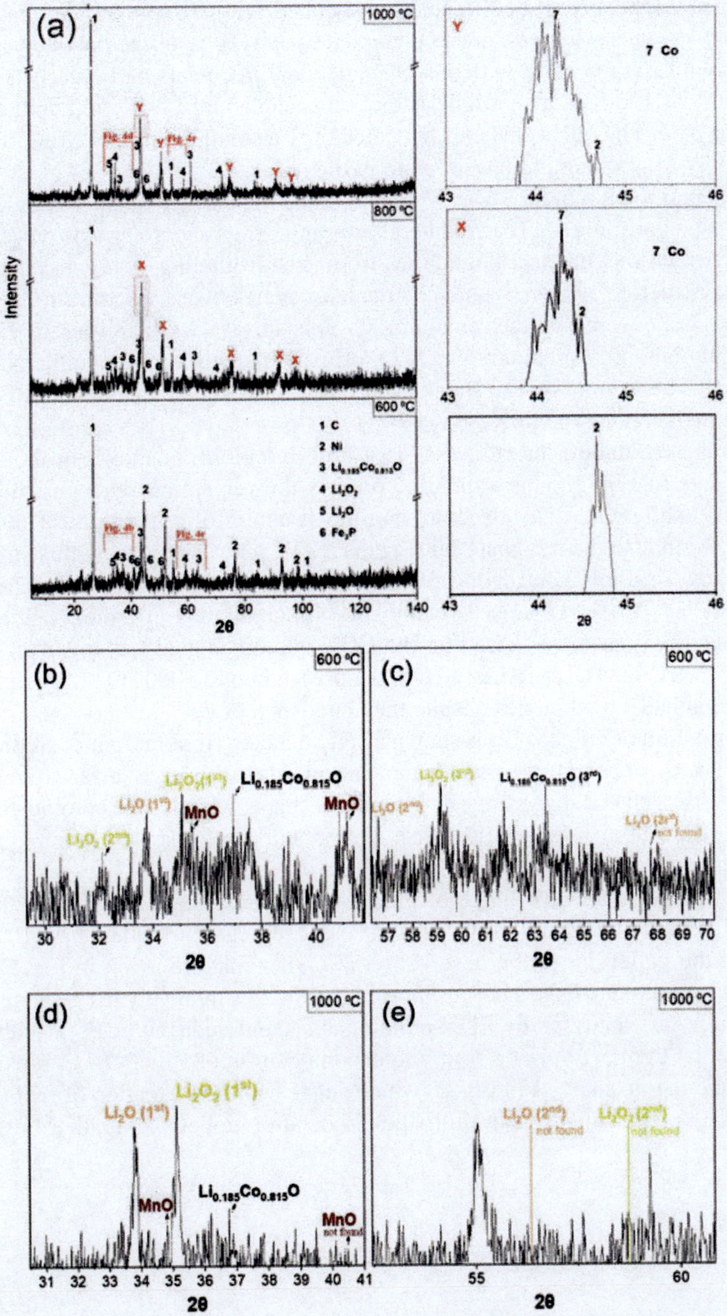

Fig. 4 **a** XRD analyses of the reduced samples at 600, 800, and 1000 °C. Magnified XRD spectra: **b** 600 °C at 2θ = 29–42°, **c** 600 °C at 2θ = 56–70°, **d** 1000 °C at 2θ = 30–41°, **e** 1000 °C at 2θ = 53–60°. First, second, third peak were denoted as 1st, 2nd, 3rd, respectively. XRD spectrum at 600 °C was representative to 800 °C

Enlarged XRD spectra of the peaks associated with Li compounds are shown in Fig. 4b–e, where Li was present mainly as Li_2O_2 at all temperatures. Li_2O appeared with less intensity, with only a single identified peak, particularly at higher temperature of 1000 °C. This suggests that most elemental Li was still contained in $Li_{0.185}Co_{0.815}O$. The XRD analyses indicated only a small amount of MnO, and this was the only Mn-containing phase found in the reduced samples.

High magnification of an SEM-SE image of the top surface of a pellet reduced at 600 °C is shown in Fig. 5. The middle phase, light-grey in colour, was present as a separated structure with clear boundaries from its surrounding. A higher magnification of this structure, shown by the red-square image, shows a smooth surface while some area was covered by the remaining C. This surface was analysed by EDX to contain only Ni and a small amount of O, primarily identified as metallic Ni (refer to Table 3). This result confirm the thermodynamic calculations (shown in Fig. 2a) that metallic Ni should be formed.

The area surrounding metallic Ni, dark grey in colour, is shown in the yellow square image in Fig. 5. This structure contains the remaining reactants of C and $LiCoO_2$. C exhibited a flaky structure and the structure of remaining $LiCoO_2$ was spherical, similar to the raw material (Fig. 1). EDX point analysis on this spherical phase is shown in Table 3 and deduced as non-stoichiometric $LiCoO_2$ with a chemical formula of $Li_{0.185}Co_{0.815}O$. Since Li cannot be measured with EDX, the consideration of Li-containing phase was based on the EDX and XRD analyses together. In the XRD analyses (Fig. 4), $Li_{0.185}Co_{0.815}O$ was observed up to 1000 °C, suggesting that the reduction had slow kinetics despite the abundance of C.

The formation of reduced Co is shown in Fig. 6, taken from the sample reduced at 1000 °C. The Co appeared as micro-grains, in contrast to the large surface of metallic Ni. It was observed that the grains exhibit cubical shape, which is the common crystal structure of Co above 422 °C [24–26]. This result confirms the thermodynamic calculations (Fig. 2a).

Figure 7a shows a SEM image of the Fe-containing phase, taken from the top surface of pellet reduced at 800 °C. There is clear segregation of the phase from the surface of the pellet. From the observation at higher magnification in Fig. 7b, the structure is observed to have a porous morphology with a relatively irregular surface. The structure was analysed by EDX point analysis and deduced as Fe_3P (Table 3). It appears that the Fe_3P formed from the decomposition of $LiFePO_4$. Compared to the thermodynamic analysis in Fig. 2 (which suggested the formation of Li_3PO_4), it appears that kinetics might be limiting the dissociation to FeO, and rather Fe_3P was formed.

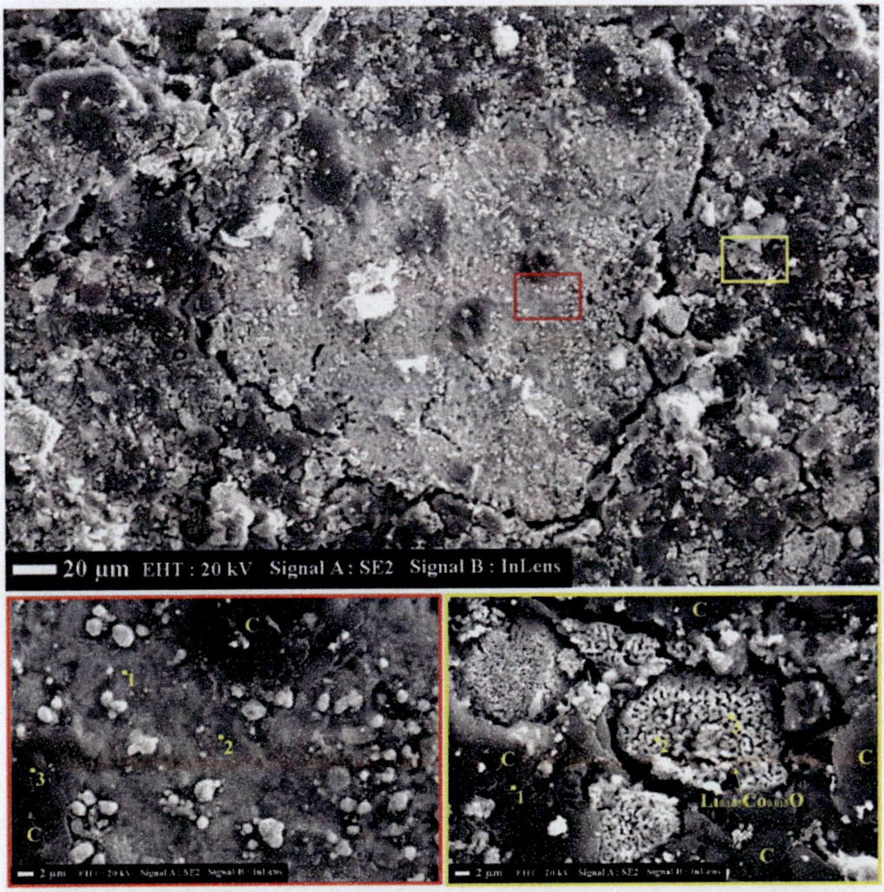

Fig. 5 SEM-SE images of the top surface of a pellet reduced at 600 °C, red: Ni-rich surface, yellow: remaining $LiCoO_2$ and C

Conclusions

Reduction of Li-ion battery waste using C and H_2 has been carried out at 400–1000 °C by thermodynamic assessment supported by experiments. The following conclusions were drawn from the study:

- Thermodynamic assessment: Metallic phases of Co and Ni were feasible to form at all temperatures. Mn and Fe would not be reduced to metallic form, rather only as far as MnO and FeO. Li phases would be generated as various Li compounds of Li_3PO_4, Li_2CO_3, and $LiFeO_2$, where $LiFeO_2$ phase was favoured at higher temperature starting around 900 °C. Full gasification of C to CO was predicted to occur above 750 °C while minor formation of CH_4 occurred below 800 °C.

Table 3 EDX point analysis of reduced black mass referring to SEM images

	Point	Wt (%)						Deduced phase
		Ni	Co	Fe	O	C	P	
Figure 5-red	1	91.02	–	–	8.98	–	–	Ni
	2	94.79	–	–	5.21	–	–	Ni
	3	–	–	–	–	100	–	C (graphite)
Figure 5-yellow	1	–	–	–	–	100	–	C (graphite)
	2	–	86.04	–	13.96	–	–	$Li_{0.185}Co_{0.815}O$
	3	–	84.92	–	15.08	–	–	$Li_{0.185}Co_{0.815}O$
Figure 6	1	–	92.15	–	7.85	–	–	Co
	2	–	96.14	–	3.86	–	–	Co
	3	–		–		–	–	Co
Figure 7	1	–	–	63.71	–	–	36.29	Fe_3P
	2	–	–	61.52	–	–	38.48	Fe_3P
	3	–	–	60.82	–	–	39.18	Fe_3P
Theoretical value		–	75.01	–	24.99	–	–	$Li_{0.185}Co_{0.815}O$
		–	–	75.00	–	–	25.00	Fe_3P

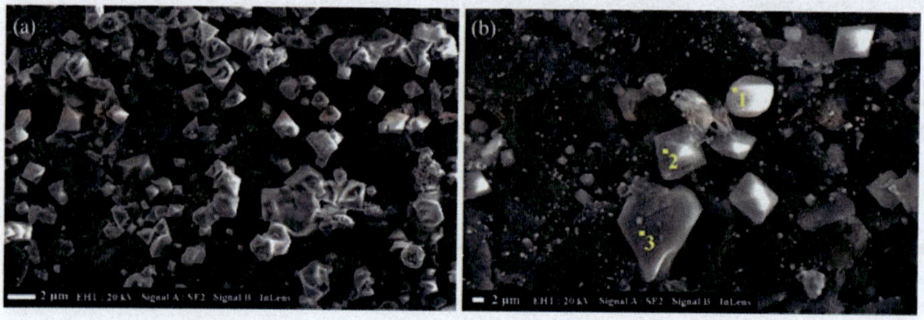

Fig. 6 SEM-SE images of the top surface of pellet reduced at 1000 °C: **a** reduced Co, **b** magnified image showing Co-crystals

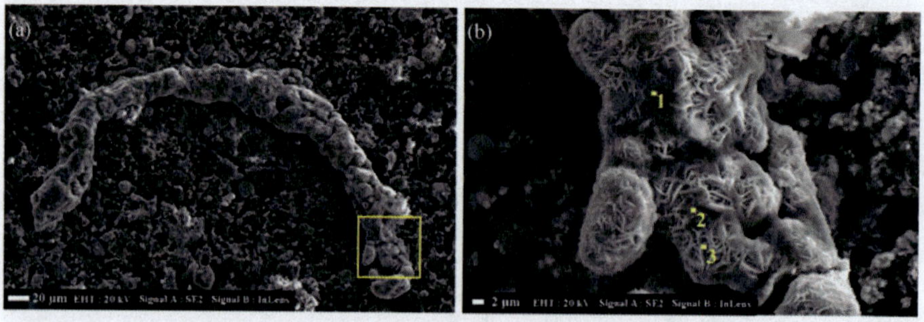

Fig. 7 SEM-SE images of the top surface of pellet reduced at 800 °C: **a** Fe-rich phase, shown by arched structure, **b** magnified image of Fe-rich phase

CO and H_2 were utilized better at lower temperatures (<750 °C) while higher temperature caused an increase in CO/CO_2 and H_2/H_2O ratio.

- Experimental results: Metallic Ni was produced at temperature as low as 600 °C, but the reduction of metallic Co required higher temperature. Fe was recovered as Fe_3P at all temperatures, while Li was recovered as Li_2O_2 and Li_2O. However, the Li compounds were found with low intensity peaks, which may indicate that most Li was still associated with non-stoichiometric $LiCoO_2$ ($Li_{0.185}Co_{0.815}O$) that was visible in the XRD spectra. The presence of excess C and remaining $LiCoO_2$ (as $Li_{0.185}Co_{0.815}O$) suggests that kinetics might limit further reduction.

Acknowledgements The authors thank and acknowledge the financial support from Swinburne University's Automotive Engineering Graduate Program (AEGP) by the Australian Government through the Department of Industry, Science, Energy, and Resources (previously the Department of Industry, Innovation, and Science) and the Commonwealth Scientific and Industrial Research Organisation (CSIRO).

Conflict of Interest The authors declare that they have no conflict of interest.

References

1. Dunn JB, Gaines L, Sullivan J, Wang MQ (2012) Environ Sci Technol 46:12704
2. Xu Y, Dong Y, Han X, Wang X, Wang Y, Jiao L, Yuan H (2015) ACS Sustain Chem Eng 3:2435–2442
3. Huang B, Pan Z, Su X, An L (2018) J Power Sources 399:274–286
4. Yao Y, Zhu M, Zhao Z, Tong B, Fan Y, Hua Z (2018) ACS Sustain Chem Eng 6:13611–13627
5. IEA (2022) Global supply chains of EV batteries. All rights reserved
6. Shin SM, Kim N, Sohn J, Yang D, Kim Y (2005) Hydrometallurgy 79:172–181
7. Chase MW (1986) JANAF thermochemical tables, 3rd edn. American Chemical Society, Washington
8. Dinsdale AT (1991) Calphad 15:317–425
9. Jung I-H, Decterov S, Pelton A, Kim H, Kang Y-B (2004) Acta Mater 52:507–519
10. Lee WK, Choy CL (1975) J Polym Sci Polym Phys Ed 13:619–635
11. Zhang YC, Tagawa H, Asakura S, Mizusaki J, Narita H (1997) J Electrochem Soc 144:4345–4350
12. Kawaji H, Oka T, Tojo T, Atake T, Hirano A, Kanno R (2002) Solid State Ionics 152–153:195–198
13. Chernova NA, Ma M, Xiao J, Whittingham MS, Breger J, Grey CP (2007) Chem Mater 19:4682–4693
14. Idemoto Y, Matsui T (2008) Solid State Ionics 179:625–635
15. Knyazev AV, Mączka M, Smirnova NN, Knyazeva SS, Chernorukov NG, Ptak M, Shushunov AN (2014) Thermochim Acta 593:58–64
16. Loos S, Gruner D, Abdel-Hafiez M, Seidel J, Hüttl R, Wolter AUB, Bohmhammel K, Mertens F (2015) J Chem Thermodyn 85:77–85
17. Cupid DM, Li D, Gebert C, Reif A, Flandorfer H, Seifert HJ (2016) J Ceram Soc Jpn 124:1072–1082
18. Gotcu P, Seifert HJ (2016) Phys Chem Chem Phys 18:10550–10562
19. Jankovský O, Kovařík J, Leitner J, Růžička K, Sedmidubský D (2016) Thermochim Acta 634:26–30

20. Cupid D, Gotcu-Freis P, Beutl A, Bergfeldt T, Giel H, Henriques D, Kozlov A, Masoumi M, Seidel J, Flandorfer H, Markus T, Mertens F, Schmid-Fetzer R, Seifert H (2017) Int J Mater Res
21. Masoumi M, Cupid DM, Reichmann TL, Chang K, Music D, Schneider JM, Seifert HJ (2017) Int J Mater Res 108:869–878
22. Wang S, Zhang G, Gao J, Wang J, Wang Y, Nan C, Huang D, Chen L, Song J, Luo P (2018) First-principles study on $LiFePO_4$ materials for lithium-ion battery. In: International workshop on materials, chemistry and engineering - IWMCE, Xiamen, China
23. Masoumi M (2019) Thermochemical and electrochemical investigations of $Li(Ni,Mn,Co)O_2$ (NMC) as positive electrode material for lithium-ion batteries
24. Kazantseva NV, Stepanova NN, Rigmant MB (2018) Superalloys: analysis and control of failure process, 1st edn. CRC Press, Boca Raton, FL
25. Knapek M, Minárik P, Dobroň P, Šmilauerová J, Celis MM, Hug E, Chmelík F (2020) Materials 13:5775
26. Schuett FM, Esau D, Varvaris KL, Gelman S, Björk J, Rosen J, Jerkiewicz G, Jacob T (2020) Angew Chem Int Ed 59:13246–13252

Selective Extraction of Vanadium from Sodium Tungstate Solution Using P507

Hanyu Wang, Guihong Han, Yanfang Huang, Shengpeng Su, Bingbing Liu, and Kunpeng Shi

Abstract Separation of tungsten (W) and vanadium (V) is of great significance for the utilization of the polymetallic complex minerals and secondary resources of these elements. In this work, selective extraction of V from sodium tungstate solution was investigated using 2-ethylhexyl phosphonic acid mono-2-ethylhexyl ester (P507) as an extractant. The extraction process was studied to optimize various parameters including initial pH, extractant concentration, phase ratio (O/A), temperature, and contact time. The simulation calculation indicated W and V exist in the form of cation $HW_7O_{24}^{5-}$ and anion VO_2^+, respectively, in solutions with pH ranging from 1 to 3, which provided a theoretical basis for the selective separation of W and V. In addition, under the optimum conditions (pH = 1.5, 20% (v/v) P507, O/A = 1/2, room temperature, and T = 20 min), the removal efficiency of V exceeds 90% with W loss below 6%.

Keywords Sodium tungstate · Vanadium · P507 · Solvent extraction

Introduction

W and V have been wildly applied in many fields, including catalysts, weapons, atomic energy, and aerospace [1–3]. Due to their remarkable physical and chemical properties, these metals have been considered strategic for the development of new technologies [4, 5]. With the continuous consumption of traditional high-quality resources, the research on the recovery of W and V from various resources is stimulated, such as nickel molybdenum ore, scrap carbides, scrap steel alloys, and the spent catalysts [6–9]. Especially, spent selective catalytic reduction (SCR) catalyst has attracted widespread attention of numerous researchers, due to its increasing discard amount [10, 11]. Hence, considering that normally both metals are located together in spent SCR catalyst, it is necessary to dispose of available procedures for their extraction and separation.

H. Wang · G. Han (✉) · Y. Huang · S. Su · B. Liu · K. Shi
School of Chemical Engineering, Zhengzhou University, Zhengzhou 450001, P.R. China
e-mail: hanguihong@zzu.edu.cn

R. G. Reddy et al. (eds.), *New Directions in Mineral Processing, Extractive Metallurgy, Recycling and Waste Minimization*, The Minerals, Metals & Materials Series,
https://doi.org/10.1007/978-3-031-22765-3_39

In recent years, the process of extracting valuable metals from spent SCR catalysts by soda roasting/water leaching has been the most studied by researchers [12, 13]. This combined processes that alkaline reagents (mainly NaOH and Na_2CO_3) are used for roasting, and the resulting sodium vanadate and tungstate can be solubilized in hot water, separating W and V from titanium (Ti) and other metallic impurities [14]. However, the main challenge is still the effective separation, at a reasonable cost, of the two metals in order to produce high-grade salts with commercial applications [13]. Compared with precipitation and ion exchange, solvent extraction possesses significant advantages including high selectivity, superior separation efficiency, and easy operation, which is considered as the most promising method recommended for separate W and V [15]. Currently, chelate extractants (LIX 63) and amine extractants (Aliquat336) are the popular choices for selective separation of W and V in mildly acidic or basic solutions [16, 17]. However, the separation of the two elements in acidic solution (pH < 2) has not been reported. P507, as an acidic extractant, possesses excellent characteristics, such as high extraction efficiency, low price, and also possesses the potential to separate W and V in acidic solution conditions [18].

The objective of present investigation is to selectively separate V from sodium tungstate solution using P507 as an extractant. The feasibility of selective separation of V was confirmed by theoretical analysis. Then various factors affecting the extraction efficiency of V including initial pH, extractant concentration, O/A ratio, and contact time were systematically investigated, simultaneously, and the key parameters affecting the stripping efficiency of V were determined. Ultimately, based on the present study, a new method for the selective separation of W and V was proposed.

Experiment

Materials

The mixed solutions of W and V were prepared from sodium tungstate dihydrate, sodium vanadate, and both the initial concentrations of W and V were identified as 1 g/L. The organic phase that was used in this experiment consisted of extractant P507 and diluent sulfonated kerosene. Hydrochloric acid (HCl) was applied to regulate the acidity of solutions. All other reagents adopted in the experiments were of analytical grade without further purification, and deionized water was used throughout the experiments.

Experimental Procedures

In each experiment, a certain concentration of organic phase was prepared by mixing P507 with sulfonated kerosene. Then the acidity of solution containing W and V was adjusted to different pH values by HCl. The aqueous phase and organic phase with different volume ratios were transferred into a conical bottle and shaken in conical flasks for a predetermined time, and the extraction experiments were conducted in a shaker with a speed of 200 rpm/min. Unless otherwise mentioned, the temperature was 25 °C with an error of less than ±0.1 °C. In addition, aqueous and organic solutions were equilibrated using a separating funnel for 10 min.

Analytical Methods

The pH value of solution was measured by a Shanghai Mettler Instruments pH meter, and the concentrations of W and V were determined by ICP-OES (Thermo Scientific). The simulation software of Visual MINTEQ was adopted to calculate the species distributions. Additionally, the concentration of metal ions was calculated by subtraction, and all experiments were carried out in duplicate to ensure accuracy. The key indexes of the separation process were calculated by Eq. (1):

$$E = \left(1 - \frac{C_{aq}}{C_0}\right) \times 100\% \tag{1}$$

where E represents the extraction efficiency of W and V, %, C_0 and C_{aq} are the concentration of W or V in the initial solution and raffinate, mg/L.

Results and Discussion

Thermodynamic Considerations

The selective separation of W and V is mainly based on the differences in physico-chemical properties between the elements [19], including the ability of W and V to polymerize and the ability of V to form cations in acidic solution [6]. Therefore, the common ion species of W and V in acidic solutions were investigated by simulation calculations of Visual MINTEQ software, and the molar rate of the W and V species as a function of pH value in W-H_2O and V-H_2O systems was given in Fig. 1. It was found that V can form cationic $VO_2{}^+$ species, while W exists in the form of oxyanion under the pH range of 0–2. Based on this difference, the use of cationic extractants is expected to selectively separate V from solution, which also provided a theoretical basis for the selective separation of W and V. The chemical reaction of solvent extraction is expressed as Eq. (2) [20]:

$$VO_2^+ + \overline{H_2A_2} = \overline{VO(A)_2} + 2H^+ \tag{2}$$

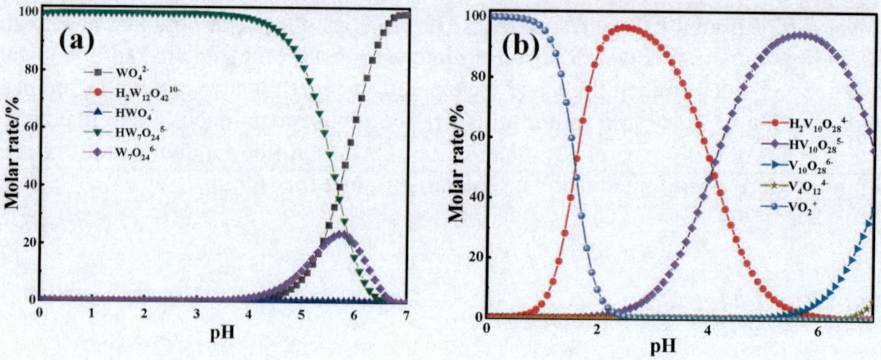

Fig. 1 Molar ratio of W (**a**) and V (**b**) species as function of pH value in W-H_2O and V-H_2O system ($C_W = 1.0$ g/L, $C_V = 1.0$ g/L, 25 °C)

where H_2A_2 represents the dimer of the phosphine extractant, and the overline represents the organic phase.

Effect of Initial pH Value

The effect of pH value on metal extraction was investigated by using 20% (v/v) P507 with a feed solution at different pH and an O/A ratio of 1/2, and the results were shown in Fig. 2. It can be observed in Fig. 2 that V maintained a high extraction efficiency when the solution pH was less than 2. However, the extraction efficiency of V significantly decreased with the pH further increased, and the extraction efficiency of W varied between 2 and 3%. Therefore, the pH value is an important parameter during the extraction process, which significantly influences the extraction efficiency of metal ions. The main reason is the gradual transformation of VO_2^+ species into $H_2V_{10}O_{28}^{4-}$ species with increasing pH value. In addition, at the solution pH value of 1.5, the single-stage extraction efficiency of W and V are 5.4% and 87.93%, respectively, and the separation factor is as high as 127.34. Thus, the optimum pH value of the initial aqueous phase is determined to be 1.5.

Effect of Extractant Concentration and Phase Ratio

To improve the separation efficiency and reduce the consumption of organic phase, the effect of extractant concentration and O/A on the extraction of W and V were investigated. As shown in Fig. 3a, the concentration of extractant has little effect on the extraction efficiency of W and V. The extraction efficiency of V increased slightly with the increase of extractant concentration. However, the extraction efficiency of

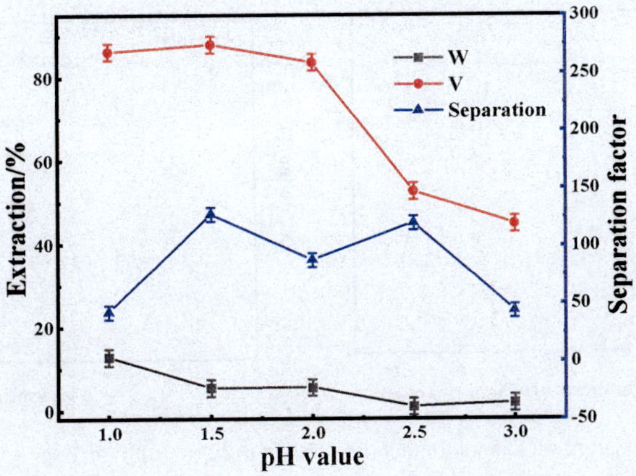

Fig. 2 Effect of initial pH on the extraction of W and V (20% P507, O/A = 1/2, T = 25 °C, and t = 20 min)

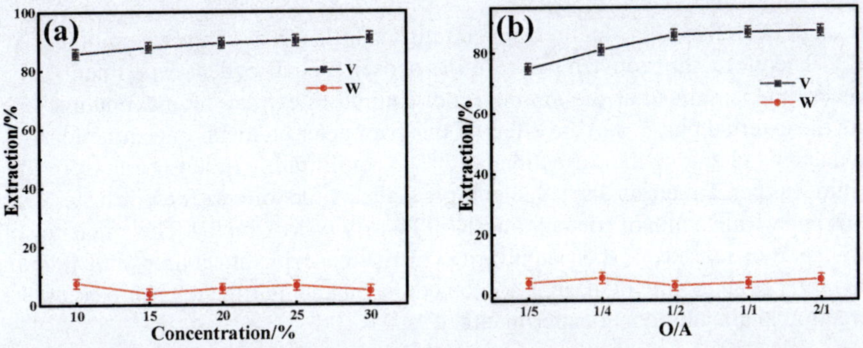

Fig. 3 Effect of organic phase on the extraction of W and V. **a** Extractant concentration (pH = 1.5, O/A = 1/2, room temperature, and t = 20 min), **b** phase ratio (pH = 1.5, 20% P507, room temperature, and t = 20 min)

V remained basically unchanged when the concentration of the extractant exceeded 20%. In light of this, the optimum extractant concentration is intended to be 20%.

The extraction experiments with O/A ranging from 1/5 to 2/1 were conducted with the solution initial pH of 1.5, and the results were displayed in Fig. 3b. As the O/A increased from 1/5 to 2/1, the extraction efficiency of V increased by 12.17% (from 74.83 to 87.00%), and the extraction efficiency of W remained basically unchanged. Additionally, the extraction efficiency of V reached equilibrium with an O/A ratio of 1/2.

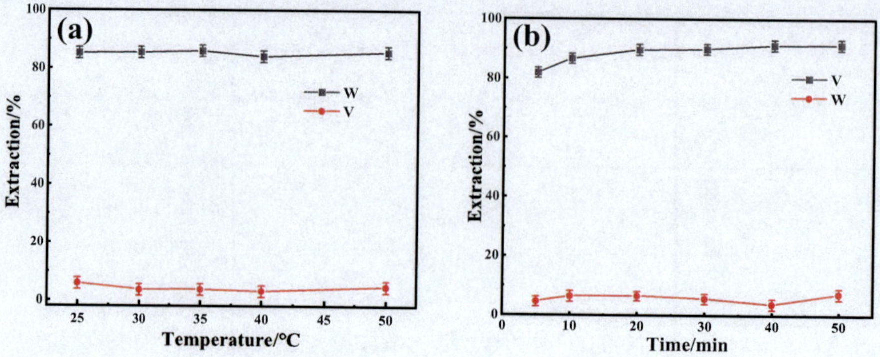

Fig. 4 Effect of temperature (**a**) and contact time (**b**) on the extraction of W and V (pH = 1.5, O/A = 1/2, and 20% P507)

Effect of Temperature and Contact Time

As can be seen from Fig. 4a, the temperature has little effect on the separation of W and V. Therefore, the room temperature was chosen for subsequent experiments. The contact time is defined as the mixing time required to extract the maximum solute from the external phase, and the effect of this parameter on metal ion extraction was evaluated by changing the extraction time. The equilibration time varying from 5 to 50 min was performed, as displayed in Fig. 4b, the extraction rate for V is faster with P507, and within 5 min of contact time, 81.90% of V was extracted. The efficiency of V transport increased and then stabilized as the increase of contact time until 40 min. Therefore, contact time of 40 min is deemed the balancing time to ensure complete extraction in all subsequent experiments.

Conclusions

The selective extraction of V from sodium tungstate solution containing 1 g/L of W and V was investigated using P507. The presence of the cationic form of V as VO_2^+ and anionic form of W as $HW_7O_{24}^{5-}$ at the strongly acidic solution pH (equilibrium pH 1.5) provides a theoretical basis for the selective separation of W and V. Based on the species' differences of W and V in acidic solutions, V is extracted into the organic phase by cation exchange with P507, while W remains in solution during the extraction process. In addition, the optimal conditions (pH = 1.5, O/A = 1/2, 20% P507, room temperature, t = 20 min) were determined by single factor experiments. After one extraction operation, the removal efficiency of V exceeds 90% with W loss below 6%. Therefore, the proposed process is promising exclusively on the suitable adoption of P507 as an extractant for effective extraction of V from sodium tungstate.

Acknowledgements This work was financially supported by the Original Exploration Project of China (52150079), the Natural Science Foundation of China (U2004215, No. 51974280, and No. 51774252), and the Educational Commission Fund of Henan Province of China (No. 20HASTIT012, No. 18A450001, and No. 17A450001).

References

1. Zhang Y, Zhang TA, Dreisinger D, Zhou W, Xie F, Lv G, Zhang W (2018) Chelating extraction of vanadium(V) from low pH sulfuric acid solution by Mextral 973H. Sep Purif Technol 190:123–135
2. Zhang WJ, Li JT, Zhao ZW, Huang SB, Chen XY, Hu KL (2015) Recovery and separation of W and Mo from high-molybdenum synthetic scheelite in HCl solutions containing H_2O_2. Hydrometallurgy 155:1–5
3. Zhang WJ, Chen YQ, Che JY, Wang CY, Ma BZ (2020) Green leaching of tungsten from synthetic scheelite with sulfuric acid-hydrogen peroxide solution to prepare tungstic acid. Sep Purif Technol 241:116752
4. Luo DS, Huang J, Zhang YM, Liu H, Hu PC (2020) Efficient and environment-friendly vanadium (V) extraction from vanadium shale leachate using tri-n-octylmethylammonium chloride. Sep Purif Technol 237:116482
5. Nguyen TH, Lee MS (2014) Separation of vanadium and tungsten from sodium molybdate solution by solvent extraction. Ind Eng Chem Res 53(20):8608–8614
6. Zhang JL, Zhao ZW (2014) Thermodynamic analysis of tungsten-vanadium separation in W(VI)-V(V)-H_2O system. Chin J Nonferr Met 24(6):1656–1662
7. Nguyen TH, Lee MS (2016) A review on the separation of molybdenum, tungsten, and vanadium from leach liquors of diverse resources by solvent extraction. Geosyst Eng 19(5):247–259
8. Choi IH, Moon G, Lee JY, Jyothi RK (2018) Hydrometallurgical processing of spent selective catalytic reduction (SCR) catalyst for recovery of tungsten. Hydrometallurgy 178:137–145
9. Luo L, Kejun L, Shibayama A, Yen W, Fujita T, Shindo O, Katai T (2004) Recovery of tungsten and vanadium from tungsten alloy scrap. Hydrometallurgy 72(1–2):1–8
10. Wang HY, Han GH, Huang YF, Su SP (2022) Solvent extraction separation of tungsten and vanadium from simulated leaching solution of spent SCR catalyst. In: Ouchi T, Azimi G, Forsberg K, Kim H, Alam S, Neelameggham N, Baba A, Peng H (eds) Rare metal technology 2022. Springer, Berlin, pp 93–100
11. Liu LJ, Wang LL, Su S, Yang T, Dai ZJ, Qing MX, Xu K, Hu S, Wang Y, Xiang J (2019) Leaching behavior of vanadium from spent SCR catalyst and its immobilization in cement-based solidification/stabilization with sulfurizing agent. Fuel 243:406–412
12. Han GH, Wang HY, Su SP, Huang YF, Liu BB (2021) Research progress and discussion on selective separation technology of dissolved tungsten and vanadium. Chin J Nonferr Met 31(11):3380–3395
13. Ferella F (2020) A review on management and recycling of spent selective catalytic reduction catalysts. J Clean Prod 246:118990
14. Zhang QJ, Wu YF, Yuan HR (2020) Recycling strategies of spent V_2O_5-WO_3/TiO_2 catalyst: a review. Resour Conserv Recycl 161:104983
15. Rout PC, Mishra GK, Padh B, Suresh KR, Ramachandra Reddy B (2017) Solvent extraction separation of molybdenum as thio-molybdate complex from alkaline tungsten leach liquor of spent HDS catalyst—a pilot study. Hydrometallurgy 174:140–146
16. Wang LP, Zhang GQ, Guan WJ, Zeng L, Zhou Q, Xia Y, Wang Q, Li QG, Cao ZY (2018) Complete removal of trace vanadium from ammonium tungstate solutions by solvent extraction. Hydrometallurgy 179:268–273

17. Truong HT, Nguyen TH, Lee MS (2017) Separation of molybdenum(VI), rhenium(VII), tungsten(VI), and vanadium(V) by solvent extraction. Hydrometallurgy 171:298–305
18. Zhang J, Zhu ZW, Chen DS, Wang LN, Wang WJ, Lei Z (2019) Preparation of highly pure vanadium electrolyte by solvent extraction and purification using P507 from V(IV) solution. Chin J Rare Met 43(3):303–311
19. Su SP, Huang YF, Liu BB, Han GH, Xue YB, Wang YZ (2021) A feasible strategy for deeply separating low concentrations of molybdenum from tungstate solutions with a high-efficiency microbubble floating-extraction concept. ACS Sustain Chem Eng 10(1):146–158
20. Xiong P, Zhang YM, Huang J, Bao SX, Yang X, Shen C (2017) High-efficient and selective extraction of vanadium (V) with N235–P507 synergistic extraction system. Chem Eng Res Des 120:284–290

Solar Thermal Application in Zn/ZnO Recovery from Spent Alkaline Batteries

Reiza Mukhlis, Deddy Nababan, Andrew Mackenzie, and Muhammad Akbar Rhamdhani

Abstract Landfilling of spent alkaline batteries is considered as more attractive than recycling if economic perspective is the main factor taken into account. The current study explores the potential use of solar thermal energy in recovering valuable Zn/ZnO from the batteries in attempt to improve the economic competitiveness of the recycling process. The black mass obtained from the batteries' anode and cathode, mixed with carbon from end-of-life lithium batteries, was subjected to carbothermal reduction at 1030 °C in a solar thermal simulator under argon. It was found that high purity Zn/ZnO powder can be recovered from the black mass, leaving valuable MnO in the reactor. The increase in argon flow was found to decrease the particle size of the Zn/ZnO powder, where the average of 2 μm obtained at 2 L/min argon. Since potassium is the main impurity in the recovered powder, neutral leaching of the black mass prior to reaction may improve product's purity.

Keywords Battery recycling · E-waste · Solar processing

Introduction

Despite the neurotoxin manganese and corrosive electrolyte content, the end-of-life (EOL) alkaline battery is not classified as hazardous waste and therefore can be thrown away as household thrash that may ended up as landfill [1]. While the consumer awareness and behaviour towards recycling are improving, the recycling rate of the used batteries is however far from sufficient, i.e. not more than 10% [2, 3]. A recent techno-economic study on alkaline battery recycling revealed that the recycling process was not considered economically feasible unless there are

R. Mukhlis (✉) · D. Nababan · M. A. Rhamdhani
Fluid and Process Dynamics (FPD) Group, Swinburne University of Technology, Melbourne, VIC 3122, Australia
e-mail: rmukhlis@swin.edu.au

A. Mackenzie
Envirostream Australia Pty. Ltd., Laverton North, VIC 3026, Australia

R. G. Reddy et al. (eds.), *New Directions in Mineral Processing, Extractive Metallurgy, Recycling and Waste Minimization*, The Minerals, Metals & Materials Series,
https://doi.org/10.1007/978-3-031-22765-3_40

significant increase in value of the recycling products and decrease in the process cost [4].

The merit of the recycling products can be increased through the recovery of valuable materials from the spent batteries. It has been previously demonstrated that upon heating beyond 900 °C, ZnO can be reduced by carbon into zinc vapor where upon condensation, the particle size of the Zn powder can be controlled by the gas flow rate [4].

The current study explored the possibility of solar thermal energy application in the alkaline battery recycling process. The process involved the production of nano-sized zinc and/or zinc oxide powders through carbothermic reduction of black mass obtained from mechanically processed EOL alkaline batteries. Carbon after hydrometallurgical recycling process of lithium batteries was used as an additional source of reductant to further reduce process cost.

Materials and Method

Battery Materials

Black mass, a mixture of cathode and anode powders produced through mechanical processing of spent alkaline batteries, was supplied by Envirostream Pty. Ltd. Based on XRD analysis, the black mass mainly consists of ZnO, Mn-oxides (MnO_2, Mn_2O_3, and Mn_3O_4), and carbon. Mn, Zn, and C content in the black mass was 28.55 wt%, 24.30 wt%, and 16.78 wt%, respectively. Carbon powder was added to the black mass to ensure complete ZnO reduction. The carbon, initially in the form of paste, was obtained from the waste of hydrometallurgical recycling process of lithium batteries supplied by Envirostream Pty. Ltd.

Sample Preparations

The carbon paste was dried in an oven at 100 °C for 5 h, grinded into powders, and sieved to 100 mesh. The resulted carbon powders were then added into black mass powder that has been sieved to 100 mesh with the molar ratio of 5 mol C to 1 mol ZnO. Subsequently, the powders were than mixed in a ball mill rotator at 50 rpm for 48 h to ensure its homogeneity. Alumina ball were added to the mixture with a ratio of 1:10 to enhance the mixing process.

The black mass-carbon mixture was then compacted by 1 tonne load into 10 mm pellets ready for the high temperature experimentation.

Thermodynamics Assessment

Thermodynamics assessment on the black mass-carbon system was carried out to determine the main parameter of the high temperature experimentation that is the carbothermic process temperature. Equilibrium reaction of black mass-carbon mixture in an inert atmosphere was simulated through *Equilib* module within Fact-sage 7.2 thermochemical package. Three databases used in present study were FactPS, FToxid, and FTmisc [5].

In order to achieve carbon addition of 5 mol C: 1 mol ZnO, the reactants for the assessment were consist of 11.2 g pure carbon (graphite) and 100 g black mass with the composition mentioned in Section a: *Battery materials*. Impurities that may exist within the carbon used for the actual experiments were excluded in the assessment.

High Temperature Experimentation

Three compacted pellets were loaded into a quartz reactor equipped with gas inlet and outlet and a thermocouple as schematically shown in Fig. 1.

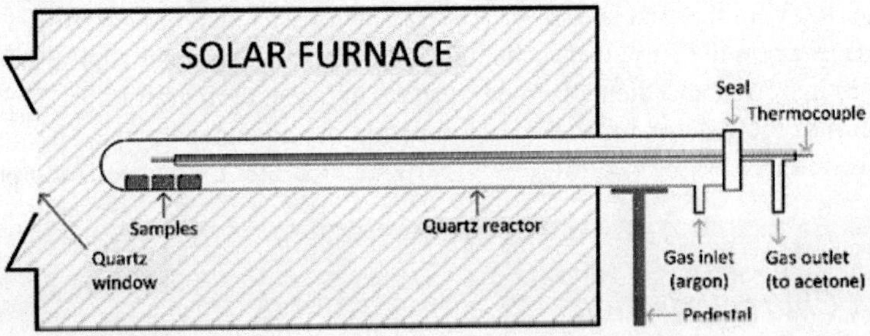

Fig. 1 Schematic diagram of quartz reactor inserted to a furnace heated by solar simulator through the quartz window

The reactor was then inserted into a solar furnace and heated to 1030 °C in a solar simulator built at Swinburne University of Technology (see Fig. 2) to enable carbothermic reduction of the black mass. The solar simulator, powered by seven 6 kWe metal halide lamps, was specifically designed to simulate highly concentrated solar energy applications in material processing [6]. A peak flux ranging from 117 to 148 kW/m^2 can be delivered by each lamp, where 1 MW peak flux can be focused to the solar furnace through the quartz window (0.175 m diameter) when all the lamps turned on. Five lamps were used in the current experiments to reach reactor temperature of 1030 °C where, according to the thermodynamic assessment, the carbothermic reduction of black mass reaches the plateau.

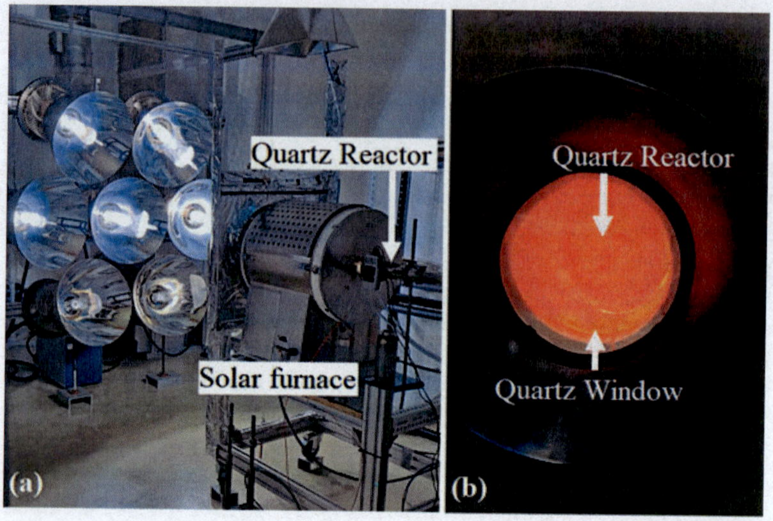

Fig. 2 Solar simulator used in the present study, **a** furnace backview showing the metal halide lamps as the source of the heat, **b** furnace front view showing the quartz windowns and the position of the quartz reactor

In order to avoid thermal shock, the quartz reactor was inserted carefully into the solar furnace when the furnace reached 800 °C. The reactor reached 1030 °C after about 2 h of insertion and then left in the furnace at target temperature for 2 h. Argon was flown into the reactor for the whole duration of the experiments. The gas flow rate was varied from 0.5 to 2 L/min to investigate its effect to the produced powder.

Material Characterisation

The X-ray diffraction (XRD) pattern analysis, scanning electron microscopy (SEM) image analysis, energy dispersive spectroscopy (EDS) elemental analysis, X-ray fluorescence (XRF) analysis, and light scattering particle size analysis were among the methods used in characterising the reactants and the products of the process.

The Bruker AXS–D8 diffractometer utilising Cu Kα radiation (1.54178 Å) operating at 40 kV and 30 mA was used to scan the sample from 20 to 90° (2θ) at a rate of 0.02° per 1.5 s time step. The ZEISS Supra 40 Field Emission SEM equipped with EDS was utilised in microstructure and elemental analysis. The particle size of the condensed powder was analysed using light scattering Malvern Zetasizer Ultra analyser. The Varian 730-ES Optical Emission Spectrometer and Olympus DELTA Professional XRF analyser were used to quantitatively identify the elements within the reactants and the products.

Results and Discussion

It is generally known that thermodynamically, zinc oxide will be reduced by carbon into zinc at elevated temperatures following Reaction (1):

$$ZnO_{(s)} + C_{(s)} \rightarrow Zn_{(g)} + CO_{(g)} \quad \Delta G_{(at\ 1030°C)} = -4687.6\,kJ \tag{1}$$

The thermodynamics assessment results on the equilibrium reaction of 100 g black mass containing ZnO with 11.2 g carbon in inert atmosphere are presented in Fig. 3. By considering the XRF, XRD, ICP, and preliminary thermodynamics calculation results on the black mass, the composition of the reactant was assumed to be: 25.22 g C, 35.93 g Mn_2O_3, 7.14 g MnO_2, 0.9 g Mn_3O_4, 29.7 ZnO, 7.94 KOH, 2.93 Mg, 1.07 Fe_2O_3, and 0.39 g Co.

One can clearly see from Fig. 3a that equilibrium reaction between black mass with carbon at the temperature ranging from 800 to 1400 °C was predicted to mainly produce gas phase, MnO, unreacted C and ZnO, K_2CO_3, Mn_7C_3, and minor amount of Fe_3C. The carbothermic reduction of zinc oxide was predicted to start becoming significant at 900 °C marked by the decrease in ZnO and C content, and the increase in the gas phase content. MnO remained constant at around 39 g where it suddenly diminished at 1400 °C where Mn_7C_3 started to form. The K_2CO_3 was predicted to constantly reduced from 9.8 g at 800 °C to 6.3 g at 1000 °C and finally diminished at 1100 °C, where all the K turned into gas phase (see Fig. 3b). It is shown clearly in Fig. 3b that zinc vapour started to reach its plateau (around 23.9 g at 1000 °C).

Considering that the expected product of the current recycling process is Zn and/or ZnO powder, the high temperature experimentation using the solar simulator was set to 1030 °C. It was expected that at the set temperature, almost all of the zinc has left the black mass as a vapour which can be recovered once condensed. It was also expected that the zinc powder recovered from the process conducted at 1030 °C will relatively be pure. At temperatures higher than the set temperature, the gas phase was predicted to increase, where all condensed vapor will be ended up as impurities

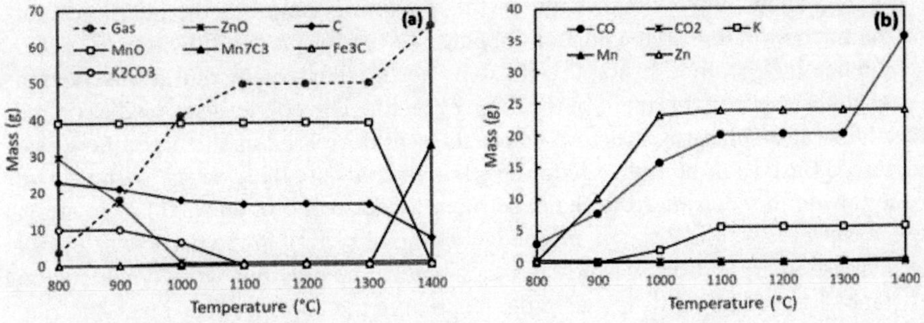

Fig. 3 Results of thermodynamic assessment: **a** major phases in the product, **b** main component of the gas phase marked by dashed line in Fig. 3a

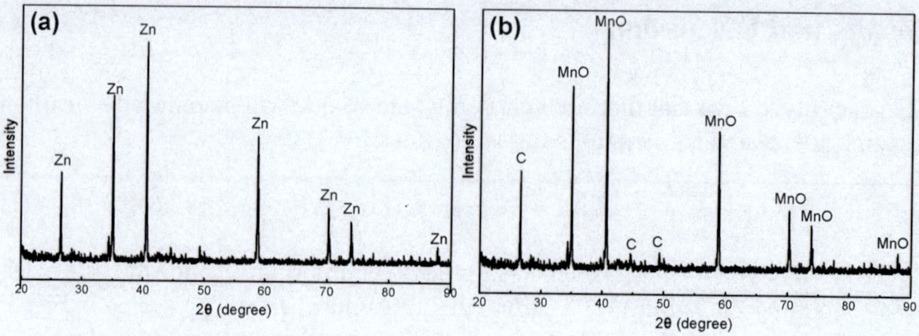

Fig. 4 XRD pattern of **a** the condensed powder and **b** sample remaining in the quartz reactor after heating at 1030 °C for 2 h under 2 L/min argon flow

in the produced zinc powder. Take potassium (K) for example; the predicted K(g) produced at 1100 °C is about 2.5 times higher than that is produced at 1000 °C.

Figure 4 shows the X-ray diffraction pattern of the condensed powder and the remaining sample (pellet) that stayed in the quartz reactor after the experiment. As mentioned in Section d: *High temperature experimentation*, the samples were heated at 1030 °C for 2 h in solar simulator under argon flowing at 0.5, 1, or 2 L/min.

As seen in Fig. 4, Zn was the only obvious phase in the condensed powder, while MnO and C peaks were found from the sample remained in the quartz reactor after heating. This is in agreement with thermodynamics assessment, which predicted that the gas phase at 1030 °C was mainly consisted of CO and Zn. At room temperature, only Zn will be condensed while CO will leave away as gas. Similar with Zn, potassium should condense at room temperature, but no potassium peaks were detected. The peaks found from the reacted sample remaining in the reactor are also in line with the thermodynamic assessment. As shown in Fig. 3, at the temperature ranging from 1000 to 1100 °C, the predicted products of reaction were MnO, minor K_2CO_3, and unreacted C. Since no K_2CO_3 was predicted to exist as a product of the reaction at 1100 °C, it is understandable that no K_2CO_3 peaks were found from the reacted samples.

Figure 5 shows the microstructure of the remaining sample in the quartz reactor and the microstructure of the condensed phase for various argon flow rate.

In general, crystalline structure with some needle like structures was observed in the reacted sample that remain in the quartz reactor. The condensed powder, on the other hand, has spherical structure where the size decreased as the argon flow rate increased. Based on the light scattering particle analysis, the average particle size of the powder condensed from the experiment under 0.5, 1.0, and 2.0 L/min argon flow was 3.72, 2.3, and 2.0 μm, respectively. The reduction in the particle size with the increase in argon flow was not as significant as the previous work conducted in resistance furnace [4]. This is due to the fact that in the current experiment, the zinc vapour needs to travel longer along the hose that connect quartz reactor gas outlet to the wash bottle that contain acetone. A 2-m hose was needed to keep the wash

Fig. 5 Microstructure of **a** remaining sample in the quartz reactor, and the condensed powder from the experiment conducted under **b** 0.5 L/min, **c** 1 L/min, **d** 2 L/min argon flow

bottle away from the solar simulator lamp, as otherwise the acetone will evaporate very rapidly. Consequently, quick-quench of the zinc vapor cannot be achieved.

The EDS elemental mapping of the sample remained in the quartz reactor and the condensed powder are shown in Fig. 6a, b, respectively. The EDS map in Fig. 6a confirmed that the sample remaining in the quartz reactor was consisted of manganese oxide and carbon. No zinc was found in the sample. This indicated that the zinc oxide in the black mass has successfully been reduced into Zn where at the set temperature (1030 °C), it is in a form of the gas which swept away from the sample by argon flow. The zinc vapor was then subsequently condensed into solid when the temperature dropped, which was clearly shown by the EDS map of the condensed powder (Fig. 6b). Traces of oxygen, however, were found in the condensed powder, which most likely be introduced during the SEM sample preparation due to prolonged contact with air.

The concentration of elements within the samples shown in Fig. 6 is listed in Table 1, along with element concentration in pure oxides.

When the carbon content in the Fig. 6a sample is not included in calculation, the ratio of Mn to O in the sample is 77.44 Mn: 22.55 O, which resemblance the element ratio of MnO instead of other manganese oxides listed in Table 1. This is in agreement with the XRD analysis that shown only MnO peaks. Initially, the black mass contains MnO_2, Mn_2O_3, and Mn_3O_4. Thermodynamically, these oxides will be reduced to MnO at 1030 °C by carbon according to the reactions below:

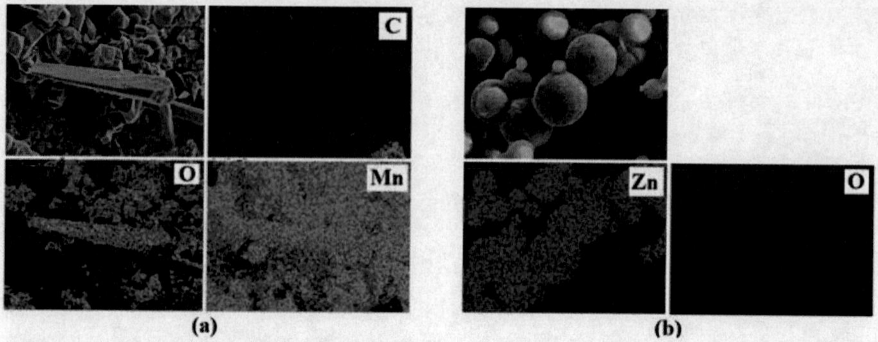

Fig. 6 EDS elemental mapping of **a** remaining sample in quartz reactor, and **b** condensed phase

Table 1 Concentration of elements of samples shown in Fig. 6 along with common oxides

Element	Figure 6a sample (wt%)	Figure 6b sample (wt%)	MnO (wt%)	MnO_2 (wt%)	Mn_2O_3 (wt%)	Mn_3O_4 (wt%)	ZnO (wt%)
Mn	71.70	0	77.44	63.19	69.6	72.03	0
Zn	0	51.68	0	0	0	0	80.34
O	20.88	7.34	22.55	36.81	30.4	27.97	19.66
C	7.42	37.98	0	0	0	0	0
K	0	0.38	0	0	0	0	0
Cu	0	2.67	0	0	0	0	0

$$MnO_{2(s)} + C_{(s)} \rightarrow MnO_{(s)} + CO_{(g)} \quad \Delta G = -231.8\,\text{kJ} \tag{2}$$

$$Mn_2O_{3(s)} + C_{(s)} \rightarrow 2MnO_{(s)} + CO_{(g)} \quad \Delta G = -175.8\,\text{kJ} \tag{3}$$

$$Mn_3O_{4(s)} + C_{(s)} \rightarrow 3MnO_{(s)} + CO_{(g)} \quad \Delta G = -159.8\,\text{kJ} \tag{4}$$

If the Gibbs free energy is the only consideration, the carbon in the black mass will easily consumed by the reduction of zinc oxide as it has significantly lower ΔG (see Reaction (1)) and then consumed by the manganese oxides reduction. In the case that carbon content in the black mass is higher than the requirement for Reactions (1)–(4) to happen, it will remain unreacted in the sample (black mass) if the process conducted under inert atmosphere.

The presence of C in the condensed powder (see Table 1) was definitely an artefact as the powder was adhered to carbon tape for the SEM/EDS analyses. CO will not condense at room temperature and will not contribute to the presence of carbon in the sample. As expected, minor amount of potassium was presence in the condensed powder. Potassium was originated from the alkaline battery electrolyte, i.e. KOH. The presence of Cu in the condensed powder is arguable. While Cu is present as impurity in the carbon used in the current experiment (i.e. waste of Li battery recycling), the amount was not that significant as it only reached 0.04 wt% maximum according to XRF analysis. Additional Cu was most likely an artefact as the carbon tape was attached to Cu plate to ensure good conductivity required for SEM/EDS analysis. Based on XRF analysis, the resulted Zn powder has a purity ranging from 99.56 to 99.79% with potassium as the main impurity.

Conclusion and Recommendations

The current study demonstrated that valuable Zn powder can be recovered from end-of-life alkaline battery by the application of concentrated solar thermal energy. Based on thermodynamics assessment, the process temperature should be in the range of 1000–1200 °C as the recovery of high purity zinc reached its maxima. Below 1000 °C, the amount of the zinc that can be recovered from the EOL battery is predicted to reduce significantly. On the other hand, at the temperatures beyond 1200 °C, the purity of the recovered zinc was predicted to reduce since many elements, e.g. potassium and manganese will be evaporated which will later condense together with the zinc and will act as impurity.

Zinc powder, up to 99.79% pure, can be recovered from the alkaline battery when the black mass with carbon addition is heated to 1030 °C. The main impurities in the zinc were found to be potassium. Through the application of pre-process, i.e. neutral leaching of the black mass, purity of the recovered powder can be increased as potassium will be washed away from the black mass prior to the carbothermic reduction reaction. The particle size of the zinc powder can be reduced by increasing the gas flow rate. Due to the nature of the current experiments, however, the size reduction was not significant. Modification to the reactor should be explored to ensure rapid cooling of the zinc vapour to accommodate size reduction.

It is required to supply enough carbon to the black mass to ensure maximum zinc recovery from the used alkaline batteries. Carbon is also required to ensure complete reduction of manganese oxides in the spent batteries into MnO, which was saleable as it has been proven to be adequately effective fertiliser [7]. The plant takes up manganese as Mn^{2+} from the soil solution, and therefore, MnO is preferable than other valence of manganese oxides, i.e. MnO_2, Mn_2O_3, and Mn_3O_4. Carbon content, however, should not be too much as the remaining will left unreacted in the sample.

Acknowledgements The authors thank Sustainability Victoria for the funding and support throughout the duration of the study.

References

1. US-EPA (2020) Universal waste. https://www.epa.gov/hw/universal-waste. Accessed 10 Sept 2020
2. ABRI (2020) Handheld battery recycling. http://www.batteryrecycling.org.au/wp-content/uploads/2014/01/Handheld-battery-recycling-QA-2014.pdf. Accessed 27 Apr 2020
3. Islam MT, Huda N, Baumber A, Hossain R, Sahajwalla V (2022) Waste battery disposal and recycling behavior: a study on the Australian perspective. Environ Sci Pollut Res 1–22
4. Mukhlis R, Mackenzie A, Rhamdhani MA (2021) Small scale recycling process for spent alkaline batteries: technoeconomic analysis and potential use of solar energy. Resour Conserv Recycl 166:105367
5. Bale CW, Bélisle E, Chartrand P, Decterov SA, Eriksson G, Hack K, Petersen S et al (2009) FactSage thermochemical software and databases—recent developments. Calphad 33(2):295–311
6. Ekman BM, Brooks G, Rhamdhani MA (2015) Development of high flux solar simulators for solar thermal research. In: Energy technology 2015. Springer, Cham, pp 149–159
7. Brennan RF, Bolland MDA (2004) Comparing manganese sources for spring wheat grown on alkaline soils. J Plant Nutr 27(1):95–109.

Study on the Application of Absorbing Colloid Flotation for Phosphate Removal from the Hydrometallurgy Waste Liquid

Lulu Kou, Wenjuan Wang, Yanfang Huang, and Guihong Han

Abstract On some situations, environmentally hazardous phosphate ions exist in hydrometallurgy waste liquids. Among the available treatment technologies, adsorbing colloid flotation (ACF) has been considered and is of particular importance. In this work, ferric chloride was used as colloid adsorbent and sodium dodecyl sulfate (SDS) as collector to remove phosphate from the hydrometallurgy waste liquid by flotation. Different parameters, including solution pH, contact time, molar ratio, and collector concentration, were comprehensively investigated. The optimum phosphate removal efficiency was 96.05% when Fe/P molar ratio of 2:1, contact time of 30 min, solution pH of 4, and collector concentration of 160 mg/L were applied. The results indicated that adsorbing colloid flotation can effectively remove phosphate from solution and has great application prospects in hydrometallurgy waste liquid treatment.

Keywords Adsorbing colloid flotation · Phosphate · Hydrometallurgy · Waste liquid

Introduction

Phosphorus is mostly present as inorganic phosphate ($H_2PO_4^-$, HPO_4^{2-}, PO_4^{3-}) in natural water or waste liquid [1, 2]. With the rapid development of the hydrometallurgy industry, a large amount of waste liquid containing phosphate will be produced. When the phosphate concentration in the water body is higher than 0.003–0.8 μg/L, it tends to cause eutrophication in water body, which can cause algal blooms, kill fish, and destroy ecosystem [3–5]. Therefore, it is important and urgent that remove phosphate from the hydrometallurgy waste liquid. Common removal phosphate technologies include: chemical precipitation [6], biological technology [7], adsorption, and

L. Kou · W. Wang · Y. Huang · G. Han (✉)
School of Chemical Engineering, Zhengzhou University, Zhengzhou 450001, P.R. China
e-mail: hanguihong@zzu.edu.cn

R. G. Reddy et al. (eds.), *New Directions in Mineral Processing, Extractive Metallurgy, Recycling and Waste Minimization*, The Minerals, Metals & Materials Series,
https://doi.org/10.1007/978-3-031-22765-3_41

ion exchange [8, 9]. These technologies often produce by-products or need pretreatment. Therefore, the development a method for effective removal of phosphate is important and necessary.

The technique of adsorbing colloid flotation (ACF) has evolved from the micro-flotation and precipitate flotation [10]. ACF is based on the adsorption of material onto colloidal sized particles which then collect at the gas liquid interface. It typically involves the formation of a hydrous metal oxide (e.g. hydrous oxides of Fe(III) and Al(III)) and the adsorption or coprecipitation of waste species onto this hydrous metal oxide [11]. Then, surfactants are added to attach them to the bubbles [12]. Compared with the traditional process, the adsorbing colloid flotation has the advantages of easy operation, low cost, and reusable adsorbent [13], which has great application prospects for the removal of phosphate in hydrometallurgy waste liquid treatment.

The main objective of this work was to use the adsorbing colloid flotation for the removal of phosphate from hydrometallurgy waste liquid. In this work, ferric chloride and sodium dodecyl sulfate were used as colloid adsorbent and collector, respectively. Different parameters including solution pH, contact time, Fe/P molar ratio, and collector concentration were comprehensively investigated.

Experimental

Materials

N, N-dimethylformamide (C_3H_7NO, 99.5%), ferric chloride hexahydrate ($FeCl_3{\cdot}6H_2O$, 99%), sodium dodecyl sulfate ($C_{12}H_{25}SO_4Na$, >92.5%), potassium dihydrogen phosphate (KH_2PO_4, 99.5%), NaOH, HCl. All the chemicals used in the experiments were of analytical grade.

Absorbing Colloid Flotation Experiments

A certain mass of potassium dihydrogen phosphate was dissolved in distilled water to prepare a 1000 mg/L solution of phosphate and used as the simulated hydrometallurgy waste liquid. Different concentration of phosphate standard solutions could be obtained using the dilution method. All adsorption experiments conducted in 100 mL conical flasks filled with 50 mL simulated phosphate ion solution and a certain amount of ferric chloride solution. The solution pH was adjusted by using 0.1 M HCI and 0.1 M NaOH. The conical flask filled with the solution was reacted at 25 °C in shaker. Secondly, a certain amount surfactant solution was added to the conical flask. The solution with phosphate ion, ferric chloride, and surfactant was transferred to flotation column. Then, the clean solution was collected and filtered using a 0.22 μm membrane filter. Finally, the residual concentration of phosphate ions was determined by ICP-OES. The phosphate ion concentration was 100 mg/L, the temperature was 25 °C, and impeller speed was 150 r/min for all adsorption experiments. The removal efficiency (R_t) was calculated by Eq. (1).

$$R_t = \frac{C_0 - C_t}{C_0} \times 100\% \quad (1)$$

where R_t is removal efficiency at time t (%); t is the contact time (min); C_0 is the residual phosphate ion concentration (mg/L); C_t is the phosphate ion concentration at time t (mg/L).

Results and Discussion

Effect of Contact Time

The contact time is one of the important factors affecting the removal efficiency. To obtain the effect of contact time on the removal of phosphate ion, the following experiment with the contact time 10–210 min was carried out. As shown in Fig. 1, the removal efficiency of phosphate declined, and residual phosphate ion concentration increased from 10 to 20 min. However, the removal efficiency increased, and phosphate residual phosphate ion concentration declined from 20 to 30 min. The removal efficiency of phosphate reached the maximum of 100% firstly when the contact time was 30 min. At the same time, the residual phosphate ion concentration achieved 0 mg/L. The optimum contact time was 30 min to improve economic benefits.

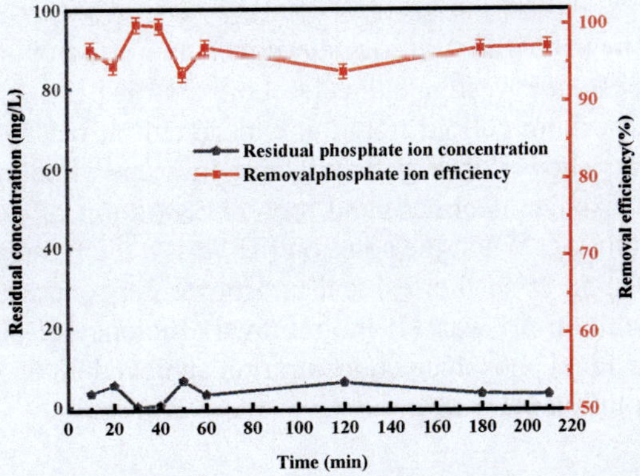

Fig. 1 Effect of contact time on the removal of phosphate (solution pH = 6; Fe/P molar ratio = 2:1)

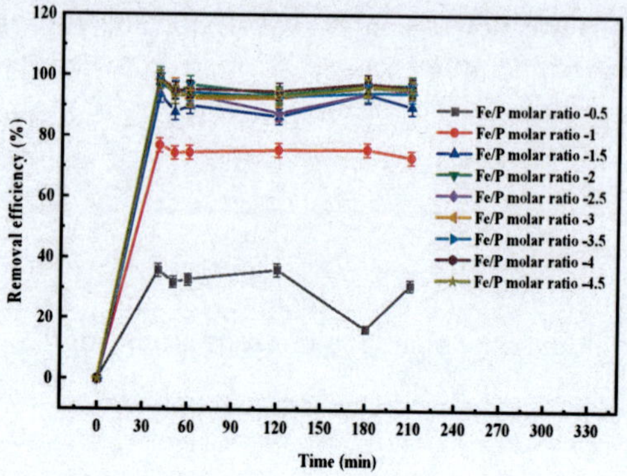

Fig. 2 Effect of different Fe/P molar ratio on removal efficiency (solution pH = 6; contact time = 30 min)

Effect of Fe/P Molar Ratio

It is well known that the dosage of adsorbent has a decisive effect on adsorption process. To explore the effect of Fe/P molar ratio on the removal of phosphate by ACF, the following experiment with the Fe/P molar ratio 0.5–4.5 was conducted.

As shown in Fig. 2, the removal efficiency of phosphate first increased and then remained constant with the increase of the Fe/P molar ratio. When the Fe/P molar ratio was 0.5:1, the removal efficiency of phosphate was 32.875%. And when the Fe/P molar ratio was 2:1, the removal efficiency of phosphate could be reached 99.8%. Therefore, the optimum Fe/P molar ratio was 2:1 to improve economic benefits.

Effect of Solution pH

The pH of the solution determines the form of phosphate ions. Therefore, it is meaningful to investigate the effect of solution pH on the removal phosphate by ACF. The absorbing colloid flotation experiment at pH 3–11 was carried out, and the result was showed in Fig. 3. With the increase of solution pH, the removal efficiency of phosphate declined, and the concentration of residual phosphate increased in the solution. When the solution pH was 4, the removal efficiency of phosphate reached 93.95%, and the residual phosphate concentration achieved 6.05 mg/L. When the solution pH was 11, the removal efficiency of phosphate reached 1.325%, and the residual phosphate concentration achieved 98.675 mg/L. Therefore, the optimum of solution pH was 4.

Fig. 3 Effect of solution pH on residual phosphate ion concentration and removal efficiency (Fe/P molar ratio = 2:1; contact time = 30 min)

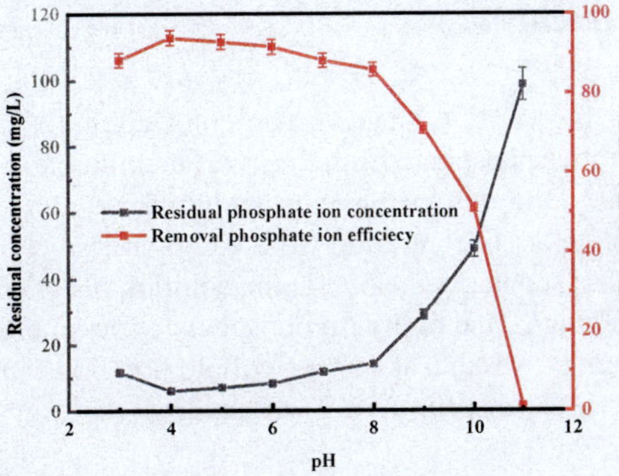

Effect of Collector Concentration

It was necessary to explore the effect of sodium dodecyl sulfate surfactant concentrations on the removal of phosphate and the solution turbidity; hence, the follow experiment with the sodium dodecyl sulfate surfactant concentrations 0–280 mg/L was conducted. The result was showed in Fig. 4.

As shown in Fig. 4, collector concentration has little effect on the removal efficiency of phosphate but has a significant effect on the solution turbidity. The removal efficiency of phosphate did not change significantly as the SDS surfactant concentration increased, with removal efficiency remaining above 90%. However, the solution turbidity changed obviously along with SDS surfactant concentration increased. The solution turbidity firstly increased and then decreased with the increase of the SDS concentration. When the SDS surfactant concentration was 40 mg/L, the solution turbidity reached the maximum of 22.52 NTU. When SDS surfactant concentration was above 160 mg/L, turbidity approximatively became 0 NUT. Hence, the optimum collector concentration was 160 mg/L.

Fig. 4 Effect of SDS surfactant concentrations on removal efficiency and turbidity (solution pH = 4; contact time = 30 min; Fe/P molar ratio = 2:1)

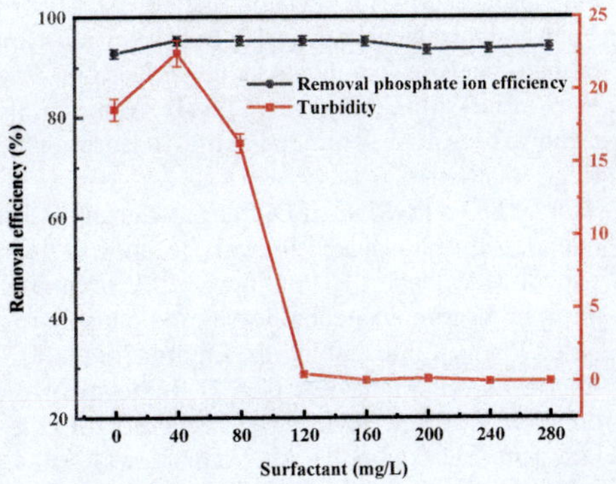

Conclusions

In this work, we used ferric chloride and sodium dodecyl sulfate to successfully remove phosphate from the hydrometallurgy waste liquid by adsorbing colloid flotation. Different parameters including solution pH, contact time, molar ratio, and collector concentration were comprehensively investigated. When Fe/P molar ratio of 2:1, contact time of 30 min, solution pH of 4, and SDS surfactant concentration of 160 mg/L, the optimum phosphate removal efficiency reached 96.05%. The experiment proves that absorbing colloid flotation can effective removal phosphate and has a bright prospect to deal with hydrometallurgy waste liquid.

Acknowledgements This work was financially supported by the Natural Science Foundation of China (U2004215, No. 51974280, and No. 51774252) and the Educational Commission Fund of Henan Province of China (No. 20HASTIT012, No. 18A450001, and No. 17A450001).

References

1. Vianna MTG, Marques M, Bertolino LC (2016) Sun coral powder as adsorbent: evaluation of phosphorus removal in synthetic and real wastewater. Ecol Eng 97:13–22
2. Mangwandi C, Albadarin AB, Glocheux Y, Walker GM (2014) Removal of ortho-phosphate from aqueous solution by adsorption onto dolomite. J Environ Chem Eng 2(2):1123–1130
3. Le C, Zha Y, Li Y, Sun D, Lu H, Yin B (2010) Eutrophication of lake waters in China: cost, causes, and control. Environ Manage 45(4):662–668
4. Shinohara R, Imai A, Kohzu A, Tomioka N, Furusato E, Satou T, Sano T, Komatsu K, Miura S, Shimotori K (2016) Dynamics of particulate phosphorus in a shallow eutrophic lake. Sci Total Environ 563:413–423
5. Zhang Y, Li M, Dong J, Yang H, Van Zwieten L, Lu H, Alshameri A, Zhan Z, Chen X, Jiang X, Xu W, Bao Y, Wang H (2021) A critical review of methods for analyzing freshwater eutrophication. Water 13(2)
6. Massey MS, Ippolito JA, Davis JG, Sheffield RE (2010) Macroscopic and microscopic variation in recovered magnesium phosphate materials: implications for phosphorus removal processes and product re-use. Bioresour Technol 101(3):877–885
7. Yuan Q, Oleszkiewicz J (2010) Selection and enrichment of denitrifying phosphorus accumulating organisms in activated sludge. Desalin Water Treat 22(1–3):72–77
8. Wu K, Liu T, Ma C, Chang B, Chen R, Wang X (2014) The role of Mn oxide doping in phosphate removal by Al-based bimetal oxides: adsorption behaviors and mechanisms. Environ Sci Pollut Res 21(1):620–630
9. Ahmed S, Ashiq MN, Li D, Tang P, Leroux F, Feng Y (2019) Recent progress on adsorption materials for phosphate removal. Recent Pat Nanotechnol 13(1):3–16
10. Saoudi MA, Dabert P, Ponthieux A, Vedrenne F, Daumer M-L (2022) Correlation between phosphorus removal technologies and phosphorus speciation in sewage sludge: focus on iron-based P removal technologies. Environ Technol 2022:1–13
11. Zouboulis AI, Matis KA (1997) Removal of metal ions from dilute solutions by sorptive flotation. Crit Rev Environ Sci Technol 27(3):195–235
12. Lehmann M, Zouboulis AI, Matis KA (1999) Removal of metal ions from dilute aqueous solutions: a comparative study of inorganic sorbent materials. Chemosphere 39(6):881–892
13. Choi SJ, Seo EJ, Shin WS (2000) Effect of activators on Cu(II) removal by adsorbing colloid flotation with $Fe(OH)_3$. Environ Technol 21(12):1357–1361

Author Index

R. G. Reddy et al. (eds.), *New Directions in Mineral Processing, Extractive Metallurgy, Recycling and Waste Minimization*, The Minerals, Metals & Materials Series,
https://doi.org/10.1007/978-3-031-22765-3

Subject Index

R. G. Reddy et al. (eds.), *New Directions in Mineral Processing, Extractive Metallurgy, Recycling and Waste Minimization*, The Minerals, Metals & Materials Series,
https://doi.org/10.1007/978-3-031-22765-3